MORE Electronic Gadgets for the Evil Genius

Evil Genius Series

Bionics for the Evil Genius: 25 Build-it-Yourself Projects

Electronic Circuits for the Evil Genius: 57 Lessons with Projects

Electronic Gadgets for the Evil Genius: 28 Build-it-Yourself Projects

Electronic Sensors for the Evil Genius: 54 Electrifying Projects

50 Awesome Auto Projects for the Evil Genius

Mechatronics for the Evil Genius: 25 Build-it-Yourself Projects

MORE Electronic Gadgets for the Evil Genius: 40 NEW Build-it-Yourself Projects

123 PIC® Microcontroller Experiments for the Evil Genius

123 Robotics Experiments for the Evil Genius

MORE Electronic Gadgets for the Evil Genius

BOB IANNINI

McGraw-Hill
New York Chicago San Francisco Lisbon
London Madrid Mexico City Milan New Delhi
San Juan Seoul Singapore Sydney Toronto

The McGraw-Hill Companies

Cataloging-in-Publication Data is on file with the Library of Congress.

2 3 4 5 6 7 8 9 0 QPD/QPD 0 1 0 9 8 7 6

ISBN 0-07-145905-7

The sponsoring editor for this book was Judy Bass and the production supervisor was Richard Ruzycka. It was set in Times Ten by MacAllister Publishing Services, LLC. The art director for the cover was Anthony Landi.

Printed and bound by Quebecor/Dubuque.

This book is printed on acid-free paper.

About the Author

Bob Iannini runs Information Unlimited, a firm dedicated to the experimenter and technology enthusiast. Founded in 1974, the company holds many patents, ranging from weapons advances to children's toys. Mr. Iannini's 1983 *Build Your Own Laser, Phaser, Ion Ray Gun & Other Working Space-Age Projects,* now out of print, remains a popular source for electronics hobbyists. He is also the author of the wildly successful *Electronic Gadgets for the Evil Genius*, an earlier volume in this series.

Contents

Contents

Contents

Acknowledgments

I wish to express thanks to the employees of Information Unlimited and Scientific Systems Research Laboratories for making these projects possible.

Their contributions range from many helpful ideas to actual prototype assembly. Special thanks go out to department heads Rick Upham, general manager in charge of the lab and shop and layout designer; Sheryl Upham, order processing and control; Joyce Krar, accounting and administration; Walter Koschen, advertising and system administrator; Chris Upham, electrical assembly department; Al (Big AI) Watts, fabrication department; Sharon Gordon, outside assembly; and all the technicians, assemblers, and general helpers at our facilities in New Hampshire, Florida, and Hong Kong that have made these endeavors possible.

Also not to be forgotten is my wife Lucy, who has contributed so much with her support and understanding of my absence due to long hours in front of the computer necessary for preparing this manuscript.

I want to acknowledge and thank Dr. Barney Vincelette for his excellent contribution of Chapter 13, the Microwave Cannon.

Also, thank you to Robert Gaffigan for the basic bark detect circuitry section of Chapter 34 titled the Canine Controller.

MORE Electronic Gadgets for the Evil Genius

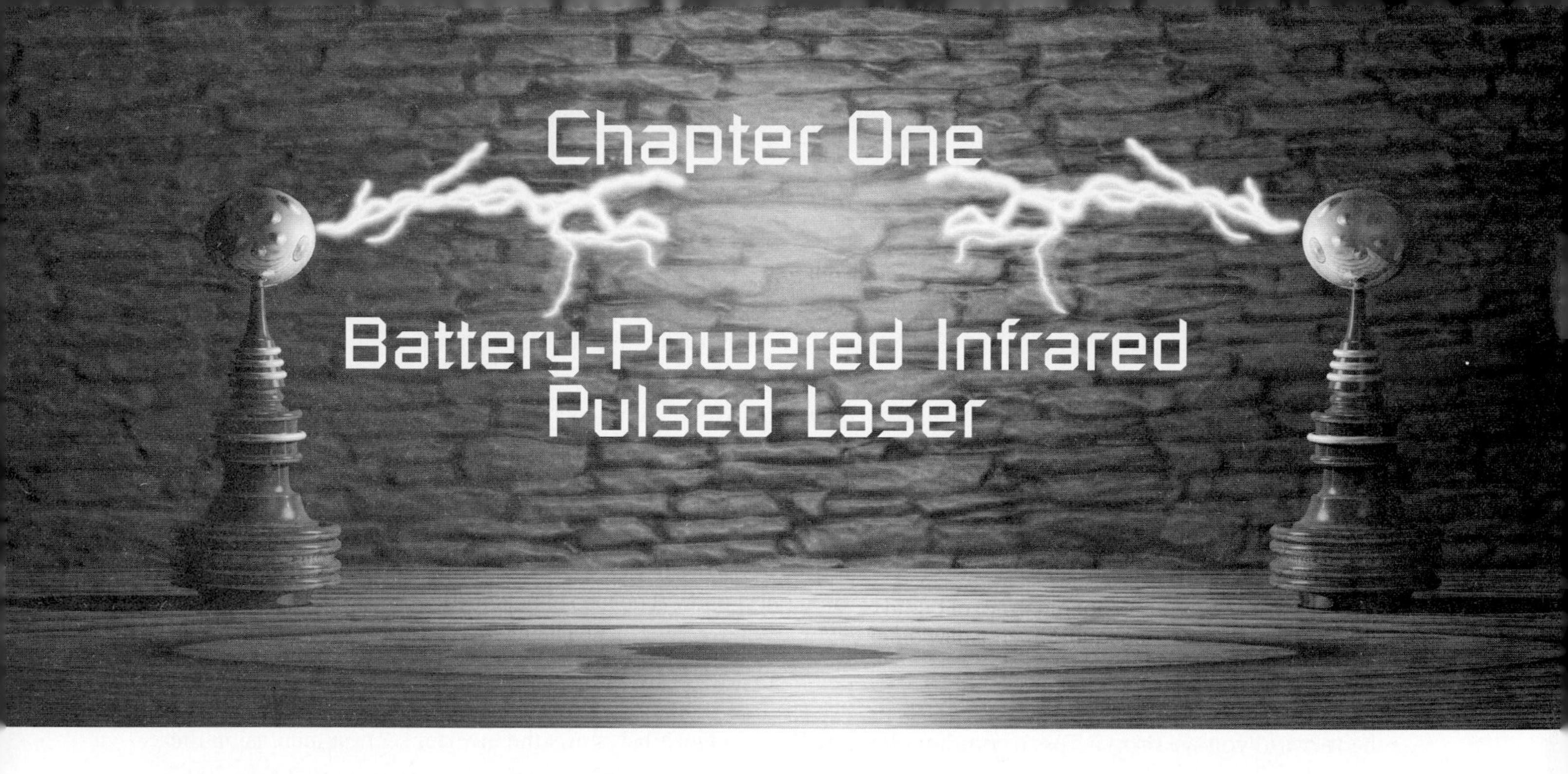

Chapter One

Battery-Powered Infrared Pulsed Laser

This project shows how to construct a low-powered, portable, solid-state, pulsed infrared laser. The system uses a gallium arsenide laser diode and provides pulse powers from 10 to 100 watts, depending on the diode used. The device operates from batteries and is completely self-contained. It may be built in a pistol, rifle, or a simple tubular configuration. The device is intended as a source of adjustable-frequency pulses from 10 to 1,000 repetitions per second of infrared energy at 900 nanometers

The system is shown in Figure 1-1 as being built on a perforated board assembly and combination copper ground plane that all fits into a tubular enclosure with a lens and holder for a collimator. The housing serves as the enclosure for the batteries and contains the control panel at its rear. It may be fitted with a handle that can hold an optional trigger switch when the device is designed in a gun configuration. A conventional sighting system is also easily adapted to the device.

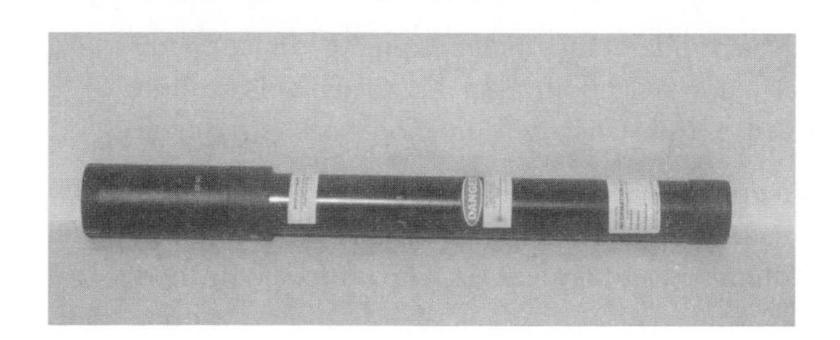

Figure 1-1 *Photograph of laser*

The laser is intended as a rifle-type simulated weapon, whose range can be several miles or as a long-range, laser-beam protection fence with a similar range of several miles. It is intended to be used with our high-speed laser pulse detector described in Chapter 2.

The laser, when assembled as shown, is a class 3B FDA certified device and requires the appropriate labeling and several included safety functions, as described in the assembly instructions. At no time should it be pointed at anyone without protective eyewear or at anything that could reflect the pulses. Never look into the unit when the power is on. The device is intended to be used for ranging, simulated weapons practice, intrusion detection, communications and signaling, and a variety of related scientific, optical experiments and uses.

This is an intermediate- to advanced-level project requiring electronic skills and basic electronic shop equipment. Expect to spend $100 to $150. All parts are readily available, with specialized parts obtainable through Information Unlimited (www.amazing1.com), and they are listed in Table 1-1.

Theory of Operation

A laser diode is nothing more than a three-layer device consisting of a *pn* junction of *n*-type silicon, a *p* type of gallium arsenide, and a third *p* layer of doped gallium arsenide with aluminum. The *n*-type material contains electrons that readily migrate across the *pn* junction and fill the holes of the *p*-type material. Conversely, holes in the *p* type migrate to the *n* type and join with electrons. This migration causes a potential hill or barrier consisting of negative charges in the *p*-type material and positive charges in the *n*-type material that eventually ceases growing when a charge equilibrium exists. In order for current to flow in this device, it must be supplied at a voltage to overcome this potential barrier. This is the forward voltage drop across a common diode. If this voltage polarity is reversed, the potential barrier is simply increased, assuring no current flow. This is the reversed bias condition of a common diode.

A diode without an external voltage applied to it contains electrons that move and wander through the lattice structure at a low, lazy average velocity as a function of temperature. When an external current at a voltage exceeding the barrier potential is applied, these lazy electrons increase their velocity so that some of them, by colliding, acquire a discrete amount of energy and become unstable. They eventually emit the acquired energy in the form of a photon after returning to a lower-energy state. These photons of energy are random both in time and direction; hence, any radiation produced is incoherent, such as that of an LED.

The requirement for coherent radiation is that the discrete packets of radiation must be in the form of a lockstep phase and in a definite direction. This demands two essential requirements: first, sufficient electrons at the necessary excited energy levels, and second, an optical resonant cavity capable of trapping these energized electrons to stimulate more electrons and give them direction. The amount of energized electrons is determined by the forward diode current. A definite threshold condition exists where the device emits laser light rather than incoherent light, such as in an LED. This is why the device must be pulsed with high current. The radiation from these energized electrons is reflected back and forth between the square-cut edges of the crystal that form the reflecting surfaces due to the index of refraction of the material and air.

The electrons are initially energized in the region of the *pn* junction. When these energized electrons drift into the *p*-type transparent region, they spontaneously liberate other photons that travel back and forth in the optical cavity interacting with other electrons, commencing laser action. A portion of the radiation traveling back and forth between the reflecting surfaces of these mirrors escapes and constitutes the output of the device.

Circuit Theory of Operation

Figure 1-2 shows the inverter section increasing the 12 volts of the portable battery pack to 200 to 300 volts performed by the circuit, which consists of a switching transistor (Q1) and step-up transformer (T1). Q1 conducts until saturated for a time until the base reverse biases and can no longer sustain it at an on state and Q1 turns off. This causes the magnetic field in its collector winding (COL) to collapse, thus producing a stepped-up voltage in the secondary (SEC) of proper phase. The variable resistor (R3) controls the charging current to the capacitor (C2), the transistor turnoff time, and consequently system power.

The stepped-up square wave voltage on the secondary of T1 is rectified by diode (D1) and integrated onto the storage capacitor (C3). The trigger circuit determines the pulse rep rate of the laser and uses a timer (I1) whose pulse rate is determined by the timing resistor and capacitor (R9 and C6). R9 is adjustable for changing the laser pulse rate. The output trigger pulse is differentiated by capacitor (C4) with negative overshoot being clipped by diode D2. This differentiated pulse is fed to the gate terminal of the *silicon controlled rectifier* (SCR) switch.

The discharge circuit generates the current pulse in the laser diode (LD1) and consequently is the most important section of the pulser. The basic configuration of the pulse power supply is shown in the system schematic. The current pulse is generated by the charging storage capacitor C3 being switched through the SCR and laser diode LD1. The rise time

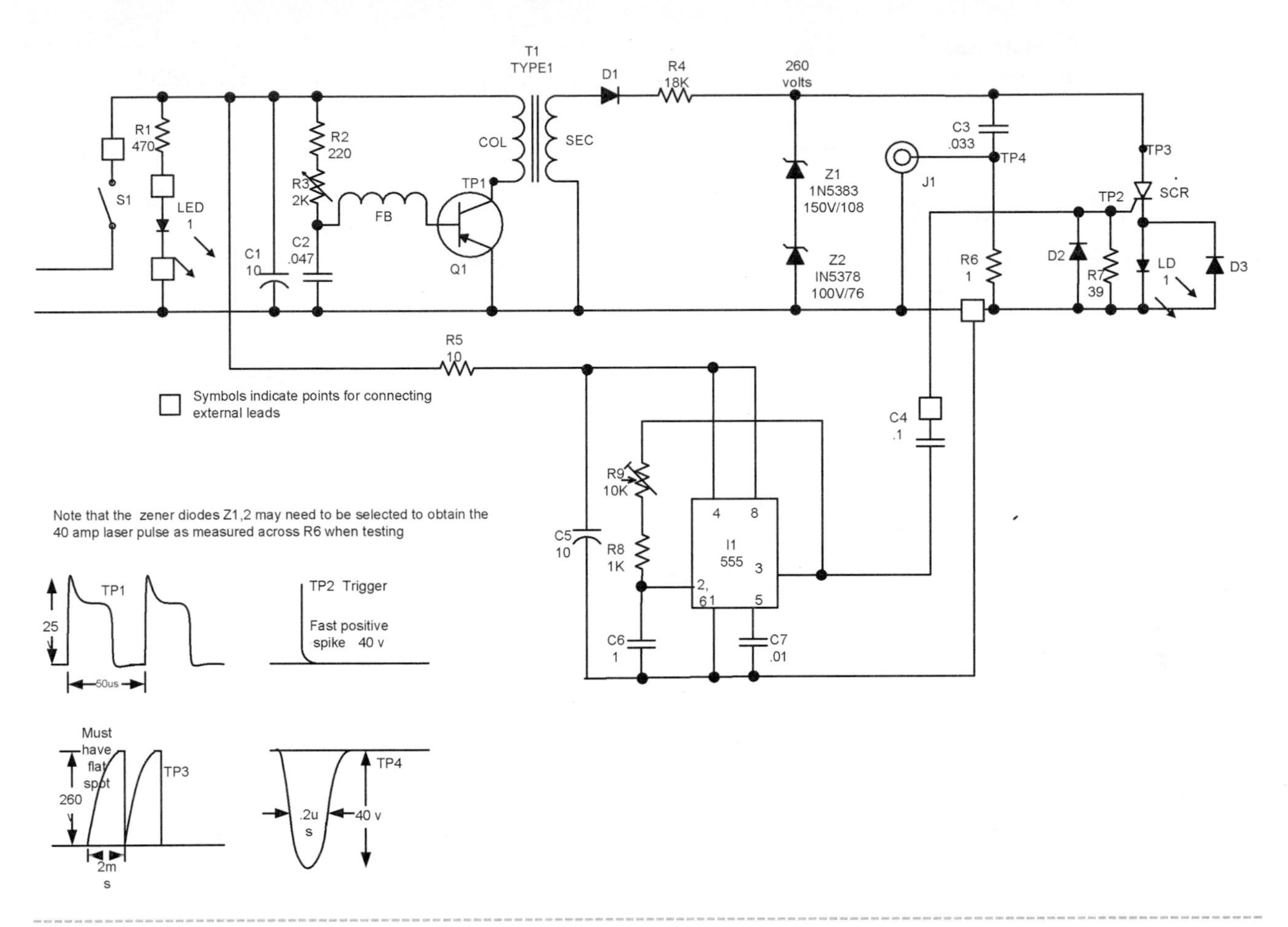

Figure 1-2 *Circuit schematic*

of the current pulse is usually determined by the SCR, while the fall time is determined by the capacitor value and the total resistance in the discharge circuit.

Figure 1-3 shows the typical anode voltage and current waveforms of the SCR during the current pulse through the diode laser. The peak current, pulse width, and voltage of the capacitor discharge circuit are related for various load and capacitance values. The peak laser current and charged capacitor voltage relationships are given for several different capacitor values and typical laser types. The voltage and current limits of the SCR are also shown. Short pulse widths provide less time for the SCR to turn on than longer pulse widths; therefore, the SCR impedance is higher and more voltage is required to generate the same current. Also shown are the current pulse waveforms for the three different values of the capacitance. The capacitor is charged to the same voltage in all three cases, that is, 400 volts.

In conventional SCR operation, the anode current, initiated by a gate pulse, rises to its maximum value in about 1 microsecond. During this time, the anode-to-cathode impedance drops from an open circuit to a fraction of an ohm. In injection laser pulsers, however, the duration of the anode-cathode pulse is much less than the time required for the SCR to turn on completely. Therefore, the anode-to-cathode impedance is at the level of 1 to 10 ohms throughout most of the conduction period.

The major disadvantage of the high SCR impedance is that it causes low circuit efficiency. For example, at a current of 40 amps, the maximum voltage would be across the SCR, while only 9 volts would be across the laser diode. These values represent very low circuit efficiency.

The efficiency of a laser array is greater due to its circuit impedance being more significant. Because the SCR is used unconventionally, many of the standard specifications such as peak current reverse

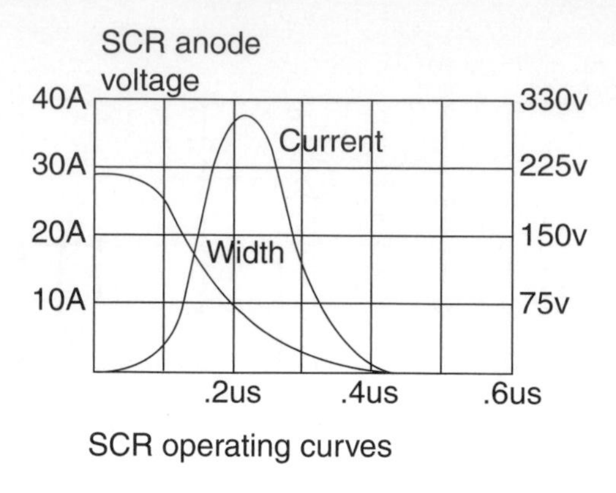

SCR operating curves

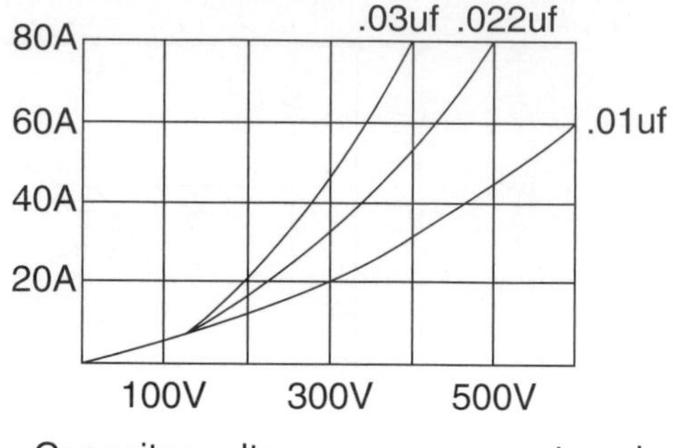

Capacitor voltage versus current peak

Typical curves and parameters for laser pulses using SCR silicon controlled rectifier and capacitor discharges

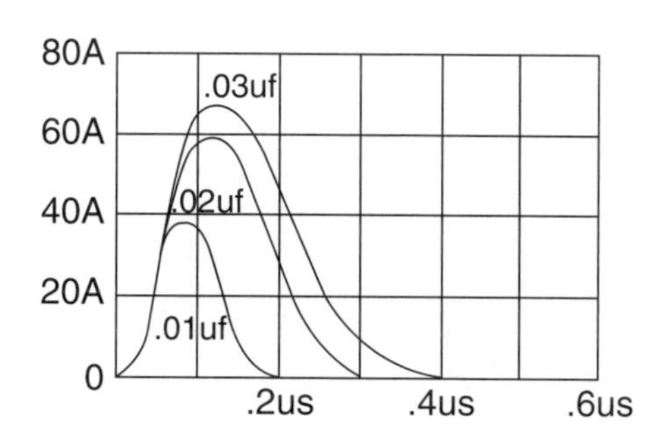

Capacitor value versus current peak

Figure 1-3 *SCR operating curves*

voltage, on-state forward voltage, and turn-off time are not applicable. In fact, it is difficult to select an SCR for a pulsing circuit on the basis of normally specified characteristics. The specifications important to laser pulser applications are the forward-blocking voltage and current rise time. A *use test* is the best and often the only practical method of determining the suitability of a particular SCR.

The voltage rating of the storage capacitor must be at least as high as the supply voltage. With the exception of ceramic types, most capacitors (metallized paper, mica, etc.) will perform well in this circuit. Ceramic capacitors have noticeably greater series resistance but are usable in slower speed pulsing circuits.

Lead lengths and circuit layout are very important to the performance of the discharge circuit. Lead inductance affects the rise time and peak value of the current, and it can also produce ringing and undershoot in the current waveform that can destroy the laser. A well-built discharge circuit might have a total lead length of only 1 inch and therefore an inductance of approximately 20 nanohenries. If the current rises to 75 amperes in 100 nanoseconds, the inductive voltage drop across this lead can be 15 volts. It is now obvious that if proper care is not taken in wiring the discharge circuits, high-inductive voltage drops will result.

A 1-ohm resistor in the discharge circuit will greatly reduce the current undershoot in single-diode lasers. Laser arrays usually have sufficient resistance to eliminate undershoot. The small resistance in the discharge circuit is also useful in monitoring the laser current, as described in the following section.

A clamping diode (D3) is added in parallel with the laser to reduce the current undershoot. Its polarity should be opposite that of the laser. Although the clamping diode is operated above its usual maximum current rating, the current undershoot caused by ringing is very short and the operating life of the diode is satisfactory.

Current Monitor

The current monitor in the discharge circuit provides a means of observing the laser current's waveform with an oscilloscope. A resistive-type monitor (R6) reduces circuit ringing and current undershoot, but the lead inductance of the resistor may cause a current reading that is higher than it actually is. A current transformer such as the Tektronix CT-2 can also be used to monitor the current and is not affected by lead inductance. Because the transformer does not respond to low-frequency signals, it should be used with fast waveforms that have a short pulse width and a fast fall time.

Charging Circuits

The second major section of the pulser is the charging circuit. This circuit charges the capacitor to the supply voltage during the time interval between the laser current pulses. It also isolates the supply voltage from the discharge circuit during the laser current pulse, thereby allowing the SCR to recover to the blocking state. Because the response times of the charging circuit are relatively long, lead lengths are not important and the circuit can be remotely located from the discharge circuit.

The simplest charging circuit is a resistor-cap combination. The resistor must limit the current to a value less than the SCR holding current, but it should be as low as practical, because this resistance also determines the charging time of the capacitor, C3.

Chassis Assembly

The following is a list of steps to construct the laser project.

1. Fabricate the copper chassis, as shown in Figure 1-4. This part is shown in Table 1-1 as being available.

2. Assemble the board as shown in Figure 1-5. Lay out and identify all the parts and pieces. Separate the components that go to the assembly board and fabricate the (PB1) perfboard from a 1.4 × 5-inch piece of .1 × .1 grid. Note the two holes for attachment to the copper chassis section from Figure 1-6.

3. Assemble the components as shown and observe the polarity of the capacitors and semiconductors.

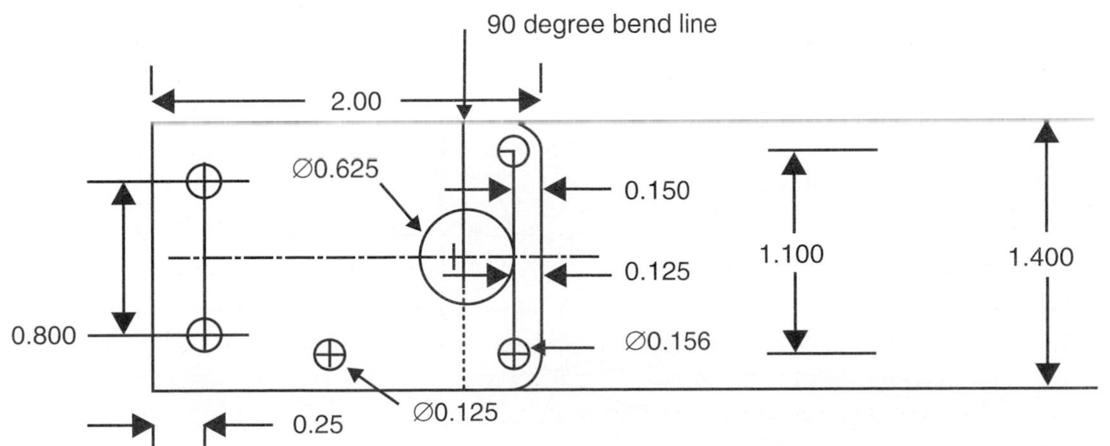

Figure 1-4 *Fabrication of laser chassis*

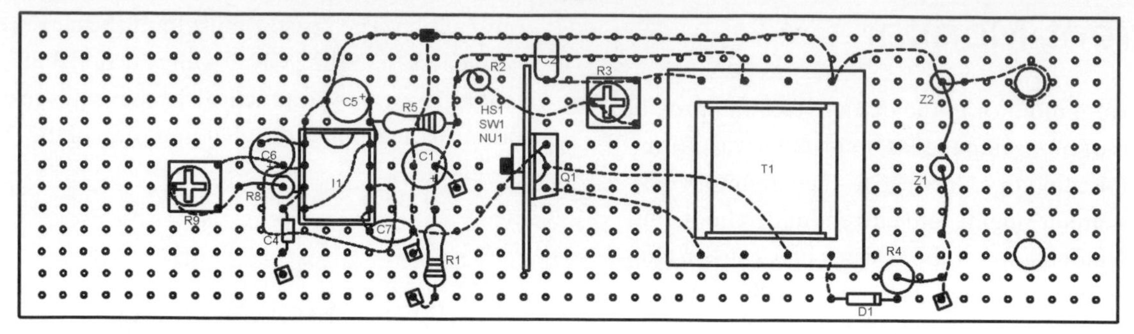

Drill two .155 holes on .8" centers for mating to copper chassis from figure 1-4

Fabricate the heatsink for Q1 from a piece of .06 aluminum 1 x .65"' with a .15 hole for SW1. Round off corners for clearance when fitting into the tubular enclosure.

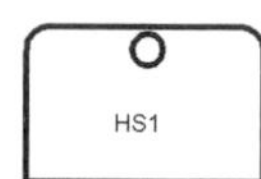

Dashed lines are connections on underside of perforated circuit board using the component leads. These points can also be used to determine foil runs for those who wish to fabricate a printed circuit board.

- Small dots are holes used for component insertion.

■ Black squares are solder junctions

▣ Squares with dots are holes that may require drilling for component mounting and points for external connecting wires indicating strain relief points. It is suggested to drill clearance holes for the leads and solder to points beneath board.

Figure 1-5 *Assembly board parts identification*

4. Connect the components using their actual leads wherever possible. Follow the layout and avoid bare-wire bridges.

5. Figure 1-6 shows the wiring of the copper chassis (COPCHA1). This type of wiring is preferred for high-frequency, sensitive, low-noise circuitry. It is used in this circuit due to the high-current pulses with fast rise times.

   Start by forming the leads and soldering to the copper plate. You will need a good hot soldering iron to do this correctly. The components themselves serve as anchor points and will be self-supporting when soldered to one another.

6. Figure 1-7 shows the interconnection of the copper chassis and the perforated assembly board via the two screws and nuts. Attach and connect the external leads and components as shown.

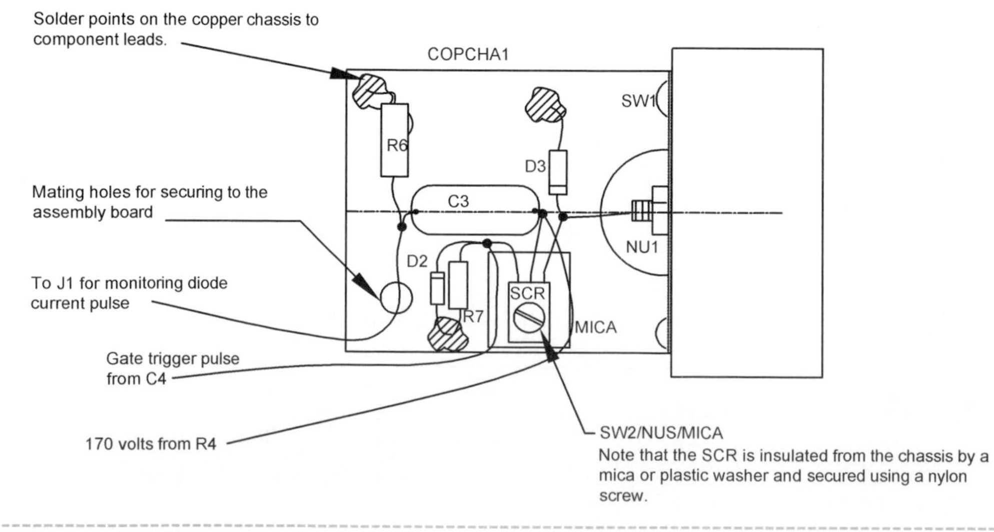

Figure 1-6 *Assembly of laser chassis*

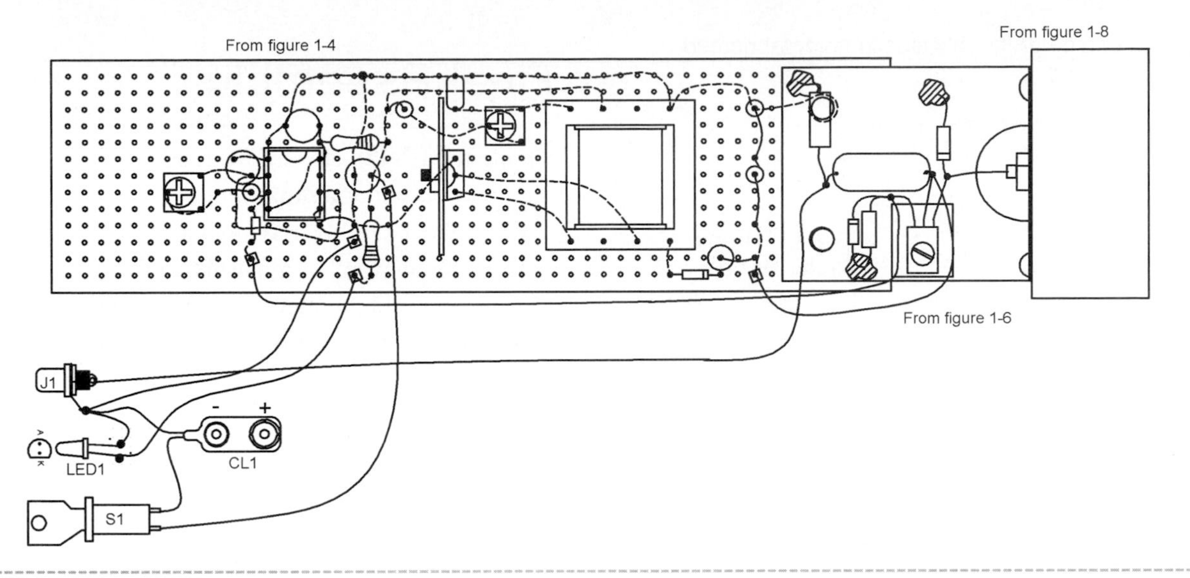

Figure 1-7 *Interconnecting the assembly board to the laser chassis*

7. Fabricate the LAHOLD1, as shown in Figure 1-8. Mount laser diode LD1 into the center hole and secure via a nut (NU1). Secure this assembly to the bent-down flange of the copper chassis. Use two 1/4-inch screws as shown.
8. A collimator adapter is shown and is optional.

Assembly Testing

The following steps will require basic testing equipment.

1. It is assumed at this point that the assembly is complete, as shown in Figure 1-7. Verify the circuit and preset the two trimpots R3 and R9 to midrange.
2. Obtain a 12-volt bench power supply and connect the 12 volts to the battery clip CL1. Turn on the key switch and note the LED coming on. A current meter on the bench supply should read approximately 300 milliamps.
3. Connect a scope to TP1, as shown in Figure 1-2, and adjust R3 for the waveshape as shown.
4. Connect the scope to TP2 and note a spiky waveform that verifies the trigger pulse to the SCR.
5. Connect the scope to TP3 and note the waveform, which is the voltage across the SCR. Note the flat spot indicating that capacitor C3 has reached full charge before the SCR is triggered. The top of this waveshape is affected by the pulse repetition rate and should support 500 repetitions per second when set by R9.
6. Connect the scope to TP4 and set it for a negative-going fast pulse, indicating the current waveform through the LD1 laser diode. If you have followed the instructions and your assembly of the copper chassis is secure with short leads, you should see a picture-perfect waveform as shown.

   Note that the current pulse is 40 amps and corresponds to 1 amp for every volt read at TP4. If you find the pulse is less, you can increase the voltage ratings on one of the zener diodes. The laser diode specified requires 40 amps for full output. Any value over 40 amps will greatly shorten the useful life of this part.
7. Verify the laser output by using a night vision scope or a video camera. You should see an indication of the output. Screw in LEN1 so it is flush with the holder and adjust for the smallest footprint at over 50 feet. Our lab model adjusted to about two turns below being flush with the holder for the best results. The output should be a sharp, narrow bar shape with the diode specified.

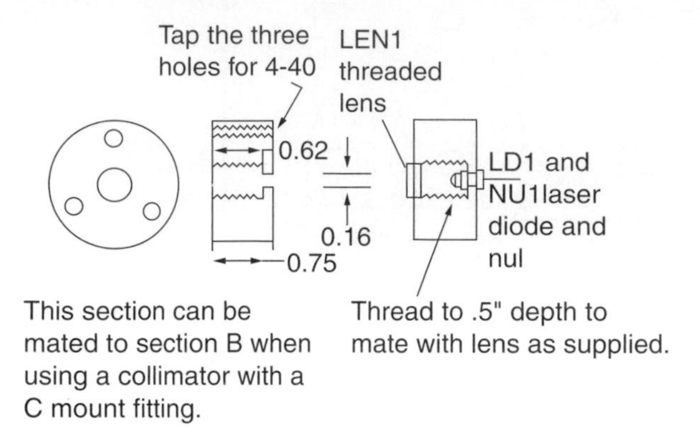

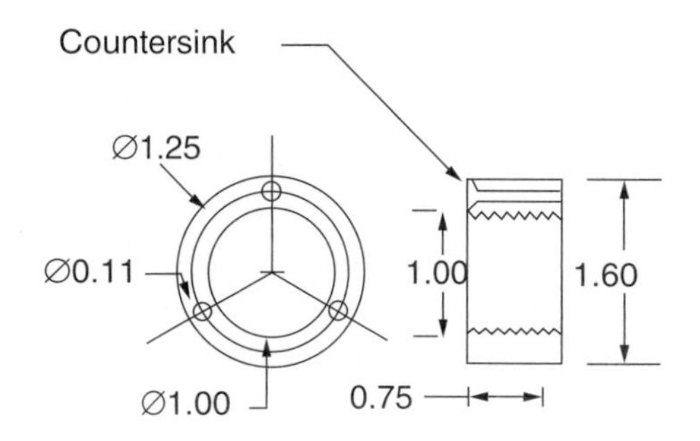

Figure 1-8 *Laser diode and lens holder*

Final Assembly

Create the EN1 enclosure from a 12 × 1.5-inch inner diameter plastic or aluminum tube (see Figure 1-9). Enclose the assembly and label it as shown in Figure 1-10. You may want to construct a mounting block or another assembly to mechanically connect the laser to a good, sturdy video tripod for future use.

Special Note

This laser is rated at 10 to 20 watts and is not to be confused with a continuous-output device rated at this power level. The advantage of this system is that the pulses are optically equivalent to 10 to 20 watts, and a suitable detection system will see them at this power level.

The equivalent energy output is related to the repetition rate times the peak power times the pulse duration and is approximately equal to joules. A joule is a watt second and is energy. If that energy is released in 1 microsecond as a pulse, the power of that pulse is now 1 megawatt.

A very suitable detector for this pulsed laser is our laser light detector described in Chapter 2, and a useful application for it as a laser property security fence is described in Chapter 5. Note that a laser pointer is 5 milliwatts, meaning the pulsed laser is 5,000 times more detectable.

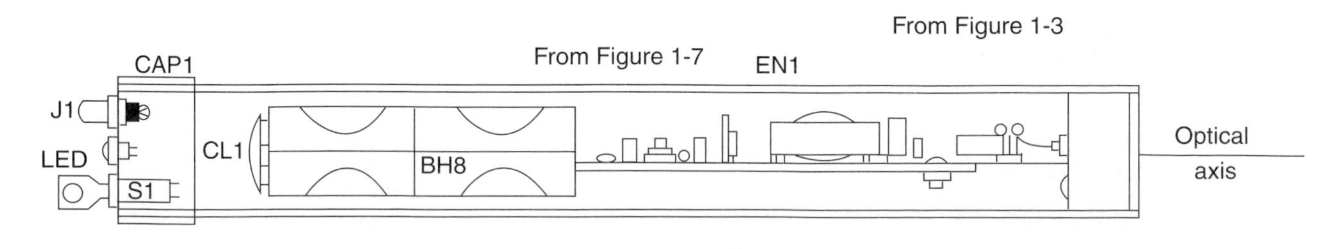

Figure 1-9 *X-ray view of enclosure*

CAUTION: Viewing of laser beam of direct reflections require the use of protective eye wear available through Information Unlimited at www.amazing1.com

Instruction

1. Turn key switch counterclockwise to verify unit is off. Note key is easily removed in the "off" position
2. Remove CAP1 and insert 8 AA batteries into holder.
3. Turn key clockwise and note LED lighting. If not, check the battery holder. Note key cannot be removed in the on position. Laser energy is infrared and can not be seen with the eye. You must use infrared detection devices such as our laser pulse detector or suitable night vision devices. Video cameras will often detect the laser.
4. Adjust collimator for desired effect if you are using.

Note that collimator will expand beam at near field but greatly reduce it at far field.

Compliance tests
1. Verify correct labels as shown
2. Key switch — nonremoval in off position
3. Beam indicator LED indicates beam emission

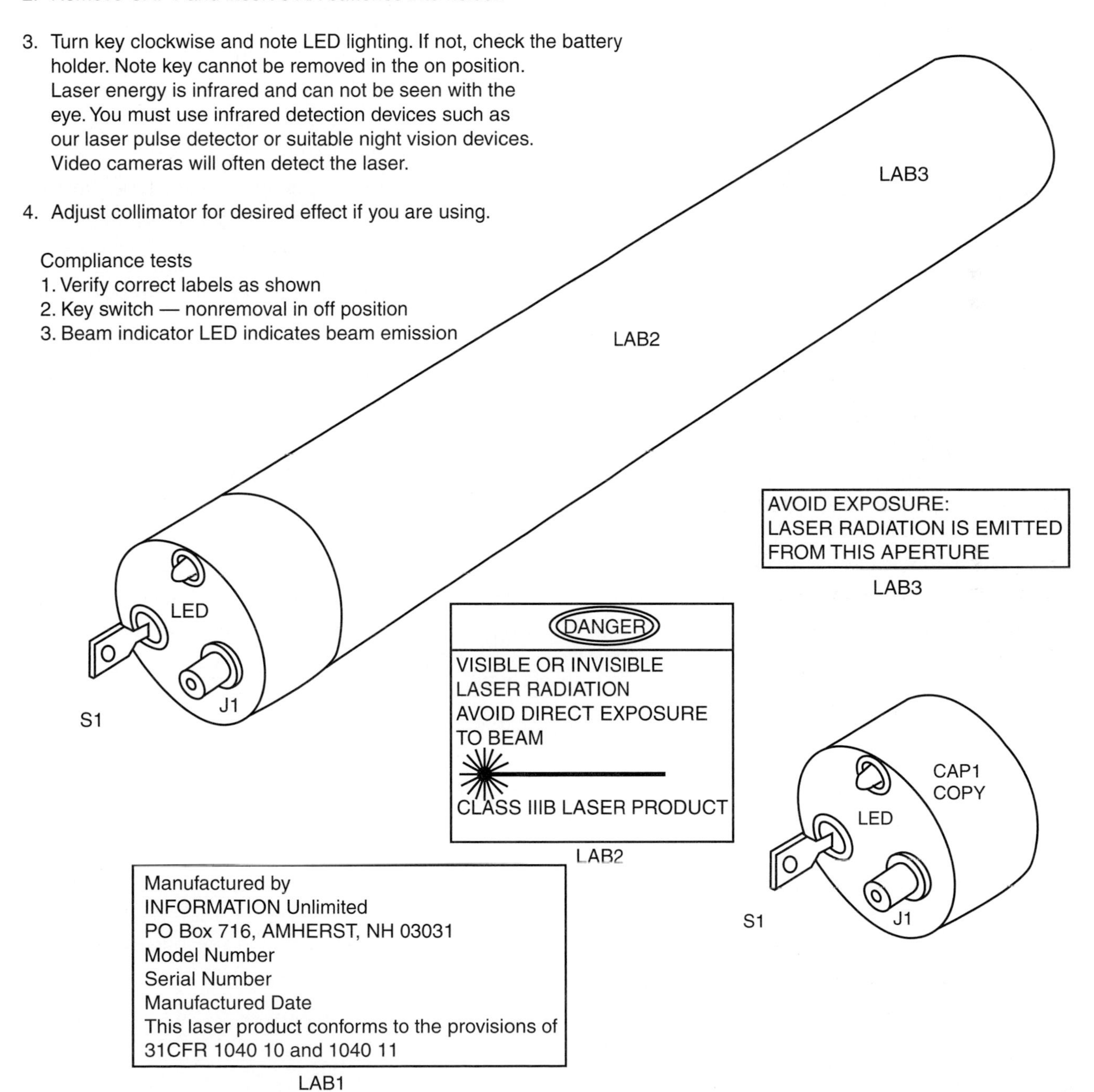

Figure 1-10 *Final assembly view and labels*

Table 1-1 Battery-Operated Infrared Laser Parts List

Ref. #	Description	DB Part #
R1	470 1/4-watt resistor (yel-pur-br)	
R2	220 1/4-watt resistor (red-red-br)	
R3	2K horizontal trimmer resistor	
R4	18K, 3-watt metal oxide resistor	
R5	10-ohm, 3/4-watt resistor (br-blk-blk)	
R6	1-ohm, 1/2-watt carbon resistor (br-blk-gold)	
R7	39 1/4-watt resistors (or-wh-blk)	
R8	1K 1/4-watt resistor (br-blk-red)	
R9	10K horizontal trimmer resistor	
C1,5	Two 10-microfarad, 50-volt vertical electrolytic capacitors	
C2	.047-microfarad, 50-volt plastic capacitor 473	
C3	.033-microfarad, 250-volt polypropylene capacitor	
C4	.1-microfarad, 50-volt plastic capacitor	
C6	1-microfarad, 50-volt vertical electrolytic capacitor	
C7	.01-microfarad/50-volt plastic capacitor (103)	
D1,3	Two 1N4937 1-kilovolt, 1-amp diode	
D2	IN914 silicon diode	
LED1	Bright green *light-emitting diode* (LED)	
LD1	40-amp, 10- to 20-watt, 904-nanometer TO18 pack laser diode	DB# LD650
SCR1	C106M 600-volt, 4-amp *silicon-controlled rectifier* (SCR)	DB# C106M
Z1	100/76-volt, 5-watt zener diode 1N5378	
Z2	150/108-volt, 5-watt zener diode 1N5383	
Q1	MJE3055T NPN power transistor in TO220 case	
I1	555 *dual inline package* (DIP) timer	
T1	400-volt square-wave-switching transformer	DB# TYPE1PC
PB1	1.4 × 5-inch, .1 × .1 grid perforated board	
S1	Key switch with non-removable key in on position	DB# KEYSWSM
J1	RCA phono jack	
COPCHAS1	Copper chassis, created in Figure 1-4	DB# COPCHAS1
LAHOLD	Laser and lens holder, created in Figure 1-8	DB# LAHOLD1
LENS13	Threaded lens	DB# LENS13
EN1	1 1/2-inch inner diameter × 1/32-inch wall × 12-inch plastic or aluminum housing tube	
CAP1,2	1 5/8-inch plastic caps, reworked as shown in Figure 1-9	
BH8	Eight AA cell holder	
SW1	Five 6-32 × 1/4-inch screws	
NU1	Four 6-32 hex nuts	
SW2	4-40 × 3/8-inch nylon screw	
MICA	Mica insulating washer	
NU2	4-40 hex nut	
HSINK	Heatsink tab for Q1	

Chapter Two

High-Speed Laser Pulse Detector

This project shows how to build a specialized device capable of detecting fast, high-speed optical pulses. A suitable laser generating these pulses is described in Chapters 1 and 4 and can have a range, limited only by the curvature of the earth. Its uniqueness is its ability to resolve these pulses with rise times in the nanosecond range. This capability makes it possible to transmit large amounts of information within a small period of time.

The heart of this device is the actual front-end detector, in this case a PIN diode device. PIN comes from *positive negative* (PN) junction with intrinsic layer (I). The characteristics of this detector are low capacitance, high resistance, and low leakage when operating in the reverse bias mode. It may also be used as a photo cell when operated in the forward mode.

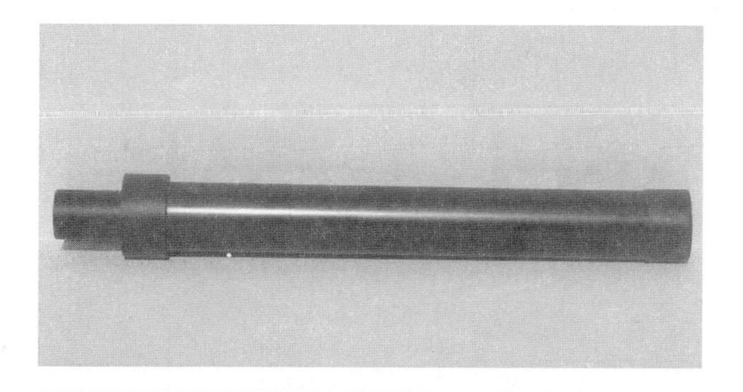

Figure 2-1 *The high-speed laser pulse detector*

The first part of the circuit is built on a piece of thin copper and uses discrete components. The builder may choose to build on a fully copper-clad circuit board if available, and a facsimile output can be set up via a phono jack or a BNC connector. A signal-conditioning circuit is shown built on a normal perforated circuit board and is designed to actuate a relay in the absence of the laser pulses. This mode of operation allows its use as a long-range laser perimeter or invisible fence intrusion detector, as described in Chapter 5.

Optics are shown for those wanting to operate at long ranges. Machined fittings are shown to mate with the 1.5-inch inner-diameter plastic enclosure used for the housing.

This is an intermediate- to advanced-level project requiring certain electronic skills, and one should expect to spend $35 to $70. All parts are readily available, with specialized parts obtainable through Information Unlimited (www.amazing1.com), and they are listed in Table 2-1.

Circuit Description

The circuit in Figure 2-2 utilizes a pin photodiode (LPIN) reverse biased by resistor R3. The reverse-biased photodiode produces a voltage across bias resistor R2. The response of the PIN photo detector is determined by the resistance of R2 and is relatively low due to the nanosecond rise time of the laser pulses detected. Capacitor C3 couples the positive-going detected laser pulses to the gate of a preamplifying *field effect transistor* (FET) Q1 where the negative-going pulse is amplified and fed into a second FET (Q2) for further amplification. NPN transistor Q3 again amplifies the positive-going pulse for Q2 to a negative pulse. This amplified signal is a reasonably good facsimile of the actual laser pulse and can be monitored or further processed via phono jack J1.

The negative-going pulses are also fed to a Schmidt discriminator that is in an untriggered state, providing that transistor Q4 continues to see these pulses. When the pulses are interrupted by an object in the laser path, Q4 turns on due to a positive bias set by trimpot resistor R15 and turns transistor Q5 off, regenerating the trigger state of the Schmidt discriminator. R15 allows the adjustment of a trigger threshold if required. The triggering time of the circuit also helps prevent false triggering such as flying birds, windblown debris, and insects.

The output of the Schmidt is DC coupled into the base of the emitter-follower transistor Q6 where the signal is inverted by transistor Q7. The negative-going level at the collector of Q7 now triggers timer I1 on for a preset time that controls an alarm or another function resulting from a qualified laser beam interruption. A relay (RE1) is shown in the circuit and can be used as normal open or normal closed contacts for control of external circuit functions. This is shown in the phasor pain field generator in Chapter 35.

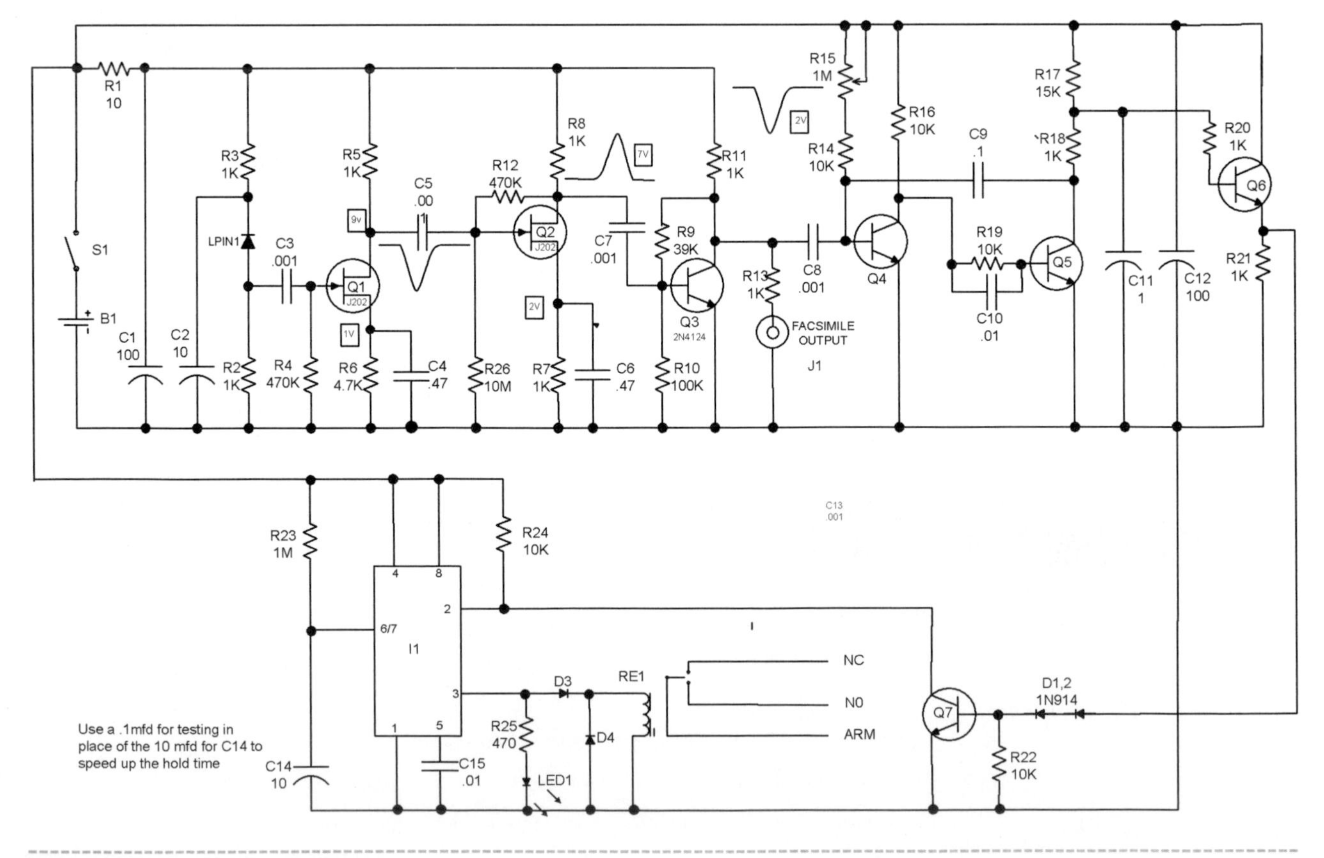

Figure 2-2 *Circuit schematic*

System Assembly

1. Assemble the board, as shown in Figure 2-3. Lay out and identify all the parts and pieces. Separate the components that go to the assembly board and create the PB1 perfboard from a 1 1/2 × 5 1/2-inch piece of .1 × .1 grid. Note the two holes for attachment to the copper chassis section.
2. Assemble the components as shown and observe polarity of the capacitors and semiconductors.
3. Connect the components using their actual leads wherever possible. Follow the layout and avoid bare-wire bridges. Note that Figure 2-3 shows pieces of #20 bus lead used as VC+ and COMMON rails. This approach is easier for wiring as it gives a direct route for the component leads.
4. Create the copper chassis as shown in Figure 2-4. This part is listed as a fabricated piece in Table 2-1.
5. Figure 2-5 shows the wiring of the COPCHA1 copper chassis. This type of wiring is preferred for high-frequency and sensitive, low-noise circuitry. It is used in this circuit due to the fast-rise-time, low-level-voltage pulses.

   Start by forming the leads and soldering to the copper plate. You will need a good, hot soldering iron to do this correctly. The components themselves serve as anchor points and will be self-supporting when soldered to one another.
6. Figure 2-6 shows the connection of the copper chassis and the perforated assembly board via the two screws and nuts. Attach and connect the external leads and components as shown.

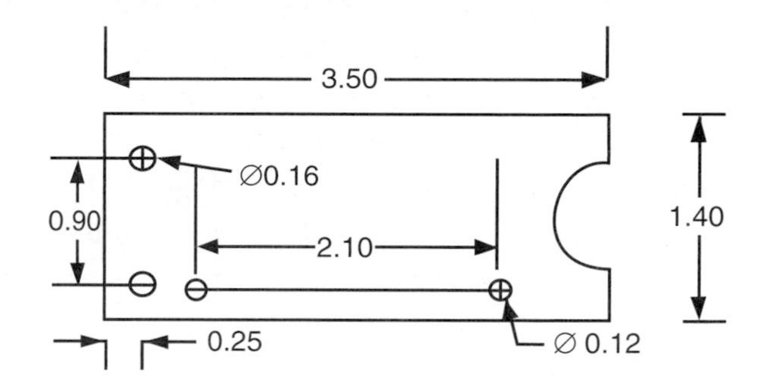

Figure 2-4 *Fabrication of the laser chassis*

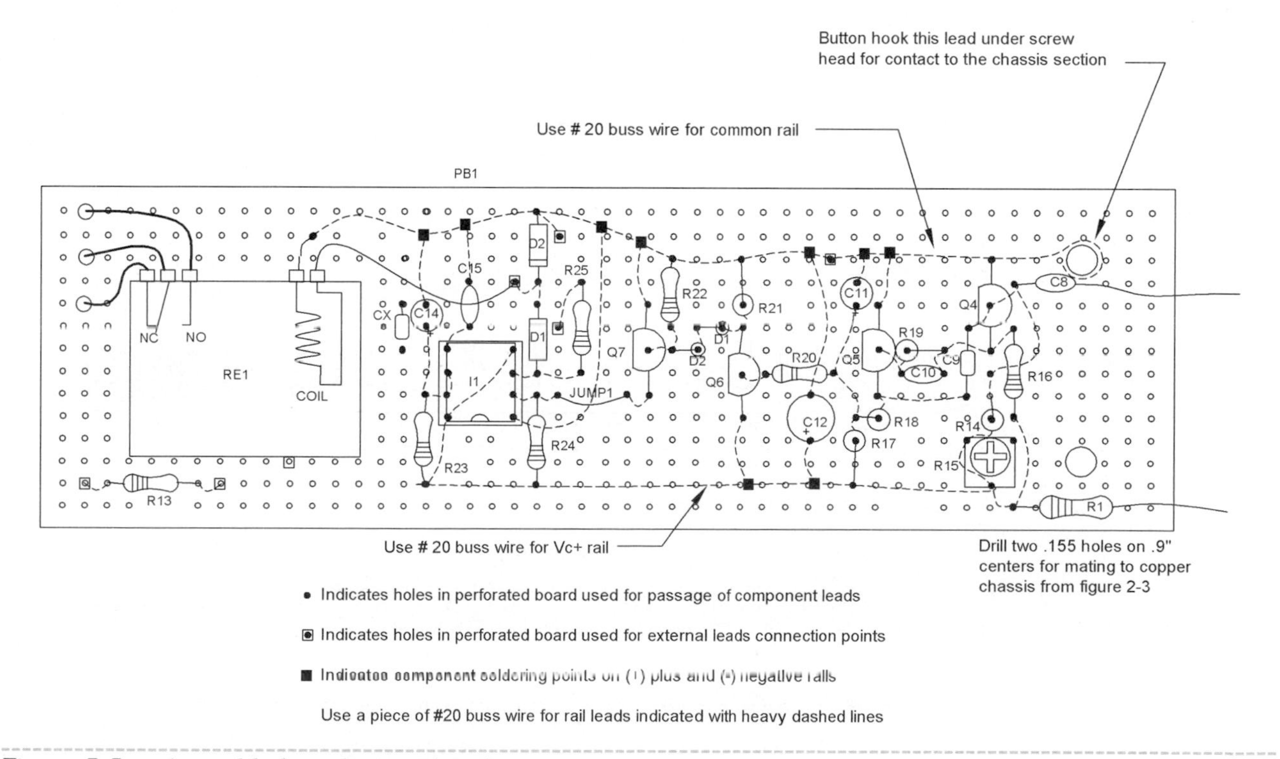

Figure 2-3 *Assembly board parts identification*

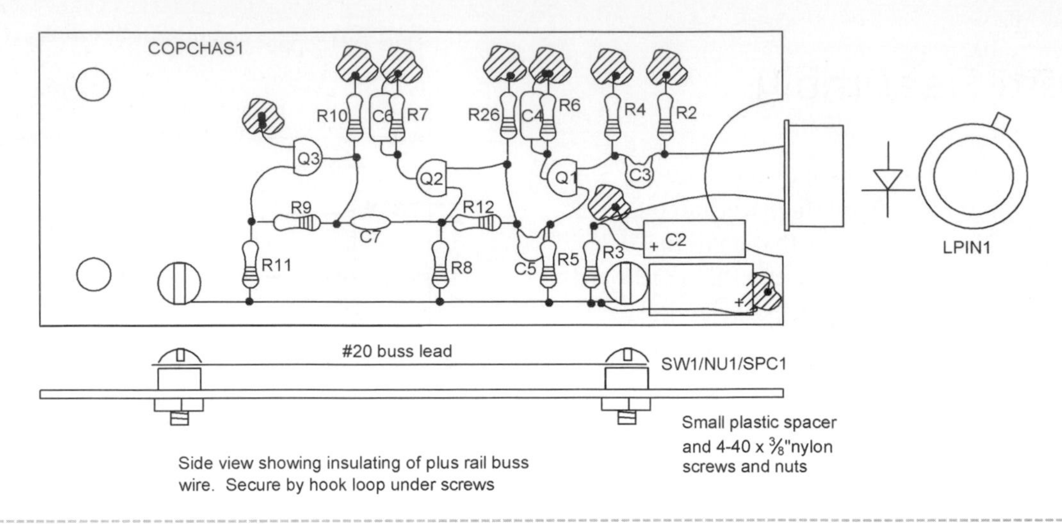

Figure 2-5 *Assembly of the laser chassis*

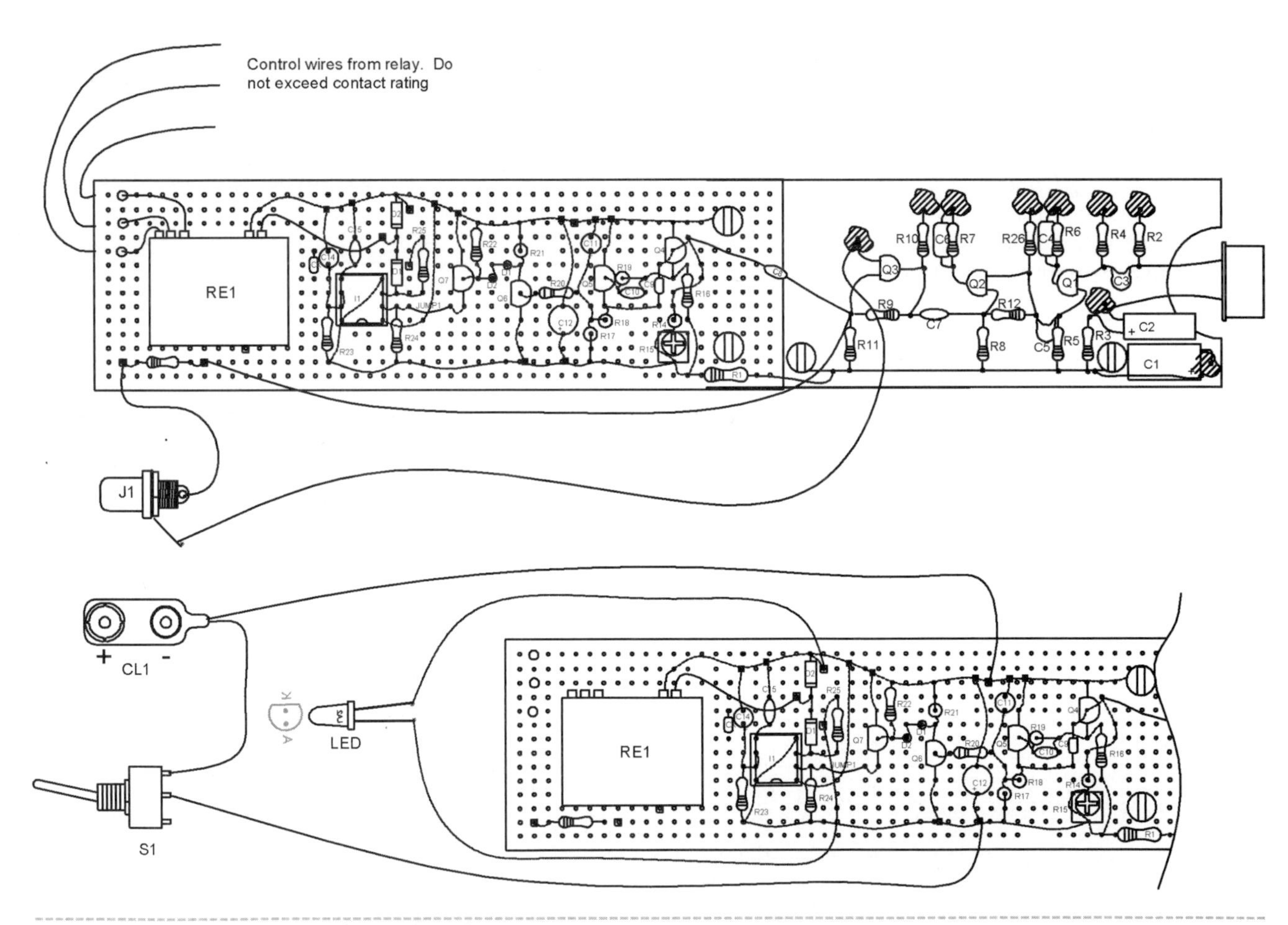

Figure 2-6 *Connection of board to chassis*

Chapter Two

Assembly Testing

Testing of the laser receiver involves the use of a compatible laser transmitter to provide the necessary high-speed pulses. The laser described in Chapter 1 is recommended.

1. It is assumed at this point that the assembly is complete, as shown in Figure 2-6. Verify the circuit and preset the R15 trimpot to midrange. This test is performed *without* the circuitry installed and without a lens. You should temporarily replace C14 with a .1-microfarad value.
2. Obtain a 12-volt bench power supply and connect the 12 volts to the battery clip CL1. Turn on the S1 switch and note the LED coming on and flashing erratically. A current meter on the bench supply should read approximately 10 milliamps.
3. Measure the DC voltage levels, as indicated in Figure 2-2, noted in the squares. These values should be within 10 to 15 percent of those indicated.

   At this point, you will need an optical laser pulse transmitter, such as that described in Chapter 1. Any suitable laser or optical transmitter will work with a fast pulse rise of 1 usec or so. The detector, as shown, uses a silicon PIN diode with a spectral sensitivity peak at 960 nanometers. This is very close to the spectral output of the lasers described in Chapters 1 and 4.
4. Position the laser pulse generator source so that it is pointing in the direction of the laser receiver being tested. You will need to connect a scope at the drain of Q1 and then attempt to position the receiver so it is receiving the pulses and is secure in this position. This may be tricky and require patience. When the laser is perfectly aligned with the receiver, the signal will block the Q1 and produce an overloaded waveshape. Carefully adjust the receiver off axis until the scope indicates the negative-going facsimile pulse as shown. Continue through the circuit and observe the waveshapes as shown, again adjusting the receiver off axis to obtain the required waveshape.

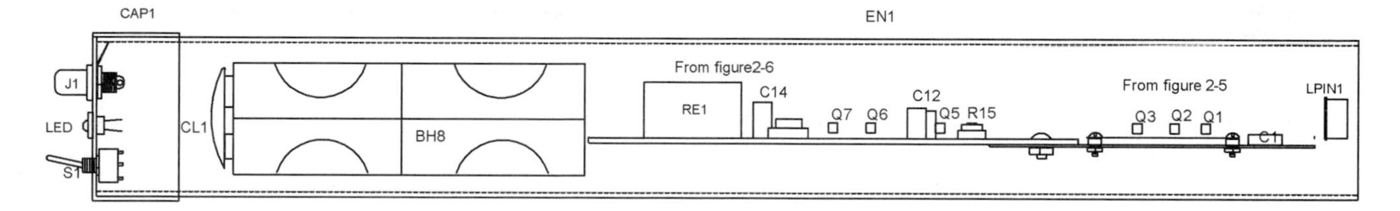

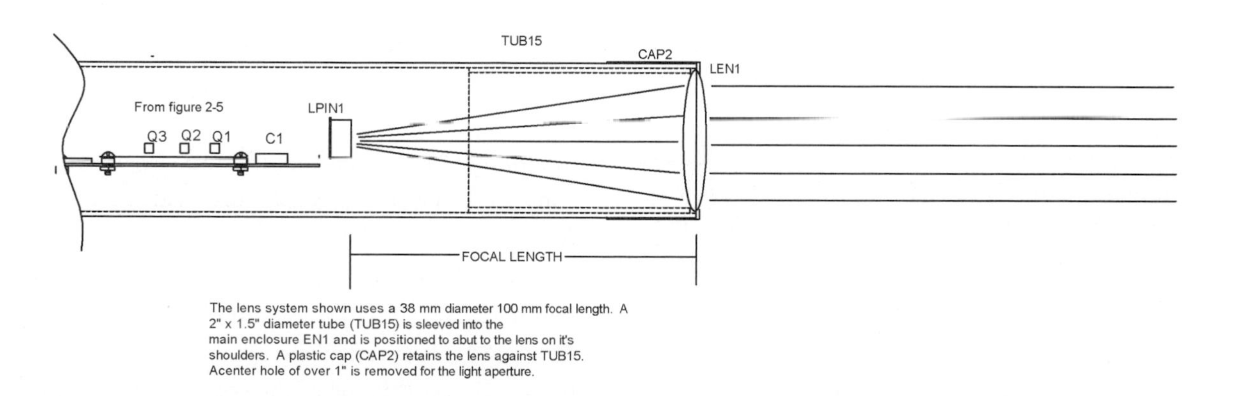

Figure 2-7 *X-ray view of enclosure*

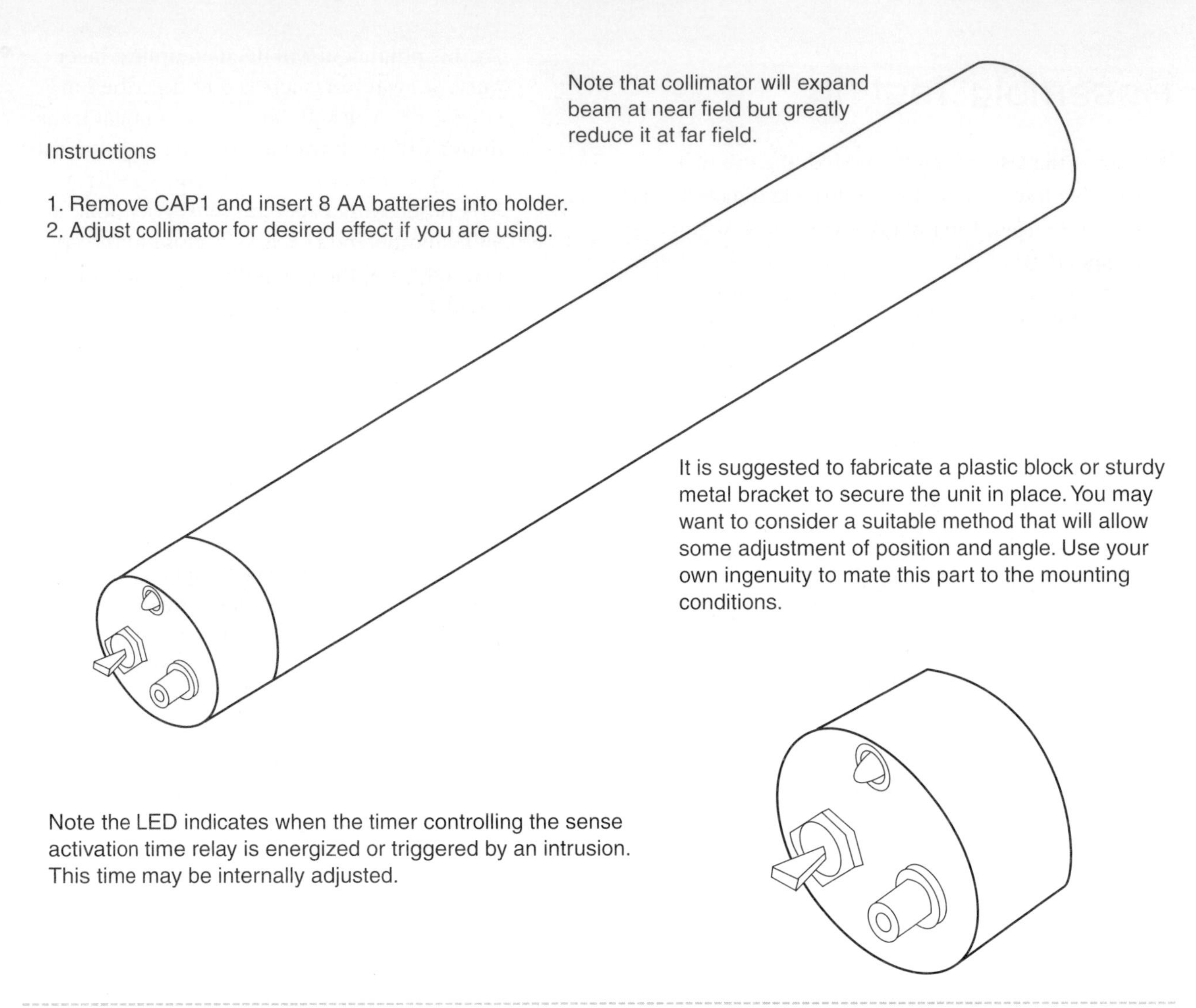

Figure 2-8 *Final assembly view*

Note that the LED will be off if the circuit is working and will come on if you block the light with a solid object.

5. Verify that the discriminator works and you can set the sensitivity trip level via R15. Verify the timer trips when the light beam is broken, activating the relay for a preset time as determined by the value of C14. This feature now allows external circuitry to be activated when the laser beam is broken, powering alarms or deterrents. The relay as shown provides a normally closed and open circuit that can be selected by the user. Do not switch loads that exceed the relay contact rating.

Final Assembly

Fabricate the EN1 enclosure from a 15 × 1.5-inch ID plastic or aluminum tube. Enclose the assembly as shown in Figure 2-7. You may want to create a mounting block or another assembly to mechanically connect the laser to a good, sturdy video tripod for future use

The system as shown is enclosed in a long tube that includes the batteries. You may separate the batteries and relay control circuits from the actual detection section consisting of the copper chassis and optics. This approach will be a two-part system that is connected via a cable or cord.

Special Note

This laser pulse detector receiver, when used with a pulsed laser like that described in Chapter 1, can have a control range exceeding that allowed by the curvature of the earth when used with suitable optics. Alignment is about as critical as sighting a high-powered rifle and requires extreme mechanical stability once set. The alignment of a long-range system requires patience, perseverance, and a little black magic.

Chapter 5 describes how to use both the laser pulser from Chapter 1 and the above assembly in a multiple-reflection property protection device with an ultrasonic shock deterrent.

Special Note on Photo Detectors

Two photo detectors are referenced in Table 2-1. LPIN is the low-cost piece that will usually provide the necessary operating parameters for most hobbyist laser projects. LPINX is more expensive and has a much lower dark current rating. It is used for lower noise and more sensitive circuitry.

Table 2-1 Laser Pulse Detector Parts List

Ref. #	Description	DB Part #
R1	10 ohms, 1/4 watt (br-blk-blk)	
R2, 3, 5, 7, 8, 11, 13, 18, 20, 21	Ten 1K, 1/4-watt resistors (br-blk-red)	
R4, 12	Two 470K, 1/4-watt resistors (yel-pur-yel)	
R6	4.7K, 1/4-watt resistor (yel-blk-red)	
R9	39K, 1/4-watt resistor (or-wh-or)	
R10	100K, 1/4-watt resistor (br-blk-yel)	
R14, 16, 19, 22, 24	Five 10K, 1/4-watt resistors (br-blk-or)	
R15	1M horizontal trimmer resistor	
R17	15K, 1/4-watt resistor (br-blk-or)	
R23	1M, 1/4-watt resistor (br-blk-grn)	
R25	470-ohm, 1/4-watt resistor (yel-pur-br)	
R25	10M, 1/4-watt resistor (br-blk-blue)	
C1, 12	Two 100-microfarad, 50-volt vertical electrolytic capacitors	
C2, 14	Two 10-microfarad, 50-volt vertical electrolytic capacitors	
C3, 5, 7, 8	Four .001-microfarad, 100-volt disc capacitors	
C4, 6	Two .47-microfarad, 50-volt plastic capacitors	
C9, 13	Two .1-microfarad, 50-volt plastic capacitors	
C10, 15	Two .01-microfarad, 50-volt plastic capacitors (103)	
C11	1-microfarad, 50-volt vertical electrolytic capacitor	
Q1, 2	Two J202 N-channel FET transistors	
Q3	2N4124 *negative positive negative* (NPN) high-frequency transistor	
Q4,5,6,7	Four PN2222 NPN general-purpose transistors	
D1,2	Two IN914 silicon diodes	
D3,4	Two IN4007 1-kilovolt, 1-amp diodes	
I1	555 *dual inline package* (DIP) timer integrated circuit	
LED1	High-brightness *light-emitting diode* (LED)	
*LPIN1	Silicon PIN photodiode dark current less than .10 nano-amp (see text)	DB# PIN
*LPIN1X	Silicon photo diode dark current less than .02 nano-amp (see text)	DB# PINX
RE1	12-volt coil, 400-ohms, 120 VAC 15-amp contacts	DB# RE12115
COPCHAS1	1.4 × 3.5-inch .031 copper plate (see Figure 2-4)	
SW1/NU1	Four 4-40 × 3/8 nylon screws and metal hex nuts	
SPC1	Two 1/4-inch plastic spacers	
PB1	5½″ × 1½″ .1 × .1 grid perforated circuit board	
J1	Chassis mount RCA phono jack	
S1	Small *single pole, single throw* (SPST) toggle switch	
CL1	Battery clip	
BH8	Holder for 8 AA cell battery	
EN1	1 1/2-inch inner diameter × 1/32-inch wall plastic or aluminum tube	
CAP1, 2	Two 1 5/8″ plastic caps reworked as shown in Figure 2-8	
LEN1	38 mm × 100 mm focal length DCX glass lens	DB# 38100
RET1	Reworked 1 1/2-inch plastic cap for retaining lens	
TUB15	2-inch length of 1.5-inch outer diameter plastic or aluminum tube	

Chapter Three

Ultra-Bright Green Laser Project

This ultra-bright green laser project, as shown in Figure 3-1, is intended for the laser experimenter who desires to add a simple optical system that provides far-field focusing, beam collimation, or beam expanding.

Construction parts such as tubing, handles, brackets, and hardware are all readily available from most local hardware stores (see Table 3-1). Our suggested lab approach is to locate some telescopic tubing of metal or plastic, or a combination of both, and fabricate the laser and lens holders from a suitable sized plastic rod or equivalent. The center holes of the holders must be as true as possible or the optical system's integrity will be greatly degraded.

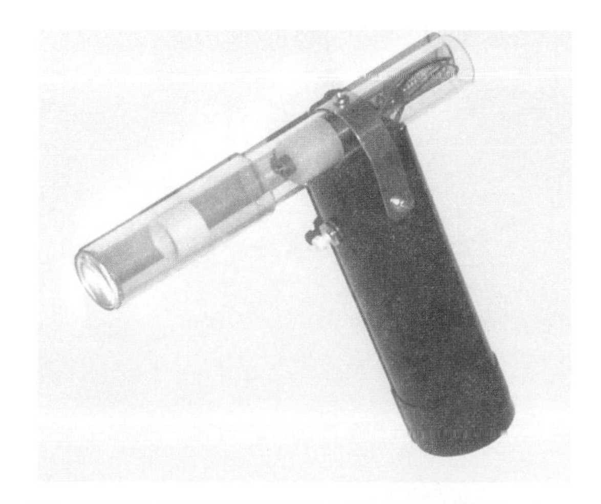

Figure 3-1 *Photo of green laser project*

The grip or handle is created from a 1-inch-thick piece of finished pine that has a section carefully removed for the battery holder and push-button switch. Passage holes for the power leads are drilled as shown. You may also use a suitable sized plastic box. A clear piece of Lexan (polycarbonate) sheet is formed with a flange at a 90-degree bend and is the surface for mounting the switch. This cover is screwed to the wooden handle and makes an attractive design if done with care and precision. The handle can be constructed with a contoured shape to match the curvature of the tube, and you may also add rings, fins, and other décor to make a space-age optical ray gun.

The operation of the laser allows usage without any optics, providing similar performance to a high-quality laser pointer. The addition of the lens system allows far-field focusing, close range expansion, and far-field collimation where the beam spot impact will be reduced by the collimating power of the system. As an example, the bare foot beam diameter (or operation without external optics) at 300 meters will be approximately a 50-centimeter diameter spot. With an ×10 collimator, the diameter can be reduced to 5 centimeters.

This is a class iiia beginner's laser project that uses a working laser module. Class iiia is a classification for lasers under 5 milliwatts of output power

Never point it at vehicles, aircraft, or directly at people. Expect to spend $40 to $50 for this very rewarding laser project.

Assembly Steps

1. Rework the laser module (LM1) as shown in Figure 3-2.
2. Wire the laser module with pushbutton switch (SW1) and battery clip (CL1) as shown in Figure 3-3.
3. Make final assembly as shown in Figure 3-4.
4. Connect batteries and activate SW1. Note a green impact point. Adjust lenses as shown for desired beam profile.

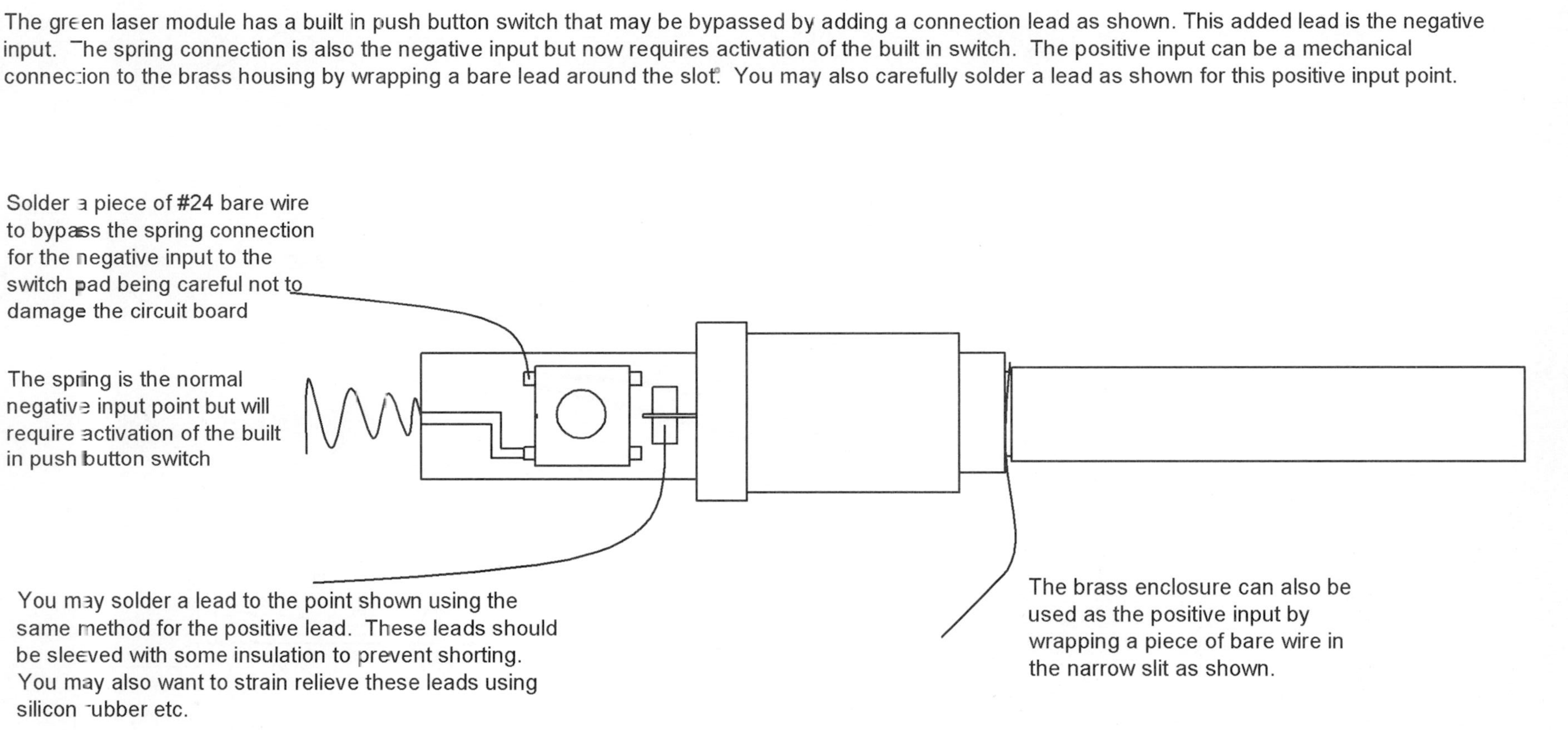

The green laser module has a built in push button switch that may be bypassed by adding a connection lead as shown. This added lead is the negative input. The spring connection is also the negative input but now requires activation of the built in switch. The positive input can be a mechanical connection to the brass housing by wrapping a bare lead around the slot. You may also carefully solder a lead as shown for this positive input point.

Solder a piece of #24 bare wire to bypass the spring connection for the negative input to the switch pad being careful not to damage the circuit board

The spring is the normal negative input point but will require activation of the built in push button switch

You may solder a lead to the point shown using the same method for the positive lead. These leads should be sleeved with some insulation to prevent shorting. You may also want to strain relieve these leads using silicon rubber etc.

The brass enclosure can also be used as the positive input by wrapping a piece of bare wire in the narrow slit as shown.

Figure 3-2 *Laser module rework*

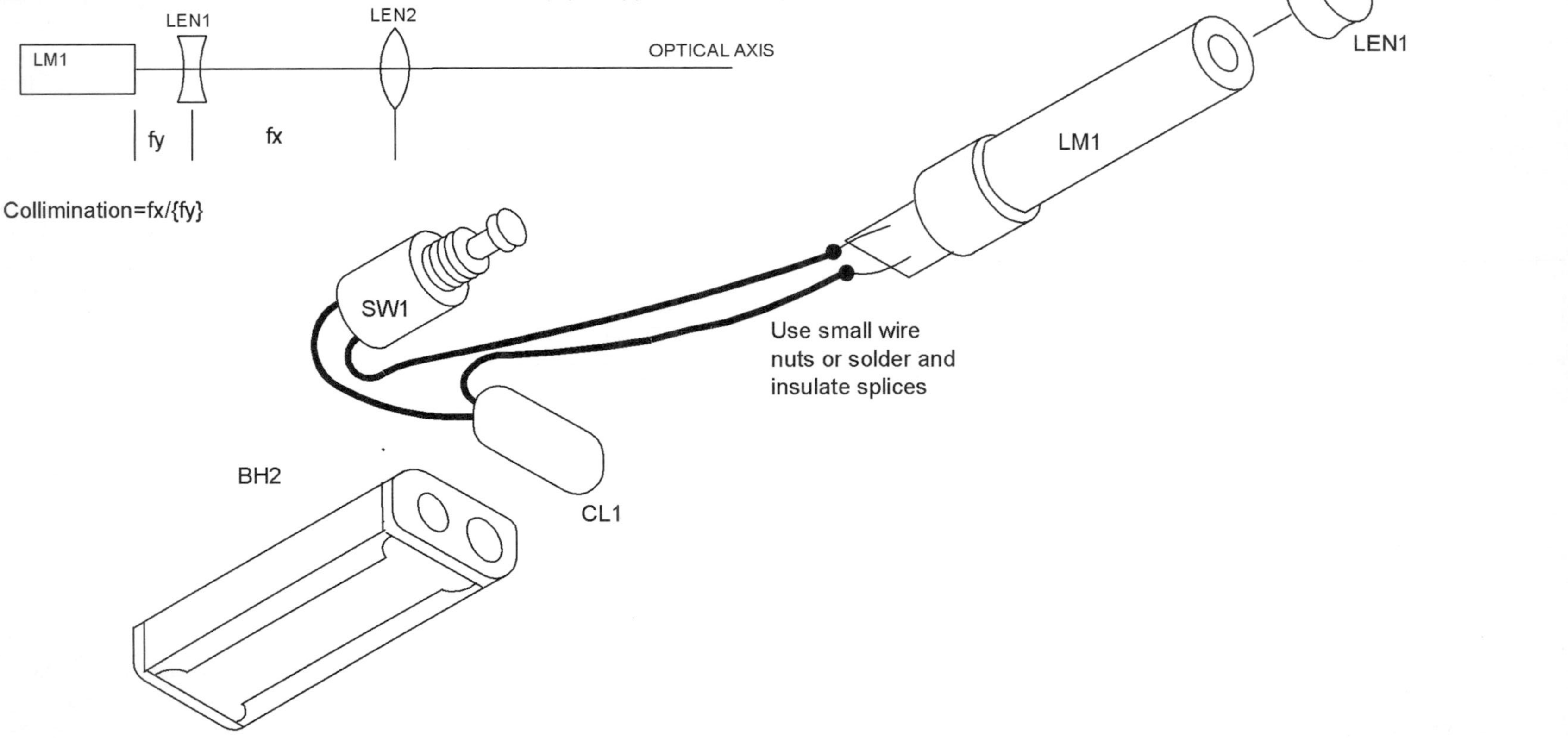
LEN2 position will depend on focal lengths and focus distance of target. LEN1 lens distance from module is not critical, but attempt to get close to module as this will limit overall length of system. Lenses are mounted inside of fabricated bushings/holders and secured in place by small trace amounts of glue. Lenses must be perfectly centered for proper operation. The magnification value of the system is determined by the focal length of the convex lens divided by that of the concave lens. This value expands the beam width but decreases the beam divergence by the same factor greatly increasing the range capability producing a much smaller spot size at far field.
Approximate position shown for lenses supplied with kit. Adjustment is by varying the seperation distance between LEN1 and LEN2. Start at a distance (fx) of approx 6.25mm (2.5")
LEN1
LEN2
LM1
OPTICAL AXIS
fy
fx
Collimination=fx/{fy}
OPTICAL
AXIS
LEN2
LEN1
LM1
SW1
Use small wire
nuts or solder and
insulate splices
BH2
CL1
The green laser module has a built in push button switch that may be bypassed by adding a connection lead as shown. This added lead is the negative input. The spring connection is also the negative input but now requires activation of the built in switch. The positive input can be a mechanical connection to the brass housing by wrapping a bare lead around the slot. You may also carefully solder a lead as shown for this positive input point.

Figure 3-3 *Ultra bright green isometric assembly*

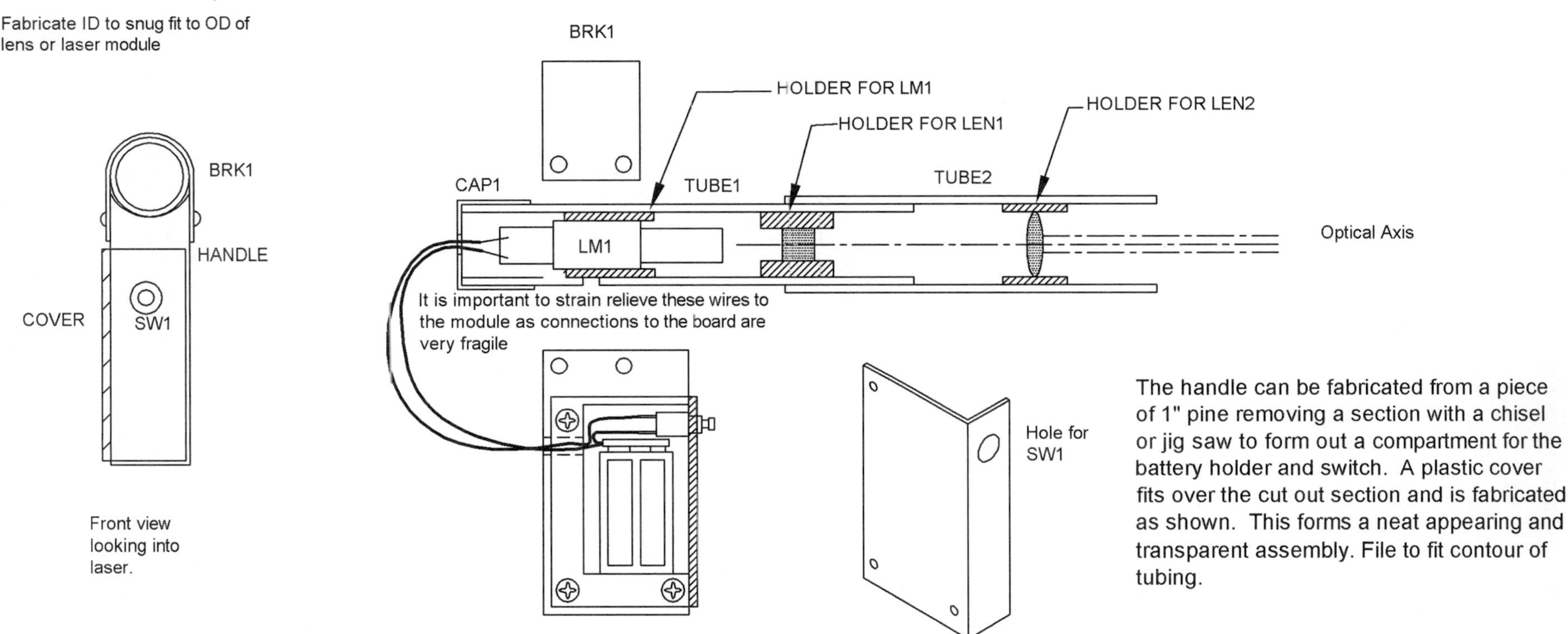

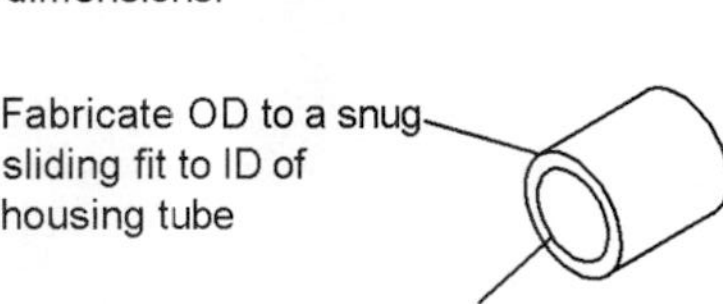

Figure 3-4 *Ultra bright green optical laser gun suggested final design*

Table 3-1 Ultra-Bright Green Laser Parts List

Ref. #	*Description*	*DB Part #*
SW1	Pushbutton NO switch	
BH2AA	Two-cell AA battery holder	
CL1	Battery snap clip	
LM1	5-milliwatt green laser module	DB# LM-532-P5
LEN1	Concave lens .6 × -.75-inch focal length	DB# LE15
LENS	Convex lens .9 × 3.5-inch focal length	DB# LE2475
LAB1	Class 3a danger laser label	
LAB2	Certification label	
LAB3	Aperture label	
The mechanical parts you get yourself; they are not critical and those shown represent our finalized lab approach.		
TUBE1	1-inch OD thin-walled plastic tubing	
TUBE2	Tube for telescope over TUBE1	
HANDLE	2 × 4 × 1-inch piece of soft pine	
COVER	Piece of 1/16-inch clear plastic sheet	
BRK1	1/32 × 1/2-inch aluminum bracket	
LENS HOLDERS	Wood or plastic dowels	
SCREWS	Small wood	

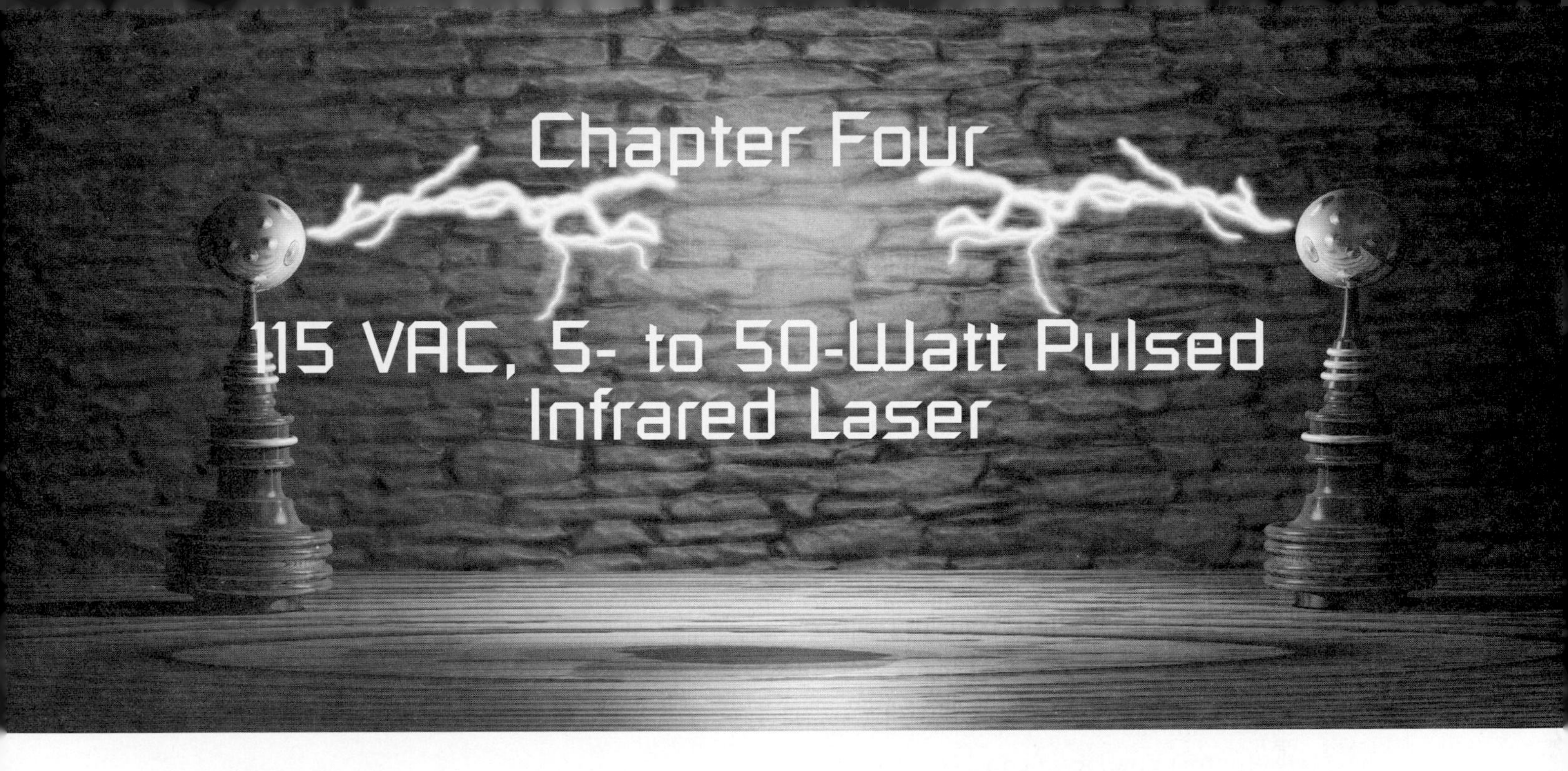

Chapter Four

115 VAC, 5- to 50-Watt Pulsed Infrared Laser

This project shows how to construct a laboratory-use, solid-state, pulsed infrared laser (see Figure 4-1). The system utilizes a gallium arsenide laser diode providing pulse powers of 10 to 100 watts depending on the diode used. The device operates from the 115 VAC line via an isolation transformer and is constructed in two parts. The laser is an excellent source of variable-frequency pulses adjustable from 10 to 2,500 repetitions per second of infrared energy.

The laser, when assembled as shown, is a class iiib device and requires the appropriate labeling and several included safety functions, as described in the assembly instructions. At no time should it be pointed at anyone without protective eyewear or at anything that could reflect these pulses. Also, never look into the unit when the power is on. It is intended to be used for ranging, simulated weapons practice, intrusion detection, communications and signaling, and a variety of related scientific, optical experiments and uses.

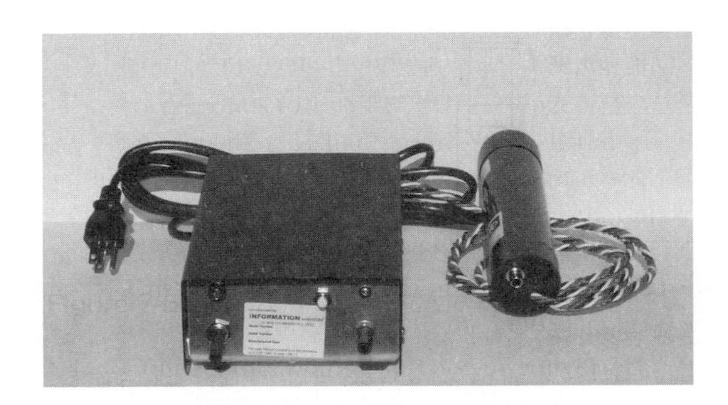

Figure 4-1 *IR laser*

This is an intermediate- to advanced-level project requiring electronic skills and basic electronic shop equipment. Expect to spend $100 to $150. All parts are readily available, with specialized parts obtainable through Information Unlimited (www.amazing1.com), and they are listed in Table 4-1.

Theory of Operation

A laser diode is nothing more than a three-layer device consisting of a *pn* junction of *n*-type silicon, a *p*-type of gallium arsenide, and a third *p* layer of doped gallium arsenide with aluminum. The *n*-type material contains electrons that readily migrate across the *pn* junction and fill the holes of the *p*-type material. Conversely, holes in the *p*-type migrate to the *n*-type and join with electrons. This migration causes a potential hill or barrier consisting of negative charges in the *p*-type material and positive charges in the *n*-type material that eventually cease growing when a charge equilibrium exists. In order for current to flow in this device, it must be supplied at a voltage to overcome this potential barrier. This is the forward voltage

Table 4-1

INFO #	Pulse width	Package	Diodes	Peak current	Emitting area	Peak power	Duty factor	Beam symmetry	Spectral width
LD660	200 ns	TO18	1	40 amps	9 × 1	12 watts	.1%	15 × 20	3.5 nm
LD780	200 ns	TO5	2	40 amps	21 × 1	20 watts	.1%	15 × 20	3.5 nm
LD1630	200 ns	TO5	2	40 amps	21 × 1	30 watts	.1%	15 × 20	3.5 nm

drop across a common diode. If this voltage polarity is reversed, the potential barrier is simply increased, assuring no current flow. This is the reversed bias condition of a common diode.

A diode without an external voltage applied to it contains electrons that move and wander through the lattice structure at a low, lazy average velocity as a function of temperature. When an external current at a voltage exceeding the barrier potential is applied, these lazy electrons now increase their velocity to where some, by colliding, acquire a discrete amount of energy and become unstable, eventually emitting this acquired energy in the form of a photon after returning to a lower energy state. These photons of energy are random both in time and direction; hence, any radiation produced is incoherent, such as that of a *light-emitting diode* (LED).

The requirements for coherent radiation are that the discrete packets of radiation must be in the form of a lockstep phase and in a definite direction. This demands two essential requirements; first, sufficient electrons at the necessary excited energy levels and, second, an optical resonant cavity capable of trapping these energized electrons for stimulating more and giving them direction. The amount of energized electrons is determined by the forward diode current. A definite threshold condition exists where the device emits laser light rather than incoherent light, such as in an LED. This is why the device must be pulsed with a high current. The radiation from these energized electrons is reflected back and forth between the square-cut edges of the crystal that form the reflecting surfaces due to the index of refraction of the material and air.

The electrons are initially energized in the region of the *pn* junction. When these energized electrons drift into the *p*-type transparent region, they spontaneously liberate other photons that travel back and forth in the optical cavity interacting with other electrons commencing laser action. A portion of the radiation traveling back and forth between the reflecting surfaces of these mirrors escapes and constitutes the output of the device.

Circuit Theory of Operation

AC power, as shown in Figure 4-2, is obtained via polarized plug CO1 through fuse FH1. Proper grounding of the green cord lead is a necessity to prevent an unnecessary shock hazard along with usage of grounded test and measuring equipment. S1 is a key switch type where the key can be removed only in the off position. Transformer T1 provides a one-to-one ratio and isolation from the power line, while step-down transformer T2 provides the low voltages necessary for the control circuits. The power indicator lamp consists of neon lamp NE1 and associated current-limiting resistor R1.

Diodes D1 and D2, along with capacitors C1 and C2, comprise a voltage doubler. The voltage across C1 and C2 is 1.4 × 230 or approximately 340 volts.

A major section of this laser pulser is the charging circuit. This circuit charges the pulse discharge capacitor (C10) to the supply voltage during the time interval between laser current pulses. It also isolates the supply voltage from the discharge circuit during the laser current pulse, thereby allowing the switching *silicon-controlled rectifier* (SCR) to recover to the blocking state. Because the response times of the charging circuit are relatively long, lead lengths are not important, and the circuit can be remotely located from the discharge circuit.

The simplest charging circuit is a resistor-capacitor combination. In this simple case, the resistor must limit the current to a value less than the SCR holding

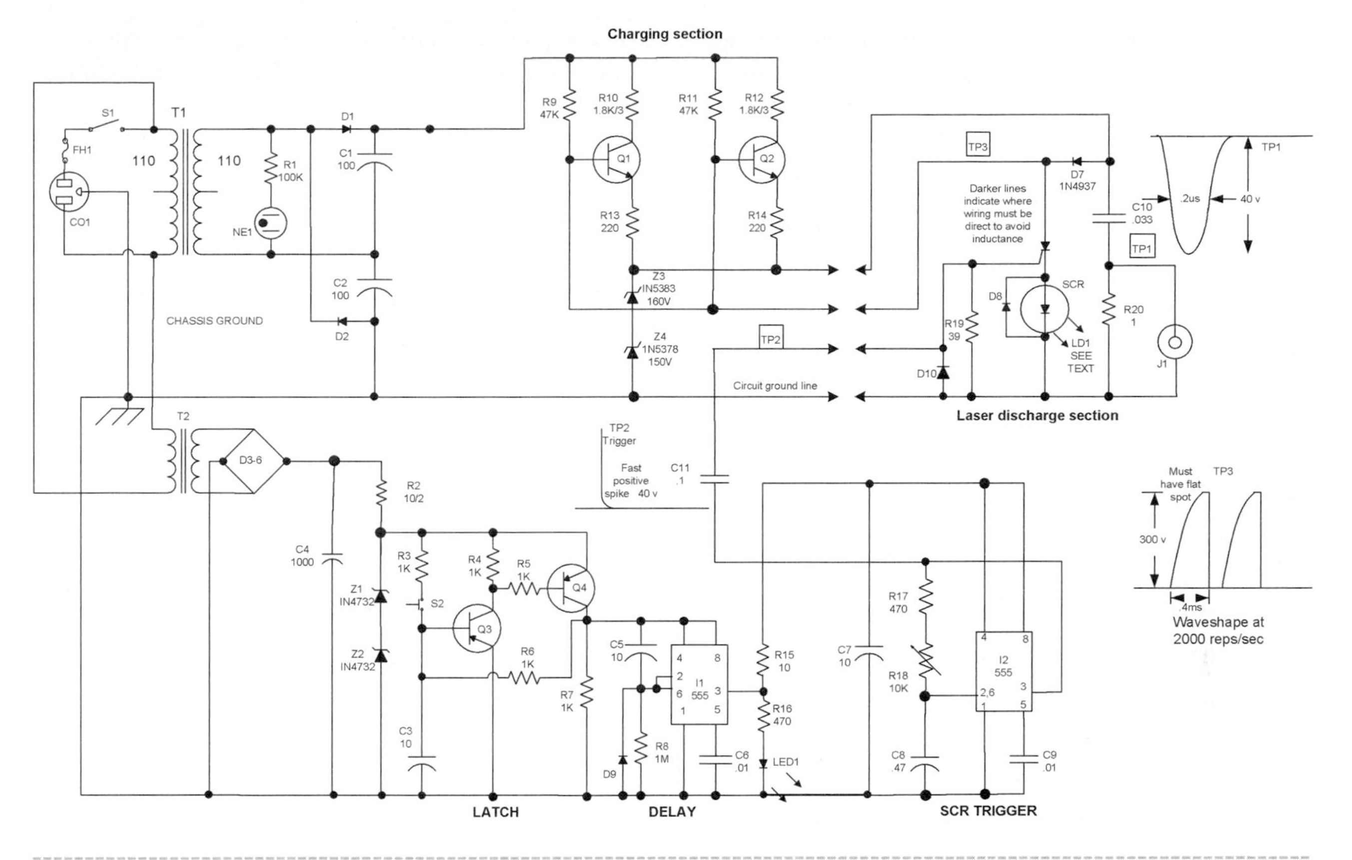

Figure 4-2 *Circuit schematic*

current but should be as low as practical because this resistance also determines the charging time of the capacitor C10, thus determining the available laser pulse repetition rate at full energy. For example, a resistance of 40 kilo-ohms limits the current to 10 milliamps from a 400-volt supply. This current value is just at the holding level of an average SCR. A time of almost 4 milliseconds is required to charge C10, a 0.033-microfarad capacitor, to the supply voltage in three time constants through a 40-kilo-ohm resistor. Therefore, the *pulse repetition rate* (PRR) of the pulsing circuit is limited to less than 250 Hz. If the PRR exceeds this value, the capacitor does not completely recharge between pulses and the peak laser current decreases with increasing PRR.

Varying the supply voltage may control the peak current in the discharge circuit, provided the PRR is low enough to allow the capacitor to fully recharge between current pulses. Both the supply voltage and the PRR determine the peak laser current. A considerable risk exists in increasing the supply voltage to compensate for an insufficient recharge time. If the PRR is decreased while the supply voltage is high, the capacitor again recharges completely, and the laser pulse current increases to a value that may damage or destroy the laser diode.

The needs for a variable voltage supply and the low PRR limit are the major disadvantages of the resistive-type charging network. Therefore, the limitations of the simple resistor drivers are that a resistor large enough to keep the current below the holding current of the SCR also limits the pulse repetition rate.

The frequency capability of the pulse power supply can be greatly improved with the charging circuits, as described. Capacitor C10 is charged to the supply voltage when the SCR is not conducting. Diode D7 is in the off state, because no current flows through it during this period. At the onset of a pulse trigger at the gate, the impedance of the SCR drops rapidly and capacitor C10 discharges toward ground potential. During this period, current is surging through diode D7, causing zero biasing of the current-control transistors (Q1, Q2) by shorting the bias resistors (R9, R11). The SCR now turns off when the current drops below its holding current. The voltage across capacitor

C10 then charges back to the supply voltage. During the capacitor charge cycle, diode D7 passes no current, and Q1 and Q2 are forward biased into the saturation region. Obviously, the charging time of C10 is now shorter due to being charged through the smaller-value resistors (R10, R12). The emitter resistors R13 and R14 balance the current through Q1 and Q2.

Zener diodes (Z3, Z4) are selected to regulate and provide the proper required current pulse for the laser diode in operation. The zener diodes provide a maximum charge voltage of 310 volts that produces 40 amps through laser diode LD1 for up to 2,500 laser pulses per second.

Other laser diodes may require a selection of other voltage-rated zener diodes. The builder may wish to provide a multiposition switch to select these various combinations of zener diodes externally via a front panel control.

The trigger circuit consists of an astable timer (I2) that derives its power through the output of timer I1, which provides a delay time required for FDA laser compliance. The trigger and delay circuitry, as well as the timers, are powered by 12 VDC from transformer T2 and rectifier diodes D3 through D6. Capacitor C4 filters the rectified wave and steps up the voltage to 15 VDC. Resistor R2 drops the 15 volts to the zener voltage of zener diodes Z1 and Z2 of regulated 12 VDC.

Operation of the laser requires activating pushbutton switch S2, which latches transistor Q4 upon supplying power to timer I1. The timer must time out to 10 seconds, as determined by resistor R8 and capacitor C5. The emission indicator LED1 comes on simultaneously with power to the trigger timer I2, which pulses the laser diode LD1. The laser pulse rate is determined by external pot R18.

Notes on SCR

In the conventional operation of an SCR, the anode current, initiated by a gate pulse, rises to its maximum value in about 1 microsecond. During this time, the anode-to-cathode impedance drops from open circuit to a fraction of an ohm. In injection laser pulsers, however, the duration of the anode-cathode pulse is much less than the time required for the SCR to turn on completely. Therefore, the anode-to-cathode impedance is at the level of 1 to 10 ohms throughout most of the conduction period. The major disadvantage of the high SCR impedance is that it causes low circuit efficiency. For example, at a current of 40 amps, the maximum voltage would be across the SCR, while only 9 volts would be across the laser diode, which represents very low circuit efficiency. The efficiency of a laser array is greater due to its circuit impedance being more significant.

Because the SCR is used unconventionally, many of the standard specifications such as peak current reverse voltage, on-state forward voltage, and turn-off time are not applicable. In fact, it is difficult to select an SCR for a pulsing circuit on the basis of normally specified characteristics. The specifications important to laser pulser applications are forward-blocking voltage and current rise time. A usage test is the best and many times the only practical method of determining the suitability of a particular SCR.

Notes on the Storage Capacitor

The voltage rating of the storage capacitor must be at least as high as the supply voltage. A high-quality metallized capacitor must be used for this part as peak currents are in the tens of amps.

Notes on Layout Wiring

Lead lengths and circuit layout are very important to the performance of the discharge circuit. Lead inductance affects the rise time and peak value of the current and can also produce ringing and undershoots in the current waveform that can destroy the laser. A well-built discharge circuit might have a total lead length of only 1 inch and therefore an inductance of approximately 20 nanohenries. If the current rises to 75 amperes in 100 nanoseconds, the inductive voltage drop will be $e = L\, di/dt$. If proper care is not taken in wiring the discharge circuits, high-inductance voltage drops will result.

A 1-ohm resistor (R20) in the discharge circuit will greatly reduce the current undershoot in single-diode lasers. Laser arrays usually have sufficient resistance

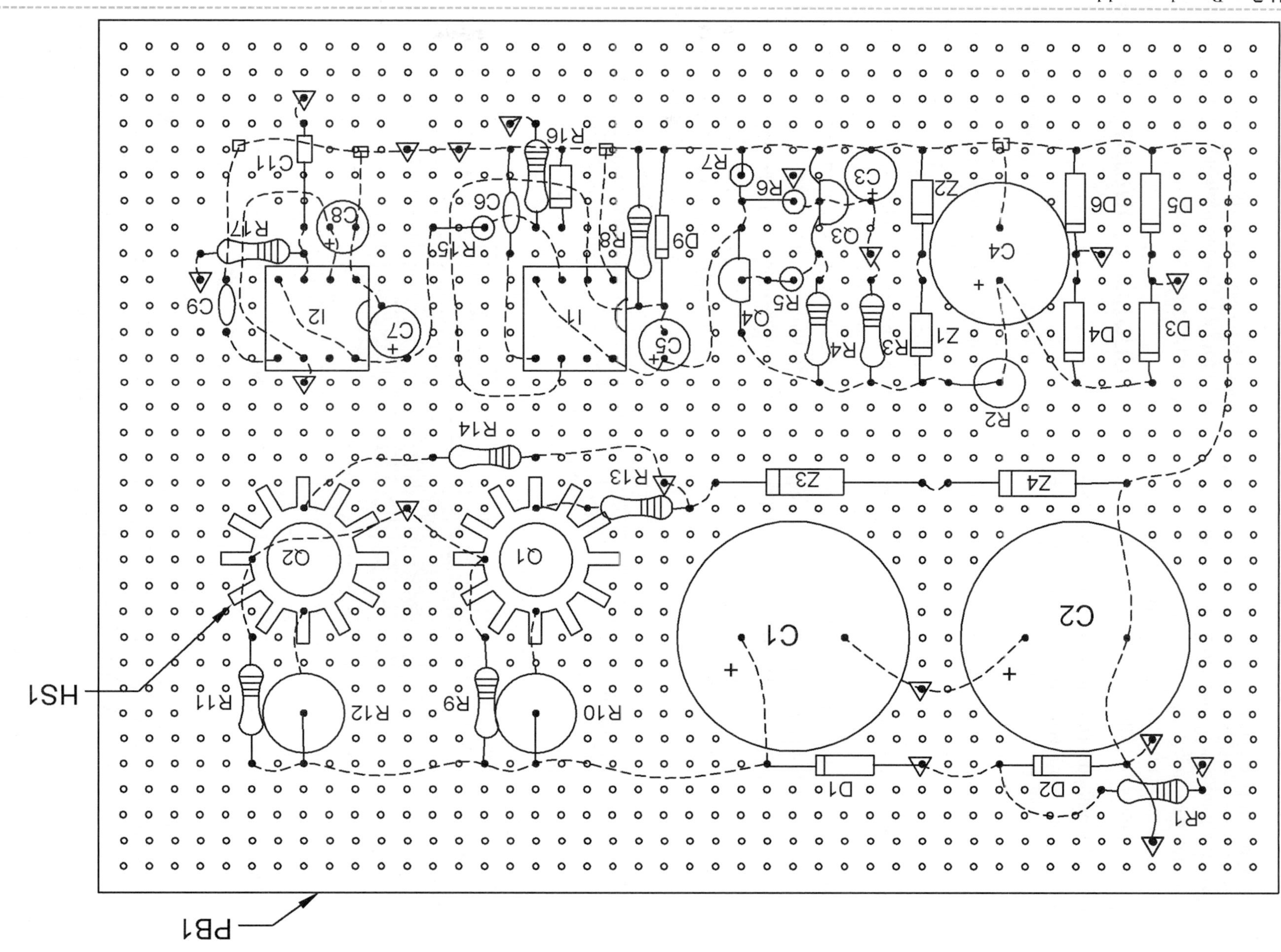

Figure 4-3 *Board assembly*

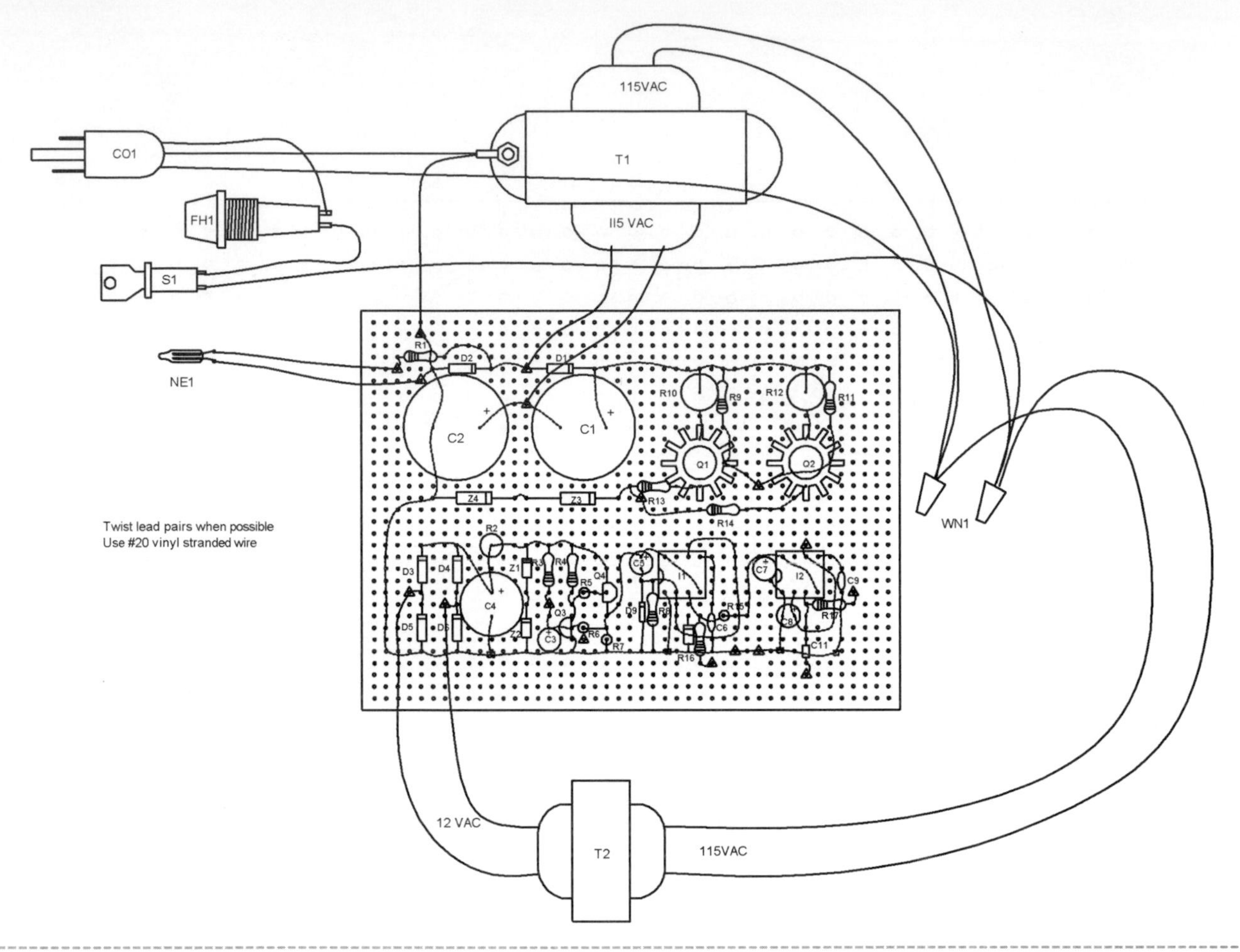

Figure 4-4 *Board input power wiring.*

to eliminate an undershoot. The small resistance in the discharge circuit is also useful in monitoring the laser current, as described in the following section.

A clamping diode (D8) is added in parallel with the laser to reduce the current undershoot. Its polarity should be opposite that of the laser. Although the clamping diode is operated above its usual maximum current rating, the current undershoot caused by ringing is very short and the operating life of the diode is satisfactory.

Notes on Current Monitoring

The current monitor in the discharge circuit provides a means of observing the laser current waveform with an oscilloscope. A resistive-type monitor (R20) reduces circuit ringing and current undershoot, but the lead inductance of the resistor may cause a higher than actual current reading. A current transformer such as the Tektronix CT-2 can also be used to monitor the current and is not affected by lead inductance. Because the transformer does not respond to low-frequency signals, it should be used with fast, short-pulse-width, fast fall-time waveforms.

Assembly Steps

Assembly of the laser system is straight forward and will require basic tools and test equipment.

1. Lay out and identify all parts and pieces. Verify them with the parts list, and separate the resistors as they have a color code to determine their value. Colors are noted on the parts list.
2. Cut a piece of .1-inch grid perforated board to a size of 5.8 × 3.5 inches, as shown in Figure 4-3.

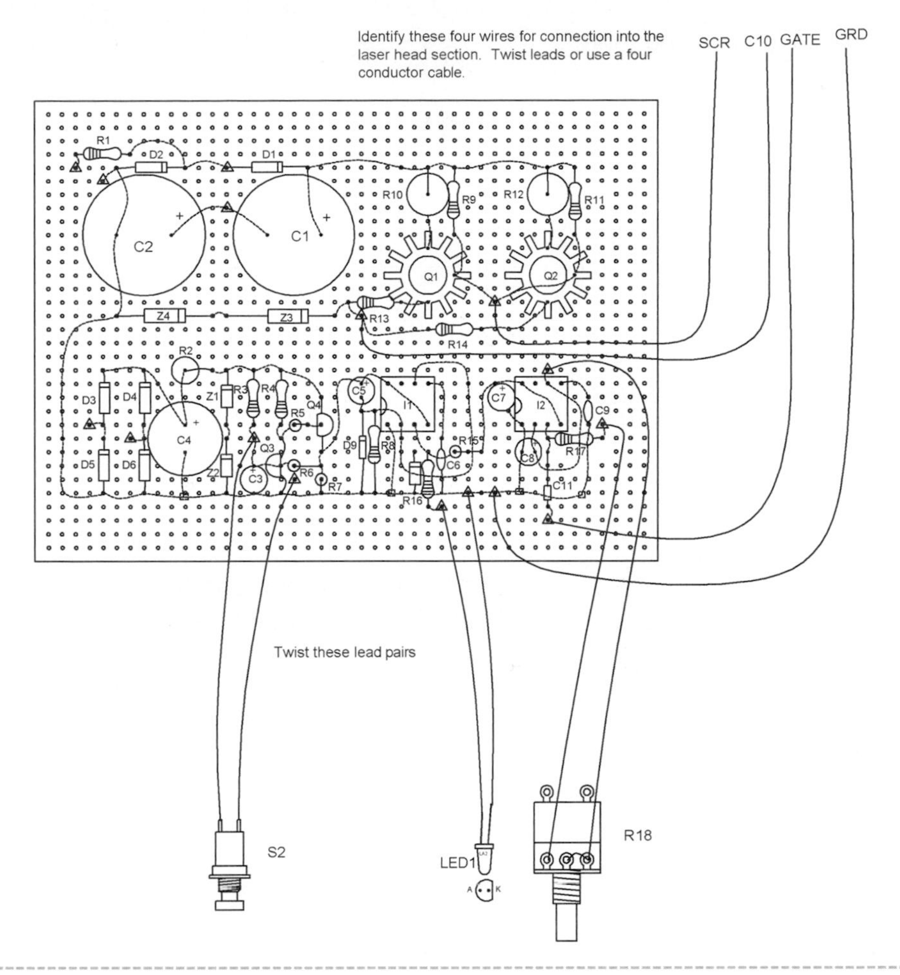

Figure 4-5 *Board output and control wiring*

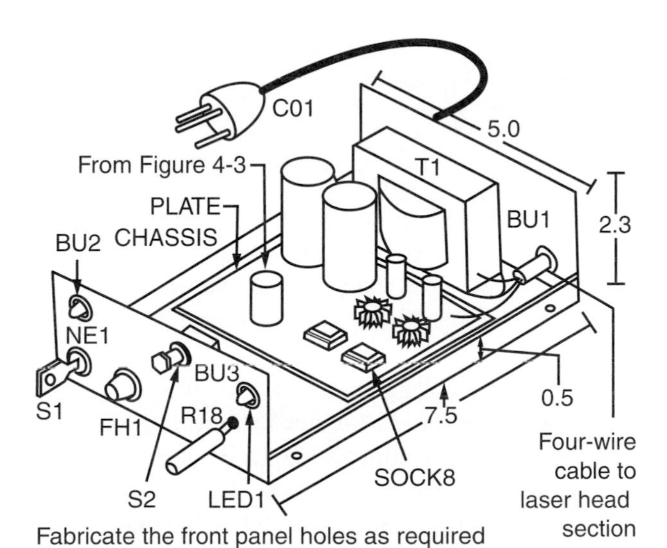

Figure 4-6 *Fabrication and chassis assembly*

3. If you are building from a perforated board, insert the components, starting in the lower left-hand corner as shown. Pay attention to the polarity of the capacitors that have polarity signs and all semiconductors. Use eight-pin sockets for I1 and I2.

   Route the leads of the components as shown and solder as you go, cutting away unused wires. Attempt to use certain leads as the wire runs or use pieces of the #24 bus wire. Follow the dashed lines on the assembly drawing as these indicate the connection runs on the underside of the assembly board. The heavy dashed lines indicate the use of thicker #20 bus wire, as this is a high-current discharge path.

4. Attach the external leads as shown in Figure 4-4 to the external components. Note the chassis grounding to the green lead of the power cord (CO1) and the use of wire nuts (WN1) for the junction of the transformer's primary inputs.

5. Attach the external leads as shown in Figure 4-5 to the external components. Note the four leads that go to the laser head assembly. These

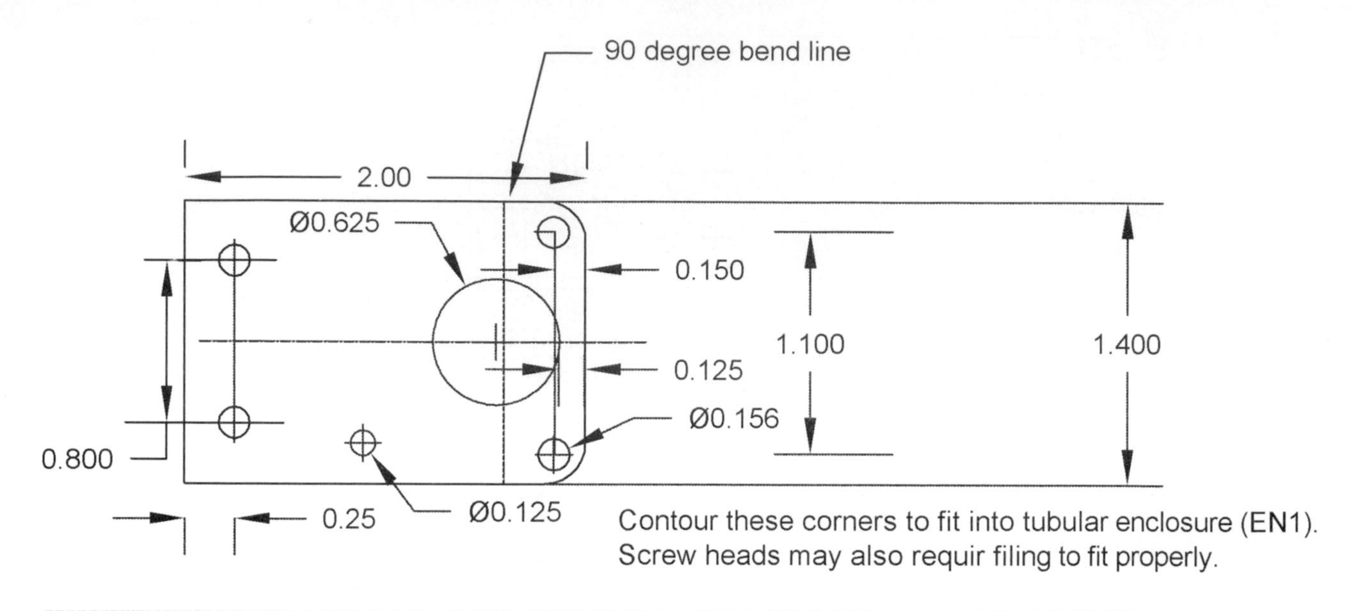

Figure 4-7 *Assembly of laser chassis*

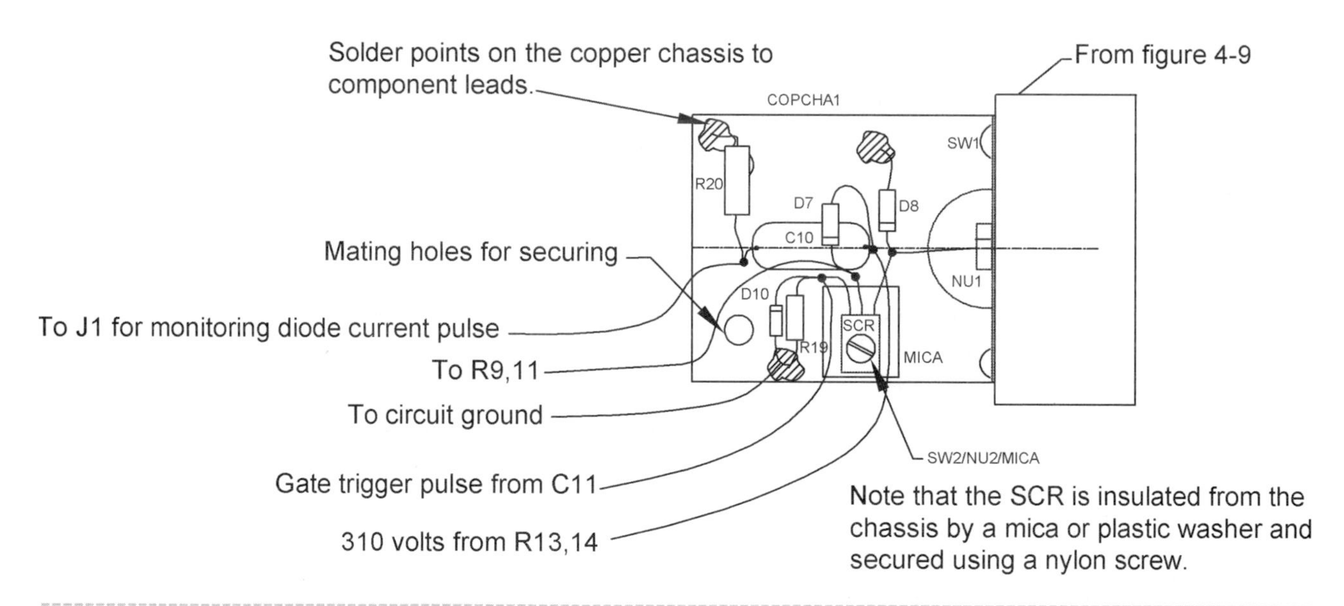

Figure 4-8 *Fabricate the laser holder*

are suggested to be 24 to 36 inches in length as well as twisted and identified. You may use a four-conductor cable.

6. Double-check the accuracy of the wiring and the quality of the solder joints. Avoid wire bridges, shorts, and close proximity to other circuit components. If a wire bridge is necessary, sleeve some insulation onto the lead to avoid any potential shorts.

7. Fabricate the chassis, as shown in Figure 4-6. Then position and drill the front and rear panel holes to fit the necessary external components. Use appropriate bushings for the power cord and wires to the laser head. Also, place an insulating piece (PLATE) under the assembly board to prevent contact with the metal chassis.

8. Fabricate the laser chassis, as shown in Figure 4-7, from a .031 piece of copper sheet metal.

9. Assemble the laser chassis, as shown in Figure 4-8. This section takes advantage of the ground plane effect of using the copper for

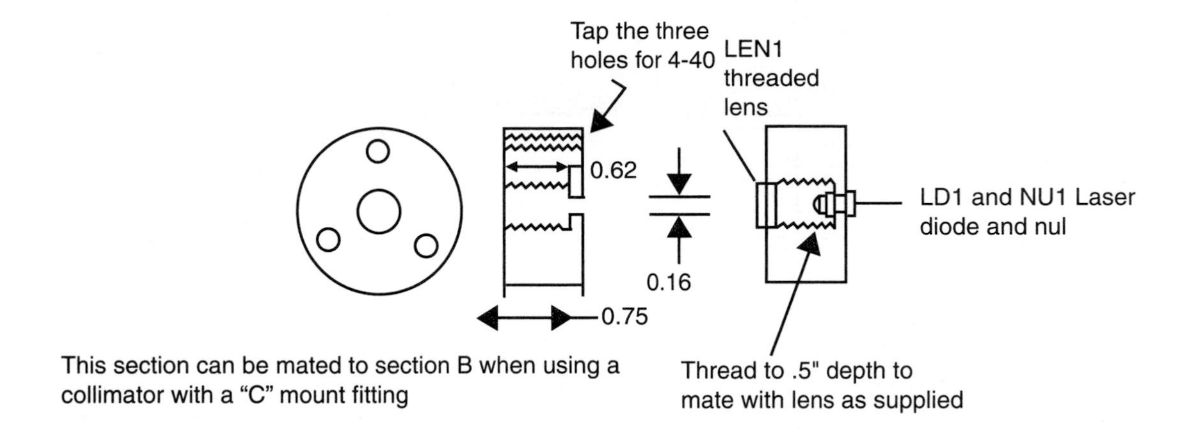

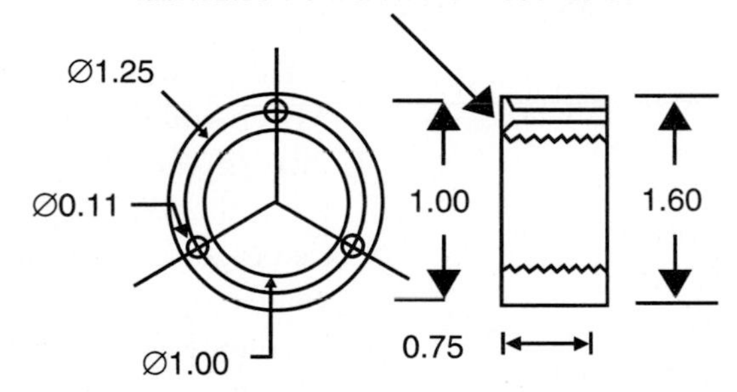

Section B adapter for fitting to the lens holder when using the optional collimator.

Three mounting holes of .11 on a 1.25" diameter 120 degres are mated to those in above.

Thread for "C" lens at 1" diameter 32 tpi to a minimum of .375".

Figure 4-9 *Laser diode and lens holder*

direct solder points. It is important to maintain a good waveshape because the switching currents are 40 amps and any undershoot is undesirable across the laser diode. Do not install the laser diode at this point.

10. Create the laser holder (LAHOLD), as shown in Figure 4-9. This assembly is attached to the bent-up lip of the chassis section by two small screws.

Assembly Testing

Testing will involve the use of an oscilloscope with a 100 MHz bandwidth and infrared detection equipment such as a night vision scope or camera.

1. It is assumed at this point that the electronic assembly is complete. Verify the circuit and preset the two trimpots, R3 and R9, to midrange.

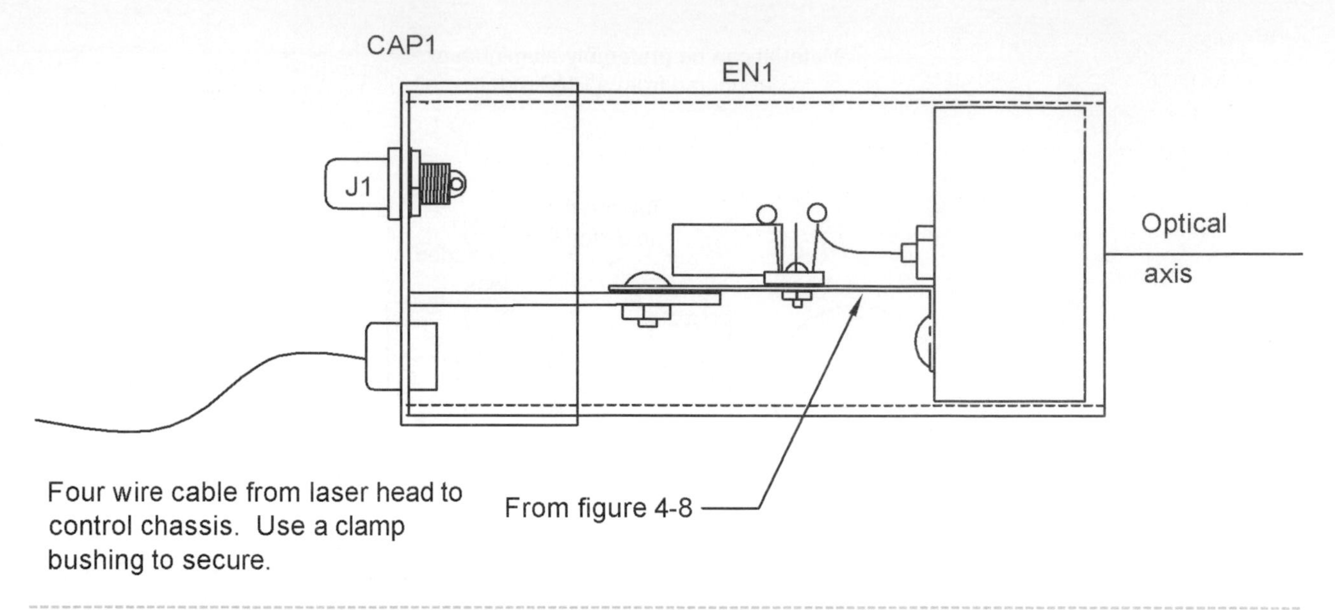

Figure 4-10 *Final head assembly*

2. Insert a .5-amp fuse into the fuse holder. Verify that the key switch is off and R18 is fully counterclockwise. Plug in a ballasted, 25-watt, 115-volt, 60 Hz power source. This is nothing more than the cord with a 25-watt lamp in series with the black lead. It protects the circuitry from the catastrophic currents of the power line should a major fault occur. Obtain a 1N4007 power diode and temporarily wire it in place of the laser diode LA1. The diode simulates the actual laser diodes and avoids costly replacement should a gross error exist. The actual laser diode is not to be wired in at this point. Turn on the key switch and note the NE1 coming on. The ballast lamp should not be lit.

13. Connect the scope to TP2 and note a spiky waveform that verifies the trigger pulse to the SCR, as shown in Figure 4-2.

14. Connect the scope to TP3 and note the waveform, which is the voltage across the SCR. Notice the flat spot indicating that capacitor C10 has reached a full charge before the SCR is triggered. The top of this waveshape is affected by the pulse repetition rate and should support 2,500 repetitions per second when set by R18 (see Figure 4-2).

15. Connect the scope to TP1 and set it up for a negative-going fast pulse, indicating the current waveform through the LD1 laser diode. If you have followed the instructions and your assembly of the copper chassis is secure with short leads, you should see a picture-perfect waveform, as shown.

    Note that the current pulse is 40 amps and corresponds to 1 amp for every volt read at TP1. If you find the pulse is less, you can increase the voltage ratings on one of the zener diodes, Z3 or Z4. The laser diode specified requires 40 amps for full output. Any value over 40 amps will greatly shorten the useful life of this part. The range of the pulse rep rate should be 200 to 2,500 pulses per second. The pulse amplitude should only vary several percent throughout this range. You will note that the amplitude will slightly decrease as the rep rate is increased.

16. At this point, you should be wearing your safety glasses. Remove the test diode D5 and connect the laser diode. Place an infrared indicator several inches from the diode and note an orange glow when the unit is turned on. Then recheck the pulse shape and amplitude.

17. Verify the laser output by using a night vision scope or a video camera. Screw in LENS13, so it is flush with the holder. Adjust it to find the smallest footprint at over 50 feet. Our lab model adjusted to about two turns below being flush with the holder for the best results. The output should be a sharp, narrow bar shape with this diode specified.

Please note that laser diodes are very easily destroyed by current pulses that are too large and overshoot. Pay attention to the correct current pulse shown in Figure 4-2.

The recommended laser diodes are available from Information Unlimited and are shown in Table 4-1.

Final Assembly Steps

Final assembly of head section as shown in Figure 4-10. For final assembly place the label as shown and verifying the compliance tests in Figure 4-11

Compliance Tests
1. Verify correct labels as shown.
2. Key switch — nonremoval in off position.
3. Beam indicator LED indicates beam emission.

Instruction
1. Turn key switch counterclockwise to verify unit is off. Note key is easily removed in the off position.
2. Plug unit into 115 vac.
3. Turn key clockwise and note NE1 lighting. Note key cannot be removed in the on position Laser energy is infrared and cannot be seen with the eye. You must use infrared detection devices such as our laser pulse detector or suitable night-vision devices. Video cameras will often detect the laser.
4. Adjust collimator for desired effect if you are using.

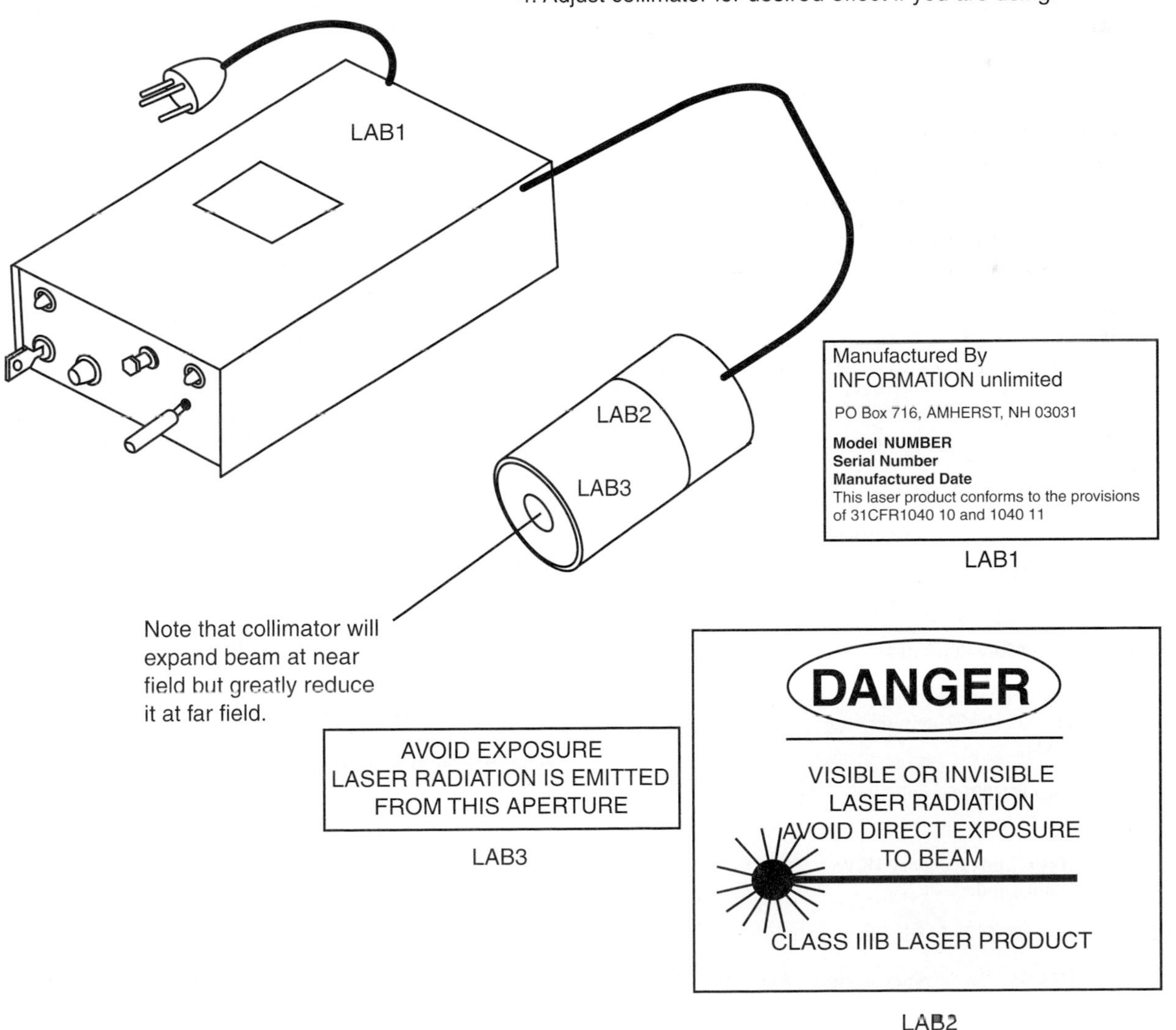

Figure 4-11 *Isometric of final view*

Table 4-2 Parts List for Pulsed Laser

Ref #	*Description*	*DB Part #*
R1	100K, ¼ –watt resistor (br-blk-yel)	
R2	10-ohm, 1-watt resistor (br-blk-blk)	
R3, 4, 5, 6, 7	Two 1K ¼-watt resistors (br-blk-red)	
R8	1M, ¼-watt resistor (br-blk-gr)	
R9, 11	47K, ¼-watt resistor (yel-pur-or)	
R10, 12	1.8K, 3-watt mox resistor	
R13, 14	220, 1/2-watt resistor (red-red-br)	
R15	10-ohm, ¼-watt resistor (br-blk-blk)	
R16, 17	470-ohm, ¼-watt resistor (yel-pur-br)	
R18	10K, 10 mm potentiometer	
R19	39-ohm, ¼-watt resistor (or-wh-blk)	
R20	1-ohm, ½-watt carbon resistor (br-blk-gold)	
C1, 2	Two 330 mfd, 16-volt vertical electrolytic capacitor	
C4	1,000 mfd, 25-volt vertical electrolytic capacitor	
C3, 5, 7	10 mfd, 25-volt vertical electrolytic capacitor	
C6, 9	.01 mfd, 50-volt disc capacitor	
C10	.033 mfd, 400-volt polypropylene capacitor	
C11	.1 mfd 50-volt plastic capacitor	
D1, 2	1N4007 1KV, 1-amp rectifier diodes	
D3–6	Four 1N4001 50-volt 1-amp rectifier	
D7, 8	1N4937 fast recovery 1KV, 1-amp diode	
D9, 10	Two 1N914 signal diodes	
Z1, 2	1N4735 6.2-volt zener diode	
Z3	1N5383 150-volt, 5-watt zener diode	
Z4	1N5384 160-volt, 5-watt zener diode	
Q1, 2	Two 2N3439 NPN high-voltage transistors	
Q3	PN2907 PNP general-purpose transistor	
Q4	PN2222 NPN general-purpose transistor	
I1, 2	Two timer dip integrated circuits LM555	
LED1	Green LED	
NE1	Neon indicator lamp	
SCR1	12-amp, 800-volt SCR S8012R	DB# S8012R
LD1	Laser diode 40-amp, 10–20-watt, 904 nm TO18 package	DB# LD650
LAHOLD	Laser and lens holder fabricated as shown in Figure 4-9	
LENS13	Threaded lens	
T1	115/115 volt, 15-volt-amp isolation transformer	DB# TR115/115
T2	115/12 volt, 1-amp transformer	DB# 12DC/.1
J1	RCA phono jack	
CO1	Three-wire #18 power cord	
S1	Compliant key switch	DB# KEYSWSM
S2	Pushbutton no switch	
FH1	Panel-mount fuse holder	
FS1	1-amp fuse	
WN1	Two small Hi3 wire nuts	
PB1	4.8 × 3.5-inch .1 × .1 perforated circuit vector board	
SOCK8	Two 8-pin integrated circuit sockets	

Chapter Five

Laser Property Guard

This project, as shown in Figure 5-1, allows the user to protect a perimeter of over 1,000 meters. The system utilizes a laser transmitter, as described in Chapter 1, and a laser receiver, described in Chapter 2. The system can easily control any deterrent to an invader, such as flood lighting, alarms, man-trap systems, or our phaser pain field generator, as described in Chapter 35.

The operation of the laser property guard is simple. The laser beam is directed through the area to be protected, using front-surface mirrors located on sturdy and secure mounts and at the points necessary to obtain the required coverage (see Figure 5-2). The receiver is normally quiescent when receiving this beam. Once the beam is broken by an intrusion, the receiver now processes the interruption as a fault and powers the desired deterrent or alarm.

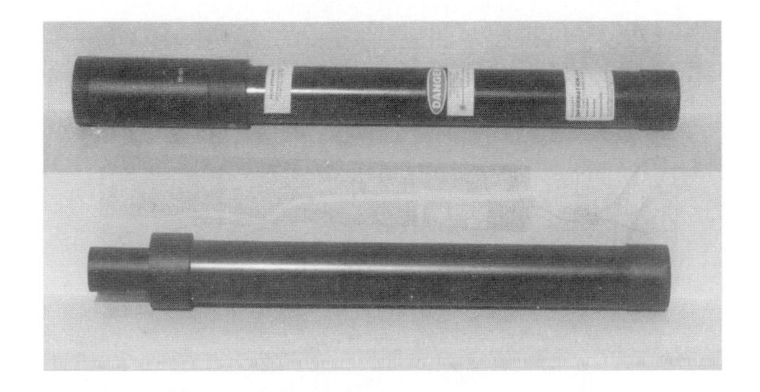

Figure 5-1 *Laser property guard protector components*

This project requires assembly of the laser devices, as described in Chapters 1 and 2. The setup requires basic mechanical skills and installation of the optical reflecting mirrors, and the hookup requires basic electrical skills. Expect to spend $25 to $50 after the assembly of the necessary components. All parts are readily available, with specialized parts obtainable through Information Unlimited (www.amazing1.com) and listed in Table 5-1.

The laser property guard system, as shown in Figure 5-3, utilizes a pulsed infrared laser to provide a perimeter path of programmed light pulses that, when interrupted, will sound an alarm, create an alert, or trigger a deterrent. The trigger fault signal actually triggers a timer or a latching circuit, which drives a switch for the external control of other functions. The timer mode allows the adjustment of the alarm period and automatically resets. The latch mode requires a manual reset.

Even though the laser pulses are rated in tens of watts, the overall energy over a period of 1 second is in the milliwatts. Installing the system at short ranges of only one to two reflection points is reasonably simple. Longer ranges requiring multiple reflections can be challenging.

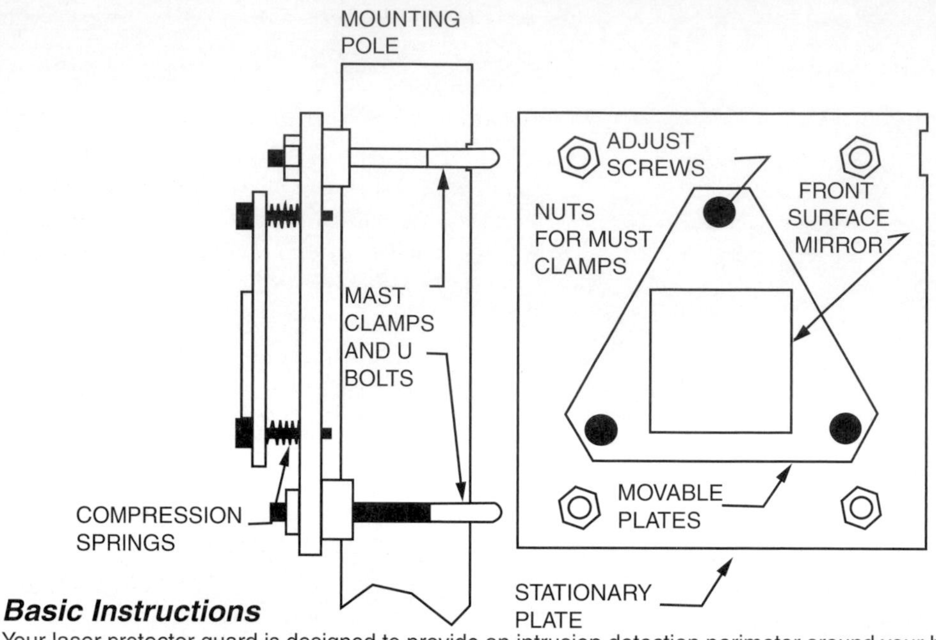

Basic Instructions

Your laser protector guard is designed to provide an intrusion detection perimeter around your home or target area. It uses a Class iiia pulsed diode laser to minimize but not eliminate any optical hazard. A low-liability sonic shock field generator such as that described in Chapter 35 is suggested as a deterrent. Caution: Check local laws for proper posting of this equipment on your property. Remember if you injure a criminal, even if he is robbing you, it can result in a stiff penalty in certain states for violating his rights.

System Setup

1. Obtain the following equipment listed in the latest Information Unlimited catalog.
   - ❑ Class iiia pulsed diode laser as described in Chapter 1.
   - ❑ Optical light detector and controller with built-in alarm as descrbed in Chapter 2.
   - ❑ Low-liability sonic shock pain field generator as described in Chapter 35.

2. Examine the area you want to protect. Attempt to position beam travel over level terrain to eliminate possible easy-to-sneak-under points.

3. Establish corner points of property where mirrors are to be positioned. Note that a clear view of adjacent mirrors, laser and receiver must be maintained in all weather conditions for reliable operation.

4. Determine the corner where you want to position the actual laser and laser detector at. Note: There should be access to these devices for powering and control.

5. Construct mirror mounts using your own ingenuity. Allow for adjustment and stability. Rough align mirror to approximate position.

6. Turn on laser and position to hit center of first mirror. Secure laser in place.

Adjust first mirror to hit center of second and repeat for remaining mirrors all the way back to the optical receiver. Secure in place as you go.

Special note: Large areas may require an optical collimator at the laser output to reduce beam width at longer ranges. The collimator can be a rifle scope, telescope, binocular, etc., positioned in axial alignment with the laser. Beam divergence will be reduced by the magnification factor of the system used, however the actual beam will be expanded by same factor. The net result is a smaller cross-section at longer ranges.

Figure 5-2 *Low-cost, adjustable mirror mount*

System Installation Suggestions

There are many ways to align an optical system of this type. The use of a visible laser module as listed in Table 5-1 can simplify the initial setup and positioning of the mirrors. Those fortunate to have night vision equipment will find this a considerable advantage.

- Determine the area you want to protect. Flat land is the most desirable, as gullies and hills are difficult to cover without using more mirrors.
- You must also consider the position of the laser transmitter and receiver, noting the power and connection into the alarm or the deterrent section. Weather exposure and convenience must also be taken into consideration, and mounting must be stable and somewhat adjustable.
- Determine the path that the light must travel to provide acceptable coverage. Always attempt to use as few mirrors as possible and consider the laser transmitter's and receiver's positions. If you are lucky, you may have some objects to mount the mirrors to. They must be stable and allow placement at the required height for your application.

  The mirrors used for this project must be the front-surface-coated type if multiple reflections are required. Regular household mirrors can be used up to two reflection points, but they will cause multiple reflections. A low, adjustable-cost mirror mount is shown in Figure 5-2. You may in certain cases use a good-quality automotive mirror, as they are easily adjustable.
- Evaluate the wildlife in your area and determine the best laser height that will provide minimal false alarms from the animal population. Obviously, deer will always be a problem, as their height may be close to that of the target intruder.
- Determine the deterrent you wish to use against the intruder. You can activate bright lights, sonic pain generators (described in Chapter 35), alarms, or just alerts so you can take further action. For more serious applications, you can activate shocking devices, such as those described in Chapter 11, nonlethal kinetic devices, or explosive charges for shock and surprise. Very serious deterrents can consist of deadly force, such as high-powered kinetic devices, shotgun traps, high-powered shocking devices, and buried explosives for lethality.

Caution: As man trapping is illegal in the United States, it may also be so in your country. Always check your laws before ever implementing a deterrent that can cause injury.

- The quality of the mirrors and their mounts cannot be overstressed when designing a high-integrity system. Here we discuss a low-cost system that may suffice in smaller installations but may be limited for use in larger, more reliable systems.

The initial alignment requires a visible, low-power laser to align the mirrors. The laser may have to be collimated for larger multiple reflection systems.

Optics on the laser light transmitter and receiver, as shown, may be sufficient for small systems. They do not have to be high-cost precision devices, as the only objective is to establish a properly aligned, position-stable beam spot.

Once the preliminary alignment is accomplished, you can put the laser transmitter in place and, with a night vision scope, recheck the alignment. The laser receiver, as constructed here, has an indicating *light-emitting diode* (LED). This function can be eliminated if required. The actual control of external functions is via a set of normally open or closed relay contacts that can handle 115 VAC at 10 amps AC.

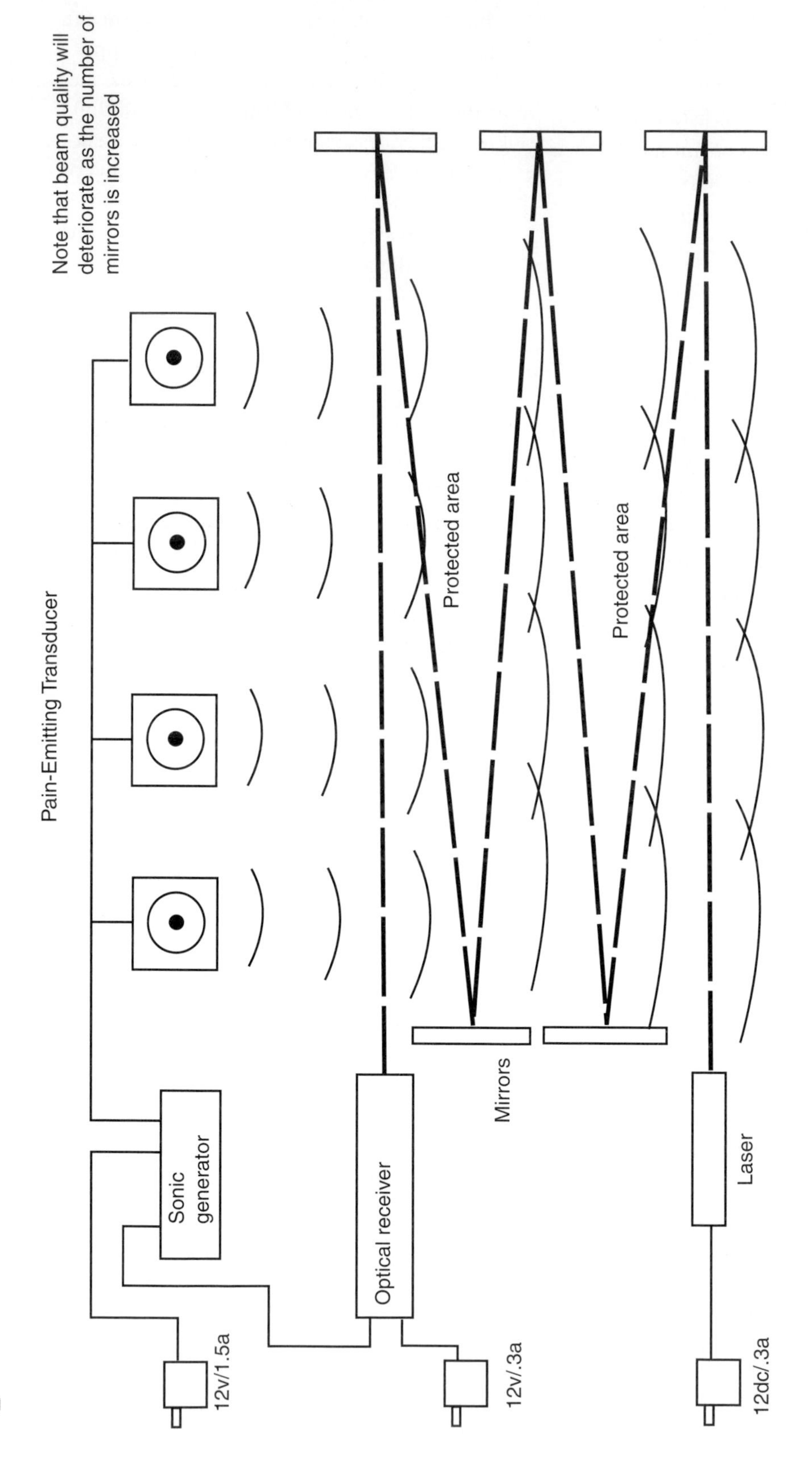

Laser beam is reflected back and forth in the protected area via strategically placed mirrors. Once the beam is broken by an intruder the optical receiver detects and sends a level to the sonic generator where the transducers are activated producing an intolerable sonic pain field discouraging the intrusion.

Even though there is no lasting effect from exposure to the transducers, it is always best to check with local laws before implementing this system.

Figure 5-3 *Example of system layout*

Table 5-1 Laser Property Guard Parts List

Ref. #	*Description*	*DB Part #*
MIRROR	3 × 4-inch front surface mirrors	DB# MIR FRSUR
LRG4K	Pulsed infrared laser kit, as in Chapter 1	DB# LRG4K
LRG40	Assembled pulsed infrared laser, as in Chapter 1	DB# LRG40
LSD4K	Laser detector kit, as in Chapter 2	DB# LSD4K
LSD40	Assembled laser detector, as in Chapter 2	DB# LSD40
PPF4K	Pain field generator kit, as in Chapter 35	DB# PPF4K
PPF40	Assembled pain field generator, as in Chapter 35	DB# PPF40
SHOCK1K	Kit of shockers, as in Chapter 11	DB# PFSHK1K
SHOCK10	Assembled shockers, as in Chapter 11	DB# PFSHK10
LM650-3	Visible red laser module for alignment aid	DB# LM650-Description P3System

The laser project shown in Figure 6-1 is a continuous infrared device providing a continuous 30-milliwatt output at 980 nanometers. The laser, when built as shown, should comply with a class 3 laser-compliant system. The completed laser can be very optically dangerous, and eye protection must be worn to prevent direct or scattered reflections to the eyes.

The project can be used for many optical applications, including illuminating a surface, such as a window, for laser listening. Laser listening is were a laser beam illuminates a window and a sensitive receiver picks up the light reflections and processes them into audio signals similar to the picked-up voices that vibrate the illuminated window glass. A system such as this is described on www.amazing1.com. It should be noted that using a class 3b laser for this application can be an optical hazard to anyone in the target area.

Converting illumination using night vision cameras and spectrally compatible viewers is another potential application, again being fully aware of a possible optical hazard. Other applications are invisible target identification, voice and data communication, and optical signaling and experimenting.

Figure 6-1 *The infrared laser module*

This is an intermediate-level project requiring basic electronic skills. Safety glasses are strongly advised, as the laser can cause eye damage if viewed directly or via direct reflections. Expect to spend $75 to $150. All parts are readily available, with specialized parts obtainable through Information Unlimited (www.amazing1.com), and they are listed in Table 6-1.

For those wanting to use this project as an illuminating laser for window bounce listeners, it is suggested that you refer to Chapter 13 in *Electronic Gadgets for the Evil Genius*, published by McGraw-Hill, ISBN 0-07-142609-4.

Circuit Description

Figure 6-2 shows the laser module LM1 and associated support circuitry. The system is powered by batteries B1 and B2 and is controlled by an internal key switch (S1) or external enabling control leads. A light-emitting diode (LED1) and a current-limiting resistor (R1) indicate the presence of laser emission.

The system will require S1 to be a key switch with a nonremovable key in the activated position if the assembly is integrated as a fully functional laser system. The power line or bench operation will require a

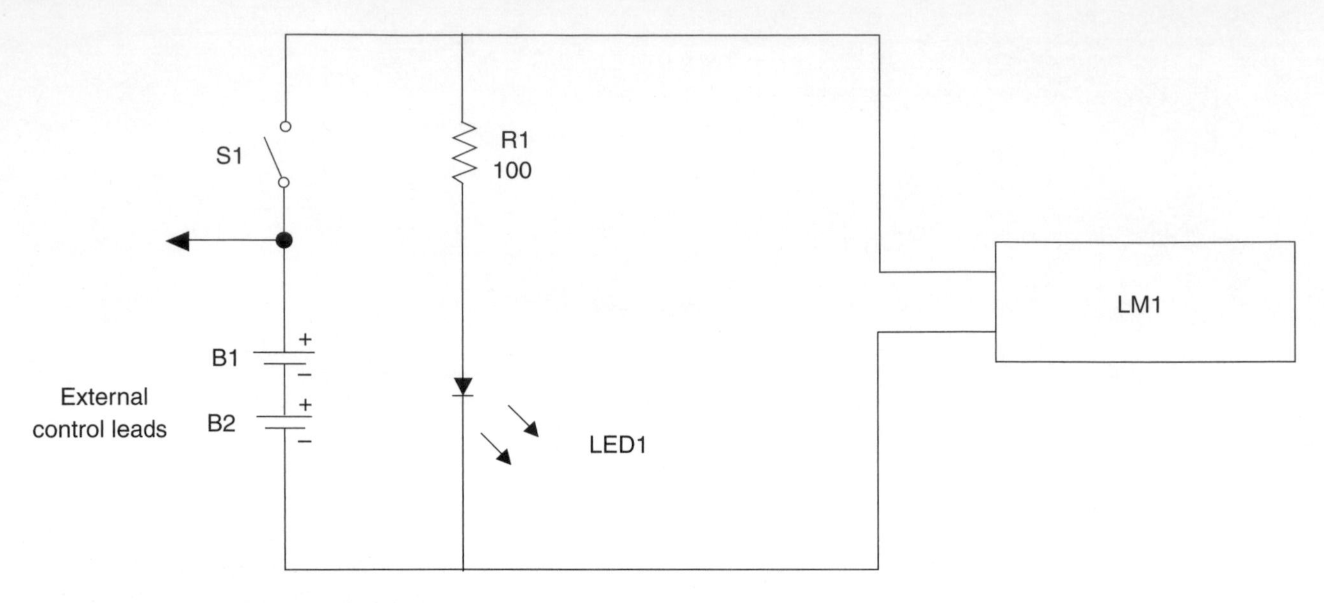

Figure 6-2 *Circuit schematic*

key switch for external enabling and associated control circuitry that may require an emission delay and automatic resetting should a loss of power occur.

Assembly

1. Rework the laser module, as shown in Figure 6-3. This step requires care as it is easy to damage the module if one is not very careful.
2. Connect the soldered leads to a source of 3 volts, paying attention to the correct polarity. If you have a current meter, you can measure 40 milliamps.

   Output can be detected using infrared photosensitive paper or a night vision scope. Certain video cameras may also detect the laser output.

   Danger! Do not view directly or indirectly without safety glasses.
3. Fabricate the laser holder (HOLD1), as shown in Figure 6-4.
4. Fabricate the laser module tube (TUB1) from a piece of 3-inch × 1-inch OD × 1/16-inch wall plastic tubing, as shown in Figure 6-5.

   Fabricate the two spacing washers (WASH1, 2) from 1/16-inch plastic. Note the inner hole is 1 inch to allow for the sleeving onto TUB1 as shown. The outer diameter fits into the inner diameter of the main enclosure (EN1). It is these two spacer washers that position and secure TUB1 into the center of EN1. Space the two washers as shown.
5. Insert the laser module LM1 as shown and wire it up to the circuitry as shown. It is a good idea to leave enough lead length to allow one to pull out the battery holder for battery replacement by removing the rear cap (CAP1). Use a piece of foam to secure the battery holder in place.
6. Verify the proper circuit operation so that when the key switch is activated, the LED emits, indicating laser emission.

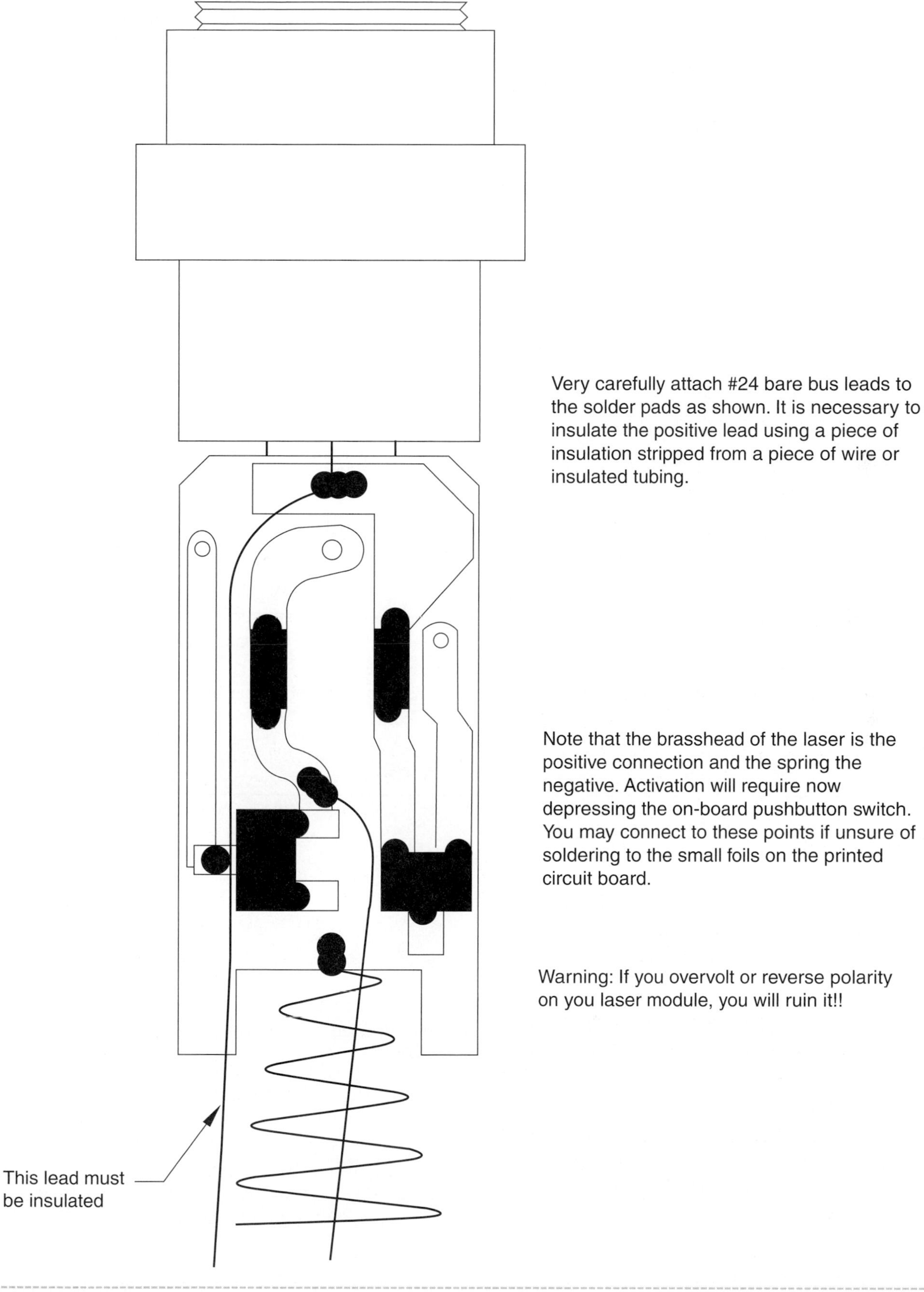

Figure 6-3 *Laser module rework*

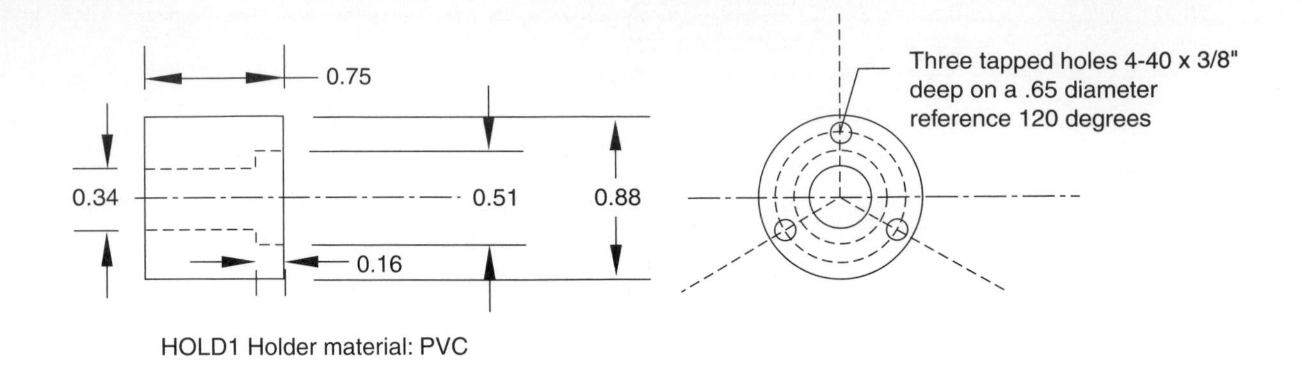

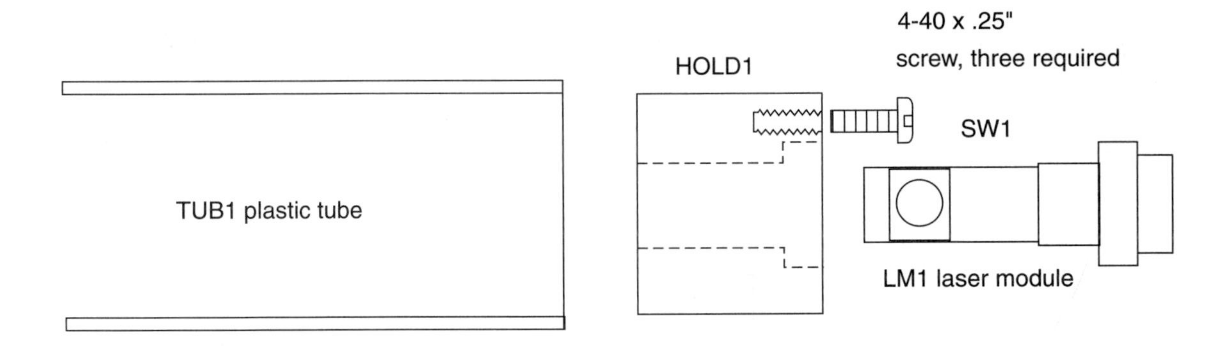

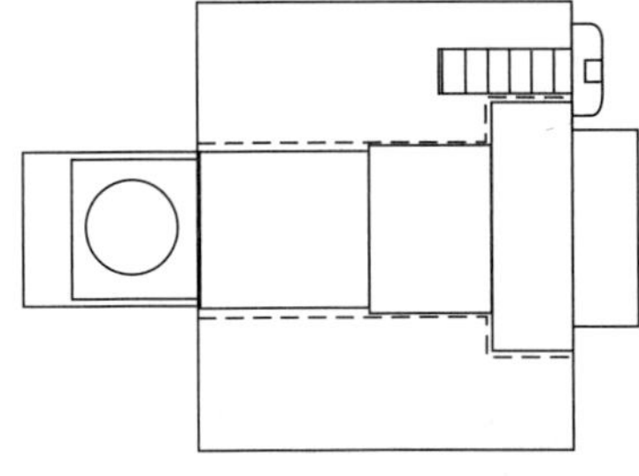

X-ray view showing assembled LM1 laser module inserted into HOLD1 holder with three retaining screws SW1

Figure 6-4 *Creation of laser module holder and retaining washer*

This is our approach to assembling this laser project with external optics. The builder may use his own ingenuity noting that the objective is optical stability and safety.

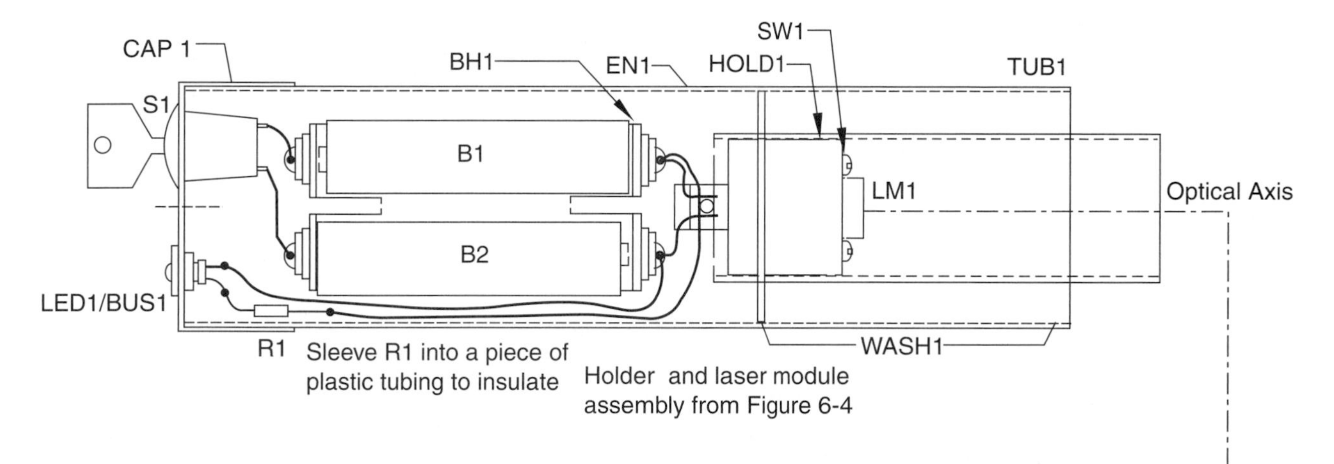

Note if the system is to be completely integrated, that the key switch will be necessary for FDA compliance in requiring this part with nonremovable key in the on position.

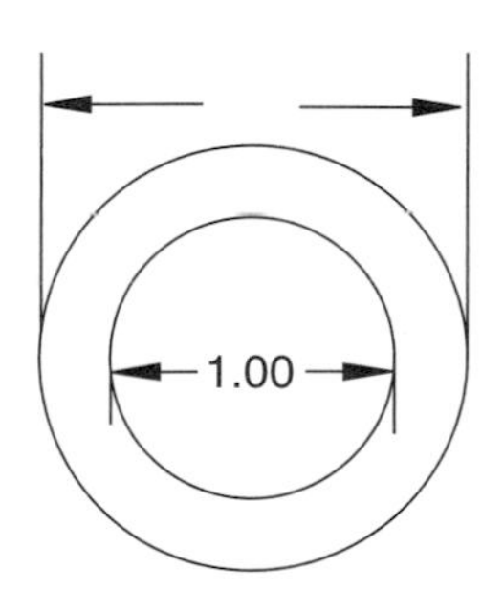

Fabrication of the two WASH1 spacing washers. Use 1/16" Lexan.

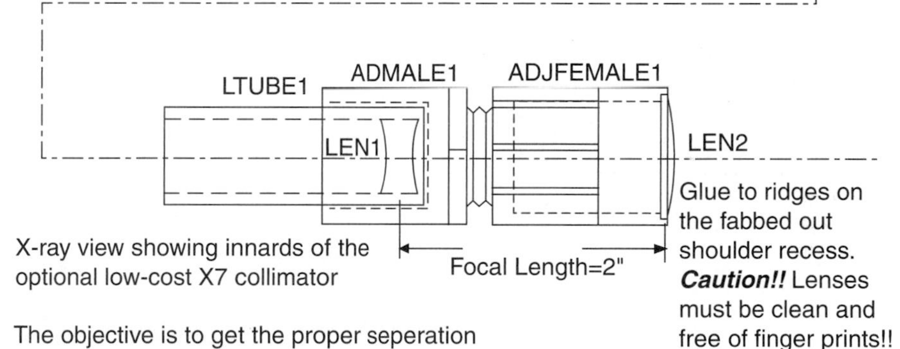

The objective is to get the proper seperation distance between the two lenses to allow proper adjustment within the range of the threaded sections of the male and female pieces.

Figure 6-5 *X-ray of complete assembly*

Collimator and Final Assembly

1. Glue the larger lens LEN2 into ADFEMALE1 as shown, using silicon rubber glue or an equivalent. Clean the lens before gluing.
2. Insert LEN1 into the laser output end of the LTUB1 assembly as shown. Do not glue at this time.
3. Fully insert the LTUB1 into the 1-inch tube TUB1. Shim it with tape for a secure but removable fit.
4. Finish the assembling and label it as shown in Figure 6-6.
5. Point the laser at a target around 100 meters away and adjust the collimator to the smallest spot. This should be done in low light, preferably at night, using a night vision device and only looking at the beam impact point. Note

Instructions

1. Turn key switch full ccw. Note switch has several positions–only two are used in this version.
2. Remove CAP1 and insert 2 AA batteries into holder.
3. Turn on key two positions cw and note LED lighting. If not check the battery holder.
4. Adjust coliminator for desired effect.

Note that coliminator will expand the beam at near field but greatly reduce it at far field. Laser module may require beam adjustment over time using the special tool.

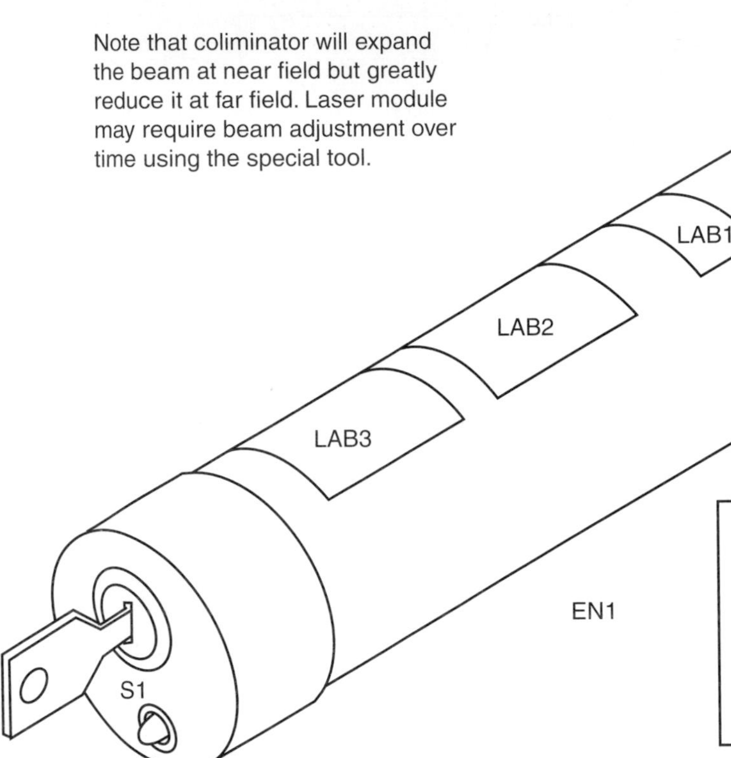

Manufactured By
INFORMATION unlimited
PO Box 716, AMHERST, NH 03031
Model Number
Serial Number
Manufactured Date
This laser product conforms to the provisions of
31 CFR 1040 10 and 1040 11

LAB3

VISIBL AND/OR INVISIBLE
LASER RADIATION
AVOID DIRECT EXPOSURE
TO BEAM

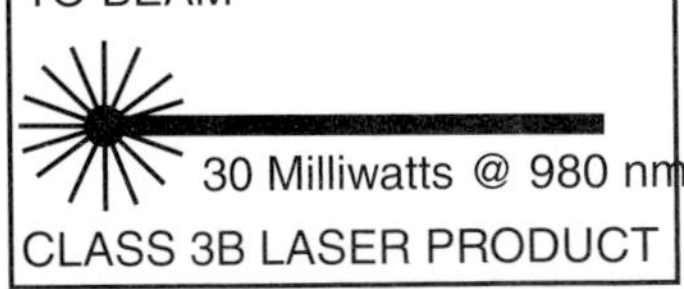

CLASS 3B LASER PRODUCT

LAB2

AVOID EXPOSURE
INVISIBLE LASER RADIATION IS
EMITTED FROM THIS APERTURE

LAB1

Compliance test

1. Verify correct labels as shown
2. Key switch–nonremoval in off position
3. Beam indicator LED indicates after delay
4. Aperture cap included

WARNING: Do not under any circumstances point this device at people, vehicles, or especially aircraft. YOU WILL BE PROSECUTED IF CAUGHT!!!

Figure 6-6 *Final assembly view*

that the adjustment should be about 1/2 to 3/4 turns from being fully tight for a mechanically stable operation. If it is too far out, the assembly will be loose. You may have to use a wire brush on the threads and lube with dry teflon for smooth action. You may compensate the distance between lenses by changing the length of TUBE1. You may now glue LEN1 in place as noted.

The collimator will expand the beam width but decrease the divergence by the same ratio. This greatly reduces the far-field beam diameter.

Note that the laser module is at a wavelength of 980 nanometers and is invisible to the eye. Also, note that most of the mechanical parts used in this project can be obtained through most hardware stores.

Table 6-1 Infrared Laser Parts List

Ref. #	Description	DB Part #
R1	220-ohms, 1/4-watt resistor (red-red-br)	
LED1	High-brightness green LED	
LM1	30-milliwatt, 980-nanometer infrared laser module	DB# LM980-25
S1	Key switch with nonremovable key in "on" position	
BH1	Dual, side-by-side two-AA battery holder	
WR24	12-inch #24 bus wire	
WR22	24-inch #22 vinyl hookup wire	
HOLD1	Fabricated holder with washer for laser module	DB# HOLD1
SWI/NU1	Two 4-40, 1.25-inch machine screws and nuts	
TUB1	3 × 1-inch OD × .875-inch plastic or metal tube	
WASH1	Two fabricated plastic spacing washers, 1.5 OD with 1-inch center hole	
EN1	6-inch × 1.625 OD × 1.5-inch ID plastic or metal tube	
CAP1	1 5/8-inch plastic slip on cap	
Parts for low-cost collimator		
LEN1	6 mm × (-1) double-concave negative glass lens	DB# LENS13
LEN2	24 mm × 75 mm plano/convex glass lens	DB# LE2475
*LTUB1	625-inch ID × 2-inch schedule 40 PVC tube	
*ADMALE1	1/2-inch PVC schedule 40 slip fit to male thread, GENOVA 30405	
*ADFEMALE1	1/2-inch PVC schedule 40 slip fit to female threads, GENOVA 30305	
*ACAP1	Plastic cap and chain for aperture cap	
LAB1	Aperture label	
LAB2	Classification label	
LAB3	Certification label	

*Available in most hardware stores

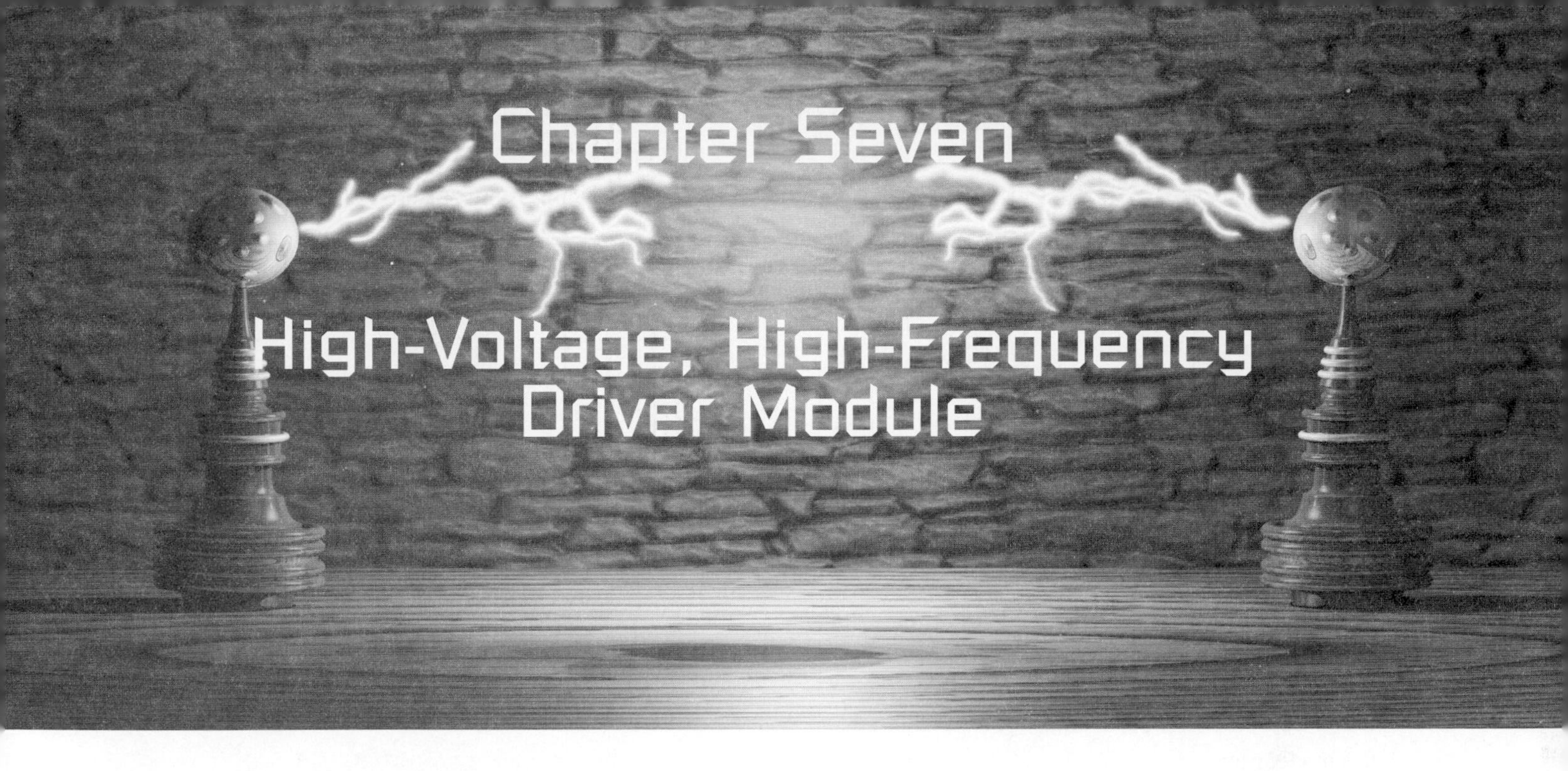

Chapter Seven

High-Voltage, High-Frequency Driver Module

Figure 7-1 shows that a universal, high-voltage, modular power supply project will provide you with many options. This "sweetheart" of a circuit has features that make it a priceless tool for the high-voltage researcher and experimenter.

The circuit operates on 11 to 15 volts DC, drawing 3 amps under a full load, allowing a portable battery or 115 volts AC via a converter. The output voltage is a 60 KHz, high-frequency current that is fully short-circuit protected. The high-frequency output also makes it possible to use low-energy voltage multiplier stacks for high-voltage DC sources, and it can also serve as an excellent plasma driver when used directly. The output current is fully adjustable via a control pot.

The unit is excellent for powering neon and all types of gas-filled vessels using one or two electrodes, or it can power objects simply by proximity. It easily retrofits to our voltage multipliers' modules that provide DC voltages up to 100 kilovolts and currents of up to .3 milliamperes. The current-limiting and control features make this combination an excellent choice for charging capacitors for low loss charging, utilizing 12-volt portable or 115 vac line operation. Also, the device is an excellent choice for powering large and small antigravity craft, ozone air purification, and other applications requiring a high-voltage current-controlled source. The module is shown built on both a rugged *printed circuit board* (PCB) and a vector perforated box. It is mounted in a plastic channel.

Figure 7-1 *Completed assembly module*

The module is used for several projects in this book. In order not to waste a lot of space, we show it as a common subassembly for the following chapters:

- Chapter 12, "Electro-magnetic (EMP) Gun"
- Chapter 32, "100-Milligram Ozone Machine"

Electrical and Mechanical Specifications

- Open-circuit voltage, 7,500 peak at 60 KHz
- Short-circuit current, 10 milliamperes short-circuit protected
- Input 11 to 15 volts DC at 3 amps fully loaded

- Adjustable current by duty-cycle-controlled pulse
- Compact size, 7 × 2 1/8 × 1 1/8 inches, weighing less than 5 ounces
- Easily retrofits to our voltage multipliers

This is an intermediate-level project requiring basic electronic skills and should cost between $25 and $50. All parts are readily available, with specialized parts obtainable through Information Unlimited (www.amazing1.com), and they are listed in Table 7-1.

High-voltage DC output is obtained from this module using a Cockcroft-Walton voltage multiplier with multiple stages of multiplication as required. Note that this method of obtaining high voltages was used in the first atom smasher that ushered in the nuclear age.

The multiplier section requires a high-voltage AC source for input supplied by the circuit transformer (T1) producing 6 to 8 kilovolts at approximately 60 KHz. You will note that this transformer is of a unique design, being owned by our company Information Unlimited. The part is very small, versatile, and lightweight for the power produced.

Circuit Description

The primary of T1 is current driven through inductor L1 and switched at the desired frequency by FET switch Q1. Capacitor C6 is resonated with the primary of T1 and zero voltage switches when the frequency is properly adjusted. (This mode of operation is very similar to class E operation.) The timing of the drive pulses to Q1 is therefore critical to obtain optimum operation (see Figure 7-2).

The drive pulses are generated by a 555 timer circuit (I1) connected as an astable multivibrator with a rep rate determined by the setting of trimpot R1 and the fixed-value timing capacitor C2.

The timer circuit is now turned on and off by a second timer, I2. This timer operates at a fixed frequency of 100 Hz but has an adjustable duty cycle (a ratio of on to off time) determined by the setting of control pot R10. I1 is now gated on and off with this controlled pulse now providing an adjustment of output power.

When the unit is interfaced with a DC voltage multiplier, an over-voltage protection spark gap is placed across the output and is easily set to break down at a preset voltage level for circuit protection. Even though the output is short-circuit protected against continuous overload, constant hard discharging of the output can cause damage and must be limited. A pulse-current-limiting resistor (R17) helps to protect the unit from these catastrophic current spikes.

Power input is controlled by switch S1, which is part of control pot R10. Actual power can be a small battery- capable of supplying up to 2 amps or a 12-volt, 2 to 3-amp converter for 115-volt operation.

Construction

The circuit is shown using the more challenging *perforated circuit board* often required for a science fair project. A *printed circuit board* is also individually available, requiring that you identify only the particular part and insert it into the respective holes as noted. The PCB is plainly marked with the part identification, and soldering is now very simple as you solder the component leads to the conductive metal traces on the underside of the board.

The perforated board approach is more challenging, as now the component leads must be routed and used as the conductive metal traces. We suggest that the builder closely follow the figures in this section and mark the actual holes with a pen before inserting the parts. Start from a corner, using it as a reference, and proceed from left to right. Note that the perforated board is the preferred approach for science projects, as the system looks more homemade.

Board Assembly Steps

If you are a beginner it is suggested to obtain our *GCAT1 General Construction Practices and Techniques*. This informative literature explains basic practices that are necessary in proper construction of electromechanical kits and is listed in Table 7-1.

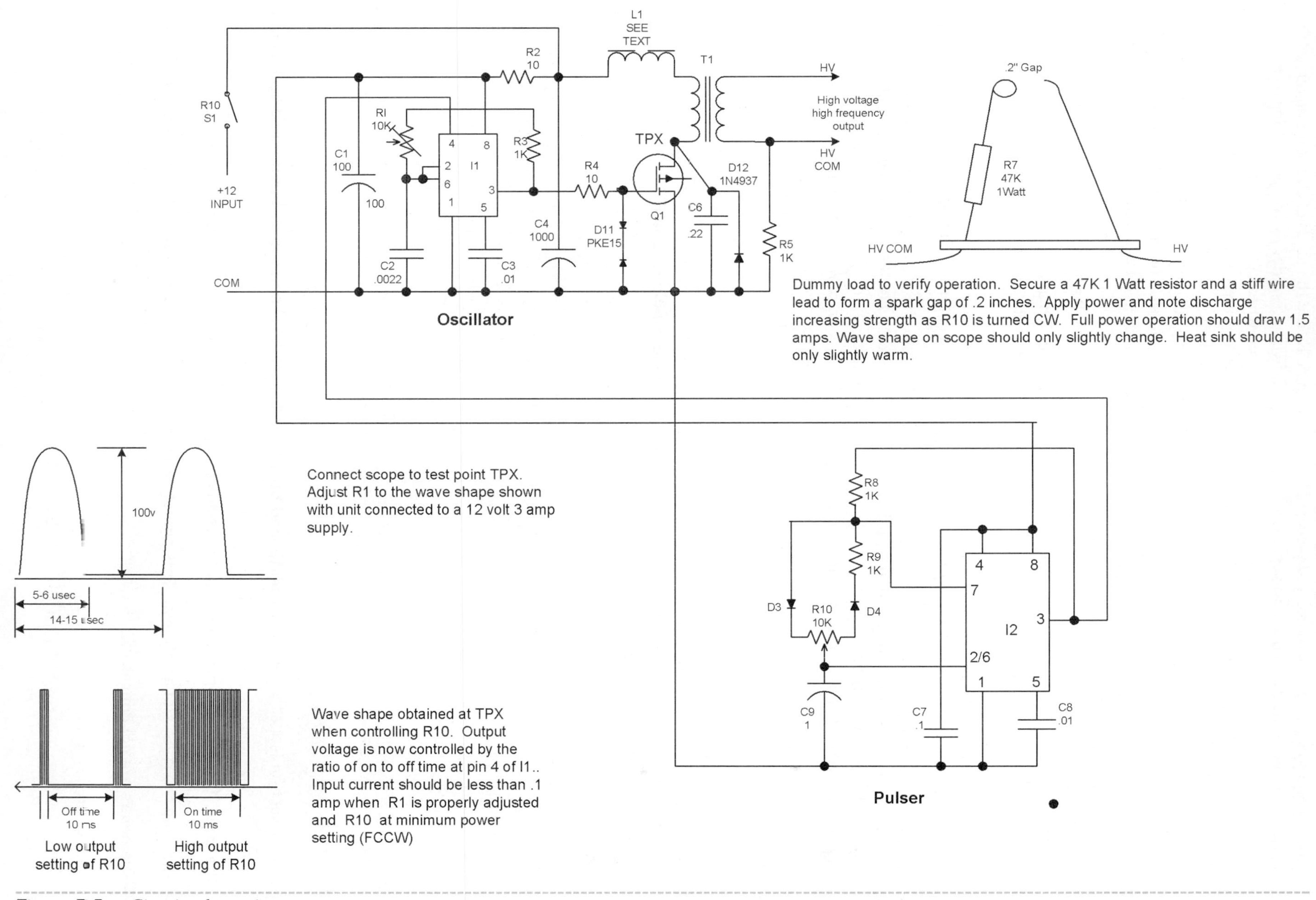

Figure 7-2 *Circuit schematic*

1. Lay out and identify all the parts and pieces. Verify them with the parts list, and separate the resistors, as they have a color code to determine their value. Colors are noted on the parts list.
2. Cut a piece of .1-inch grid perforated board to a size of 5 × 3 inches. Locate and drill the holes, as shown in Figure 7-3. An optional PCB is individually available (refer to Table 7-1).
3. Fabricate a metal heatsink for Q1 from a piece of .063 aluminum, at 1.5 × .75 inches, as shown in the inset of Figure 7-4.
4. Assemble L1 as shown in the Figure 7-4 inset. This part is individually available and listed in Table 7-1
5. If you are building from a perforated board, it is suggested that you insert components starting at the lower left-hand corner, as in Figure 7-5. Pay attention to the polarity of the capacitors with polarity signs and all semiconductors. Route the leads of the components as shown and solder as you go, cutting away unused wires. Attempt to use certain leads as the wire runs or use pieces of the #24 bus wire. Follow the dashed lines on the assembly drawing, as these indicate connection runs on the underside of the assembly board. The heavy dashed lines indicate the use of thicker #20 bus wire, as this is a high-current discharge path.

   Figures 7-6 and 7-7 show the available PCB, which is also listed as an available part in Table 7-1.

6. Attach external leads as shown in Figure 7-5.
7. Double-check the accuracy of the wiring and the quality of the solder joints. Avoid wire bridges, shorts, and close proximity to other circuit components. If a wire bridge is necessary, sleeve some insulation onto the lead to avoid any potential shorts.

These three holes for attaching optional multiplier board

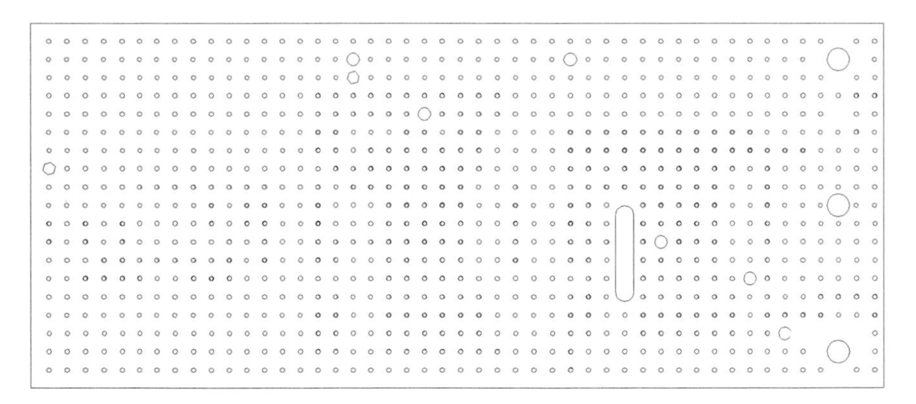

The circuit section is 4.8" x 2.9" .1 x .1 perforated board. Drill eight .063" holes in this perforated section and eleven in the

Drill the three .125" holes for attaching to the optional multiplier board used for high voltage dc applications.

Drill and drag the .125" slot as shown. This cutout and the enlarged holes are for mounting transformer T1.

Hole diameters are not critical.

Always use the lower left hand corner of perf board for position reference.

Figure 7-3 *Perforated board*

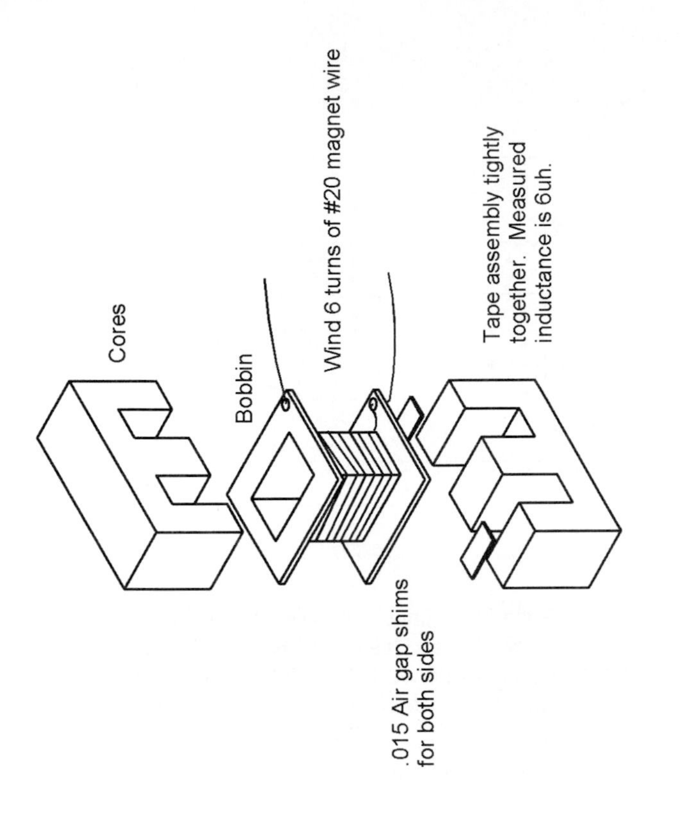

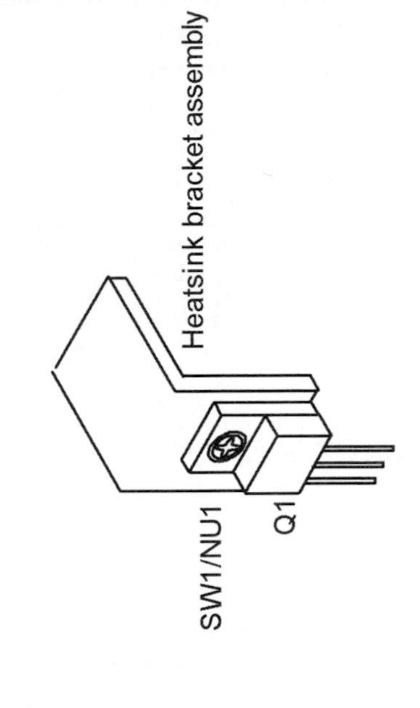

Note polarity of C1,C4,C9,D3,D4, D12 and D20A-D20J

Note position of I1,I2,Q1

Figure 7-4 *Parts assembly and fabrication*

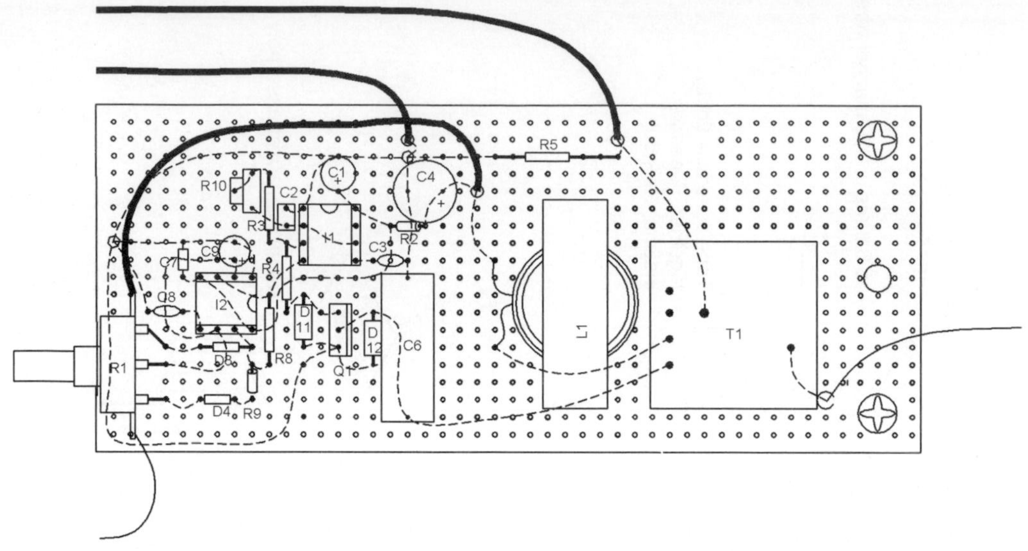

Thick dashed lines are direct connection runs beneath board of #20 buss wire (WR20BUSS) and are extended for the spark switch electrodes.

Thinner dashed lines are #24 buss wire (WR24BUSS) and component leads wherever possible.

Triangles are direct connection point junctions.

Solid black lines are external leads for input and output lines. Use red (WR20R) for +12 input. Use green (WR20G) for lifter - connection. Use black (WR20B) for com -12 input

Use 1/8-3/16 wide smooth globular solder joints for connections to C20A-J and D20A-J, R7 and HV output points. Thisis contrary to normal soldering but is necessary to prevent corona leakage.

Figure 7-5 *Wiring connections and external leads*

8. Fabricate a channel from a piece of 1/16-inch plastic material. Place it in the assembly and secure it at the corners using silicon rubber adhesive. You may also enclose it in a suitable plastic box.

   This step may not be required when used for another project referenced in this book.

Testing Steps

This step will require basic electronic laboratory equipment including a 60 MHz oscilloscope.

1. Preset the trimpot R1 to midrange and R10 to full counterclockwise. Short out the output leads using a short clip lead.
2. Obtain a 47K, 1-watt resistor and construct the spark gap "dummy load," as shown in Figure 7-2.
3. Obtain a 12-volt DC, 3-amp power converter* or a 12-volt, 4-amp rechargeable battery.*

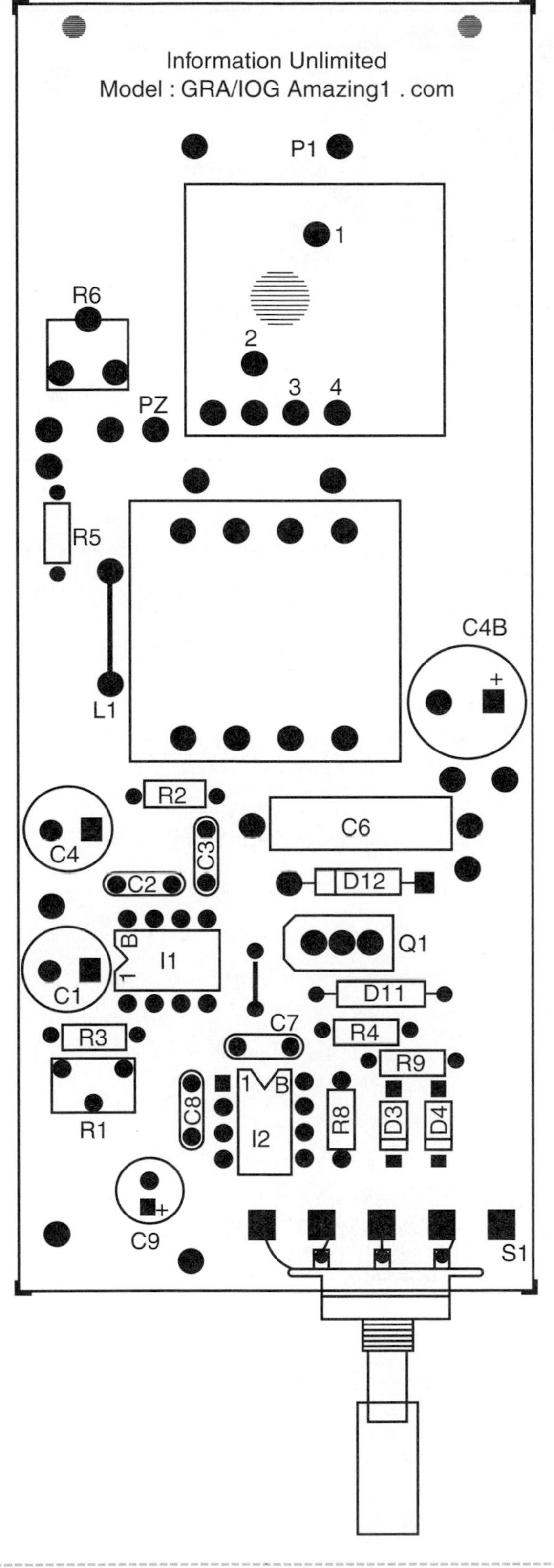

Figure 7-6 *PCB part identification*

4. Connect the input to the power converter and connect the scope, set to read 100 volts and with a sweep time of 5 usecs, to the drain of Q1.

5. Apply power and quickly adjust R1 to the waveshape shown in the inset of Figure 7-2. Note an input current dip of 100 milliamps or less. Check the heatsink of Q1, noting only warm to touch or slightly above 110 degree F.

   Note our laboratory-built units usually tune in with the pot set to 11 o'clock.

6. Remove the clip lead short in step 9 and connect the dummy load. Apply power and rotate R10 clockwise, noting the input current smoothly increasing to 1.5 amps and the spark increasing in energy. This control varies the ratio of off to on time and nicely controls the system current. Note that this system can easily provide 30 watts of usable power.

Table 7-1 High-Voltage, High-Frequency Driver Module Parts List

Ref. #	*Description*	*DB Part #*
R1	10K vertical trimpot	
R2, R4	Two 10-ohm, 1/4-watt resistor (br-blk-blk)	
R3, 5, 8, 9	Four 1K, 1/4-watt resistor (br-blk-red)	
R7	47K, 1-watt resistor (yel-pur-or)	
R10/S1	10K, 17 millimeter pot and switch	
C1	100-microfarad, 25-volt vertical electroradial leads	
C2	.0022-microfarad, 50-volt green plastic cap (222)	
C3,8	Two .01-microfarad, 50-volt disks (103)	
C4	1,000-microfarad, 25-volt vertical electrolytic capacitor	
C6	.22-microfarad, 250-volt metallized polypropylene	
C9	1-microfarad, 25-volt vertical electro cap	
C7	.1-microfarad, 50-volt cap, INFO#VG22	
D3, 4	Two IN914 silicon diodes	
D11	One PKE15, 15 volt transient suppressor	

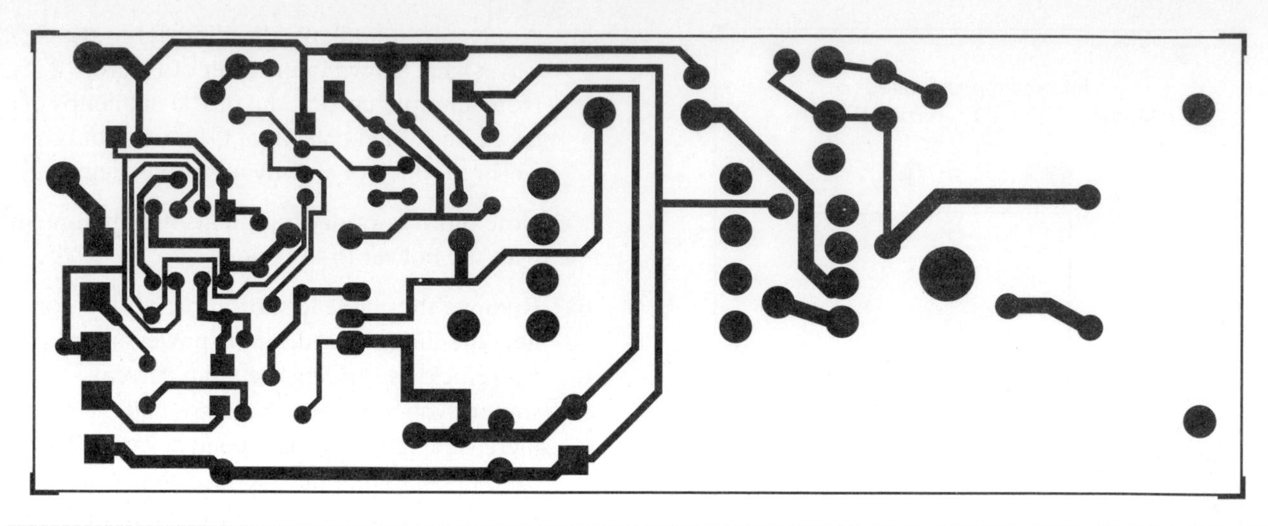

Figure 7-7 *PCB foil traces*

D12	1N4937 fast-switching 1-kilovolt diode	
Q1	IRF540 *metal-oxide-semiconductor field effect transistor* (MOSFET) transistor TO220	
I1, 2	Two 555 DIP timers	
T1	7-kilovolt, 10-milliamp mini-switching transformer	DB# IU28K089
L1	6 uH inductors (see text on assembly)	DB# IU6UH
PBOARD	5 × 2.8 × .1-inch grid perforated board; build to size per Figure 7-2	
PCGRA	Optional PCB	DB# PCGRA
WR20R	12-inch #20 vinyl red wire for positive input	
WR20B	12-inch #20 vinyl black wire for negative input	
WR20G	12-inch #20 vinyl green wire for output ground to craft return	
WR20KV	Four 20 kilovolt silicon high-voltage wires for output	
WR20BUSS	18 inches of #20 bus wire for spark gap and heavy leads (see Figure 7-5)	
WR24BUSS	12 inches of #24 bus wire for light leads (see Figure 1-5)	
HSINK	1.5 × 1-inch .063 AL plate created as per Figure 7-4	
SW1	One 6-32 1/2 Philips screw	
NUT	Four 6-32 keep nuts	
12DC/7	115 VAC to 12 VDC, 3-amp converter	DB# 12DC/7
BAT12	12-volt, 4-amp, hour rechargeable battery	DB# BAT12
BC12K	Battery charger kit for above BAT12	DB# BC12K
GCAT1	*GENERAL CONSTRUCTION PRACTICES*	DB# GCAT1

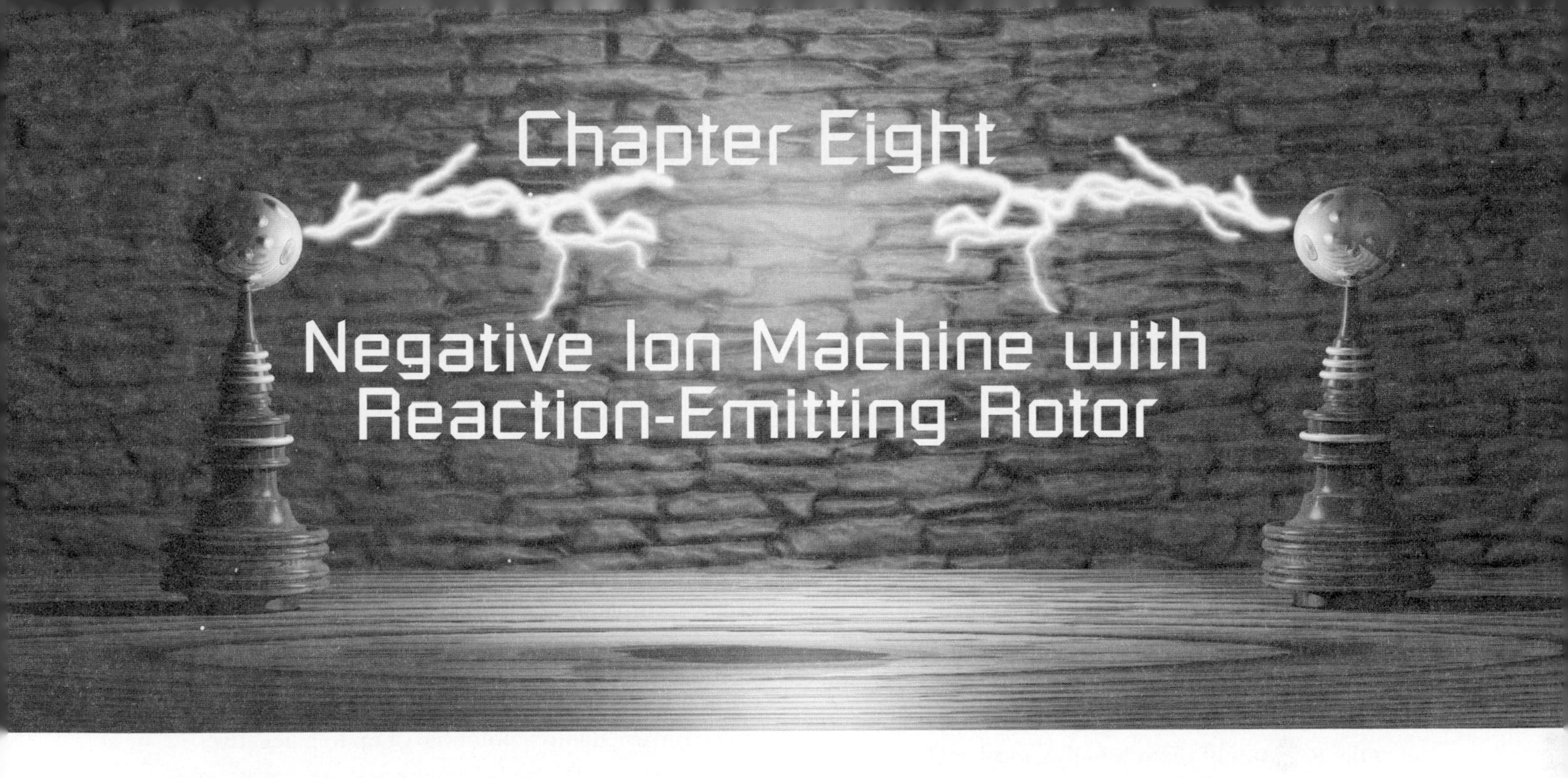

Chapter Eight

Negative Ion Machine with Reaction-Emitting Rotor

Figure 8-1 shows a new concept in negative ion generation. A spinning rotor powered by the emitted ions themselves generates copious amounts of beneficial ions. Large volumes of ions can now be produced, as the area circumscribed by the rotor greatly reduces the local charge flux. This feature allows many times more ions without ozone-producing voltages.

Negative ions are highly beneficial and cause a sense of well-being, as well as rid the air of pollen and other breathable irritants. This is an excellent device for people suffering from asthma and other bronchial problems. It is also a nice conversation piece with a desktop unit's spinning rotor silently reaching a relatively high rotational velocity. The device operates from 115 VAC/12 VDC or a battery. The project utilizes a 25-kilovolts negative output power module requiring 12 volts DC at 500 milliamps along with our proprietary rotating emitter.

This is a beginner- to intermediate-level project requiring basic electronic skills, and one should expect to spend $25 to $50. All the parts are readily available, with specialized parts obtainable from Information Unlimited (www.amazing1.com), and they are listed in Table 8-1.

Figure 8-1 *The ion machine*

Precautions: Do not use this device in explosive or flammable atmospheres, as spark discharges may cause dangerous ignitions.

Benefits of Negative Ions

In the last several decades, a medical controversy has evolved pertaining to the beneficial effects of these minute electrical particles. As with any device that appears to affect people in a beneficial sense, there are those who sensationalize and exaggerate these claims as a cure for all aliments and ills. Such people manufacture and market these devices under these false pretenses and consequently give the products an adverse name. The Food and Drug Administration now steps in on these claims, and the product, along with its beneficial facets, goes down the tubes.

People are affected by the property of these particles, which increase the activity rate of cilia, whose function is to keep the trachea clean from foreign objects, thus enhancing oxygen intake and increasing the flow of mucus. This property neutralizes the effects of cigarette smoking, which slows down this activity of the cilia. Hay fever and bronchial asthma victims are greatly relieved by these particles, and burn and surgery patients are relieved of pain and heal faster.

Tiredness, lethargy, and general worn-out feelings are replaced by a sense of well-being and renewed energy. Negative ions destroy bacteria and purify the air with a country-air freshness. They also cheer people up by decreasing the serotonin content of the blood. As can be seen in countless articles and technical writings, negative ions are a benefit to humans and their environment.

Negative ions occur naturally from static electricity, the wind, waterfalls, crashing surf, cosmic radiation, radioactivity, and ultraviolet radiation. Positive ions are also produced from some of these phenomenon and both of them usually cancel each other out as a natural statistical occurrence. However, many manmade objects and devices have a tendency to neutralize the negative ions, thus leaving an abundance of positive ions, which create sluggishness and most of the opposite physiological effects of its negative counterpart.

One method for producing negative ions is by obtaining a radioactive source rich in beta radiation (negative electrons). Alpha and gamma emissions from this source produce positive ions that are neutralized electrically. The resulting negative ions are directed by electrostatic forces to the output exit of the device and are further dispersed by the action of a fan (this method has recently come under attack by the Bureau of Radiological Health and Welfare for the use of tritium or other radioactive salts). This approach appears to be the more hazardous of the two according to the Consumer Product Safety people. The second method is to produce a potential level of negative high-voltage electricity without ozone and allow the creation of negatively charged particles by the high voltage.

Your negative ion generator produces ions via a high-potential, low-current source of DC power. The high potential causes negative charges to be produced at the sharp emission points of the rotor blade. The reactionary force of the emitted particles now causes the rotor to spin, allowing these negative ions to escape into the air stream due to the high-charge density at the sharp, pointed rotor tips. (The point of a pin will have a much higher charge density than a larger diameter spheroid for the same potential.) Ozone is kept at a minimum by keeping the voltage relatively low while allowing a sufficient charge density lets the negative particles escape into the air.

Circuit Description

The circuit (MOD25KV) consists of a high-frequency, high-voltage oscillator being fed into a multistage Cockcroft-Walton multiplier. This energy is converted into a potential of up to a negative 25,000 volts. The high-frequency stage consists of a transistor connected as a simple oscillator where its collector drives the primary winding of the resonant transformer (T1).

The high-frequency output of T1 is fed into a voltage multiplier stage consisting of a diode string (D1 to D5) and a capacitor string (C1 to C5). A 12-volt DC wall adapter (T2) powers the unit. A separate ground is required for optimum performance and involves connecting the negative 12 volts to earth ground. This is done by connection to the AC receptacle plate or ground pin. This ground provides a virtual ground for the ions.

A charge concentration occurring at the ends of the rotor blade's points now produces ions, and reactionary forces move the rotor at a high speed. Ions are produced at a high rate as the effective emitter area is equal to that of the rotating rotor. Also, charge neutralization is minimal using this method and a high ion current is generated.

The output of the unit can be approximated by the fact that a Coulomb charge equals the current multiplied by the seconds. There are also $6.25 \times E18$ negative chargers per Coulomb and the maximum current to the emitters is 400 microamps. Therefore, the unit can produce $6.25 \times E18 \times 200 \times E\text{-}6 = 2.5 \times 10E15$

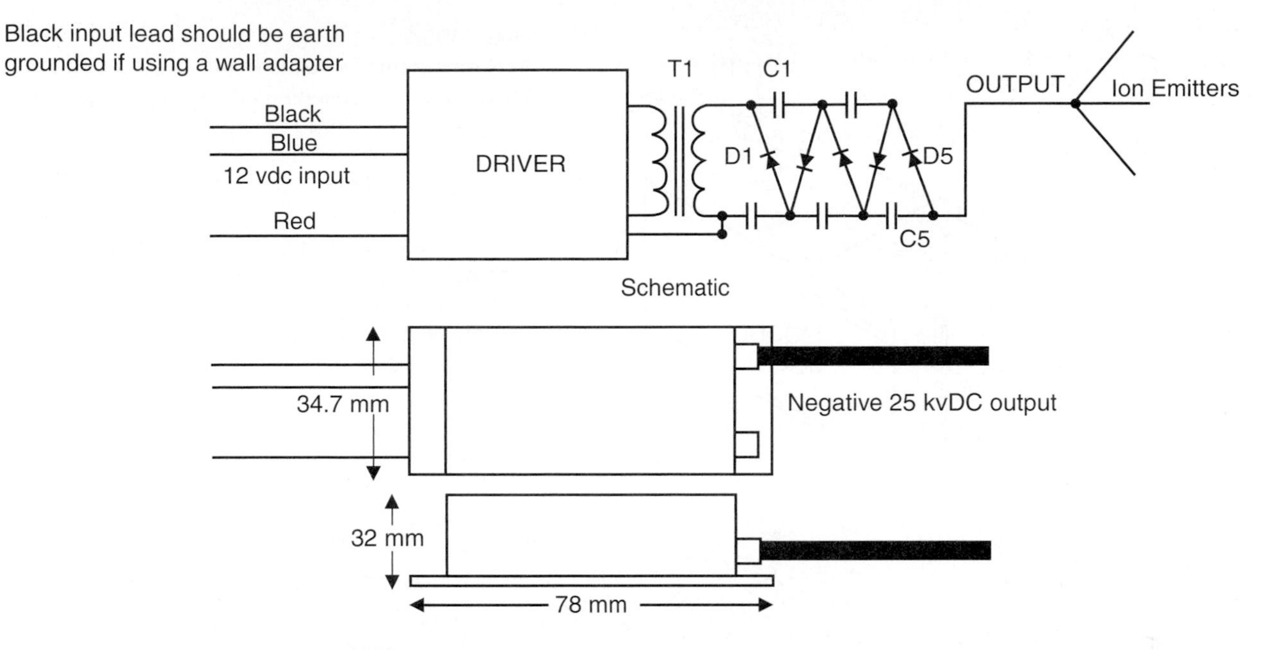

Function	Value	Notes	Other
Input Voltage	12 vdc		
Input Current (No load)			
Input Current (load)	400 ma		
Input Watt (No load)			
Input Watt (load)			
Output Volts	–20 kvDC		
Output Configuration	Single Output		
Internal Operating Frequency	35 KHz		

Use a 12 vdc/.5 amp DC wall adapter or other suitable power supply.
Always check entire system for any excessive heating.

Figure 8-2 *Circuit schematic*

negative charges per second. This is usually sufficient for a large room.

Assembly Steps

1. Lay out and identify all parts and pieces. Verify them with the parts list.
2. Carefully fabricate the rotor blade as shown in Figure 8-3 and note that it must be balanced for optimum rotational velocity and ion emission.
3. Position a common plastic pushpin (PIN1) and hot glue or epoxy to the module as shown. Glue is more forgiving if should you ever wish to replace the pin.
4. Carefully strip about 3/4 of an inch of insulation from the output lead. Tin and solder it to the bottom of the actual pin. This connection point must allow complete clearance of the entire bearing section of the rotor assembly when in place.
5. Wire up the wall adapter (T2) and ground the lead using (WN1) wire nuts, as shown in Figure 8-4.

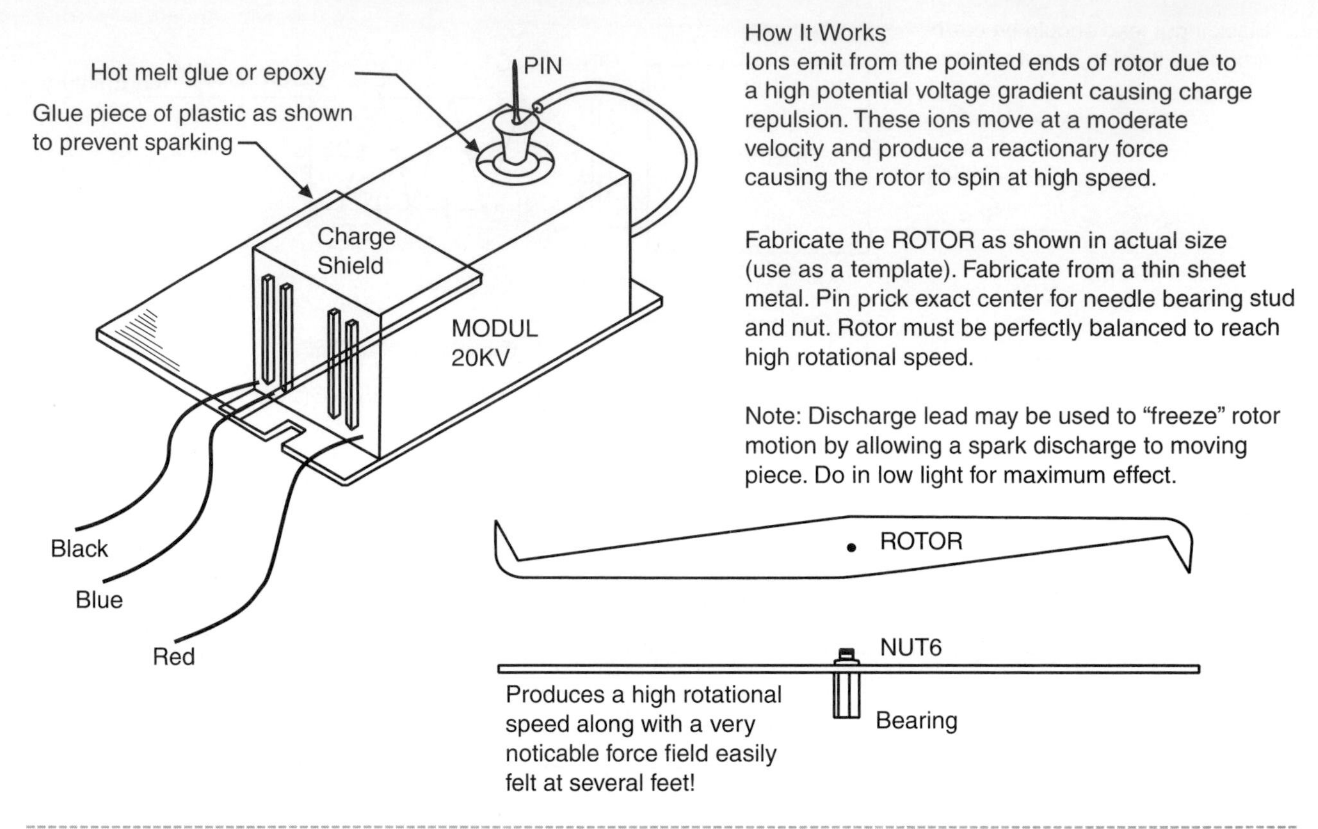

Figure 8-3 *Isometric view of the overall unit*

6. Glue on the CS1 charge shield plate to prevent premature sparking to the metal heatsink of the module. Also, stick on the rubber feet.
7. The unit is perfectly usable. However, you may want to enclose it in some sort of plastic box. Always use good ventilation and make sure the box is large enough or about 1½ times the diameter of the rotor.
8. Find a suitable location, such as a nonmetallic table, to place the unit on. Place the rotor blade on the pin and verify it is unobstructed. Then connect the grounding lead to the receptacle mounting screw and plug in the T2 adapter. The rotor should begin rotating and quickly reach a high speed.

Notes

Ions are now being emitted into the air and can easily be detected by holding a fluorescent tube or a small neon indicating lamp (LA1) near the unit, noting a flashing. This must be done in the dark. You can also place a piece of paper near the base of the unit and note the force field generated that tries to pull the paper.

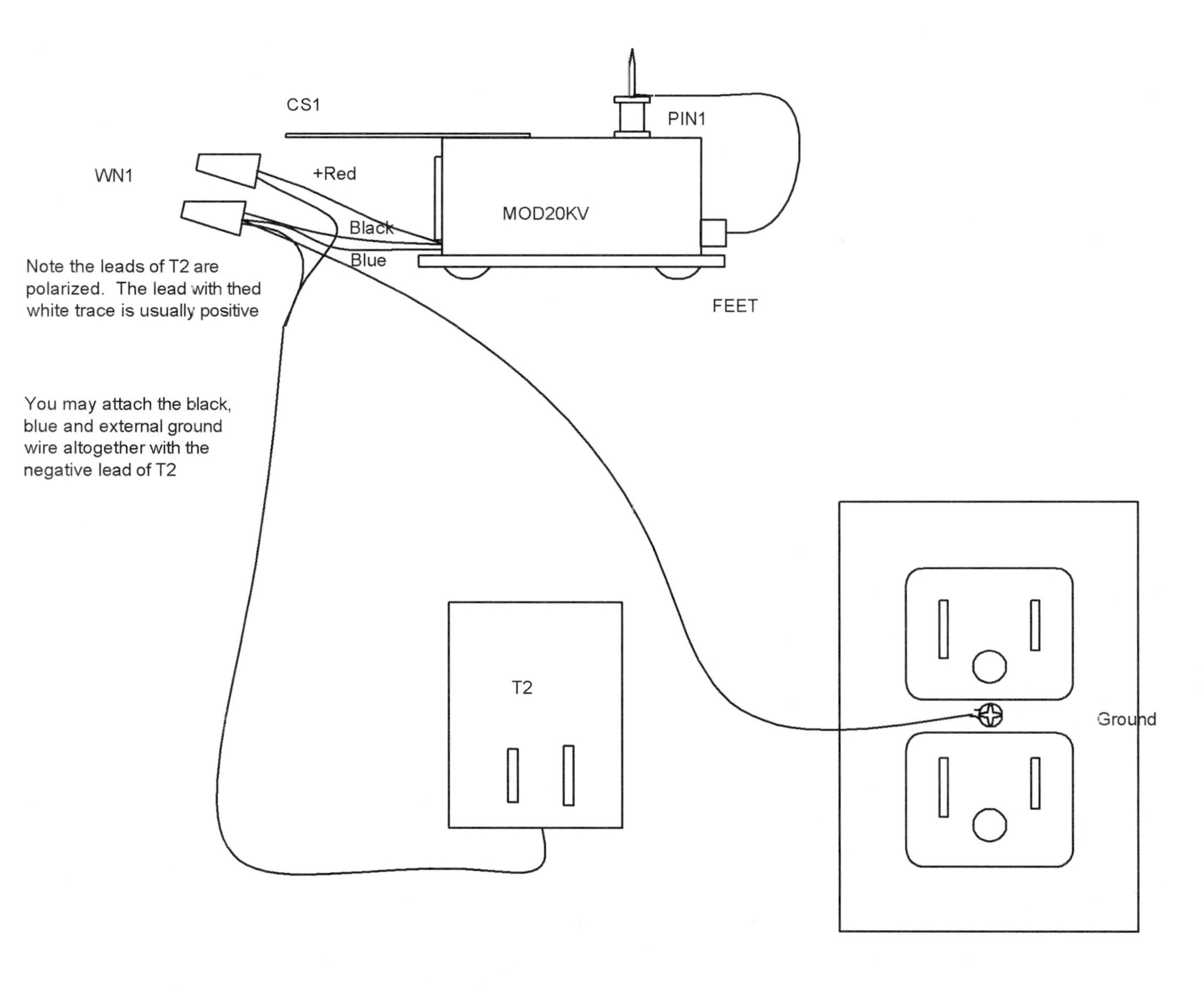

Figure 8-4 *Side view showing power connections*

Table 8-1 Negative Ion Machine Parts List

Ref. #	*Description*	*DB Part #*
MODULE	Negative 20-kilovolt module reworked	DB# MOD20KV
T2	12-volt DC, 500-milliamp wall adapter	DB# 12DC/.5
ROTOR	Metal fabricated rotor	DB# ROTOR
PIN	Common plastic pushpin	
BEARING	#6-32 threaded standoff	
NUT6	6-32 hex nut	
CS1	$2^1/_4 \times 1^1/_2$-inch-thin, $^1/_{16}$ to $^1/_8$-inch-thick plastic	
WN1	Small wire nuts	
FEET	Four stick-on rubber feet	

Chapter Nine

Kirlian Imaging Project

Kirlian photography uses *cold electron emission* to expose a photographic film. Cold electron emission is the result of electrons accelerating in an electric field. These accelerated electrons ionize the air and recombine producing ultraviolet light that is mostly invisible to the naked eye. This phenomenon is commonly known as *corona*. Many living objects are claimed to emit this radiation that will vary according to the health and condition of the object being exposed. This in itself is now subject to scientific research in material analysis. A more "psychic application" can be seen in the practice of radiation being referred to as an aura and being dependent on other more metaphysical functions. This project describes a simple method of producing Kirlian images of small objects and is an introduction to those who want to explore this interesting field. Kirlian imaging ranges from the psychic to materials analysis. Often it is used for experiments in which it is claimed that objects all possess a unique aura signature. This depends on the subject's mental makeup and condition at the time. Basically, it is a form of corona discharge that occurs as a precursor to a dielectric breakdown by most conductive objects when placed in a high-frequency, high-voltage field. Corona is a form of electrical leakage occurring without actual air breakdown such as a spark discharge. A corona discharge possesses ultraviolet emissions that easily develop film and thermographic paper, producing an image print.

This is an intermediate-level project requiring basic electronic skills and one can expect to spend $25 to $50. All parts are readily available, with specialized parts obtainable through Information Unlimited (www.amazing1.com). The parts are listed in Table 9-1.

The system is simple and built from readily available parts (see Figure 9-1). The builder may deviate from what is shown as long as the basics are adhered to. The system requires a high-voltage, high-frequency module, as shown in Chapter 10 where it will be used to supply the burning plasma for the plasma pen project.

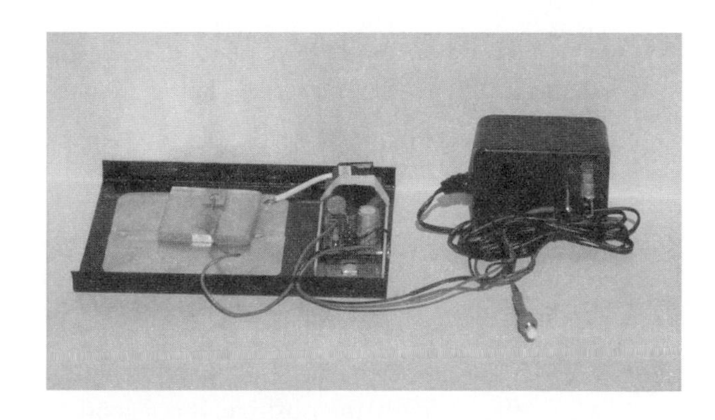

Figure 9-1 *A complete unit*

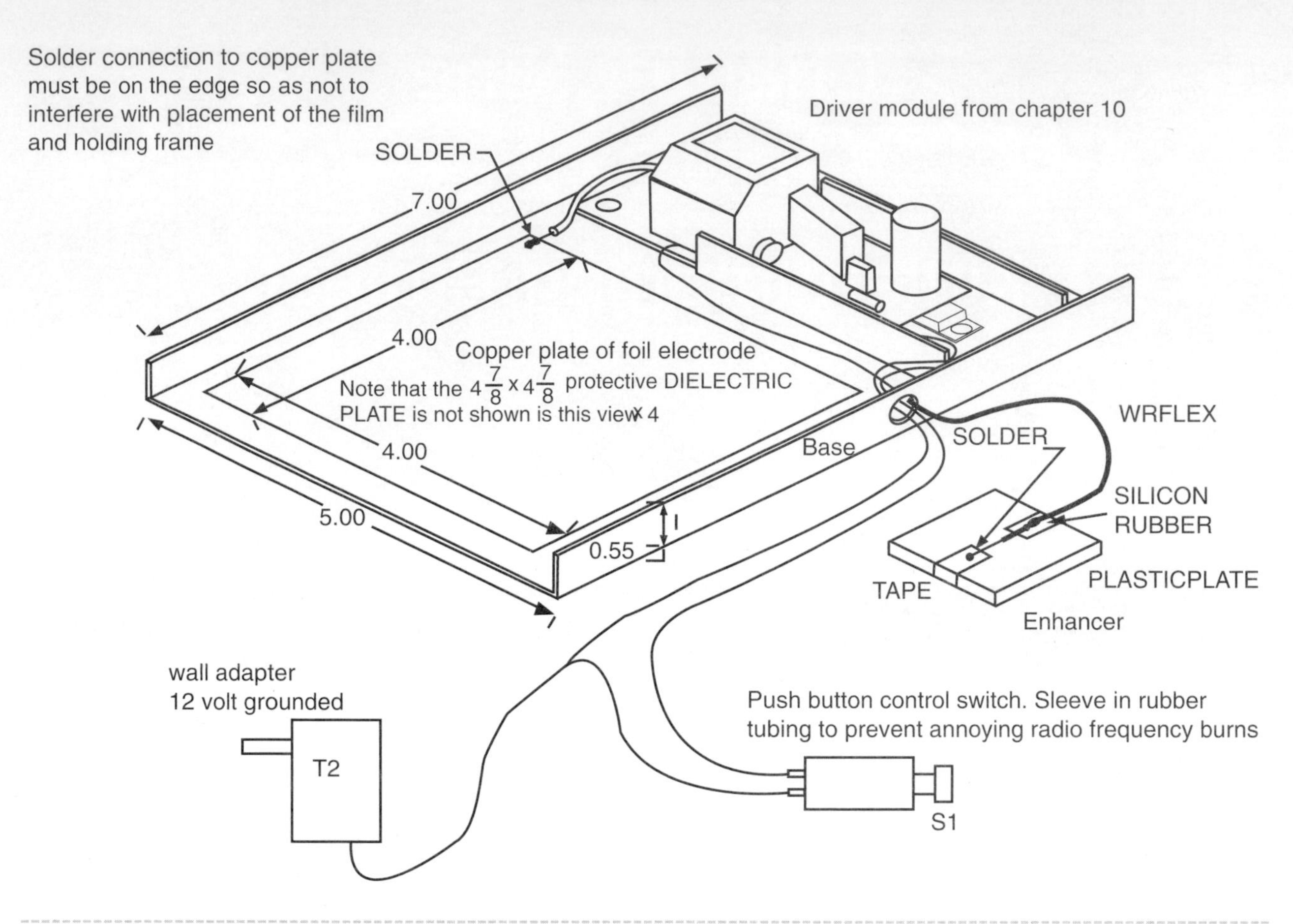

Figure 9-2 *Isometric of the image base and charge plate*

Assembly Steps

If you are a beginner, it is suggested to obtain our *GCAT1 General Construction Practices and Techniques*. This informative literature explains basic practices that are necessary in proper construction of electromechanical kits and is listed in Table 9-1.

1. Assemble the driver module, as shown in Figures 10-2 and 10-3 in Chapter 10. Verify proper operation before using.
2. Assemble the image base, as shown in Figures 9-2 and 9-3. The dimensions as shown are not critical.
3. Mount the driver module and solder the output lead as shown. Route the input and high-voltage return lead to the ENHANCER.
4. Wire the pushbutton switch (S1) to the wall adapter transformer (T2). The output wires of T2 are polarized with the lead containing the white trace, usually the positive. It is a good idea to cover the entire body of S1 except the plastic actuator; any contact to metal will result in small, very annoying burns. The wall adapter recommended has a third pin being an isolated ground for the high-frequency return.

Testing Steps for the Circuit

These steps usually will not require any special testing equipment.

1. Power up the unit and activate S1. Place your finger on the insulated plate and note a bluish corona discharge occurring under your finger. You may feel some heat or a tingling sensation. Try this for only up to 20 seconds.
2. Obtain some thermographic fax paper and place it on the dielectric plate. Place your finger on the paper and energize the circuit for 10 to 20 seconds. You will note an image on the paper.

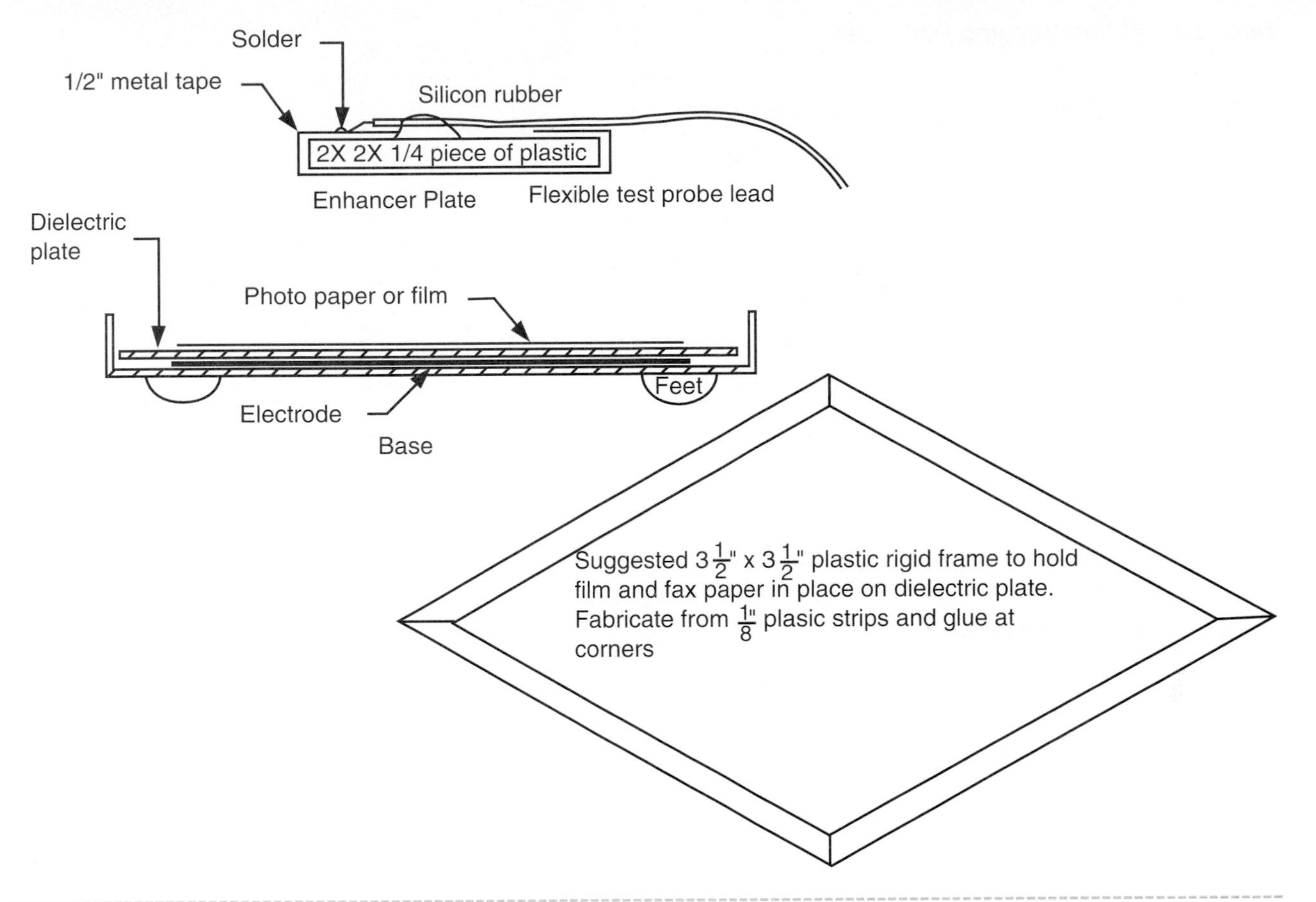

Figure 9-3 *Image base and charge plate*

Try it again with a metal object such as a coin on the dielectric plate and place the enhancer on the object. Push the button for 5 to 10 seconds and note an image forming.

3. You can upgrade your image to an actual photograph by using actual camera film. The old Polaroid 600 or other similar instant film will work. You must work entirely in the dark to avoid film exposure to any ambient light. The objective is to place the film on the dielectric plate along with the object to be photographed and the enhancer placed on top.

Notes

You will note that a few variables here can affect performance: the output voltage of the driver module and the dielectric plate's thickness and dielectric properties. The dielectric plate is important because it must form a high enough capacity between the object and the contacts to allow an adequate corona current to flow to create an image. The thinner the plastic dielectric plate, the more corona current. However, you must not allow a breakdown to occur by using material that is too thin, as this can produce a painful burn.

Many different sources of information, including some excellent books, are available on the applications, methods, and uses of Kirlian photography. You are encouraged to continue to investigate this unusual electronic phenomenon.

Table 9-1 Kirlian Imaging Parts List

Ref. #	*Description*	*DB Part #*
DRIVER	Driver module from Chapter 10	DB# MINIMAX7
BASE	6 × 7-inch plastic piece, formed as shown in Figure 9-2	
COPPER ELECTRODE	4 × 4-inch, thin piece of copper	
DIELECTRIC PLATE	5 × 5 × .02-inch piece of polypropylene sheet	DB# POLYP20
PLASTIC PLATE	2 × 2 × 1/4-inch plastic plate	
TAPE	Four inches of 1/2-inch solderable metal sticky tape	
FEET	Four sticky rubber feet	
T2	12-volt, 1.5-amp groundedadapter with third pin	DB# 12DC/1.5G
FRAME PIECES	Four strips of 1/8-inch plastic to fabricate the frame as shown in Figure 9-3	
WRFLEX	12 inches of flexible test lead wire	
S1	Push-button switch reworked as shown to prevent burns	
GCAT1	General construction practices	DB# GCAT1

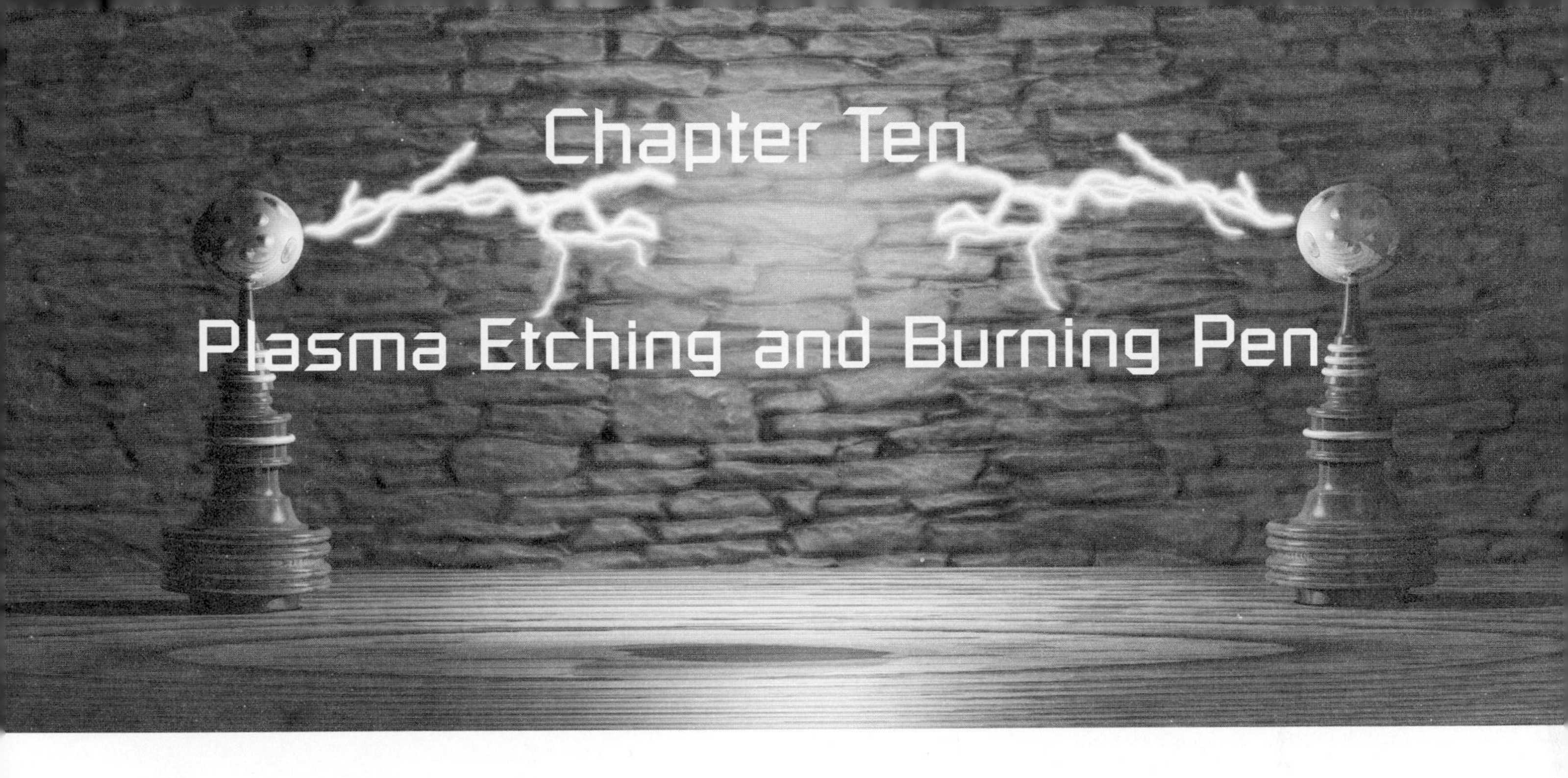

Chapter Ten

Plasma Etching and Burning Pen

A plasma pen uses high frequency, high voltage to burn or etch permanent markings into many different materials (see Figure 10-1). With a little practice, the user can create many intriguing designs of fine, lace-like etches. Names and other text can easily be etched into almost any organic material. Wood, paper, plastics, pumpkins, gourds, eggs, and leaves can all be marked with fine and fancy designs or inscriptions. The pen's other applications include Kirlian photography, aura enhancing, and magically lighting household fluorescent and neon tubes, all without any connecting wires, providing a great magic act. You can also use this for plasma experimentation or even removing warts.

This is an intermediate-level project requiring basic electronic skills. Expect to spend $25 to $35. All parts are readily available, with specialized parts obtainable through Information Unlimited (www.amazing1.com), and all of them are listed in Table 10-1.

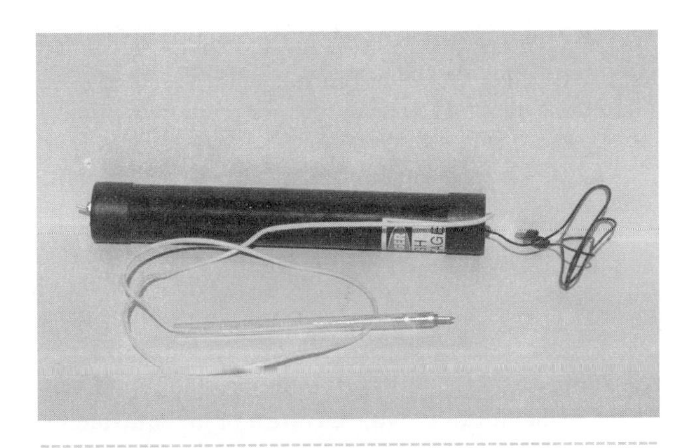

Figure 10-1 *Photo of assembly*

The system is shown in two parts: the power driver and the pen or etching section. These are connected together via a suitable, flexible, high-voltage wire. The power driver consists of the electronics module and the internal batteries for portable use. A jack is provided for external use from a 12 VDC wall adapter or another suitable source. This section is housed in a tubular enclosure with plastic end caps as shown.

Driver Circuit Description

The circuit is a frequency, high-voltage oscillator that consists of transistor Q1 connected as a simple oscillator where its collector drives the primary winding PR1 of the resonant transformer T1. Feedback is obtained via a second winding (FB) and fed to the base of Q1 through a current-limiting resistor (R1). Resistor R2 biases Q1 into conduction and initiates the oscillation (see Figure 10-2).

Capacitor C3 speeds up the turn-off time of Q1, while resistor R3 and capacitor C5 provide a filter to prevent oscillation at the self-resonant frequency of T1. Resonating capacitor C4 resonates the transformer

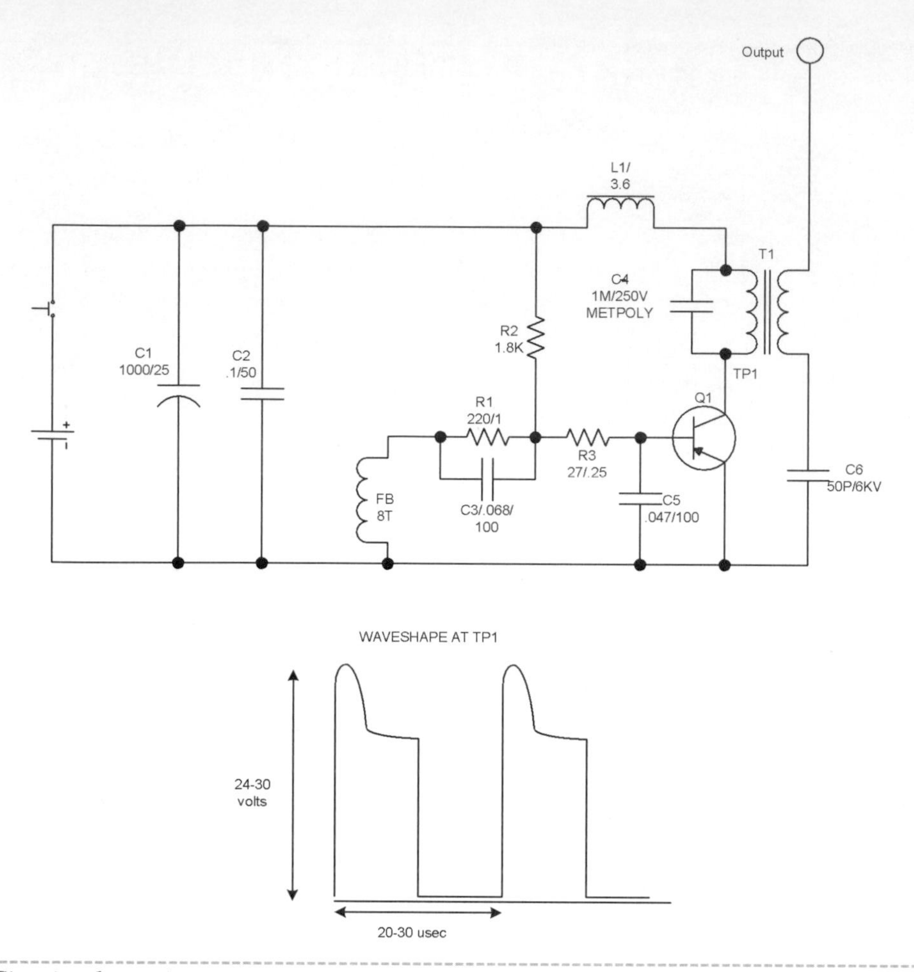

Figure 10-2 *Circuit schematic*

to around 35 KHz. The current-limiting inductor L1 limits the short-circuit current to a noncatastrophic value, and capacitors C1 and C2 bypass any signal to ground.

Driver Circuit Assembly

1. Lay out and identify all the parts and pieces. Verify them with the parts list, and separate the resistors as they have a color code to determine their value. Colors are noted on the parts list.
2. Obtain the available *printed circuit board* (PCB), as shown in Figure 10-3, or fabricate a piece of perforated circuit board to the PC as laid out on the PCB. Note the size of the PCB is 3½ × 1⅜ inches and contains the silk screening that shows the positioning of the mounted parts.
3. If you are building from a perforated circuit board, it is suggested that you insert the components starting in the lower left-hand corner. Pay attention to the polarity of the capacitors that have polarity signs and all semiconductors. Route the leads of the components as shown and solder as you go, cutting away unused wires. Attempt to use certain leads as the wire runs and follow the foil traces on the drawing as these indicate the connection runs on the underside of the assembly board.
4. Insert the components as indicated by the silk-screen printed identification numbers and compare this with the bill of materials. Attach three 6-inch, #22-20 leads as shown for the input power (P1, P2) and an external high-voltage return. Also attach a high-voltage lead to the output as shown. Note that this lead must be selected for the required length. Use a silicon, 20-kilovolt DC wire or an equivalent.

Note the ground lead that connects to the metal frame under the screw head as shown.

5. Double-check the accuracy of the wiring and the quality of the solder joints. Avoid wire bridges, shorts, and close proximity to other circuit components. If a wire bridge is necessary, sleeve some insulation onto the lead to avoid any potential shorts.
6. Cut the metal bracket (BRK1) from a piece of .062 sheet aluminum and make the final assembly, as shown in Figure 10-4. Note that the metal tab of Q1 must be insulated from the bracket as shown on the mounting scheme. This piece is also the heatsink for Q1.

Driver Board Testing

This step will require basic electronic laboratory equipment. A 60 MHz oscilloscope is suggested.

1. Obtain a 12-volt DC source, preferably with a volt and ammeter. Leave the output leads as open circuit and apply power. Note a current draw of less than 250 milliamps. If you have a scope, observe the waveshape at the collector of Q1, as shown in Figure 10-2.
2. Contact the high-voltage output lead to the bracket and note that you can draw a 1/2-inch arc with the input current going to 1 amp. This completes this module.

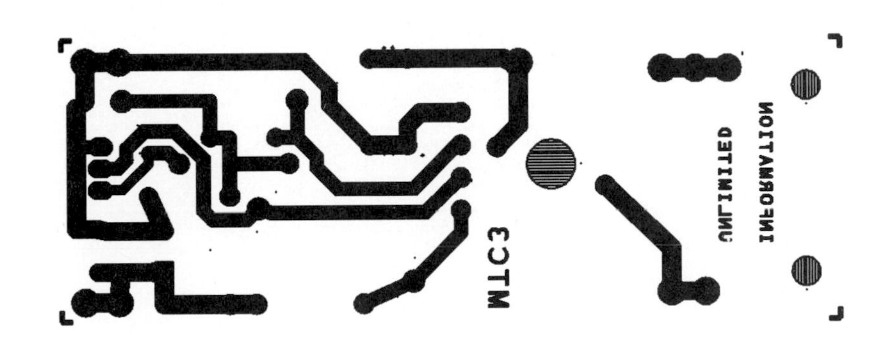

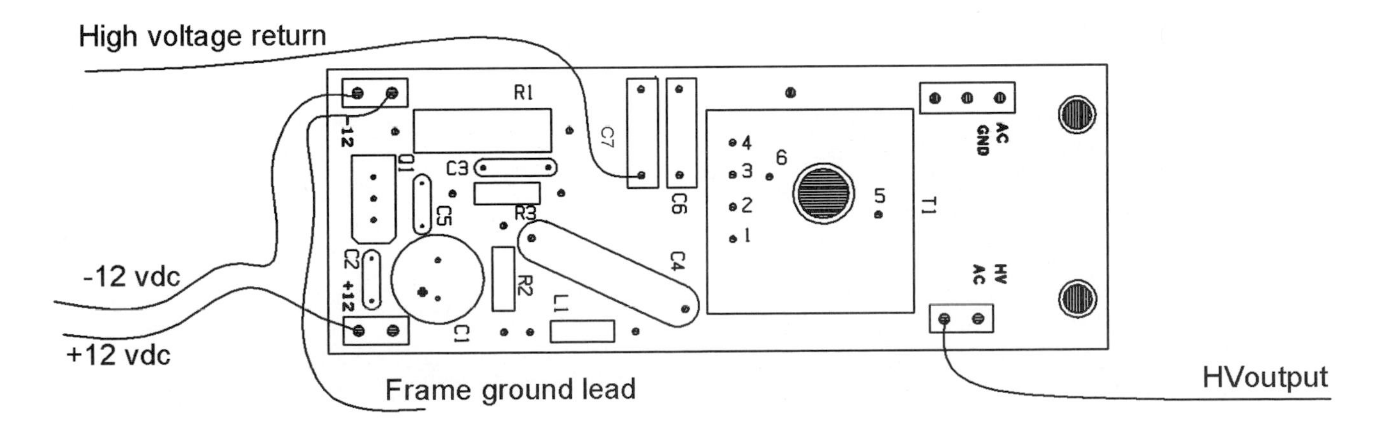

Figure 10-3 *Assembly board*

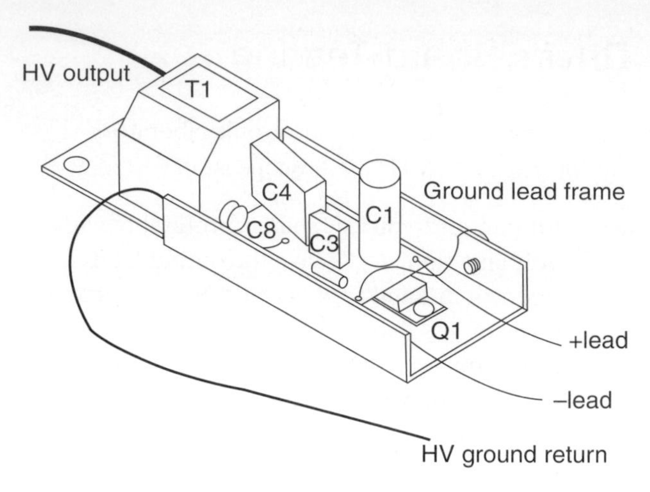

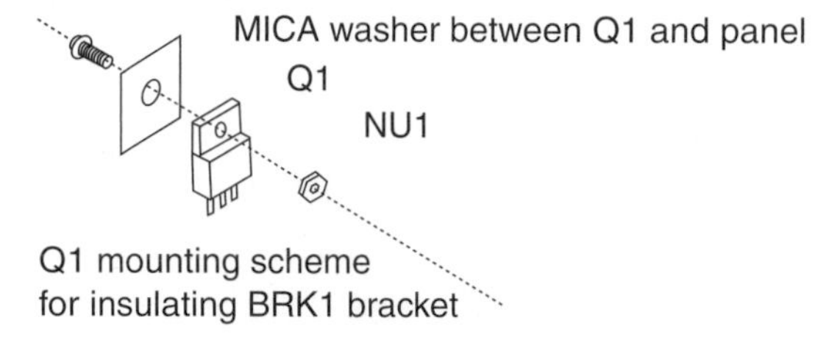

Figure 10-4 *Isometric of assembly*

Mechanical Assembly

You will need some basic cutting tools for the following steps.

1. Fabricate TUB2, as shown in Figure 10-5, and then make the caps CA1 and CA2 for passage of the output leads and for switch S1, jack J1, and indicator lamp LA1.
2. Insert the components and wire them up as shown. Note the current-limiting resistor R10. The exiting leads should be tied in a knot to relieve any strain.
3. Assemble the pen section as shown and solder the lead to the brass POINT. Shim the lead and POINT, and mechanically secure them in place.

Sample Demonstration

1. Obtain a 4 × 6-inch square piece of 1/4-inch masonite plywood, as shown in Figure 10-6. Moisten it with a sponge and place it on top of an approximately equal-size metal plate attached to the ground lead.
2. Contact the surface with the plasma pen and note the burning action. Experiment for the best results with other materials.

Note that your pen will light fluorescent lamps in your house just from its proximity to them, and it can also be used for Kirlian photography.

Switch S1 selects either internal batteries or external 12 volts, jack J1 is for the external 12 volts. LED indicates when the system is powered up. Resistor R1 limits the LED current. These components are mounted on the rear cap CAP2.

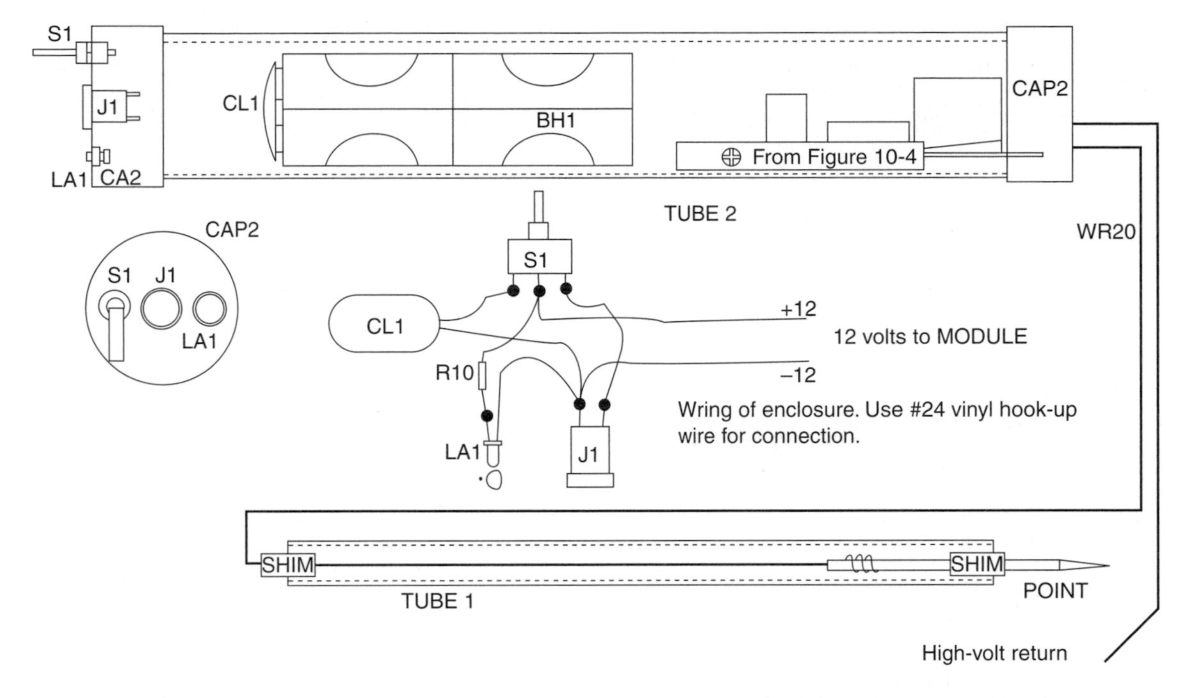

The POINT can be a 1/5 to 1/4 brass rod that can be soldered to. The SHIMS can be rubber tubing that strain relieve and position the feed lead and burning point. TUBE1 can be a 3/8" diameter plastic tube.

Figure 10-5 *Final enclosure and wiring*

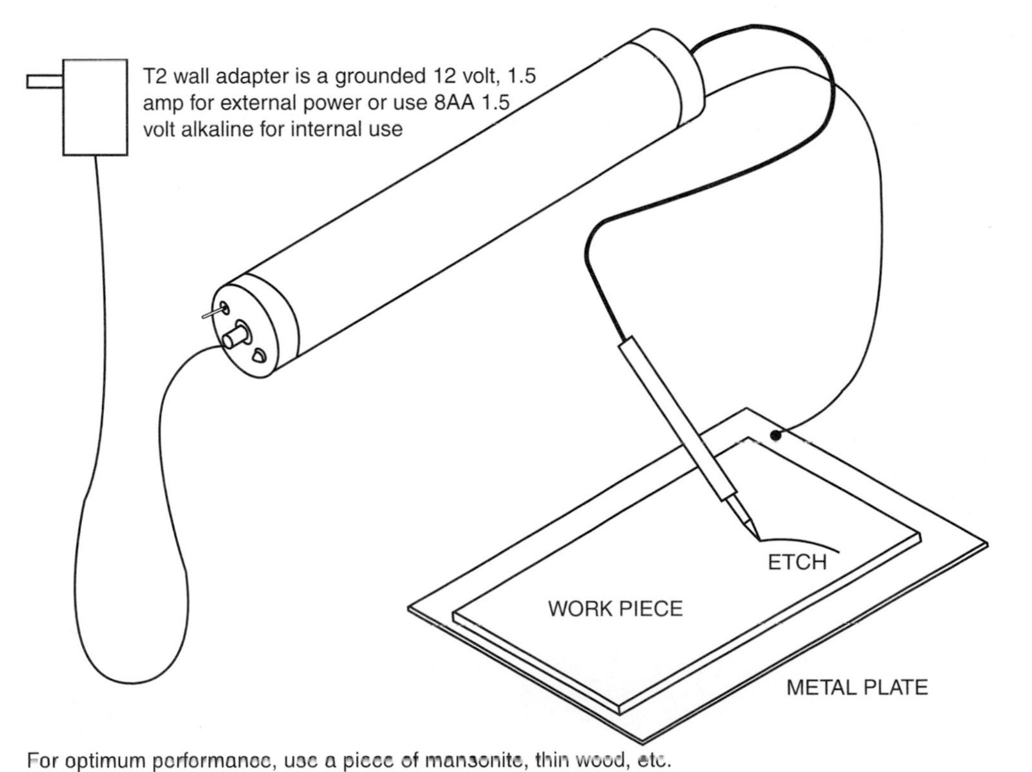

For optimum performance, use a piece of mansonite, thin wood, etc.
Place on top of metal plate connected to the high-voltage ground lead.
You may moisten the beginning area to start the etch.

Figure 10-6 *Sample demonstration*

Table 10-1 Plasma Etching and Burning Pen Parts List

Ref. #	*Description*	*DB Part #*
R1	220 ohms, 1 watt (red-red-br)	
R2	1.8K, 1/4-watt (br-gray-red)	
R3	27 ohms, 1/4-watt (red-pur-blk)	
C1	1,000-microfarad, 25-volt vertical electrolytic capacitor	
C2	.1-microfarad/50-volt small, plastic, capacitor	
C3	.068-microfarad, 50-volt plastic capacitor (683)	
C4	1-microfarad, 250-volt metallized foil large, blue, plastic capacitor	
C5	.047-microfarad, 100-volt plastic capacitor (473)	
C6, 7	Two 25-picofarad, 6-kilovolt ceramic capacitors or a single 50-picofarad, 6-kilovolt unit	
Q1	MJE3055T TO220 power tab NPN transistor	
T1	Special high-frequency transformer	DB# 28K089
WR20	24-inch #22-20 vinyl stranded hookup wire	
PCMTC	MTC3 printed circuit board or use a 3 1/2 × 1 1/2 × .1-inch grid perf board	DB# PCMTC
BRKT1	Create as shown in Figure 10-4	
TUBE2	12 × 2-inch ID schedule, 40 PVC tubing	
TUBE1	3/8 × 1/4-inch ID length of plastic tubing	
CA1, 2	Two mating flat-end caps for TUBE1	
POINT	1/8 × 3-inch brass rod filed to shape	
LA1	*Light-emitting diode* (LED), any color	
R10	220 1/4-watt resistor (red-red-br)	
CL1	Battery snap clip	
BH8	Eight-AA-cell battery holder	
S1	*Single pole, double throw* (SPDT) small toggle switch	
J1	Chassis-mounting DC power jack	
WR24	48-inch, #24 vinyl hookup wire	
WR20	24-inch, 20-kilovolt, silicon, high-voltage wire	
NEONX21	Assembled high-voltage module (see Figure 10-3)	DB# NEON21

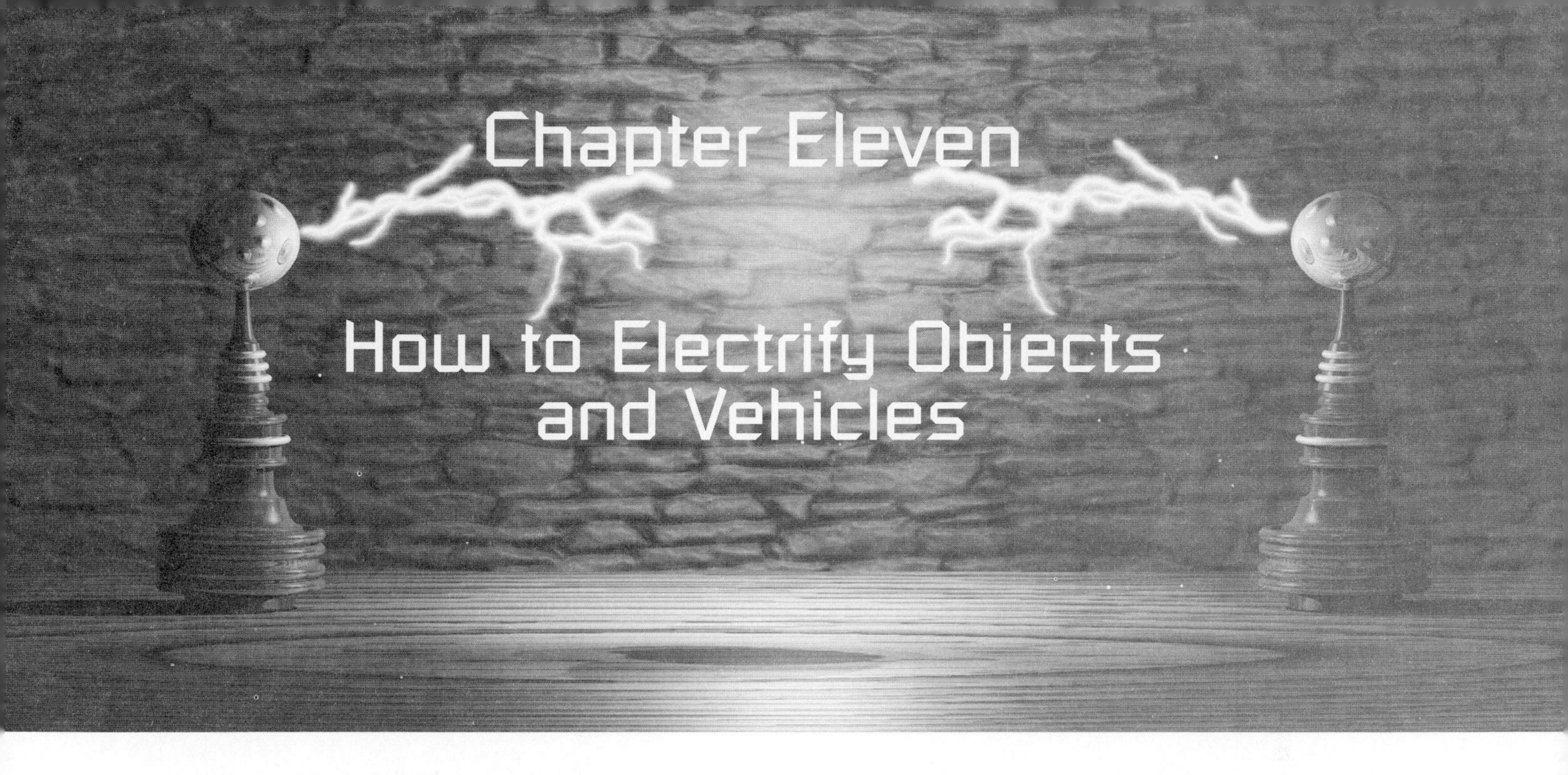

Chapter Eleven

How to Electrify Objects and Vehicles

Figure 11-1 shows two board-level devices capable of producing electrical shocks. The larger circuit generates low-energy, 20,000-volt pulses similar to an automotive ignition system. Although these pulses are nonlethal, they can still be very painful when contacted. The smaller circuit generates a steady 5-milliamp, low current at 2,000 volts. By itself, it is not lethal but again can produce a painful shock. If the output is allowed to charge a capacitor to a sufficient energy level, then the rules change and a potentially lethal situation exists. This level usually is any value over 25 joules as computed from W (joules) = half of C (capacitance in farads) × V^2 (in volts) that the capacitor is allowed to charge too.

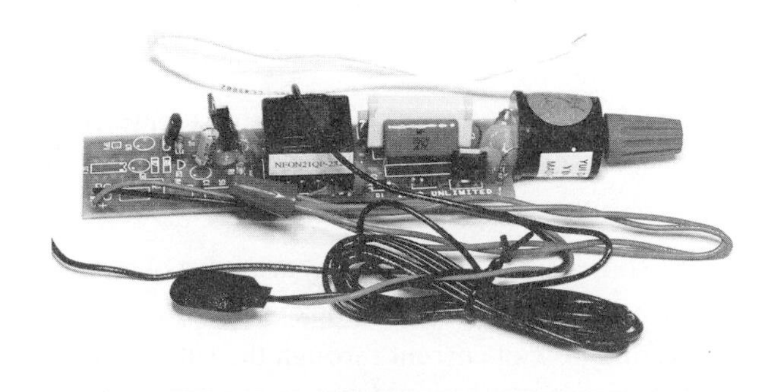

Figure 11-1 *Photograph of devices*

Caution: The use of electrification to prevent human intrusion is illegal. This application is legally referred to as "man trapping." Consult legal advice before using this if human contact is intended!

Warning: Never expose any person wearing a pacemaker or having a known physical condition to any form of electrical shock.

Data shows how to use several shocking systems with different types of outputs and applications. Even though the energy levels may be well below dangerous levels, secondary reactions can cause severe injury. Use caution if intended for human contact, and never connect the DC output version to a capacitor as a lethal charge may exist.

The following information is intended to serve as a guide to those contemplating using our equipment or other devices capable of producing a mild, severe, or possibly lethal electric shock. Any device of this nature, if used against people, may be considered a potential weapon. Remember that what may be a mild shock to one person could be harmful to another. Secondary effects, resulting from involuntary action, must also be considered.

The devices sold are intended for laboratory experiments or for use against animals. They should only be used for intimidation purposes to prevent an encounter. Human contact must be totally justified, as serious consequences can result from the illegality of their use.

Even though our electronic devices provide an average current below what is considered lethal to the average person, circumstances such as a person's heart condition or other physical parameters can influence both the immediate effect and after contact reaction.

Table 11-1 shows certain effects from various electrical currents. Remember that the indicated current must have an appropriate voltage behind it to flow through a given resistance, such as the human body. When working with sources that have storage capacitors (especially dangerous sources), one must take into account the high peak current that can result for a period of time and result in possible electrocution if this duration of time is sufficient.

Table 11-1 Electrical current scale

Voltage	*Degree of pain*	*Examples*
.01 to 1 ma	Tingling to annoying	Static electric shocks
1 to 5 ma	Annoying to painful	Spark plugs, TV picture tubes, trip current of GFI
5 to 20 ma	Painful to very painful	Oil burner ignition, bug killers, stun guns
20 to 50 ma	Very painful to possibly lethal	Neon sign transformer, old tube radios
50 to 100 ma	Possibly lethal to lethal	Low-powered transmitters, capacitor charges for lasers
100 to 500 ma	Lethal to deadly	Medium-powered transmitters, laboratory power supplies
500 ma to 1amp (+)	Deadly, with usually no second chance	Electric chair, 220-volt house current under certain circumstances

Supplementary Electrification Data and Information

The following information explains how to place an electrical charge on an object to protect against animal damage, intrusion, or theft, and the construction of several energy sources is referenced. These must be treated with caution and never be used on human beings without proper discretion due to potential injury or electrocution. Accidental contact must also be avoided. Many laws unfortunately are made to protect the criminal, and therefore any attempt made to protect oneself or property using these devices could result in a suit by the criminal if he or she can prove injury.

The methods for electrifying objects and areas will be shown as several examples that the user may adapt to his or her own needs. To successfully electrify, an object requires the following conditions to exist. Note that the following pertains to flowing electric current and not to static or super-high voltages. Figure 11-2 provides the basic explanation.

- The object of contact must be of a conductive nature such as metal, conductive paint, or wet ground.
- The subject (the intruder or animal) must supply the link to complete the circuit; that is, his or her physical extremities must come in contact with the object, along with a second common surface such as the ground, water, or another object anticipating simultaneous contact. No current flows unless contact is made to complete the circuit by the subject physically touching the object.
- The electric current flows from the +lead of SOURCE to the −lead. In doing so, it flows from the object to the point of contact by the subject to the ground and back to SOURCE via the −lead. This produces a current flow through the subject, developing an electrical shock. The severity of this shock is dependent on:
    - How well the ground will conduct this current (wetness, mineral content, and so on)
    - The electrical parameters of the SOURCE, that is, the voltage and available current to support this voltage

The source of electrical energy is a form of a power supply made to generate a sufficient voltage to force the desired flow of current through the subject when contact is made. This can take on many different forms depending on the applications and the desired results. If a system was built to kill a human being, a power supply with a current supply capability of 1/2 to 20 amps would be necessary to support a voltage high

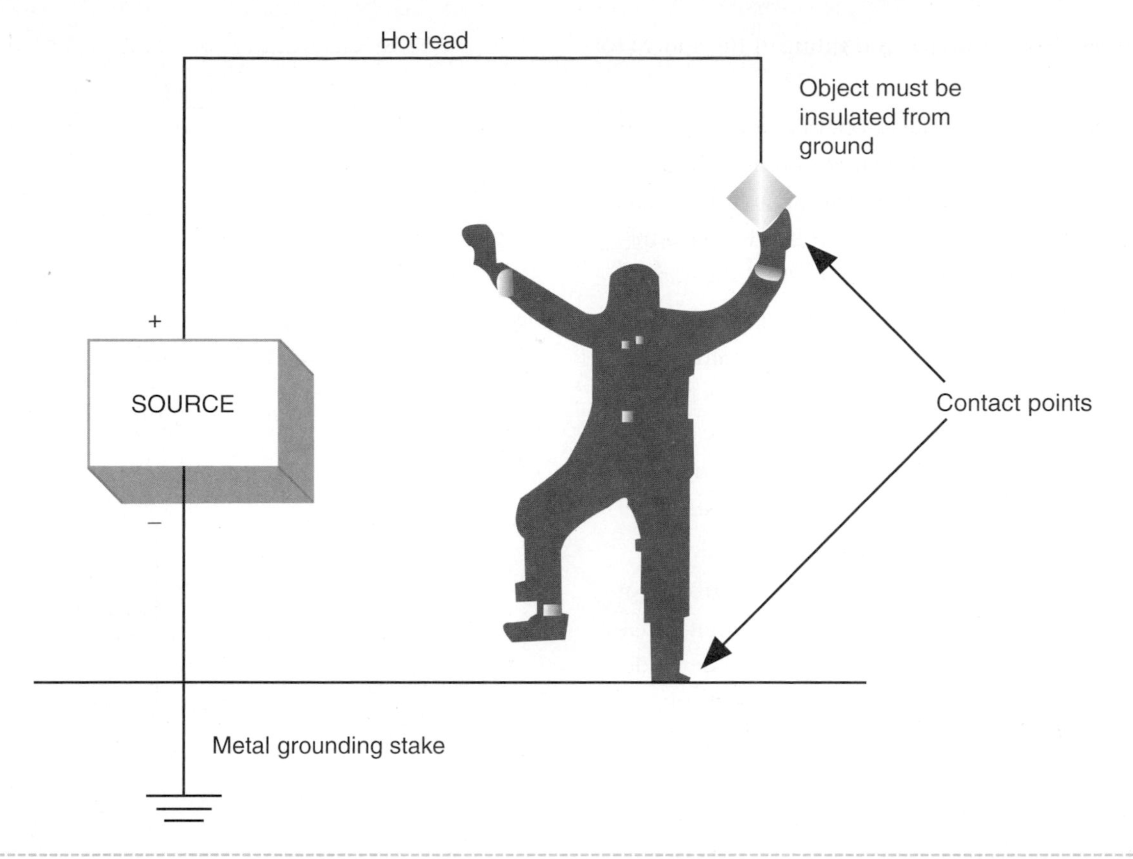

Figure 11-2 *Electrification principle*

enough to cause this current to flow, usually several thousand volts. (These parameters are similar to those used in electric chairs.) Currents of more than 1/10 of an amp are considered lethal to the average human under certain circumstances, 1/100 of an amp produces intense pain, and 1/1,000 of an amp is felt with no difficulty and is very annoying.

+
SOURCE
–
Metal can
Ground should be moist, with vegetation such as grass or can be sections of wire pieces (simulated ground) for dry areas.

Figure 11-3 *Electrification of a garbage can or similar object*

Special pulse-type energy sources, similar to capacitor discharge ignitions, can be used relatively safely as peak currents are high, yet the average is low. A source of this type is the pulsed model described in this project. This source is capable of generating the high-peak current pulses and maintaining a substantial voltage on the body of a vehicle in spite of leakages due to steel-belted tires and so on. The source is also relatively safe to exposure but will produce moderate shocks. With these parameters, one

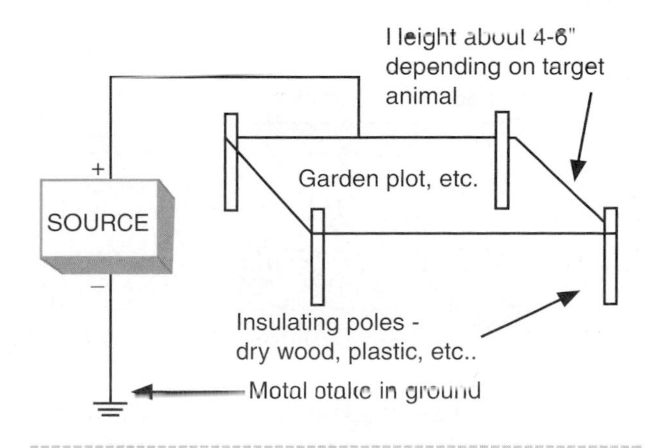

Figure 11-4 *Electrification of a small area*

can get an idea of the size and rating of the source for a particular application.

The following examples, hopefully, will apply to most situations. Figure 11-3 shows how to electrify garbage cans against annoying animals. One can see that the condition of a common contact or ground system is important because the current must also flow through this medium.

Figure 11-4 shows how to electrify a small garden area using plastic stakes with slots for wires that are pushed into the ground. Try to prevent wire droop and possible contact with the ground. Watch for fire hazards and proximity to dry vegetation if using a high, average current source.

Figure 11-5 shows how to electrify the ground of an area and protect it against annoying animals. Use #20 wire as shown with the space and size dependent on the animal and area. Anchor the wires in place with small stones. The ground must be dry and non-conductive, or use an insulated rubber or plastic mat. Please be aware that this project should only be used with a low-energy pulsed system or a hand-cranked magneto. Do not create this system if you have a heart condition.

Figure 11-6 shows a prank, but this is only to be done using a low-energy pulsed system or a hand-cranked magneto. The victim stands on a rug to open a door and zap! You turn the power on.

Figure 11-7 shows the electrification of a fence. This method is for educational use only, as serious injury can result either from the electric shock or the physical reaction. The intended use would be for military or other high-security situations.

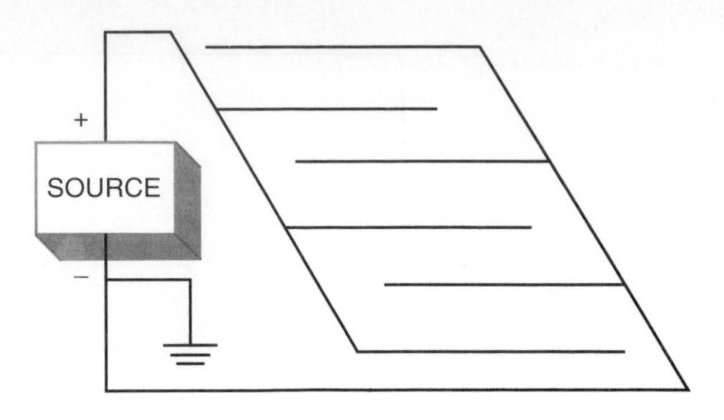

Figure 11-5 *Electrification of the ground or protected areas*

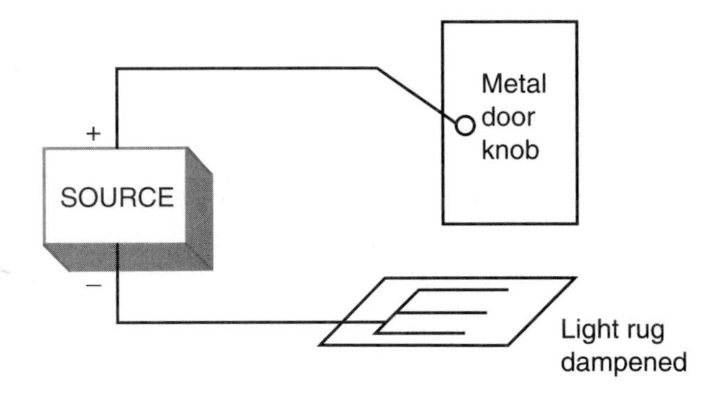

Figure 11-6 *Door knob prank*

The object here is to electrify the top-most wire of a chain-link fence, as anyone attempting to crawl over will no doubt come in contact with the top wire. This is a very effective method and can cause severe injury and even death from electrocution or physical injury from falling. The best method is to simply replace the original top-most wire with a standard, electrical farm fence insulator, as shown in Figure 11-7. The wire is

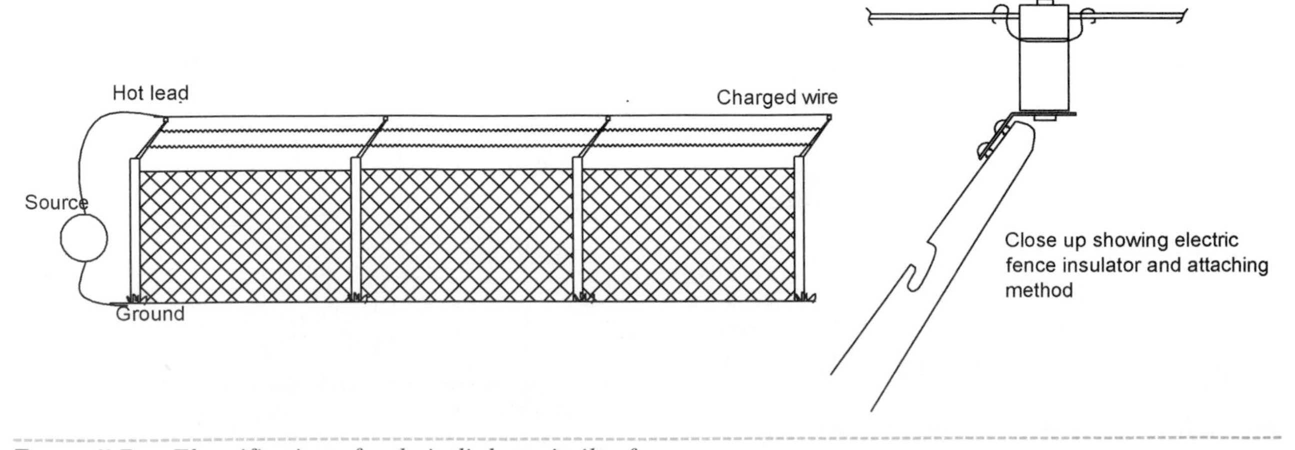

Figure 11-7 *Electrification of a chain-link or similar fence*

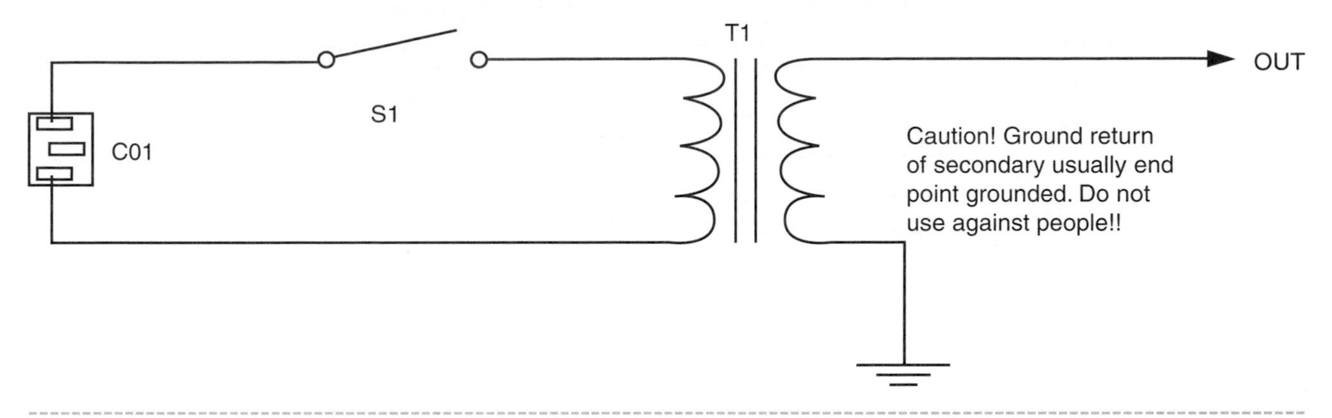

Figure 11-8 *Circuit of power supply source for animals only*

then just strung to these insulators using standard, electric farm fence techniques and is either connected to a farm-type power source such as an electric fence charger or a higher-powered source. The latter presents a dangerous choice and should never be used in a normal situation.

Danger! Do not use a transformer with a short-circuit current in excess of 9 milliamps.

Figure 11-8 shows using a current-limited transformer powered by 115 VAC. The following are suggested power sources for the circuit shown in Figure 11-8.

> 4KV/009: T1 −4 KV, 9 ma, 115 VAC, current-limited floating secondary Database #4 KV/.03 A
>
> 6KV/02: T1 6.5 KV, 20 ma, 115 VAC, current-limited floating secondary Database #6 KV/.02 A

Figure 11-9 *Electrification of a steering wheel or similar object*

You will need a three-wire grounded power cord and a switch available from most hardware stores. Connect the −connection to common using standard PVC or plastic hookup wire. Connect the +connection via a high-voltage ignition wire available from an auto supply house. A weatherproof system is required when installed outside.

Figure 11-9 shows how to electrify a steering wheel or a similar object. The project is intended as a theft or unauthorized use deterrent. Use a low-energy, pulsed, high-voltage source.

The subject sits in the seat and activates the pressure switch to the source. When he grabs the steering wheel, he experiences a mild pulse-type shock between his hand and rear end. The voltage return is via metal springs in the seat. To create this project, please follow these steps:

1. Tape some thin, bare wire on the underside of the steering wheel. Secure it with small pieces of tape, wax, or silicon rubber.

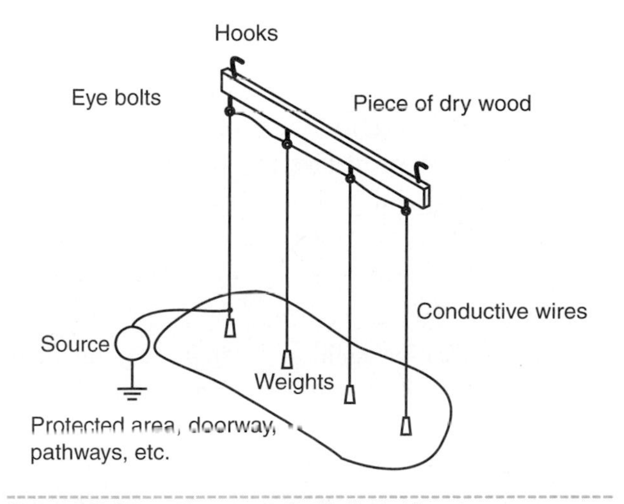

Figure 11-10 *Space or field electrification*

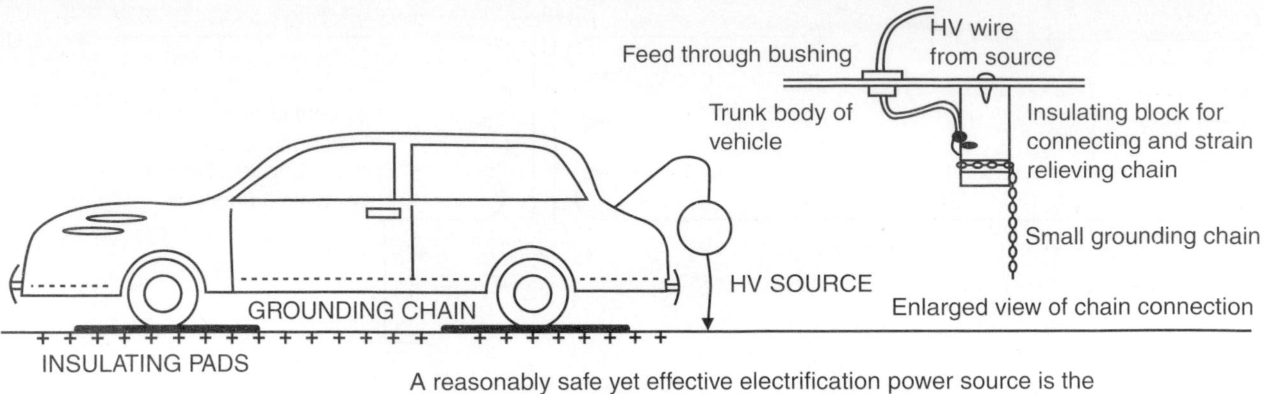

Figure 11-11 *Electrification of a vehicle or similar object*

2. Run a concealed connecting wire up the steering column to the bare wire. This wire should be insulated and connect to the output of the source.
3. Conceal an activation switch as desired and don't forget to de-energize it before entering the car. You may want to rig up a pressure-sensitive switch on the seat so when the thief gets in the unit it activates before he starts the car. Important caution: The unit must be activated without the car being in motion or serious accidents may result.
4. The use of our pulsed low-power source is suggested.

Figure 11-10 shows how to protect an area, doorway, or paths using space or field electrification. Hanging supports use ultra-thin steel wire for strength or weaker magnet wire.

Hang weights 8 to 12 inches above the ground, and use elastics or O-rings to secure connections to the ground points. The actual installation will depend on the environment and the voltage used. The source should be a low-energy, pulsed source that can be used against animals when their paths of travel are known.

Figure 11-11 shows how to electrify any vehicle using an automotive ignition coil in place of the pulse coil. Coils are available through any automotive supply house.

A remote switch must be mounted so that it is accessible from the outside of the vehicle. This may be actuated via an insulating object such as a plastic comb or something similar. An override switch may be installed to actuate this system while one is inside the vehicle in order to discourage certain encounters involving contact while driving or stopped at a light.

The suitable method for connecting to earth ground as the common terminal is accomplished via a grounding chain. This is shown strain relieved via a plastic insulating block. The user is asked to use his or her own ingenuity to create other methods of implementation.

Insulated platforms for wheels should be used if the tires have steel-belted radials or if roads are salty.

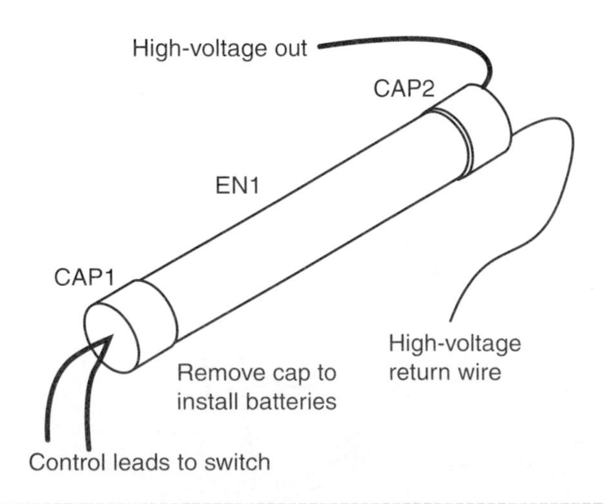

Figure 11-12 *Enclosed pulsed electrification device*

These platforms are intended for storage or long-term parking situations.

Contact the factory for information regarding a serious automotive deterrent using a high-energy discharge system to be able to support the voltage necessary to overcome tire leakage and so on. A system of this type was sold to a Central American country to protect personal when in hostile areas. Very painful 5 joule pulses were generated between the vehicle body and ground. It would be very "politically incorrect" if it were used in this country. This system produces high-current pulses of several tens of amperes for short durations. This is necessary as most tires offer considerable leakage at these high voltages, and a steady-state supply would no doubt burn off the tires and be very lethal to the perpetrator.

Assembly of Suggested Power Sources

The *pulsed version* shown in Figure 11-12 can also be used as a high-voltage trigger for triggering spark gaps, flash tubes, chemical igniters, Kirklian photography, and for electrification in animal control. The unit is housed in a tubular enclosure and powered by built-in batteries or by an external 12 VDC wall adapter. The device produces 20,000-volt pulses for electrifying objects, fences, and so on. Dual-mode operation produces low-energy pulses of .02 joules at a repetition rate of 30 per second or medium-energy pulses of .2 joules at a repetition rate of 4 per second.

The *continuous version* generates a steady, high-voltage device intended for research, capacitor charging, shocking animal pests, and small insect killers.

Caution: The use of electrification against human intrusion is illegal. Consult legal advice before using if human contact is intended.

Warning: Never expose any person wearing a pacemaker or having a known physical condition to any form of electrical shock.

Danger! Never charge a capacitor and use it for shocking humans. Energy storage over 5 joules can produce extremely painful shocks and over 25 joules can kill! Even though the energy levels may be well below dangerous levels, secondary reactions can cause a severe injury.

Circuit Description of Pulsed Shocker

A high-frequency, self-oscillating inverter circuit comprised of switching transistor Q1 and stepup transformer T1 produces a high-voltage high frequency at the secondary winding. This high-voltage AC is rectified by diode D1 and charges up storage capacitor C3 or C4 through isolation resistor R3. When this voltage charges up to the breakdown potential of SIDAC (SID1), the energy stored in the capacitors is "dumped" into the primary of the high-voltage pulse transformer T2, producing a high-voltage pulse at the output terminals (see Figure 11-13).

The oscillator circuit utilizes a winding on T1 to produce the necessary positive feedback to the base of Q1 to sustain oscillation. Resistor R1 initiates Q1 to turn on while resistor R2 and C2 control the base current and operating point.

A charging circuit consisting of the current-limiting resistor R4 and rectifier diode D2 allows external charging of battery B1 when *nickle cadmium* (NiCad) or other rechargeable batteries are used.

Construction Steps for Pulsed Shocker

If you are a beginner it is suggested to obtain our *GCAT1 General Construction Practices and Techniques*. This informative literature explains basic practices that are necessary in proper construction of electromechanical kits and is listed on Table 11-2.

1. Lay out and identify all the parts and pieces, and check them with the parts list in Table 11-2. Note that some parts may sometimes vary in value. This is acceptable because all components are 10 to 20 percent tolerant unless otherwise noted.

   Some kits contain a length of insulated wire.

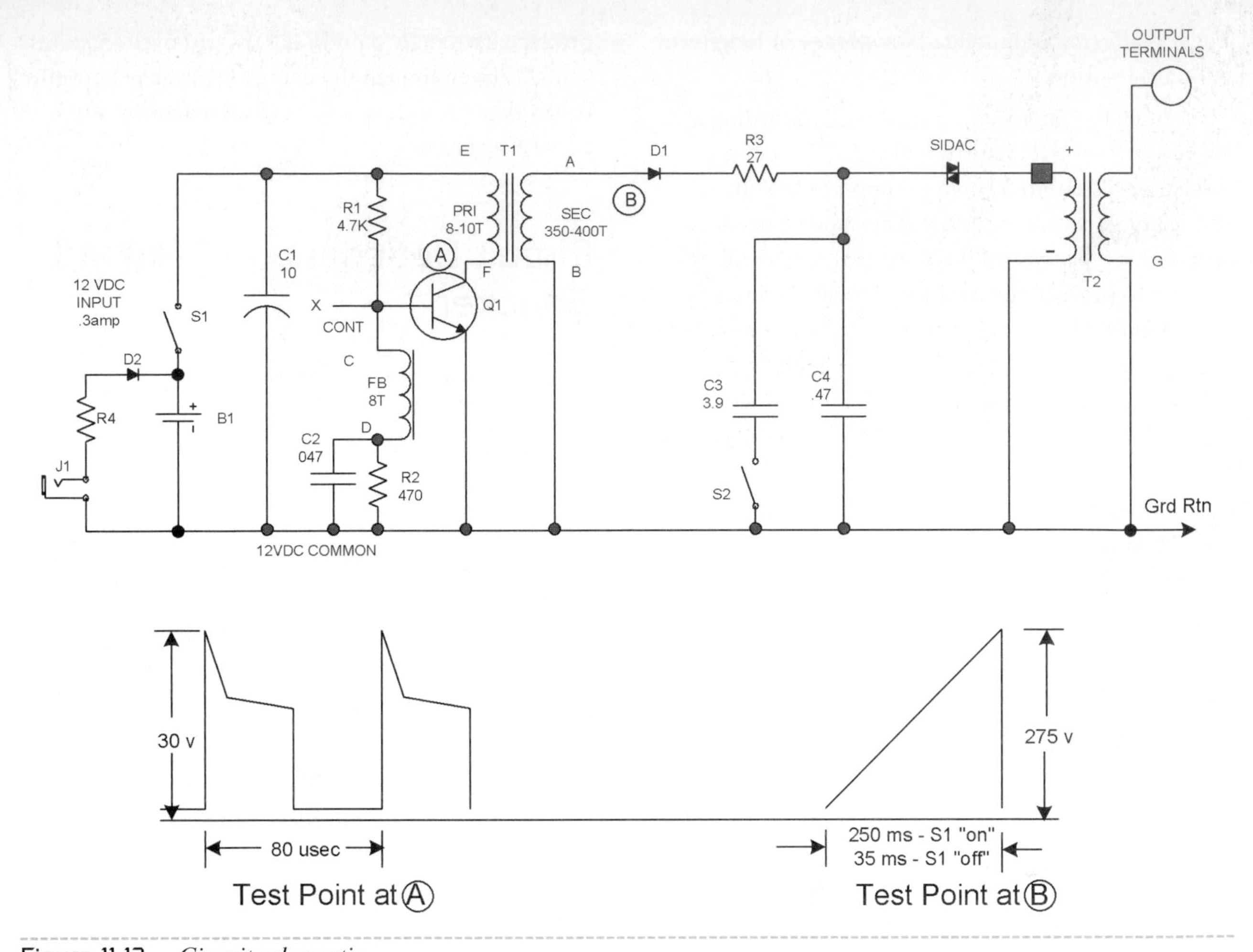

Figure 11-13 *Circuit schematic*

This must be cut, stripped, and tinned according to where it is used.

2. Assemble the board as shown in Figure 11-14. Insert the components into the board holes as shown, proceeding from left to right. Note the polarity of the components.

Note: Certain leads of the actual components will be used for connecting points and circuit runs. Do not cut or trim them at this time. It is best to temporarily fold the leads over to secure the individual parts from falling out of the board holes for now.

3. Verify the wiring, proper components, and polarity to the schematics. Check for cold solder, excess solder, and bare-wire bridge shorts. Now the unit is ready to test.

4. Test the unit as per the following:

A. Position the bare end of the ground return lead (GRD RTN) to allow a $\frac{1}{2}$- to $\frac{3}{4}$-inch air gap between the output pins of T2.

B. Connect 12 volts from a bench power supply or use eight AA NiCads at 1.25 volts each or eight AA alkalines at 1.5 volts for 12 volts.

C. Note with S2 open a fast pulsing action will produce a thin, bluish discharge. This can cause a very mild electrical shock and could be safely used within reason as a prank.

D. Note a thick, slow pulsing discharge when S2 is closed. This can produce a painful shock and is intended for use against animals.

E. The current draw with the unit properly operating should be approximately 250 milliamps.

F. Check the power tabs of both Q1 and SIDAC. These should be cool to warm to the touch.

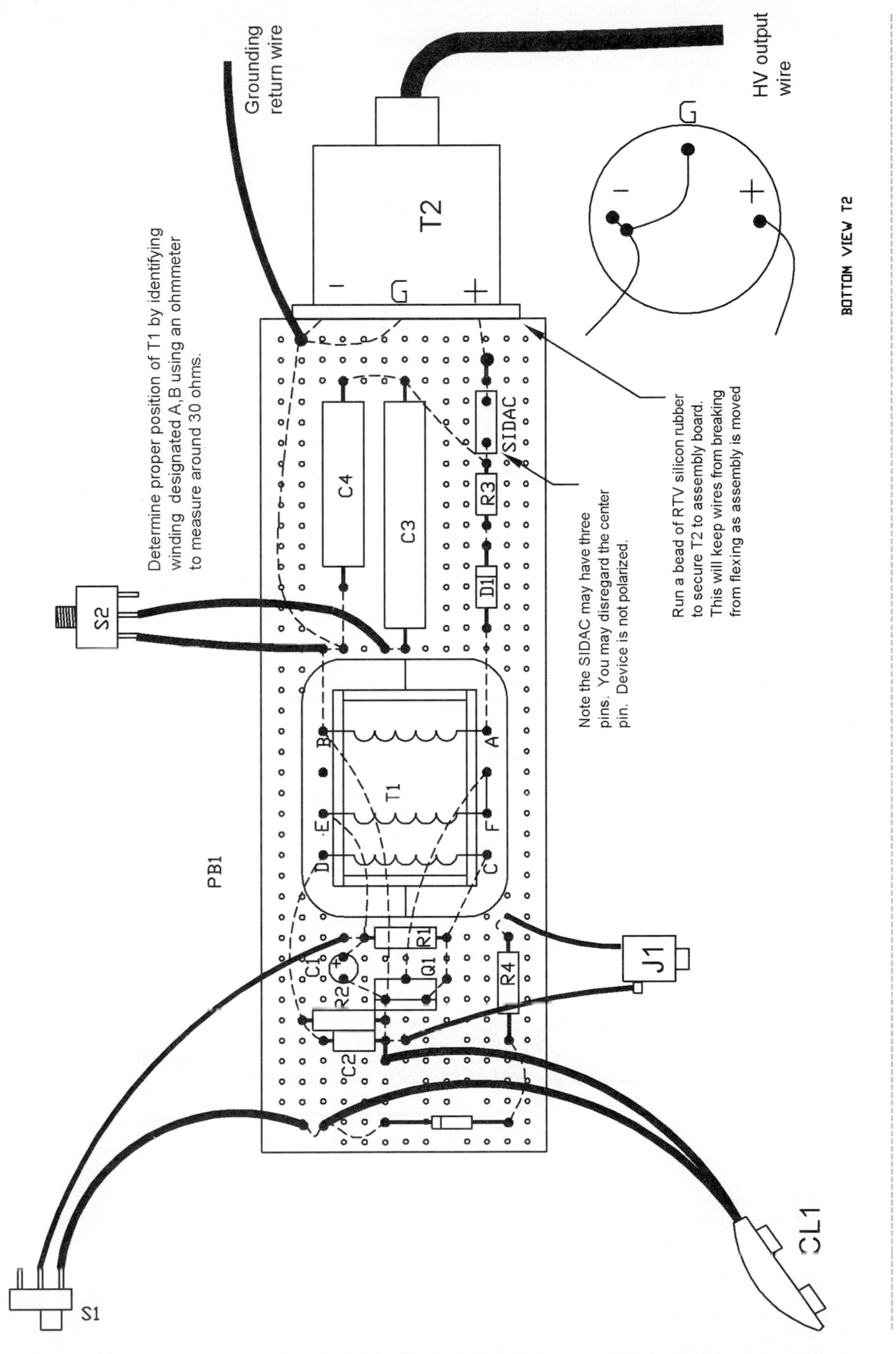

Figure 11-14 *Assembly board*

This device can produce up to 10,000 volts when used with the optional multiplier circuit shown from a 9 to 12 volt battery pack. Output is continuous and can be used to charge external capacitors. Output is 1.5KV with the single capacitor and diode at a current of 5 ma.

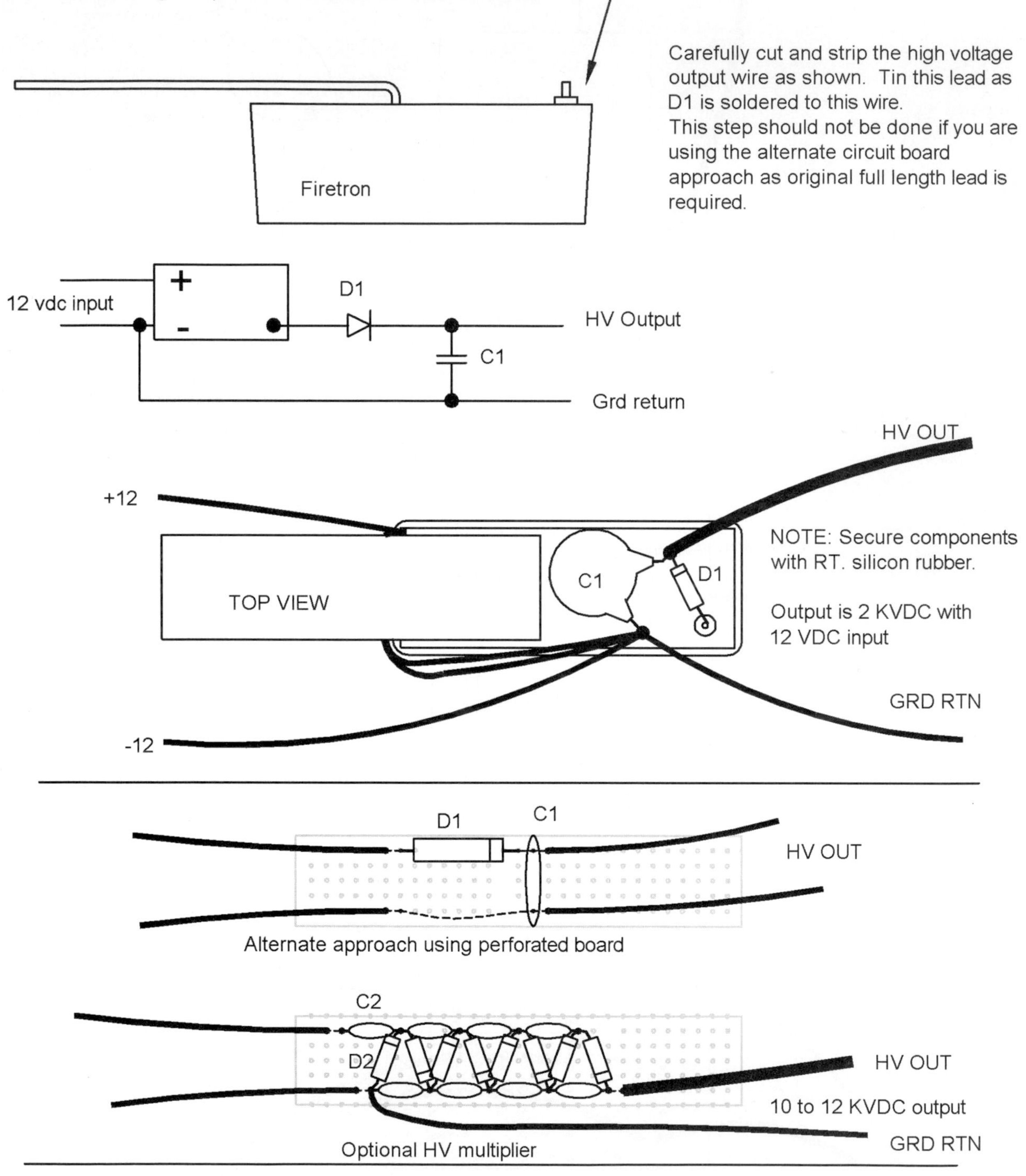

Carefully cut and strip the high voltage output wire as shown. Tin this lead as D1 is soldered to this wire.
This step should not be done if you are using the alternate circuit board approach as original full length lead is required.

Firetron

12 vdc input

\+

\-

D1

HV Output

C1

Grd return

HV OUT

+12

TOP VIEW

C1

D1

NOTE: Secure components with RT. silicon rubber.

Output is 2 KVDC with 12 VDC input

GRD RTN

-12

D1

C1

HV OUT

Alternate approach using perforated board

C2

D2

HV OUT

10 to 12 KVDC output

GRD RTN

Optional HV multiplier

WR1 24" #20 Hook up wire
PB1 1 X 4" .1 Grid perfboard
Note 4 stages shown can produce 10 to12 KV!!

Figure 11-15 *Construction of continuous shocker*

G. You may verify proper operation using a scope, noting the waveshapes as shown in Figure 11-13.

Notes Pulsed Shocker

Input is shown operating with eight NiCad cells in series for a total of 10 VDC. The unit, however, reliably operates within 8 to 14 VDC.

A simple remote switch, S1, may be installed in series with the negative or positive of the battery lead. This is left to the builder to fit his or her needs when finally packaged.

The unit may be housed in any suitable enclosure. A suggested method is to place it inside a PVC tube You may fill the enclosure with melted paraffin wax. This will allow operation in moist environments while still allowing the unit to be easily unsealed should a problem or change occur

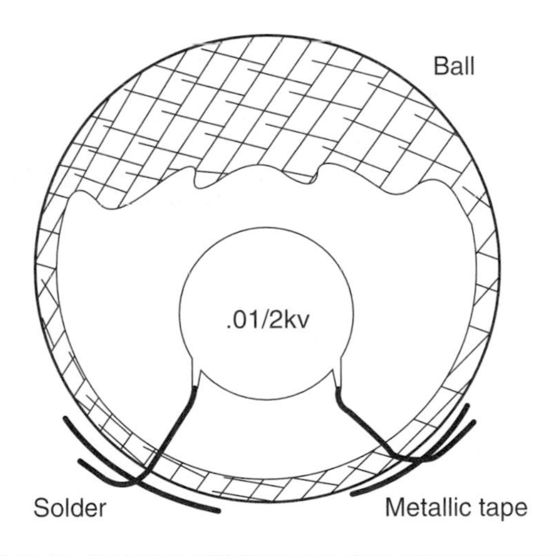

Figure 11-16 *Shock ball joke*

Construction of Continuous Shocker

A continuous shocker device is shown in Figure 11-15, and shock balls can be made by obtaining a small rubber ball or a tennis ball, as shown in Figure 11-16.

To create a shock ball, use a .01-amp capacitor at 2 kilovolts and note that a larger capacitor would produce a worse shock. Pieces of metallic tape must be used to sandwich the capacitor leads. Cut a slit in the ball to insert and position the capacitor, and charge the ball by contacting a metallic strip with the continuous 2-kilovolt source. Hold the ball without touching the metallic strips, toss it to a buddy, and watch the results.

For a shock wand application, position a unit inside a plastic or nonconductive tube with batteries and a switch. Connect the output leads by sandwiching them under metallic tape for contact with the target subject.

Table 11-2 Parts List for Shockers

Ref. #	Description	DB Part #
R1	4.7K, 1/4-watt carbon film resistor (yel, pur, red)	
R2	470 ohm, 1/4-watt carbon film resistor (yel, pur, br)	
R3	27-ohm, 1/4-watt carbon film resistor (red, pur, blk)	
R4	100-ohm, 1/4-watt carbon film resistor (br, blk, br)	
C1	10-microfarad, 25-volt electrolytic capacitor vertical mount	
C2	.047-microfarad, 50-volt polyester capacitor	
C3	3.9 to 4-microfarad, 350-volt polyester capacitor	
C4	.47-microfarad, 250-volt polyester capacitor	
Q1	MJE3055 NPN transistor TO220	
D1, 2	Two IN4007 1-kilovolt rectifier diode	
SIDAC	300-volt Sidactor switch	SIDAC
T1	400-volt switching square-wave transformer	DB# TYPE1PC
T2	25-kilovolt pulse transformer	DB# CD25B
S1, 2,	Two *single pole, single throw* (SPST) 3-amp toggle switch or equivalent	
PB1	5 × 1.5 × .1-inch grid perforated circuit board	
PCLITE	Optional *printed circuit board* (PCB)	DB# PCLITE
WR20B	36-inch #20 vinyl stranded hookup wire, black	
WR20R	36-inch #20 vinyl stranded hookup wire, red	
WRHV20	12-inch 20-kilovolt wire	
CL1	Battery clip	
CAP1, 2	Two 1 5/8-inch plastic caps	
EN1	12 inches of 1 5/8 × 1 1/2 × 12 plastic tube	
BH1	Eight-AA-cell battery holder	
J1	3.5-millimeter mono jack	
12DC/.3	Optional 12 VDC, .3-amp wall adapter	DB# 12DC/.3
Parts for the continuous version		
FIRETRON	Firetron 2,000-volt module	DB# FIRETRON
C1	.01-microfarad, 2,000-volt disc capacitor	DB# .01M/2KV
D1	6-kilovolt diode	DB# VG6
C2	270-picofard, 3,000-volt disc capacitor	DB# 270P/3KV

Chapter Twelve

Electromagnetic Pulse (EMP) Gun

Figure 12-1 shows the construction of a low-power pulse gun that will provide kilowatts of peak power at frequencies up to 100 MHz with harmonics. The unit shown is battery powered and uses a modified version of the high-voltage plasma generator shown in Chapter 7 of this book. A higher powered, more functional device is being developed in our labs and will be ready to be copied at the time the third book of this series is available.

This is an advanced level project requiring basic high-frequency electronic skills. A spectrum analyzer can be a very valuable tool in setup but is not necessary. Expect to spend $50 to $100. All parts are readily available, with any specialized parts being available through Information Unlimited (www.amazing1.com) as listed in the parts list at the end of this chapter.

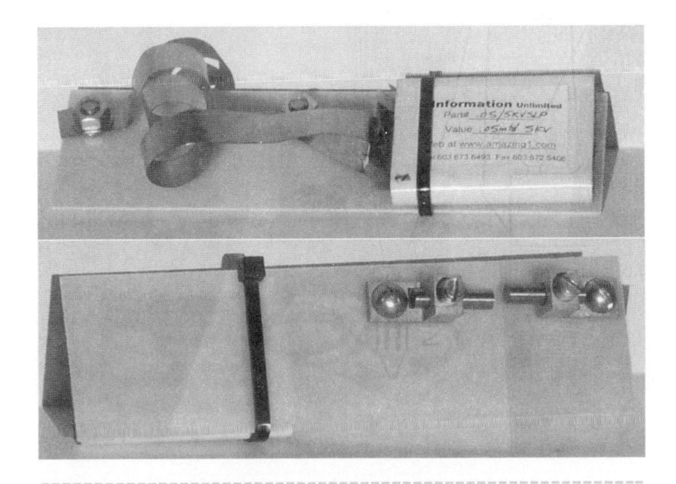

Figure 12-1 *Electromagnetic pulse gun*

Basic Theory Simplified

The ability of a signal to disrupt sensitive circuitry requires several properties. Most microprocessors consist of *field effect transistors* (FET) operating at very low voltages. Once these voltages are exceeded, catastrophic failure becomes imminent. Forgiveness to an overvoltage fault is practically nonexistent due to a microthin metal oxide between controlling elements. Any overvoltage generated across these elements consequently produces permanent damage or, in some less severe cases, causes deprogramming. To generate these damaging voltages from an external source requires a wave that can produce a standing wave of energy across circuit board traces, components, and other key points. The external signal energy therefore must be high enough for the circuitry geometry to be a significant part of energy at the wavelength. Therefore microwaves having fast rise times (high Fourier equivalents) and burst duration help to maximize the effect. The energy required is great, as it must be sufficient to enhance damage. A good measure would be the quotient of energy/wavelength. A high-power pulse of microwaves can be generated in several ways.

Explosive flux compression-driven virtual cathode oscillators and their generic relatives can produce gigawatts of peak power from only hundreds of kilojoules. This is where a seed current is pulsed into an inductor and at its peak is compressed by a shaped explosive charge thereby trapping the flux and creating a source of high energy. The coil must be compressed both along its axis and along its radius using a high-detonation velocity explosive such as the cyclotrimethyltrinitramine, its derivatives PETN, or some other equivalent energetic explosive. This trapped flux now produces an energy gain that is conditioned into the final peak-powered pulse of microwave power (HEPM). Flux compression like that of a nuclear initiation requires precise timing of the explosive chargers. For flux compression, Krytron switches or similar can be used instead of the more radiation-hardened Sprytrons that are used in nuclear initiations where ionizing radiation is produced from the inherent fissionable materials.

A virtual cathode oscillator can also easily be energized from a small Marx impulse generating 200 to 400 kilovolts. The fast current rise and high peak power can produce a powerful burst of microwaves.

Other methods include exploding wires where energy is allowed to flow into a LCR circuit and then is rapidly disrupted by the explosion of the feed wire as it vaporizes at near the peak injection current. A very fast and energetic pulse is produced, which is capable of generating an *electromagnetic pulse* (EMP).

Microwave pulses are excellent candidates for damaging sensitive electronic circuitry. But much lower frequencies are better for disrupting power grids and other similarly sized systems, as now the lengths of the conductive elements are more conducive to generating the high standing voltage waves. Obviously more energy is now required, as breakers, switches, and transformers require more energetic pulses

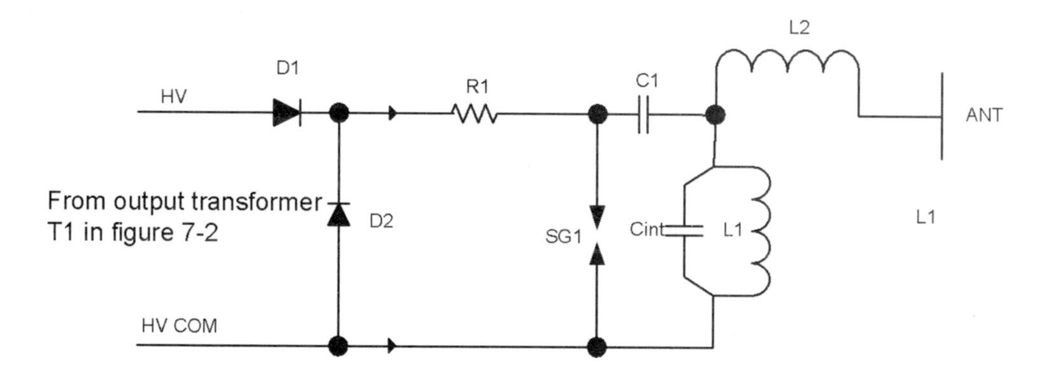

The ideal peak power of this circuit can be approximated by evaluating the product of the charge voltage across C1 x the peak current as determined by Ipk = Epk x the square root of the capacitance of C1/the total inductance of the discharge circuit. This expression implies no resistance (R) in the discharge circuit that now is only in an ideal situation. In all actuality the resistance in the real world will now be a factor in where e to the -(pi/square root of L/C) / R must be a factor in the above ideal condition. This now implies a damped waveform.

Diodes D1,2 are 10 kv 10 ma fast recovery
Resistor R1 is three 47k 1 watt resistors in series and isolate the diodes from the dv/dt of the discharge.
L2 inductor tunes out the capacitive reactance of the antenna at the desired resonant frequency

Capacitor C1 is a "slapper" capacitor. These are used to produce a very fast rise time necessary to detonate initiators necessary for initiating high explosive. Its claim to fame is the high peak current discharge current. The capacitor is constructed as strip lines with the connection leads exiting a common end.

Figure 12-2 *Circuit schematic*

Circuit Description

The circuit shown in Figure 12-2 shows a simple method of obtaining a high-power pulse with Fourier equivalents above 100 MHz. Even though the power and frequency are relatively low, close range effects are possible on many target circuits.

The project uses a high-frequency plasma source, which is converted to a direct current charging source and is short circuited functionally by the use of lossless reactive ballasting. This means that the capacitor can be charged without the use of an energy-robbing resistor, as only the real part of the complex current is seen by the battery supply.

The modified source now supplies a current charge to the reservoir capacitor (C1) to a value where the spark discharges across SG1. Current through L1 rapidly rises and rings along with the circuit and lumped capacitance (Cint). Spark gap SG1 must turn off and allow the energy to circulate in the discharge in order to generate a resonant peak of power that is now coupled into the system emitter. Experimentation of the gap settings is necessary to obtain an optimum effect.

Circuit Assembly

1. Fabricate a piece of Lexan plastic or G10 circuit board, as shown in Figure 12-3, to a 7 × 2 inch plate for PL1. Locate the two 1/4-inch spark-gap-holder (SGH1, 2) screws holes as shown. The holes are spaced 2 3/4 inches from one another.

2. Form a three-turn 1-inch diameter coil (L1) from 1/2-inch sheet copper or #14 solid copper wire, as shown in Figure 12-4. Note the leads as attached to the capacitor (C1) and the end gap holder.

3. Connect an output port at the junction of C1 and L1 as shown. This point is now considered the output and can be connected to the radiating emitter.

4. Connect it to the converted plasma driver in the manner shown in Chapter 7 to the generator (see the schematic Figure 12-2). Note the added diodes D1 and D2 for converting the output to direct current

5. You will note a coil (L2) connected in series with the output lead. This inductor tunes out the capacitive reactance of the lead and the end capacitance. Experiment using a radio wave or absorption wave meter to determine the resonant frequency of L1/C1. Select a value for L2 to provide maximum radiation from a distance.

6. Experiment on various electronic devices and observe the effects at various distances.

Notes

A low-cost spectrum analyzer would be a great aid in setting up this system. Please note this is a low-power device intended as an introductory project for those desiring to experiment with shock pulse and EMP research. Several more functional and sophisticated devices will be featured in the next book in this series.

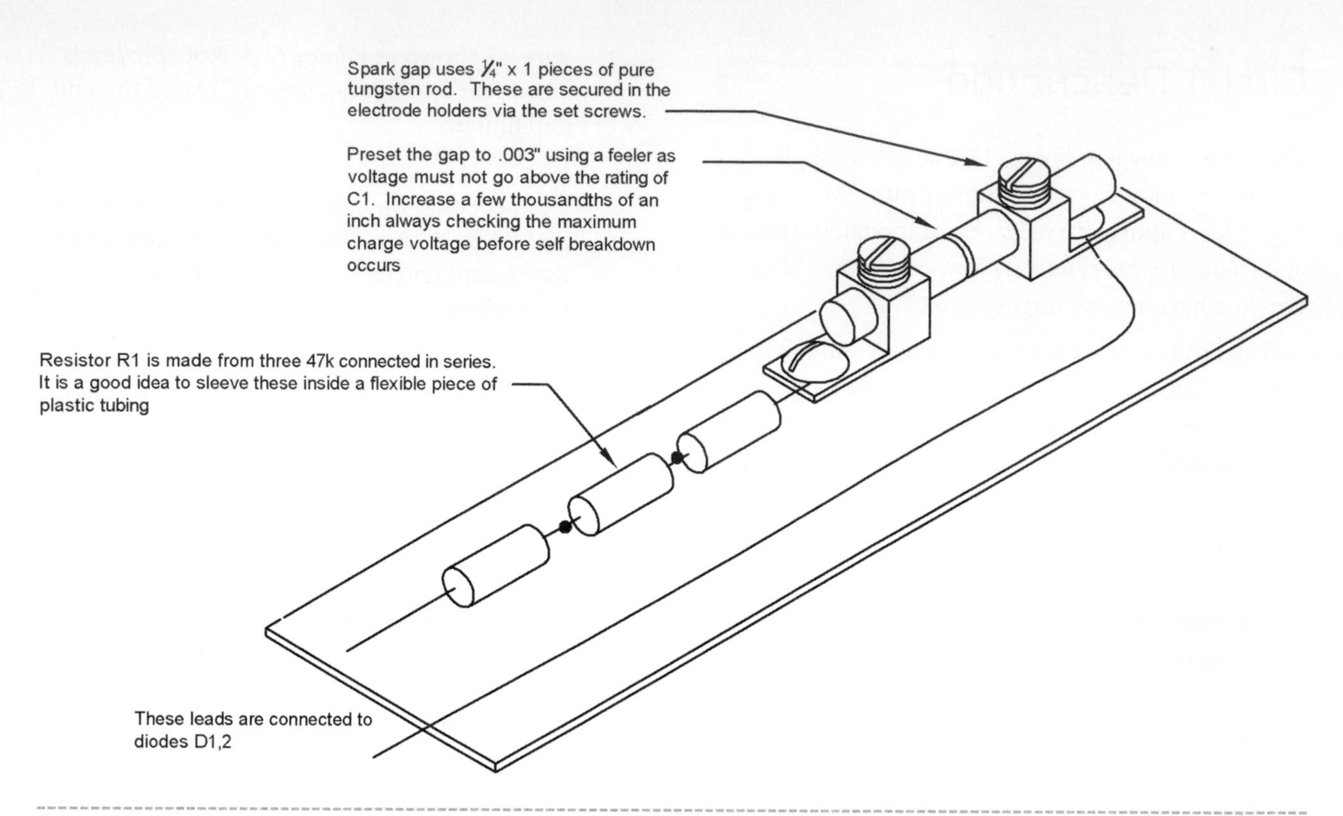

Figure 12-3 *Generator board bottom view*

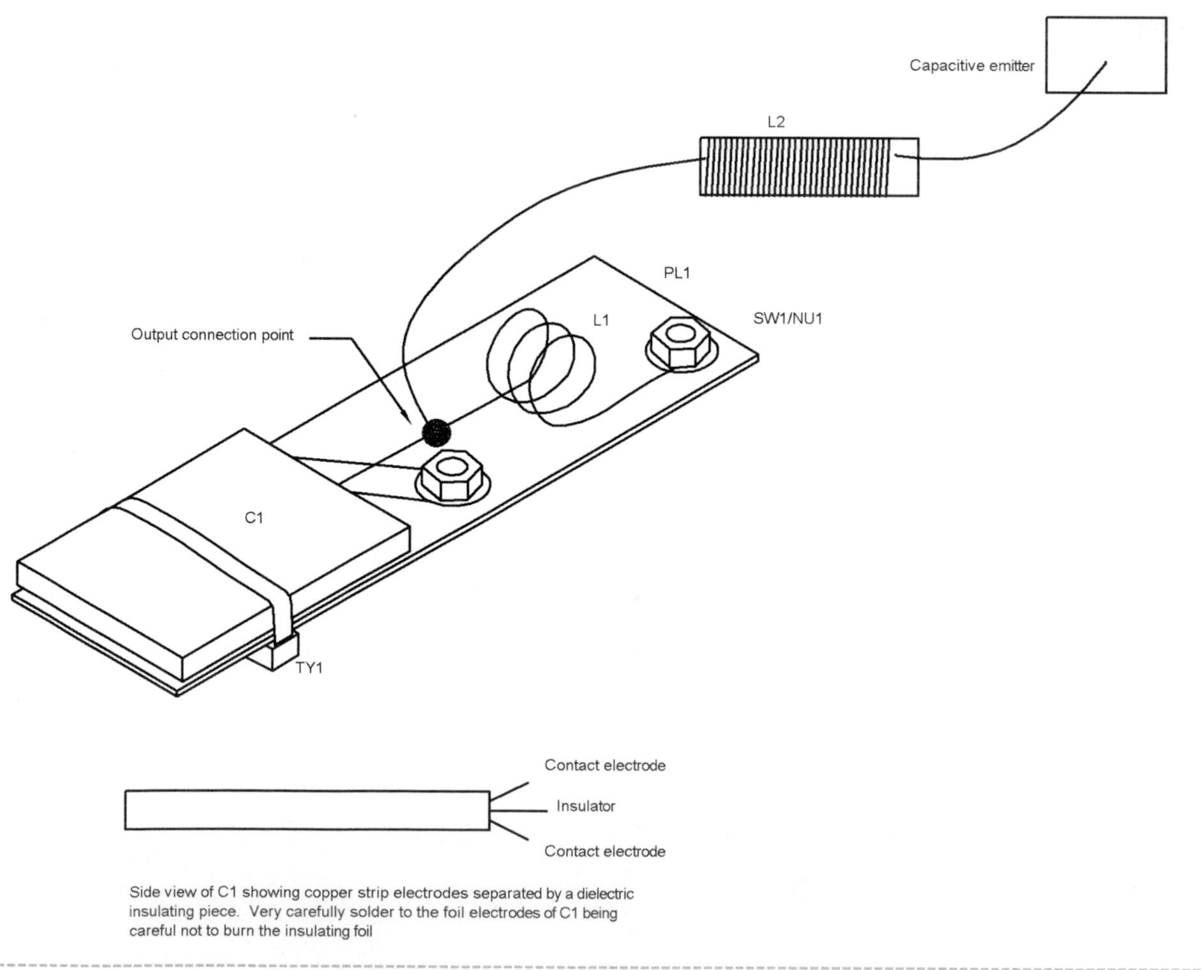

Figure 12-4 *Generator board bottom view*

Table 12-1 EMP Gun Parts List

Ref. #	*Description*	*DB part #*
R1	Three 47K, 1-watt resistors (yel-pur-or) connected in series	
D1, D2	Two 16 KV, 10 ma fast-recovery high-voltage rectifiers	DB# VG16
C1	.05 mfd, 5 KV stripline capacitor	DB#.05M/5KV
L1	Inductor 3-turn, 1-inch diameter use #12 copper wire	
L2	Inductor wind, as directed in text	
ELECTRODES	Two ½ × 1 inch pure tungsten electrodes	DB# TUNG141B
PL1	7 × 2 inch × .063 Lexan or G10 plastic	
TYE1	10-inch nylon tie wraps	
SW1/NU1	Two ¼ × 1 inch brass screw and nuts	
GRAVDRIV1K	Plasma driver kit	DB# GRAVDRIV1K
GRAVDRIV10	Plasma driver assembly	DB# GRAVDRIV10

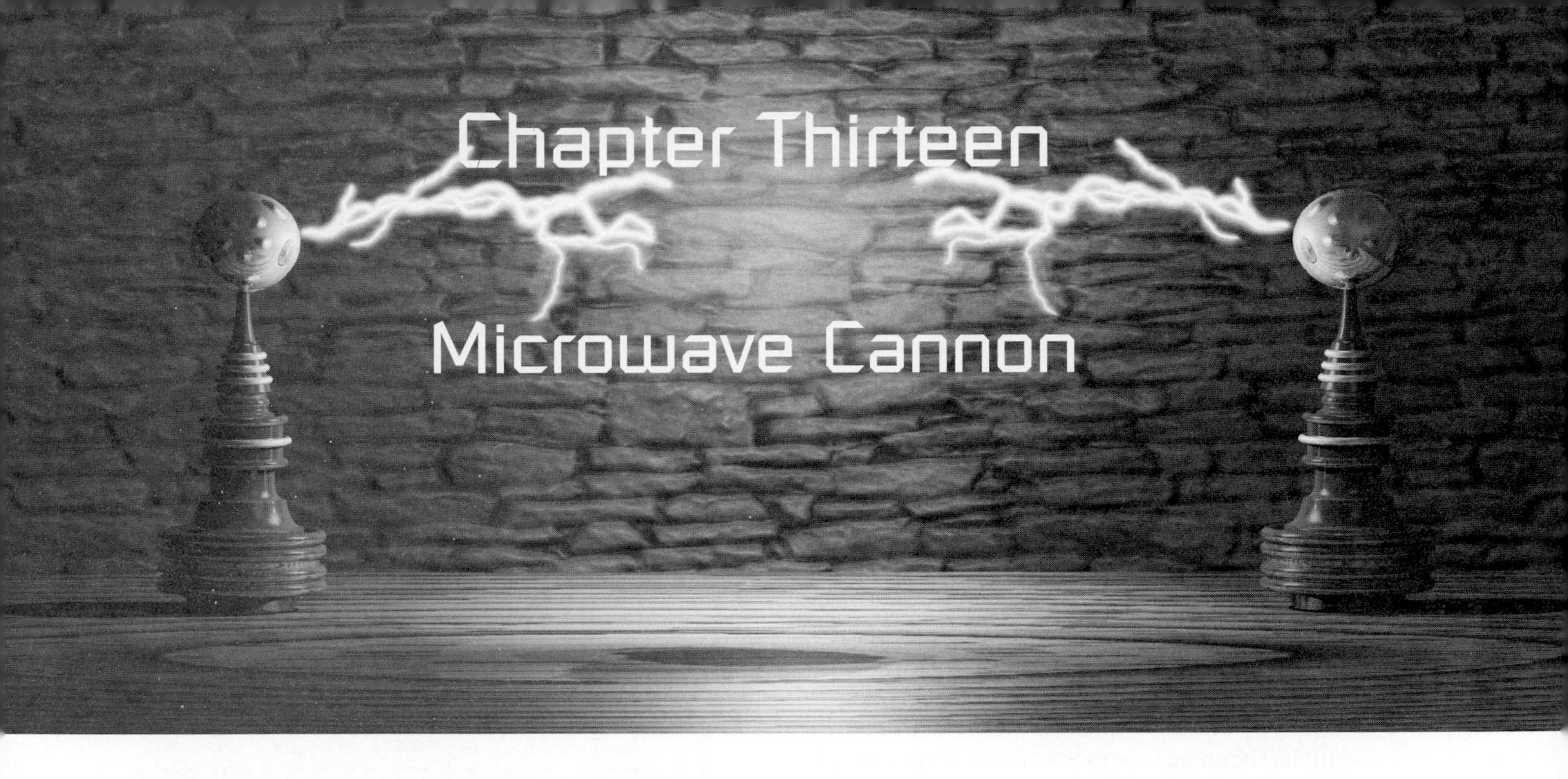

Chapter Thirteen

Microwave Cannon

This project shows how to build a powerful microwave transmitter intended to be a useful laboratory resource or as a device for disrupting bothersome stereos, boom boxes, and other annoyances perpetrated by inconsiderate hip-hop miscreants. This chapter was prepared by Dr. Barney Vincellete, who holds several doctorate degrees in the electrical sciences. It is a very useful tutorial involving the design of a microwave system, going into basic mathematical details for those who are interested. The system can be built from the parts of a microwave oven.

Warning: The project uses dangerously high voltages and can be a radiation hazard at close range if not used with protective shielding. This is an advanced project intended for experienced builders and must not be attempted by anyone unfamiliar with high voltages or high-power radio frequency circuitry.

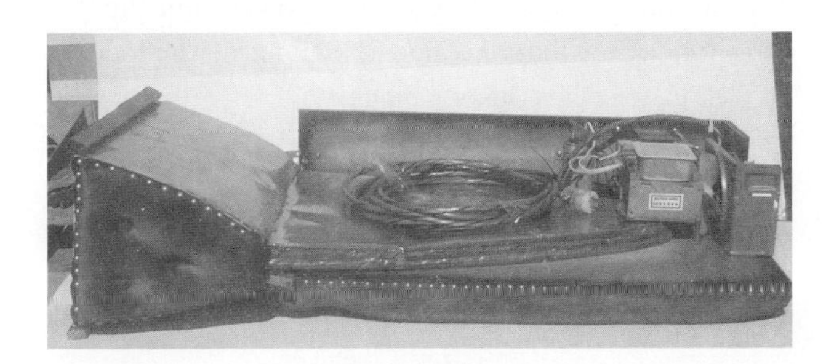

Figure 13-1 *Microwave cannon*

The microwave cannon consists of three parts: the parabolic reflector horn, the electric field focusing lens, and the magnetron. The horn focuses the magnetic oscillations that circumscribe the magnetron antenna into a narrow beam in order to concentrate the radiation into a two-dimensional space. It also prepares the electric field oscillations for the lens to concentrate the radiation in the third dimension. Because the electric field and magnetic field are orthogonal, calculations upon them and the modifications that we will perform upon them can be done on one without altering the other.

In designing the horn, we will assign the Cartesian coordinate system with the z axis along the longitudinal axis of the horn; that is, the z axis will occupy the center of the shaft of radiation that will be fired into your cultural tyrants' stereos. The Poynting vector will be on the z axis, and the x axis will be directed horizontally to the left and right of the z axis. The x axis will, during the calculations for the shape of the horn, be given an absolute value; that is, it will be endowed with a positive value whether it increases to the left or right of the z axis. This is because its values will be the square root of x squared, and it will be symmetrical to the left and right of the z axis. The origin will be located at the vertex of the parabola that will define the sides of the horn.

The y axis will be in the vertical direction to complete the basic vectors in our discourse and calculations. It will pass through the z axis at the mouth of the horn, rather than through the vertex, as the x axis does. It too will have only positive values because, due to its symmetry about the z axis, there is no need to consider negative values of y (see Figure 13-2).

The dimensions of the sides of the horn will be performed first. In building the wooden mold upon which the sheet copper will be formed, an adjustment will be made to correct for the flare that can be seen in Figure 13-3.

The horn will be made of 20-inch-wide sheet copper that can be purchased from a roofing supply house. Because the dimensions of the parabolic sides shown in Figure 13-2 are fixed by the top and bottom pieces of copper that will be soldered to the copper side pieces, the top and bottom pieces will be cut to a parabolic shape with an extra 1/4-inch margin on the sides that will be bent into a ridge that will be soldered to the side piece. The actual parabola will have a maximum width of 19.5 inches, or the maximum value of x will be 9.75 inches. Also, for structural rigidity, a half-inch lip will be bent along the front edge of the top and bottom pieces.

The parabola will contain the magnetron antenna aligned parallel to the y axis and positioned at the focal point of the parabola. The focal point will be one-eighth of the wavelength of the magnetron frequency, which, when the wavelength is computed by dividing the speed of light by the frequency or 2.45 GHz, gives us a wavelength of 2.45 inches and a focal length of 0.6 inches. The parabola is described by the equation x squared = 4 pz, where *p* equals the focal length. When *p* equals .6 inches, *r* equals 2.4 z. From this, we construct Table 13-1.

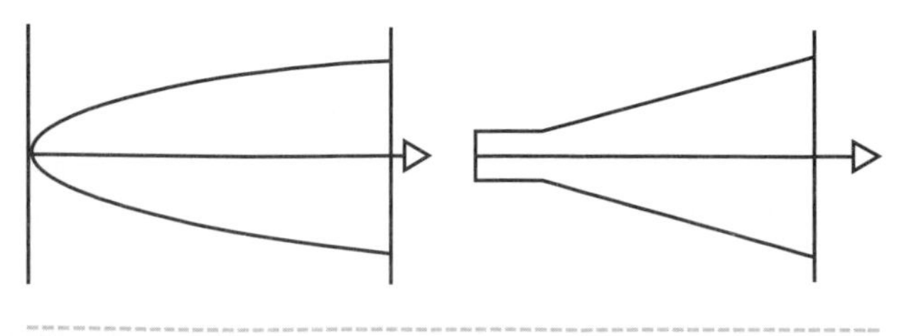

Figure 13-2 *Top and side horn templates*

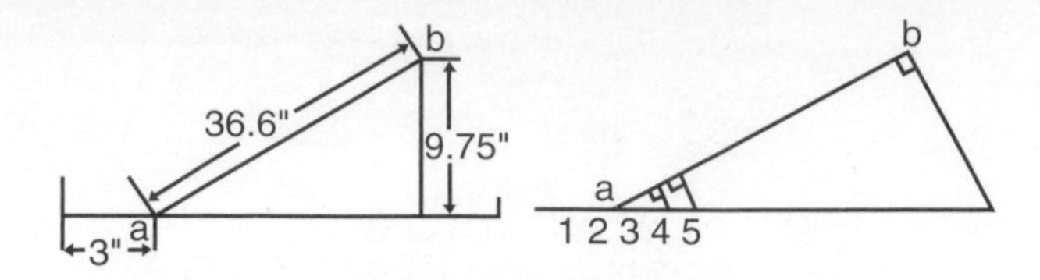

Figure 13-3 *Flair angle template*

Table 13-1 Parabolic Dimensions (measurements in inches)

z	*x*	*z*	*x*	*z*	*x*	*z*	*x*	*z*	*x*
.125	.55	3	2.68	8	4.38	16	6.2	29	8.34
.25	.77	4	3.1	9	4.65	18	6.57	32	8.76
.5	1.1	5	3.46	10	4.9	20	6.93	35	9.17
1	1.55	6	3.79	12	5.37	23	7.43	38	9.55
2	2.19	7	4.1	14	5.8	26	7.9	39.6	9.75

When regarding the side view of the horn (Figure 13-2), it can be seen that plotting the parabola, which is done by cutting the wooden sheets used for making a mold of the horn, will require an adjustment so that the sides of the parabola are not distorted where z is greater than 3 inches. The dimensions of z, if projected from the z axis to the top and bottom of the horn where z is greater than 3 inches, will have to be enlarged by dividing each inch by the direction cosine of the angle between the z axis and the top (also the bottom) of the horn. This can be done by drawing upon a sheet of 2 by 4 foot particle board, as shown in Figures 13-3.

After you have marked the board, as shown in Figure 13-3, you will begin plotting dots measured above and below the z axis at distances of x from Table 13-1 measured with a carpenter's square. You will connect these dots with a ruler, and when you cut out the parabola, the saber saw that you use will make a curved path that will very closely approximate the parabolic shape needed. After you cut out the parabola, you will make another cut along the locus of points where z equals 3 inches.

You will trace these pieces and cut duplicates from another piece of particle board, as shown in Figure 13-4. With nails, screws, glue, and two trapezoidal-shaped pieces of particle board, you will put together the form that you will use to build the copper horn (see Figure 13-5).

Stand the wooden form on its end, where z is 39.6 inches against the floor, and make a template from poster paper that has been taped together to form large enough pieces. Make a template that will wrap about the sides, and trim it to fit so that it matches the sides. Make another template for the top (also the bottom) piece the same way, as shown in Figure 13-6.

You will need 15 feet of copper roofing sheet metal to cut the pieces that you will solder together. Trace and cut the long piece from the template you made from the sides and rear of the wooden form. Add a half-inch to the length of each side so that you can bend a lip that will make the front of the sides rigid. This lip will also provide a place where you can attach the sides of a wooden frame that will make the horn more stable. Then trace the top and bottom pieces against the copper and draw a 1/4-inch additional margin surrounding the parabola you just traced. Also add a half-inch to the front that you can bend into a lip to make the top and bottom more rigid. Cut out two of these pieces, one for thc top and one for the bottom.

With a pair of pliers, bend a 90-degree lip from the 1/4-inch margin you traced. You can cut notches in this lip to make this bending easier. This lip will be soldered to the piece that will form the sides of the horn. Bend the half-inch lip at the front edges of the pieces that you will solder together, and clean the insides of the lip that you will be soldering to the sides of the horn using steel wool or sand paper. Clean the exterior 1/4-inch sides of the horn to which you will be soldering the top piece. Apply a thin coat of soldering paste and clamp the sides and top securely into place on the wooden form you built. Hammer the side lip against the sidcs of the horn for a tight fit. You do not need a sharp crease where the horn begins to flare out where z is 3 inches; a modest bend will do. You are now ready to begin soldering the seam, as shown in Figure 13-7.

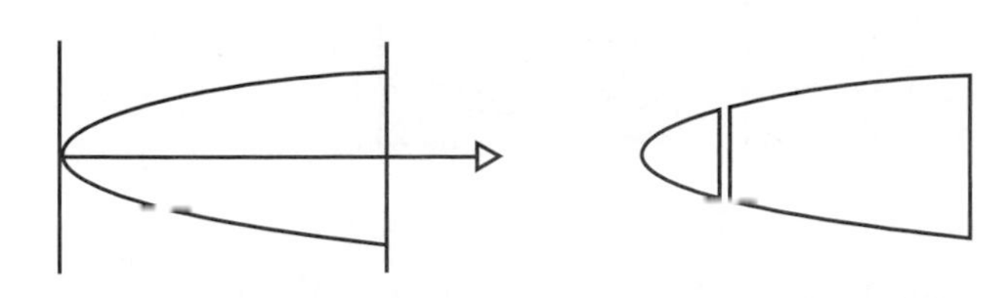

Figure 13-4 *Templates for forming the horn pieces*

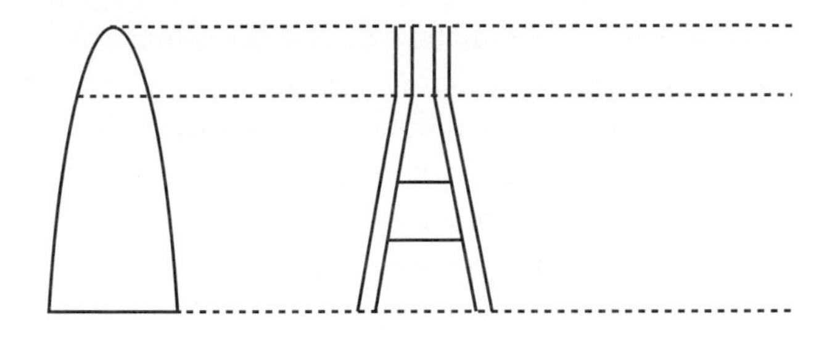

Figure 13-5 *Poster paper template*

Begin with a propane torch and plumber's solder. Apply the flame to a spot and, as soon as it is hot enough to draw in some solder, let just enough solder draw into the seam to fill the gap. Use more hammer taps to close any open spots where the lip is not tightly against the side piece. As soon as you have an inch or two soldered, remove the flame and solder another place a few inches away. Do not use enough solder to cause it to form drips or beads on the inside of the horn. Continue until the entire seam is soldered. After this is done, remove the clamps and sheet copper structure you soldered together. If you see any lumps of solder on the interior seam, you can melt it away by applying a flame.

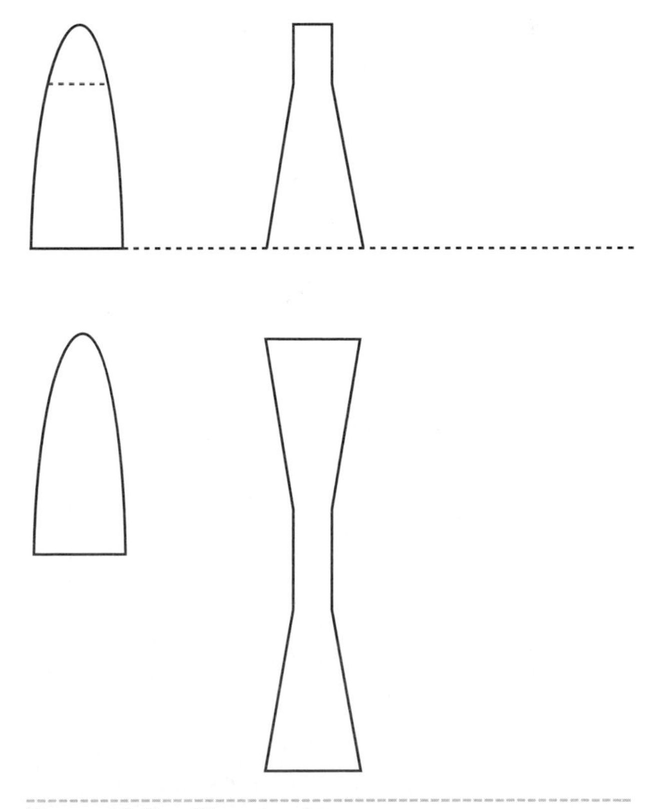

Figure 13-6 *Top template*

Now you must locate the point on the center axis at the top of the horn. This can be done by measuring the center of the front of the top section of the horn, measuring 2 feet along each side of the horn's interior, drafting a line between these points that are 2 feet back from the front, and bisecting it. You bisect the width of the front to find one point. You then measure 2 feet back on each side and draw a line between the ends. You bisect the line for the second point. This will provide two points to extend back to the vertex of the parabola's interior.

Now measure, as punctiliously as you can, 0.6 inches from the vertex. Drill a pilot hole and cut a circle that is $1^{3}/_{16}$ inches in diameter. With a hammer and a piece of pipe, tap on the interior of this hole so that its edges flare outward to make a good seal with the brass wool gasket on the magnetron you will be using (see Figure 13-8).

At this point, solder the bottom piece into place and complete the construction of the copper horn. A wooden frame must be built to make the front edges of the horn more rigid, to provide a structure for mounting the lens, and to provide a place for the front of an aluminum rail where a telescopic rifle sight will be mounted.

Here you have great artistic liberty in how you will build this cabinetry. Half-inch plywood is the most practical material to use, and cedar has the most pulchritudinous finish to stain, varnish, and paint. You can paint or carve artwork upon this frame, and you can also upholster the rear or sides with velvet and sunken buttons. In the most unlikely event the cannon becomes an exhibit in court, these characteristics will humor the jury, many of whom will be as hostile as you against the music the neighbors were forcing into your home. If it becomes newsworthy enough to attract a television crew, it will be a splendid show for the millions of people whom you will inspire to follow your example and join the revolution against cultural tyranny.

The rear of the frame will protect the operator from the deadly 4,500-volt wires between the power supply and the magnetron. It will also provide a place to mount the rear of the aluminum rail for the telescopic rifle sight.

Let us digress to a disquisition on the flair angle formed by the sides of the cannon. In order to form the best possible cancellation of standing waves, due to the impedance vicissitudes where the horn begins to flair at the mouth of the horn, the following equation must be emulated within approximately 10 percent. The dimensions of the horn must match this within a percent.

$$0.25 = [a/(2\lambda)]\tan(\theta/2)$$

where a is the dimension of the mouth in the horn in the y direction: 20 inches. λ is the wavelength in

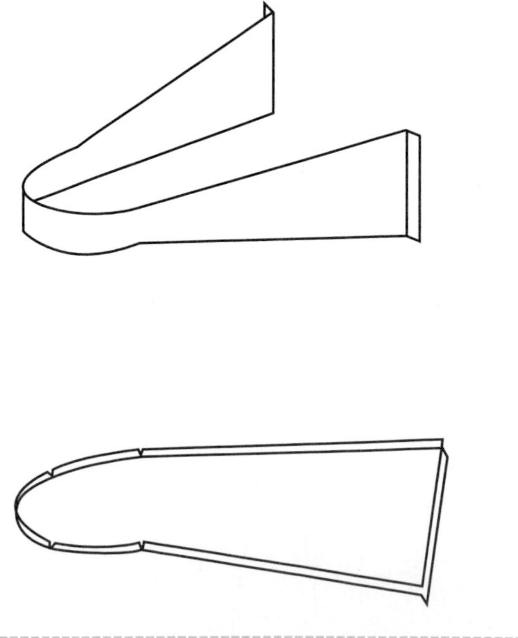

Figure 13-7 *Copper side with top and bottom pieces*

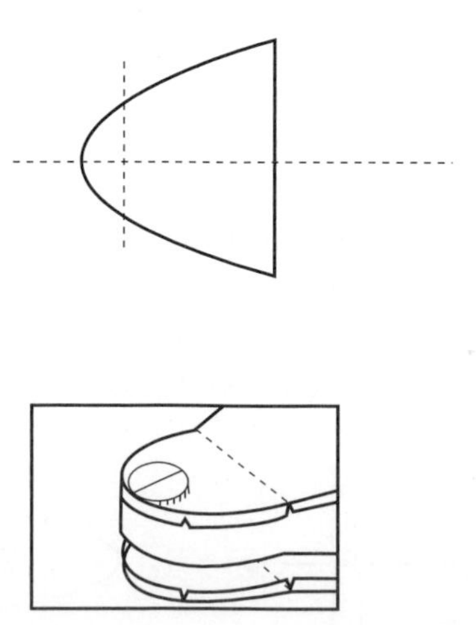

Figure 13-8 *Position of aperture hole for magnetron probe*

inches. θ is the angle between the z axis and the top (or bottom) side of the horn.

Let us now examine the efficiency of the parabolic sides of the horn for their focus of the magnetic field. The antenna is circumscribed by a magnetic field that propagates from a line source at the center of the antenna. The radiation either reflects off the sides of the parabola formed by the horn or it radiates directly from the antenna out the mouth of the horn. The latter is not useful because it dissipates rapidly with distance from the cannon. The former, the reflected radiation, is projected parallel to the zy plane, and because it originates from a line source parallel to the y axis, no diffraction occurs off the sides of the horn's mouth. The fraction of the radiation that is reflected is 92.2 percent of the total from the antenna. This is far superior to the typical 40 percent efficiency given by radar dishes that use large sizes to achieve gain.

The focusing of the electric field can now be performed by placing a lens in front of the cannon's mouth, and a description of the electric field must be provided. On the surface of a conductor there can be no voltage parallel to the surface of that conductor, because any voltage would be short-circuited out. This makes the components of the gradient of any voltage parallel to that surface zero. Since the electric field is the negative gradient of the voltage, the electric field near a conducting surface must perforce be perpendicular to that surface. As the electric field propagates forward, it forms a cylindrical-shaped wavefront in the flared part of the horn. At the mouth of the horn, this cylinder has a radius of 41 inches. A lens will convert this cylindrical wavefront to a flat wavefront, and that will complete the focusing of the radiation.

The lens will be constructed of metal plates that will be aligned parallel to the y and z axis 4 inches apart. Between these plates, any wave vector, $k = 2n/\pi$, can be the vector sum of two components, kx and kz. In the x direction, the sine wave of the wave vector will be zero every 4 inches, because at the plates the electric field in the y direction must be zero. Taking recourse to the Pythagorean theorem, the wavelength in the z direction is dilated by the inverse of the refraction index, that is, the refraction index, n = [1 − (λ/8 inches)E2]E−.5 = 0.8 (λ = 4.8 inches). See Figure 13-9.

The cylinder formed by the cylindrical-shaped wavefront plus the refraction index times the distance parallel to the z axis between the lens elements to a flat plane in front of the lens is to be constant so that the electric field will be in phase on the flat front of the lens. The y axis is shifted to the most concave depth of the lens elements, and the z axis is directed rearward. The algebra for calculating the concave curve to cut into the rear of the lens plates is rudimentary. From the following equation, the dimensions of the lens are shown in Figure 13-10.

[(47″ - z)E2 + y]E.5 + 0.8 z = 47″ from which z = {18.8 - [(18.8)E2 - (1.44y)E2] E.5} {1/.72}

The lens plates mounted against the far right and far left sides of the lens frames should be cut from leftover sheet copper and glued to wooden sides, also cut to the shape of the lens element. The other lens elements should be made from 0.025-inch sheet aluminum. They should be 21 inches long in order to mount them to the wood lens frame you will build. Small wooden blocks can hold the ends of lens plates in place. Pieces of wood with grooves cut into them will also work.

The efficiency of the lens can also be calculated. This efficiency is due to the reflection of radiation that takes place when radiation passes from a medium of one refraction index into a medium of a different refraction index where the space impedance is changed. Two reflections occur, one when the radiation enters the rear of the lens and a second reflection when the radiation exits the front of the lens. The second reflection is in opposite phase with the first reflection, so it is 2π radians out of phase with the first reflection. The appropriate Fresnel equation for calculating the power reflected when the electric field

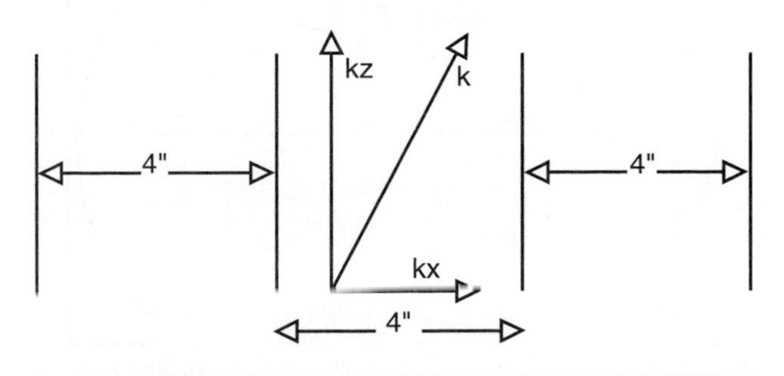

Figure 13-9 *Focus plate geometry*

is parallel to the plane containing the angle of incidence, θi, and the angle of transmission, θt, is as follows:

$$R = \{[-\cos\theta i + (n1/n2)\cos\theta t] \div [\cos\theta i + (n1/n2)\cos\theta t]\}E2$$

θt is arctan dz/dy, as can be seen in Figure 13-11. θi is θt − arctan $[y/(47 - z)]$.

In the second reflection, θi and θt are both zero. n1 is 1 and nt is 0.8. The second reflection that must be subtracted from the first one is cos $[2\pi$ radians $2(z + .25'')\lambda z]T$, where T is the fraction of the power transmitted into the lens and λz is 4.8 inches/0.8. Since the total reflected power is almost everywhere less than a percent, we can make the approximation that T equals 1. If the reflected power is calculated every inch for y, the average power reflected is approximately half a percent. Thus, the gain of the horn is

$$4\pi A/\Lambda E2 = 218 \text{ or } 23.4 \text{ db}$$

With the ideal horn and lens system having been designed so that the common man can build it without having to know all the electromagnetic laws of physics that make it work, we direct our attention to the choice of a microwave source. The cardinal rule of electronic warfare is that the radiation must convey noise in order to disrupt the signals in the circuits you will be attacking. This requires modulation that will drown out enough of the unwanted music and interfere with the digital signals in disc players to make them useless. Half-wave rectification of the current through the radiation generator is subjectively the most disrupting. The lower frequencies promoted by stereo manufacturers and music producers seem to be designed to most effectively penetrate the walls of apartments and homes for the purpose of destroying the privacy and cultural autonomy of all but the wealthiest who can afford the enormous cost of solitude.

The 60-cycle voltage from a household wall socket offers a nearly ideal source of modulation to create this disruption, because in a microwave oven it is half-wave rectified by a voltage-doubling circuit. This provides a splendid spectrum of Fourier frequencies that concentrate upon the lower frequencies but distribute themselves upward through the audio range, and it gives the utmost modulation of amplitude. It also reduces the duty cycle of the magnetron while preserving the maximum power in each pulse it shoots into the neighbor's stereo for the best range.

The first inclination is to choose a magnetron that will have the highest output power. The best choice would be a radar magnetron, which might offer 30 to 100 kilowatts in pulses that are only a microsecond or two in duration. However, tests using radar against a compact disc player yield disappointing results. The Fourier series expansion on a series of microsecond-long pulses modulates less than 1 percent in the audio band. The series of ticking sounds that it introduces into a stereo amplifier is scarcely noticeable, and not enough pulses occur to put sufficient errors in the digital circuits of the compact disc player to do any good. Also, audio circuits are not as sensitive to the

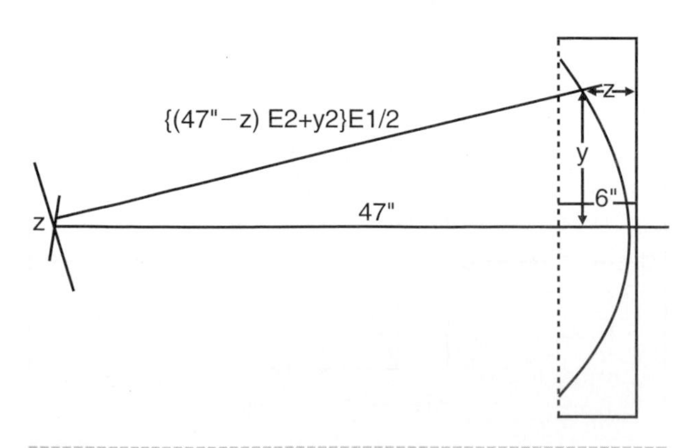

Figure 13-10 *Wavefront and electric lens geometry*

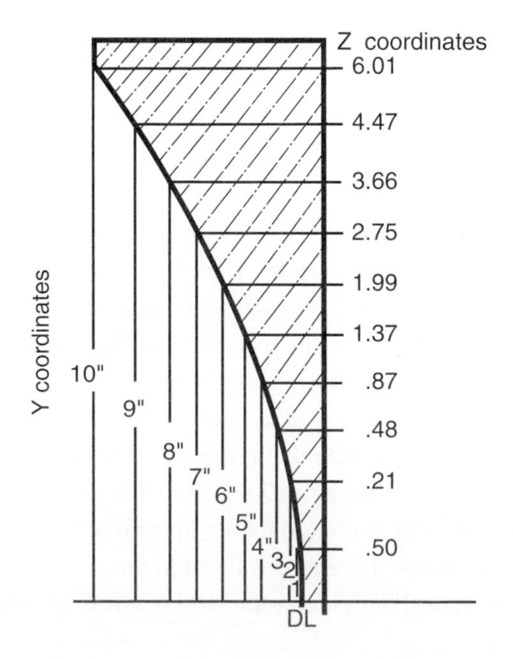

Figure 13-11 *Calculated lens dimension*

higher frequencies of radar as they are to microwave oven frequencies.

We might also be inclined to consider the higher-powered magnetrons used in industrial heaters and plasma generators, but this can be problematic. First, higher power is achieved more by increasing the duty cycle than by increasing the peak output power we desire. The 3,000-watt Panasonic 2M265 magnetron has a 7-kilowatt peak output power, but it runs on pulse-modulated direct current for a longer duty cycle than the half-wave voltage doubler of a microwave oven. Its operation is more critical; the cathode heater must have its voltage adjusted to the power output, and the cathode is more brittle in these more powerful magnetrons. It also costs as much as $1,000 and the power supply is even more expensive. The 6,000-watt Toshiba E3328 magnetron runs on filtered direct current and it has even more expensive computer controls that adjust an electromagnet to control the current. The peak output power is less than that of the 3,000-watt magnetron.

Another problem occurs with these magnetrons as well. They are so unforgiving of such things as a large metal object near the cannon in which they might be installed that reflections could mismatch the impedance enough to shorten the service life. The manufacturers recommend that they have a branch off of a waveguide to a dummy load that will absorb any energy due to such an impedance mismatch in order to protect the magnetron. This makes the parabolic horn impossible to use because it is not practical to feed with a waveguide. A pyramidal horn would be needed and it only has two-thirds of the gain of the parabolic horn. Thus, the 6,000-watt magnetron would be as good as a 4,000-watt one in a parabolic horn, and the 3,000-watt magnetron would be as good as two-thirds of 7,000 watts in a parabolic horn. Several vastly cheaper household oven magnetrons can easily outclass the more expensive and difficult-to-operate choices that appear more attractive until they are examined with greater perspicacity.

Another difficulty with purchasing industrial magnetrons also exists. Companies will not sell them to private individuals out of fear that they might be used for negative or hateful enterprises, such as in machines designed to sabotage overly loud stereos. Although almost any magnetron and power supply from any microwave oven will work, the best oven magnetron for our purposes is the 2M121A, which has its number conflated with the number 53 or 57. These last two numbers make no difference in the choice of editions of the 2M121A you procure. They are used in the following ovens: the Panasonic 1030, the Brother MF5000 and MF7000, and the Merrichef models 136M, 165M, and 206M. It has a peak output power of 5,800 watts, better than the 2,000-watt magnetrons that cost more than twice as much. It can be ordered from the Expert Appliance site (www.expertappliance.com) under the part number Z9-Panasonic NEI 030-412 if you type in Panasonic NEI 030 for the type of oven for which you are purchasing parts. The transformer number is 210 Panasonic NEI 030-5414. Global Microwave Parts has voltage-doubling capacitors that are 0.85 microfarads rated at 2,500 volts AC and diodes. You should use one rated at half an ampere. Magnetrons and parts are available and listed in Table 13-2. You will need copper and aluminum sheeting, along with mat-erials available from hardware or building suppliers.

The magnetron is attached to the antenna, as shown in Figure 13-12. Follow the instructions closely and observe that the brass washer is flush with the copper conductive surface.

The power supply is a simple voltage-doubler circuit, shown in Figure 13-13. It must be understood that the voltages in the wires will be in the thousands and can produce a spark that can jump through the air and through an insulator, conveying enough current that it would probably cause electrocution. Accordingly, the wires should be positioned where they will not be approached or touched. The magnetron and horn must be securely grounded or it will be 4,500 volts above ground potential. The frame that you must build around it will keep people's hands away from these wires. Also, after the cannon is shut off, four of the capacitors can hold a charge that would inflict a shock unless it is discharged, either through a bleeder resistor that may be built into the capacitor or it can be discharged with a safety discharge probe.

The magnetron must also have a fan to cool it during operation. The 2M121A requires at least 85 cubic feet per minute of air directed through its fins for best operation.

When your microwave cannon is completed, testing it is little more than a matter of switching on the power and aiming it at a fluorescent tube or a small 120-volt lightbulb. The lightbulb should be no more than 25 watts and should glow slightly when it is within a few feet of the mouth of the cannon. Fluorescent lamp tubes should light up as far as 10 feet away.

You should not put any body parts directly in front of the cannon at close range because, after a few seconds to a few minutes, you could burn yourself, but a distance of over 50 feet away would provide safety. However, prolonged overheating of internal organ tissue is remotely possible and the cornea of the eye can be damaged by overheating. Also, testicles can overheat, and at close enough range for the *root mean square* (RMS) microwave field to exceed 4,000 watts per square meter, a 5-minute uninterrupted exposure is supposed to build up enough internal steam pressure to cause the testicles to explode. However, the possibility of having a hot spot this intense a few inches in front of the cannon is marginal.

When testing or using your microwave cannon, you can do so with utmost safety to yourself by constructing a gown out of metal screen from a hardware store. The gown should include a hood that completely covers the head and should extend over the torso down to just below the vaudevillian parts of the body. The sex organs should be stuffed into a tin can, the mouth of which can be lined with fake fur to make it more comfortable. (Obviously, the ladies need not take recourse to this expedient.) This is more than enough to prevent any harm when performing experiments that last half an hour or more and is probably hundreds of times the protection you will need.

It can be advantageous to test the cannon in your backyard. Sooner or later the neighbors will see it, and if they see a mysterious-looking machine that lights up fluorescent tubes 10 or more feet away and they see you dressed in a wire screen gown complete with a tin can, it will frighten the emunctory indiscretions out of them. Further, when they complain to the authorities they will sound much the same way mental patients sound as they describe exotic macrunes, stereos malfunctioning, strange costumes complete with tin cans, and the like. It is even slightly possible

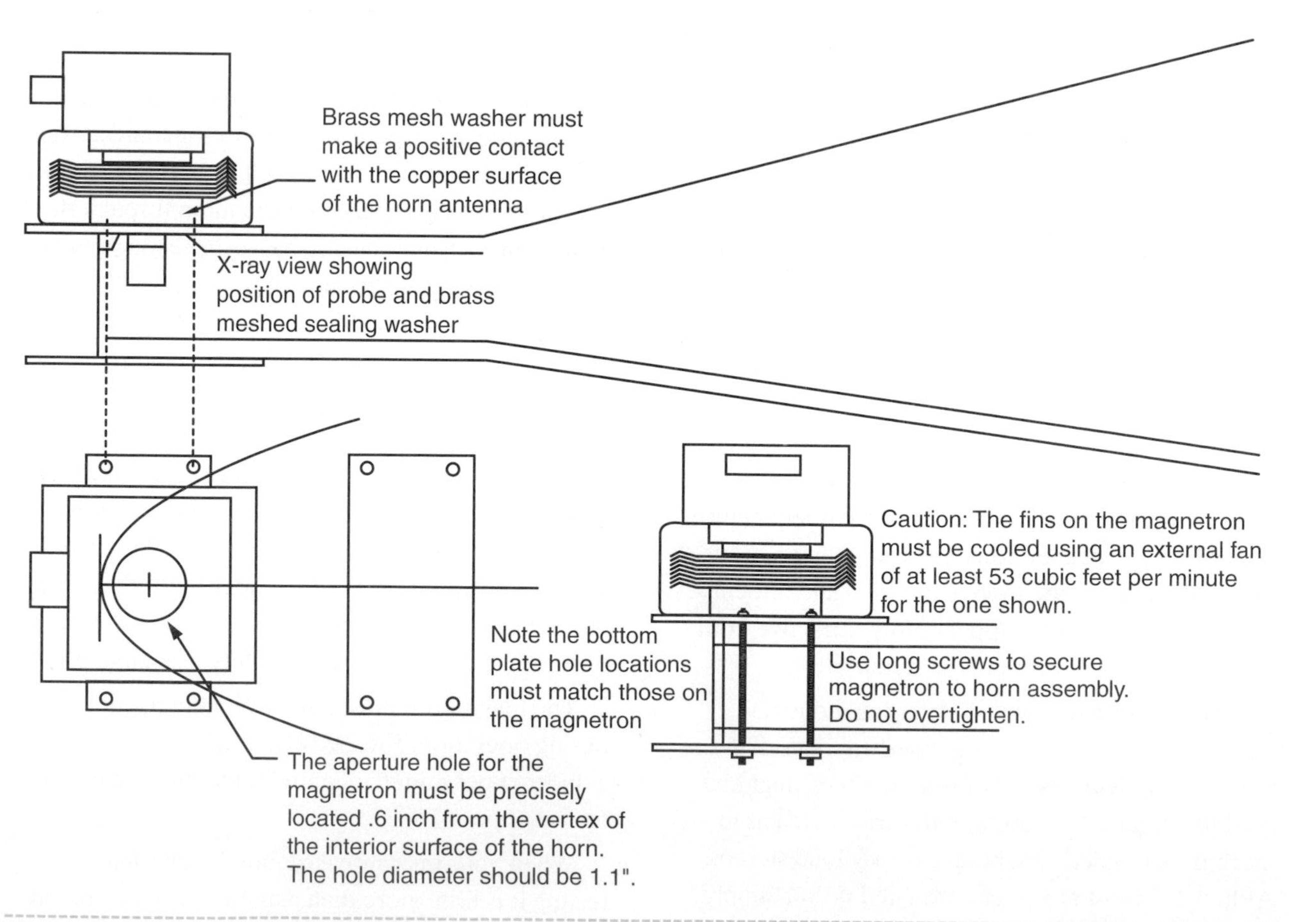

Figure 13-12 *Attaching the magnetron*

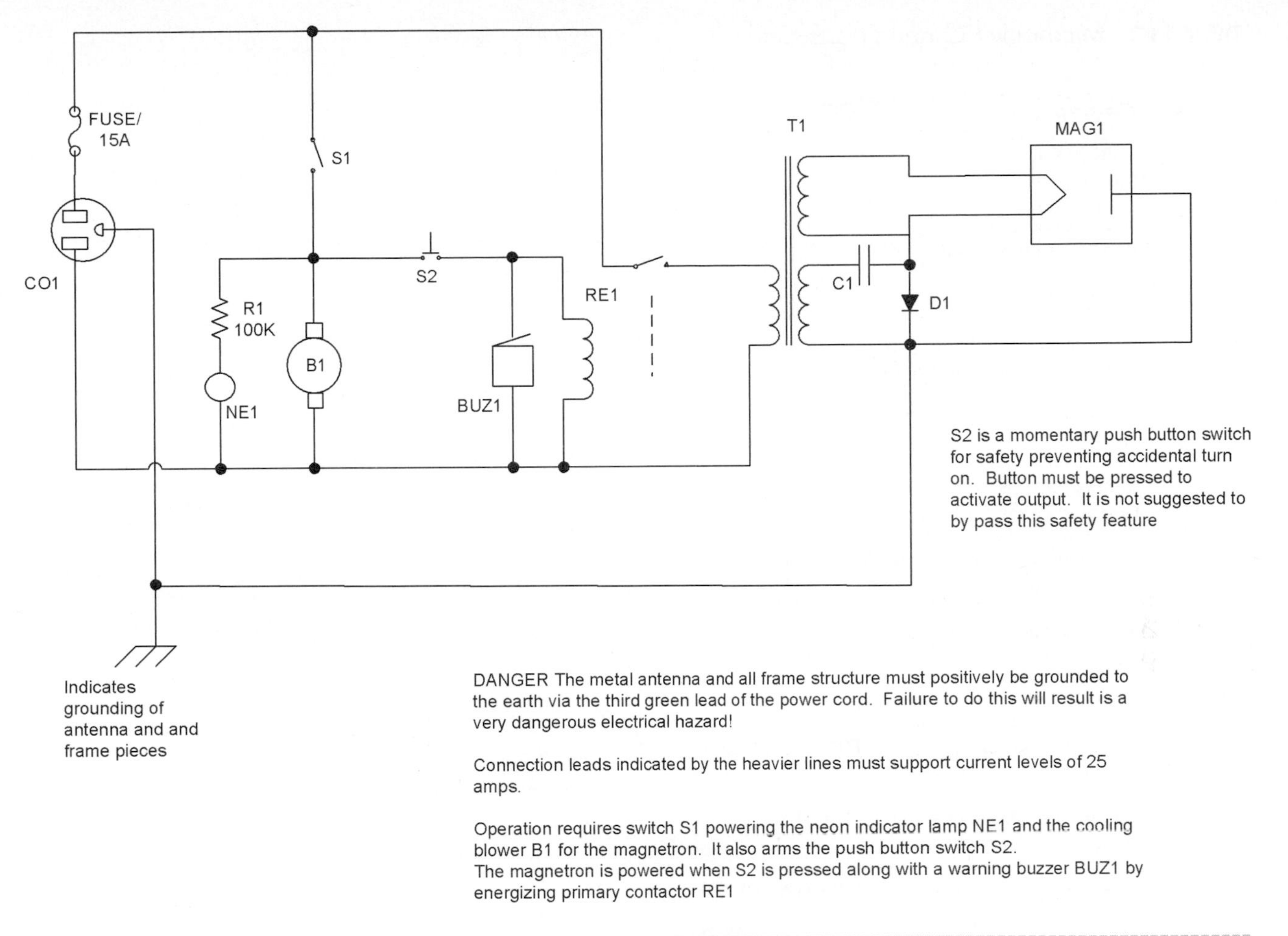

Figure 13-13 *Microwave cannon schematic*

that they might be invited to undergo observation at the local state psychiatric hospital, where, if you are really ambitious, you could locate their room, aim your microwave cannon, and make their stereo buzz whenever the psychiatrist's back is turned and cause a diagnosis that will keep them out of the neighborhood for a longer time. But in the remote case you get caught doing this, it will be almost impossible to prosecute you because no state legislature has the time to write and process special laws against shooting microwaves into peoples' stereos. The television publicity possible if the state attempts to prosecute you will be a disaster for the system that supports cultural tyranny because it will inspire countless television viewers to join in the revolution to make the world a better place by building more microwave cannons and shooting down more obnoxious stereos. And even if they can win a battle in court, they will ultimately lose the war that they started against human dignity.

Table 13-2 Microwave Cannon Parts List

Ref. #	*Description*	*DB Part #*
R1	100K 1/4-watt resistor (br-blk-yel)	
CO1	Heavy-duty, 15-amp, three-wire power cord	
FUSE/15	15-amp fuse and holder	
B1	85 cubic feet/minute or better cooling fan for magnetron fins	
S1	*Single pole, single throw* (SPST), 5-amp, 115 VAC toggle switch	
S2	Momentary push-button switch (see note on Figure 13-13)	
NE1	Neon indicator lamp with leads or equivalent	
BUZ1	120 VAC high output safety buzzer	
RE1	10- to 20-amp contacter or relay	
T1	Magnetron power transformer NEI030-5414	DB# MAGTRAN
C1	.85-microfarad, 2500 VAC voltage-doubler capacitor	DB# MAGCAP
D1	5 kilovolt, 1/2-amp diode	DB# MAGDIODE
MAG1	2M121A magnetron	DB# MAGNETRON

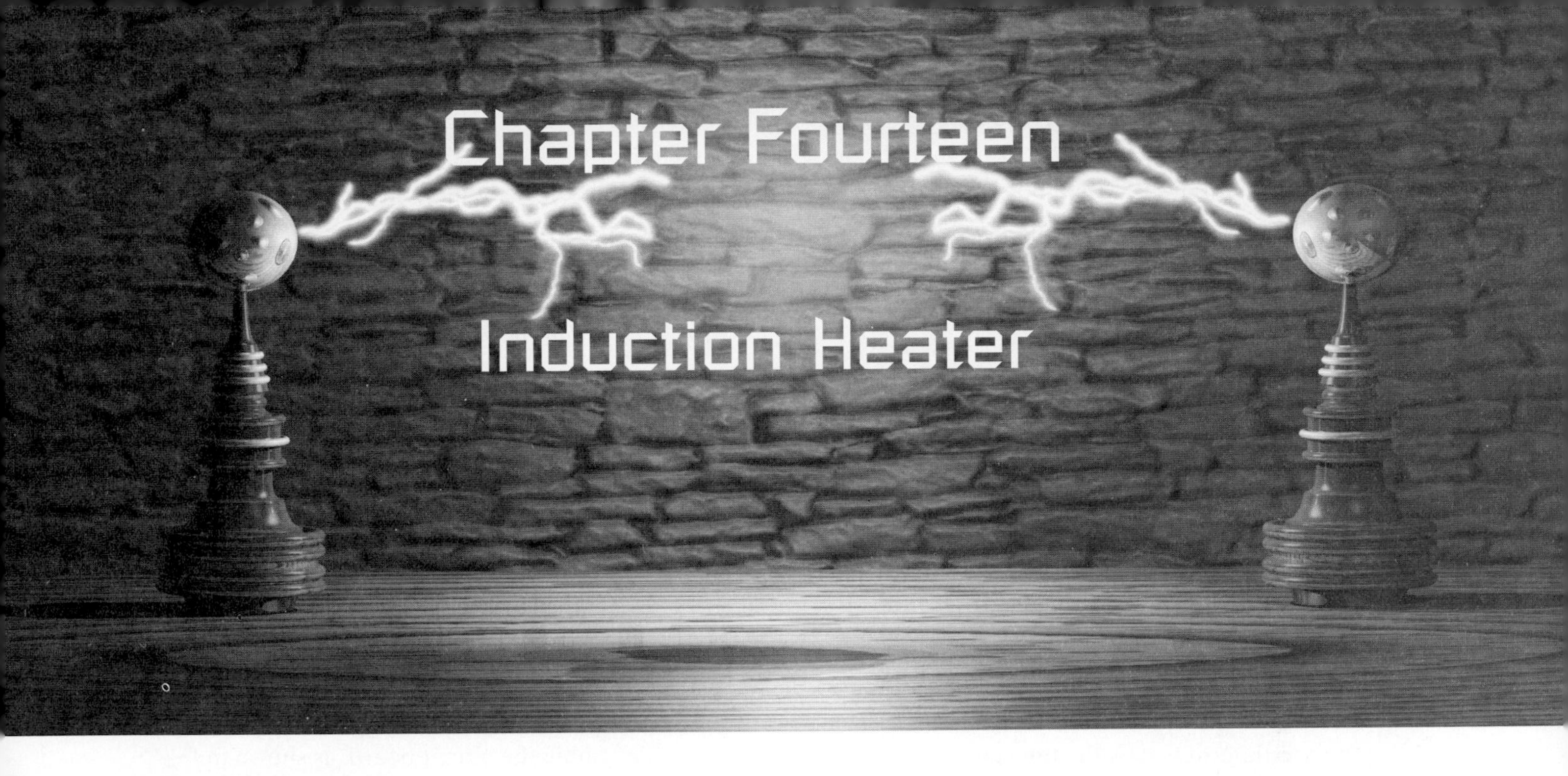

This device shown in Figure 14-1 shows an induction heater capable of heating metals to over 1,500 degrees Fahrenheit. It can be used in the processing of gas discharge tubes often used for plasma, neon displays, laser discharge tubes, and spectrum tubes. These tubes usually require electrodes that must be heated to drive out the impurities that are now sucked out by a vacuum system. The induction heater heats these electrodes directly through the glass.

The completed system consists of a donut-shaped heating coil head connected via an umbilical cord, which is in turn connected to the power-conditioning and control box. The system is powered directly from 115 VAC and requires caution when building and using. The head is a coil of wire that surrounds the target area to be heated. As shown, it is an air-cooled system that requires a waiting time for processing between operations because the coil will get hot. The common circuit ground is above ground and requires an isolation transformer or an ungrounded scope to view the waveforms from a grounded scope.

This is an advanced-level project requiring electronic skills and the use of an oscilloscope. Expect to spend $35 to $75. All the parts are readily available, with specialized parts obtainable through Information Unlimited (www.amazing1.com), and they are listed in Table 14-1.

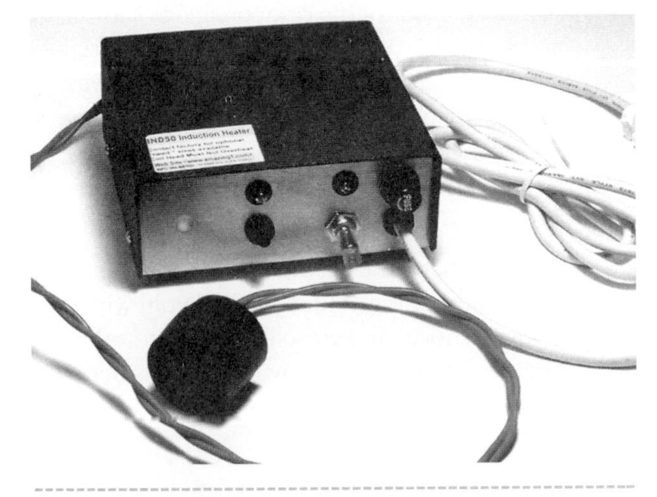

Figure 14-1 *Induction heater*

Circuit Operation

The power cord CO1 supplies 115/220 VAC from a wall receptacle where the third, green wire, grounding lead must be securely attached to the metal chassis (CHASSIS) for safety (see Figure 14-2).

Switch SW1 energizes the primary circuits, and fuse FUS1 protects the main circuit from any catastrophic faults. Indicator lamp NEON1 requires the current-limiting resistor R12 and lights up when SW1 is turned on. The peak charging current to capacitors C10 and C11 is limited by in-rush resistor Rx. The rectifiers D1, D2, D3, and D4 are shown in a full-bridge configuration when used for 220 VAC input. For 115 VAC operation, only D1 and D3 are used along with the JUMP jumper in a voltage-doubler configuration. The output voltage across C10 and C11 is 300 volts DC and is referred to as the positive and negative rails.

Dropping resistors R2 and R3 provide voltage to the oscillator driver IC1 with zener diode Z1 regulating the voltage at 15 volts. Switch SW2 is normally closed, shorting out this voltage across Z1. Capacitor C2 provides the high instant of current for the output pulses and must be physically close to IC1. Operation requires SW2 being pressed, allowing 15 volts to be applied to the oscillator driver generating the drive pulses.

These drive pulses turn the main MOSFET switches, Q1 and Q2, off and on. The switching frequency is determined by the time constant of power control pot R10 and timing capacitor C3. A limit resistor, R11, prevents driving the circuit beyond the intended limits.

The switching circuit is in a half-bridge configuration where the MOSFET connected to the positive end (+ rail) must be driven with its source pin referenced at 150 volts above the common end (− rail). This is accomplished by biasing a bootstrap capacitor (C4) through an ultrafast diode (D5), providing the correct DC level to fully control Q1. Resistors R6 and R7 eliminate the high-frequency parasitic oscillation that occurs as a result of rapidly switching the capacitive load of the MOSFET gates. Capacitor C40 limits the current flow to a second neon indicator light, NEON2, which is connected in series and is energized by the output frequency when the heating cycle is activated by SW2.

A network consisting of capacitor C7 and resistor R8 slows down the transition time of the switched pulses across Q1 and Q2. The resultant time constant limits the rate of voltage rise, dv/dt, that could cause a premature turn-on at the wrong switch, creating a catastrophic fault mode.

Capacitors C5 and C6 provide a voltage midpoint and produce the necessary storage energy to maintain the voltage level of the individual pulses. The output is taken at the junction of Q1/Q2 and C5/C6 and is a square wave fed to the induction heating coil L1. A current rise through L1 now commences as a function of Et/L where E is the voltage of the drive pulse, t is the applied time, and L is the inductance of L1. This changing current now induces eddy currents in the target piece to be heated as a function of $E = L\, di/dt$. These circulating currents quickly heat up the target piece.

Project Assembly

1. Assemble the heating head coil, as shown in Figure 14-3. This coil must be tightly and evenly wound for optimal performance. It may be difficult for the beginner and is available in three sizes, as indicated in Table 14-1.
2. Assemble the circuit board, as shown in Figure 14-4. Note that you should allow a half-inch for the Q1 and Q2 pins. This is necessary to allow these parts to be placed flat against the chassis when mounting as they exit at the midsection of the part.

   It is also strongly suggested that you use an eight-pin IC socket (SO8) because IC1 is very sensitive to static electricity and possible errors during the assembly.

   Note that if you are building from a perforated or vector circuit board, it is suggested that you use the indicated traces for the wire runs and insert components starting in the lower left-hand corner. Pay attention to the polarity of capacitors with polarity signs and all semiconductors.

   Route the leads of the components as shown and solder as you go, cutting away the unused wires. Attempt to use certain leads as the wire runs or use pieces of the #24 bus wire. The heavy foil runs should use the thicker #20 bus wire because these are the high-current discharge paths.
3. Double-check the accuracy of the wiring and the quality of the solder joints. Avoid wire bridges, shorts, and close proximity to other circuit components. If a wire bridge is necessary, sleeve some insulation onto the lead to avoid any potential shorts.

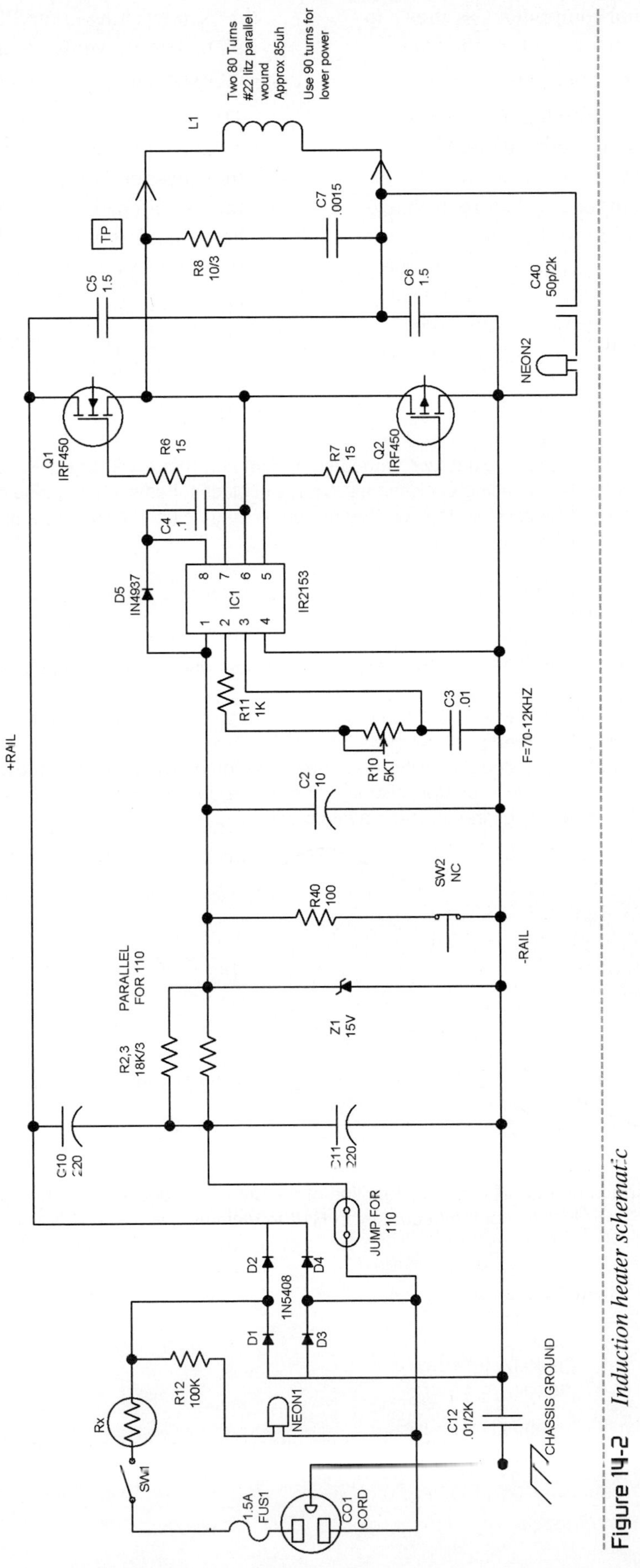

Figure 14-2 *Induction heater schematic*

4. Connect the external components as shown in Figure 14-5. Use 6-inch lengths of the wire size as noted.

5. Create the chassis, as shown in Figure 14-6, from a piece of .065-inch 5052 bendable aluminum. Note to trial-fit and measure the diameters of the components before drilling the holes. Pay attention to the holes for mounting Q1 and Q2 because these must be reasonably accurate to align with the layout of the PCB. Also, there must be at least 1/8 inch of clearance between the board and the chassis for sandwiching in the PLASTPL plastic plate. This prevents grounding out of the board wiring traces to the metal chassis.

6. The final assembly is shown in Figure 14-7. Note the mounting scheme for Q1 and Q2, as they must be insulated from the chassis using the thermo pads, THERM1. These pads allow heat transfers and provide electrical insulation from the metal chassis.

   Drill the hole for grounding lug LUG1 and solder the green grounding lead from the power cord and circuit board's earth ground.

The heating head coil assembly can be built to accommodate a diameter from a 1/2" to a 1" object. The assembly with the material shown is intended for using enclosed electrodes in glass tubes such as those used in neon tubes, laser tubes, and spectrum tubes. High-temperature materials may be used if you are anticipating using for other applictions.

The coil should be wound with high-frequency wire to limit heating losses. Use #22 litz wire consisting of multiple strands of #36 magnet wire.

1. Fabricate the winding bobbin from 1.25" lengths of 1/16" wall GT fiber glass tubing of your diameter choice of 1/2, 5/8, 3/4, 7/8 or 1".
2. Wind an even 1" layer of the #22 litz wire. Glue the start wire with a fast-curing glue. Tape the finished winding using a single layer of high-temperature glass tape.
3. Continue winding layers and repeat step 2 until there are 80 turns. You may truncate the successive layers by several turns at each end to keep from slipping off or use a quick-curing glue.
4. Attach the input leads and attach by gluing and taping into place.

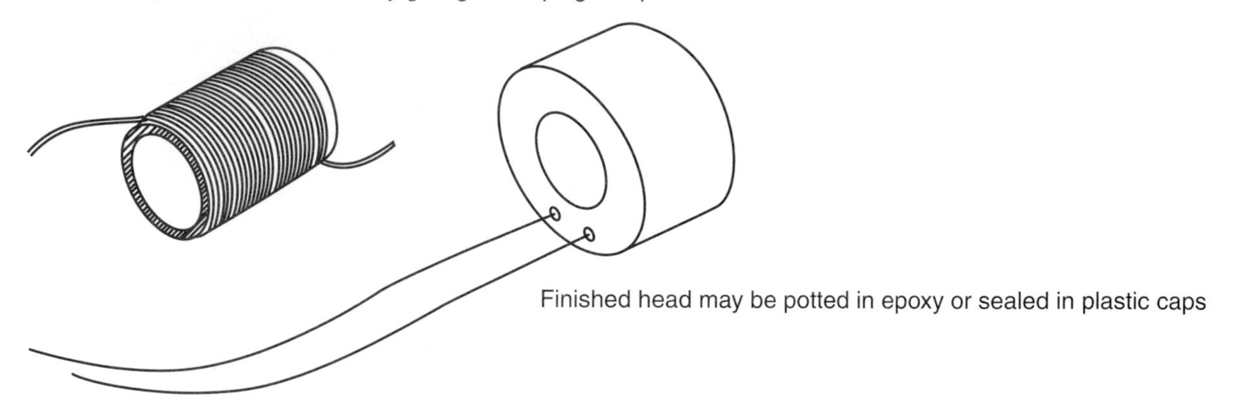

The inductances of the various heads will vary as a function of the diameter. Ideal current through the head is 20 amps peak. You should therefore adjust the frequency of R10 to maintain this value when changing the head sizes as the following:

1/2" head	50 uh	75 KHz
5/8" head	75 uh	50 KHz
3/4" head	100 uh	40 KHz
7/8" head	125 uh	35 KHz
1" head	150 uh	30 KHz

These heads are available assembled for those not wanting to attempt winding their own.They are listed in Table 14-1.

Figure 14-3 *Heater head coil assembly*

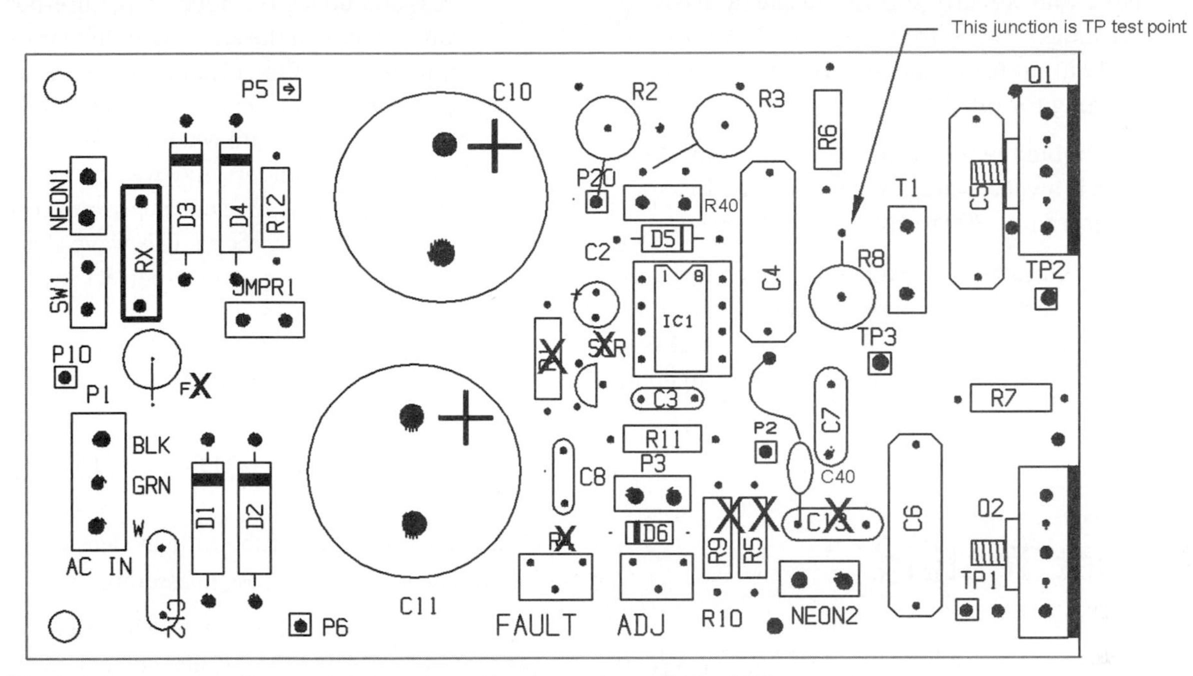

1. Black dots are connections points to external leads as shown figure 14-5

2. The PC board as shown is used in several projects and does not use all the components as indicated. Those indicated by an X are omitted for this project. R40 and C40 are added to the pads as shown and are not identified on the PC board printing

3. This circuit uses a 8 pin socket for IC1 as it is prone to failing if there are circuit faults

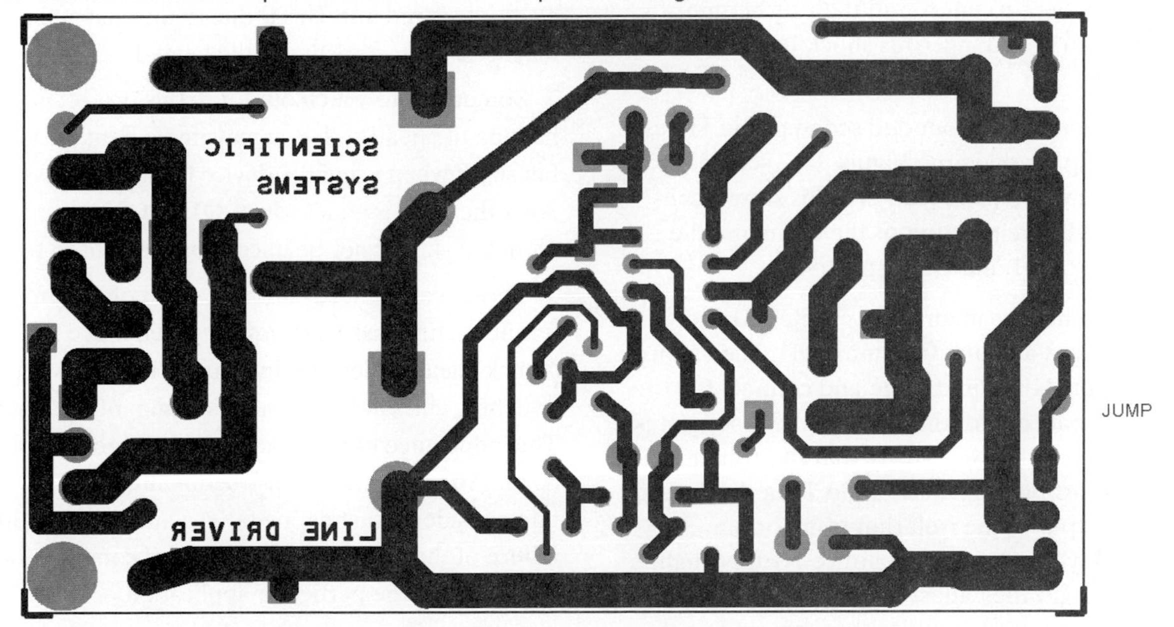

Assembly of the board assembly is shown using a printed circuit board available through www.amazing1.com Builders may use a vector board and use the component leads to duplicate the foil traces as shown.

On all IND40 assembly boards this jump lead is to be the first part on the board. Once the unit is final assembled with the heating head connected in place the jump now can now be disconnected. This step prevents static electric damage to the semiconductors.

Figure 14-4 *Assembly of the circuit board*

Note that we have shown the use of plastic bushings (BU38) for the neon indicator lamps and output leads, and also BU58CL clamp bushing for the line cord.

7. Assemble the test jig as shown in Figure 14-8. This is a very useful fixture for any 115 VAC project up to 300 watts.
8. Now fabricate the cover to fit the chassis section. Use bendable plastic or a thin, perforated aluminum sheet. Secure it to the chassis via four screws (SW1) on the folded-up lip in this section.

Project Testing

Caution: The circuit commonly noted as −RAIL is electrically above earthground by over 150 volts. It is strongly recommended that you obtain a 200-watt isolation transformer before proceeding and testing this or other similar direct line powered circuits. Experienced experimenters working on dry, wooden floors may choose to unground their test equipment. This can be a dangerous shock hazard!

1. Connect the ungrounded scope to the TP test point, as noted in the Figure 14-2 schematic or in Figure 14-4. The lead of R8 is shown connected to the junction of the drain and the source of Q1 and Q2 respectively.
2. Verify that the main power switch S1 is off and insert a 2-amp fuse into FH1. Then adjust trimpot R10 to midrange and connect CO1 to the female end of the cord on the test jig. It is important that you verify that the test jig is not shorting the ballast lamp. Plug the jig into the output of the isolation transformer. Quickly turn on S1 and notice NEON1 lighting up and the ballast lamp not lighting. If it lights, you must troubleshoot and find your error before proceeding.

Press and hold PB1, noting a picture-perfect square wave on the scope with the lamp only dimly lighting. Quickly adjust R10 for a 20-usec pulse period. This corresponds to 50 kHz. NEON2 will indicate when PB1 is activated. Experienced builders may wish to monitor the current to the head using a current probe attachment to the scope. This wave shape should be another picture-perfect sawtooth of 20 amps maximum.

Using the Heater Project

Obtain a gas discharge tube with an electrode to fit into the head size you chose in assembly.

At this point, it is a good idea to have a meter capable of measuring the AC line current, as this value is very dependent on the object that you place into the head and should not exceed 2 amps. You can drop this value by increasing the frequency, which is done by adjusting R10. When heating electrodes, you will note the ampmeter dropping in level when the electrode is red hot. This is due to the change in the magnetic characteristics of the heated material.

Something to watch out for in this project is overheating the head coil. You may implement a cooling fan stand when using the device for multiple tasks. Also, the chassis panel where Q1 and Q2 are mounted should not be uncomfortably hot to touch.

Those experienced can create other heads for different heating tasks and you can even make a "quick" heat solder pot. Just make sure you keep the switching current below the maximum of 20 amps. The inductance of the head should not go below 80 uh. Alternate head construction must always take into consideration heating fatigue and the eventual failure of the head wires, bobbin, and construction materials for the particular application.

Notes:
1. The parts on the printed circuit board with an X are not used for this project.
2. C40 and R40 are added components not indicated on the board
3. Lead lengths can be shortened in final assembly

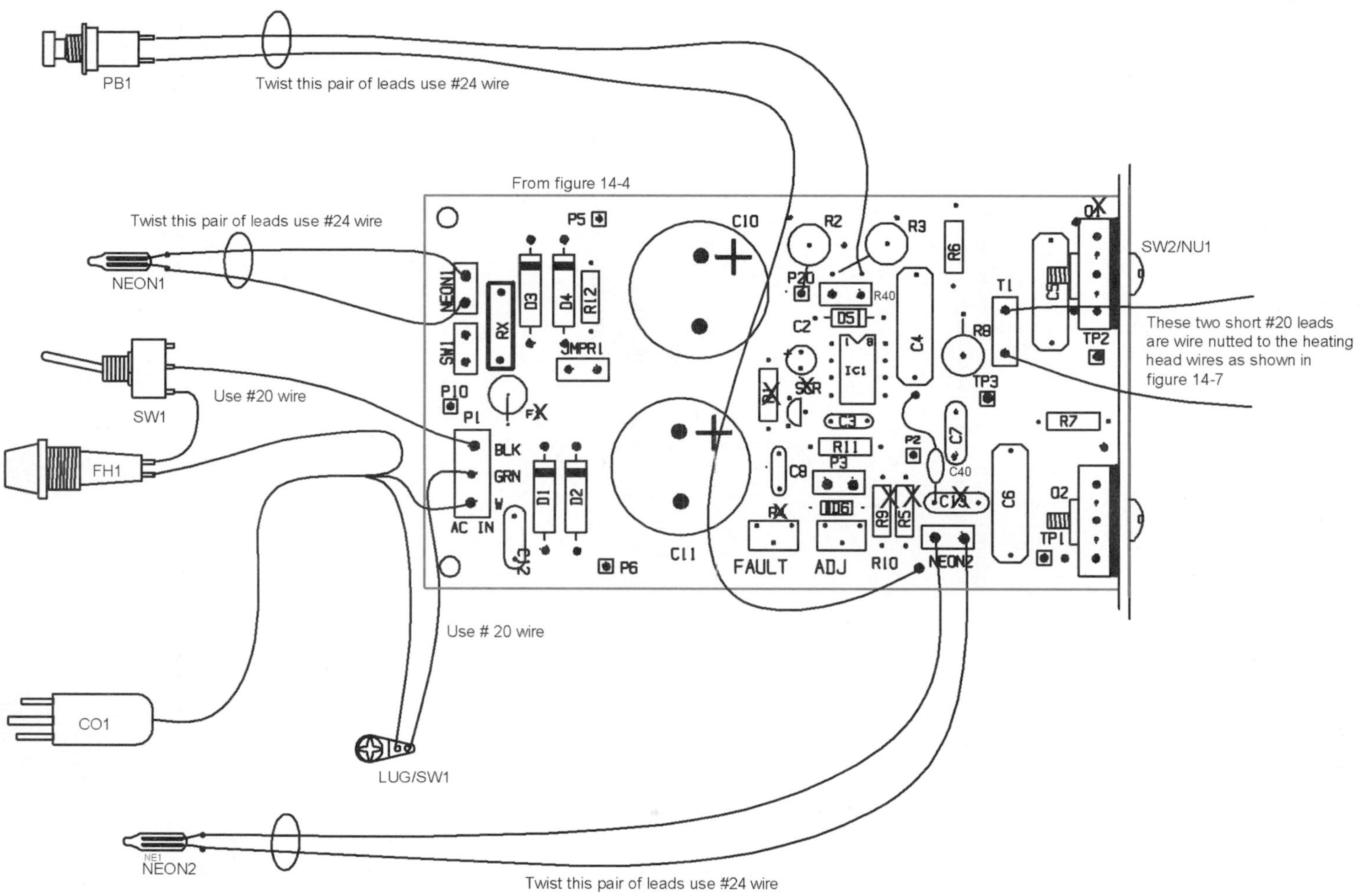
PB1
Twist this pair of leads use #24 wire
From figure 14-4
Twist this pair of leads use #24 wire
NEON1
SW1
Use #20 wire
FH1
Use # 20 wire
CO1
LUG/SW1
NE1
NEON2
Twist this pair of leads use #24 wire
SW2/NU1
These two short #20 leads are wire nutted to the heating head wires as shown in figure 14-7
P5
C10
R2
R3
R6
P20
R40
C2
D51
IC1
C4
T1
R8
TP2
TP3
R7
C3
R11
P2
C7
C40
P3
C8
D6
R9
R5
C13
C6
Q2
TP1
FAULT
ADJ
R10
NEON2
C11
P6
D1
D2
C12
AC IN
BLK
GRN
W
P1
P10
SW1
NEON1
RX
D3
D4
R12
JMPR1

Figure 14-5 *External wiring*

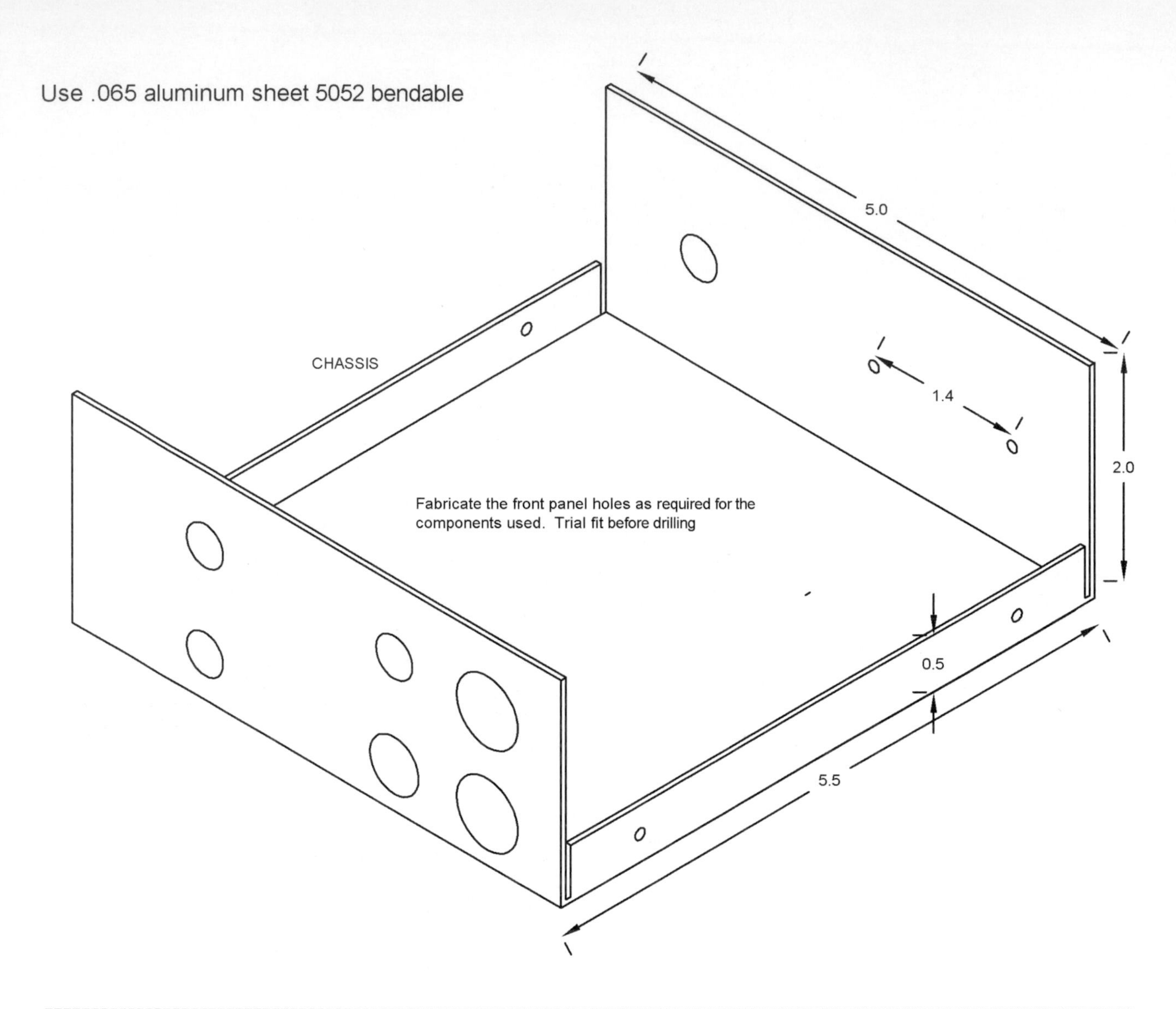

Figure 14-6 *Chassis fabrication*

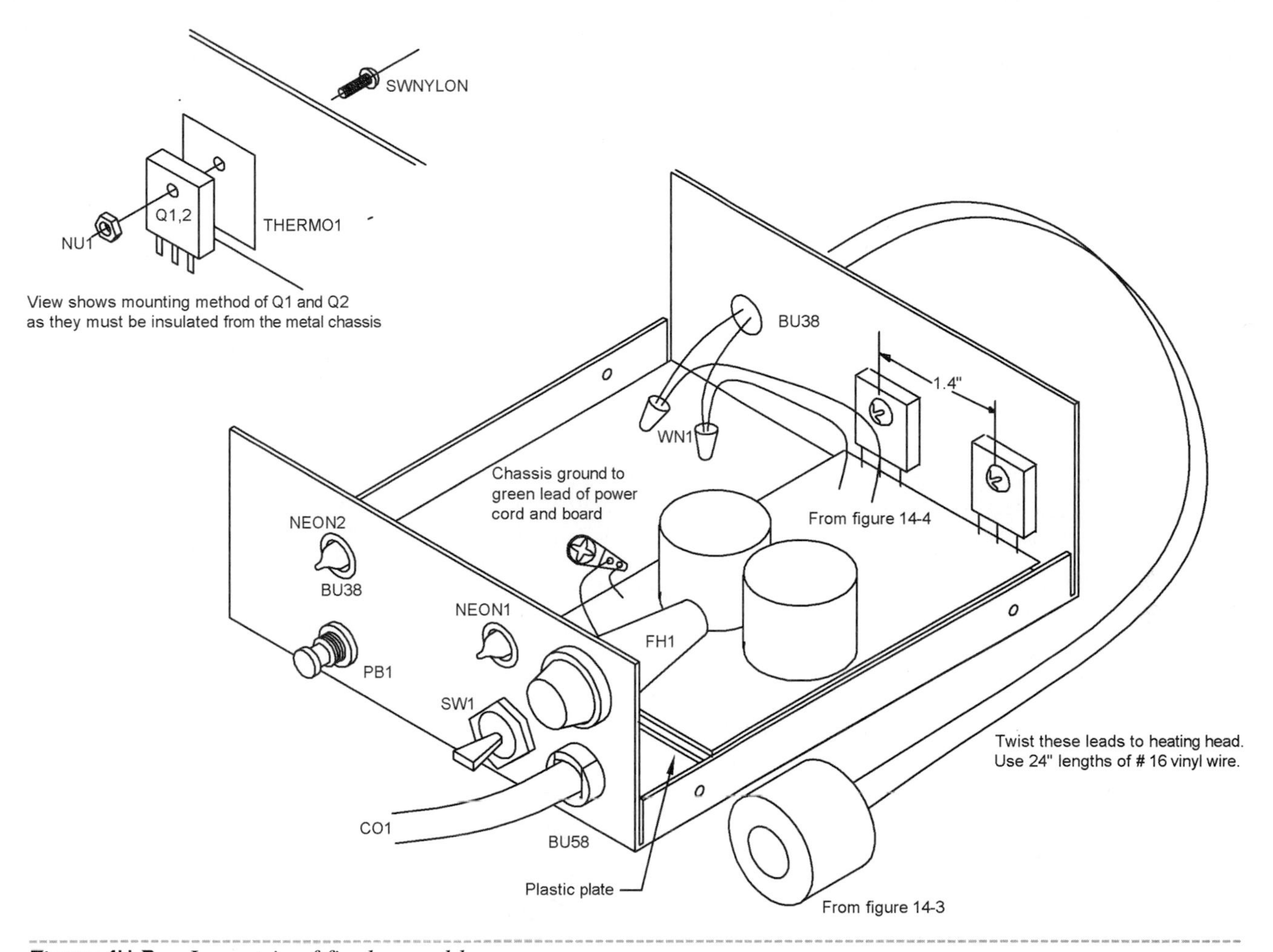

Figure 14-7 *Isometric of final assembly*

This test jig as shown prevents catastrophic damage to AC powered circuitry in the event of a fault when first assembled. Basically it uses a 60-watt lamp as a ballast, connected in series with the hot wire of the AC line. The lamp will usually allow ample current to flow for operational circuit verification yet limit short circuit current to that of the lamp should the circuit have a catastrophic fault. It is suggested that the serious experimenter construct this test jig and enclose it in a suitable container with receptacle for plugging in any AC circuit under test.

80-watt lamp
lamp socket
switch
black
green
white
male plug
Short three-wire power cord
female receptacle

Figure 14-8 *Useful test jig for line-drive circuits*

Table 14-1 Induction Heater Parts List

Ref. #	*Description*	*DB Part #*
R2, 3	Two 18K 3-watt resistor	
R4	100-ohm, 1/4-watt resistor (br-blk-br)	
R6, 7	Two 15 1/4-watt resistor (br-grn-blk)	
R8	10-ohm, 3-watts metal oxide resistor	
R10	5K vertical trimmer	
R11	1 K 1/4-watt resistor (br-blk-red)	
R12	100 K 1/4-watt resistor (br-blk-yel) or 10-ohm 3-watt metal oxide resistor	
Rx	5-amp in-rush limiter (KC006L-ND)	
C2	10-microfarad, 50-volt vertical electrolytic capacitor	
C3	.01- microfarad, 50-volts plastic capacitor (103)	
C4	.1-microfarad, 600-volt plastic capacitor	
C5, 6	Two 1.5-microfarad, 400-volt metal polyester capacitors	
C7	.0015-microfarad, 600-volt polypropylene capacitor	
C10, 11	Two 200- to 800-microfarad, 200-volt vertical electrolytic capacitors	
C12	.01-microfarad, 2-kilovolt disc capacitor	
C40	50-picofarad, 6-kilovolt ceramic capacitor	
D1, 2, 3, 4	Four 1,000-volt, 3-amp rectifiers (1N5408)	
D5	1-kilovolt, 1-amp fast diode (1N4937)	
Z1	15-volt, .5-watt zener diode (1N5245)	
Q1, 2	IRFP450/460 *metal-oxide-semi-conductor field effect transistors* (MOSFET)	
IC1	IR2153 half-bridge driver	
NEON1, 2	Two neon indicator bulbs with leads	
FH1	Fuse holder panel mount	
FUS1	2-amp slow blow 3AG fuse	
PB1	Normally closed push-button switch	
SW1	*Single pole, single throw* (SPST) toggle switch	
CO1	Three-wire line cord	
LUG1	#6 solder lug	
SO8	Eight-pin *integrated circuit* (IC) socket	
THERM1	Two thermo mounting pads for Q1 and Q2	
PCLINE	*Printed circuit board* (PCB)	DB# PCLINE
WN1	Two small wire nuts	
CHASSIS	Chassis fabricated from .065-inch aluminum sheet per Figure 14-6	
COVER	Cover fabricated from plastic sheet to fit chassis (see Figure 14-6)	
BU38	Three 3/8-inch plastic bushings	
BU58CL	5/8-inch clamp bushing for cord CO1	
SW1	Five #6 × 3/8-inch sheet metal screws for cover and LUG1	
SW2/NU1	Two 6-32 × 1/2-inch nylon screws and nuts	
WR24	24 inches of #24 vinyl stranded hookup wire	
WR20	24 inches of #20 vinyl stranded hookup wire	
WR16	48 inches of #16 vinyl stranded hookup wire	
PLASTIC	2 × 4-inch plastic insulating plate for under PC board	
HEAD12	Head with 1/2-inch clearance hole	DB# HEAD12
HEAD58	Head with 5/8-inch clearance hole	DB# HEAD58
HEAD34	Head with 3/4-inch clearance hole	DB# HEAD34
HEAD78	Head with 7/8-inch clearance hole	DB# HEAD78
HEAD10	Head with 1-inch clearance hole and head parts for those wanting to attempt assembly	DB# HEAD10
WR22LITZ	#22 LITZ of multiple strands of #36 enamel wire	
BOBBIN	1.25 inches of 5/8-inch GT tubing	

Chapter Fifteen

50-Kilovolt Laboratory DC Supply

This project, as shown in Figure 15-1, provides the high-voltage experimenter or researcher a DC source of 5 to 60 kilovolts at 5 milliamps. The system is short-circuit protected and current limited by inductive ballasting. This circuitry is excellent for high-energy capacitor charging intended for high-power applications. The charging current is fully adjustable and easily converts to a voltage source by the addition of load resistors. It is an excellent device for large gravity and lifter research, as well as normal, high-voltage laboratory functions.

This is an advanced-level project requiring electronic skills and high-voltage experience. Expect to spend $100 to $250. All parts are readily available, with specialized parts obtainable through Information Unlimited (www.amazing1.com), and they are listed in Table 15-1.

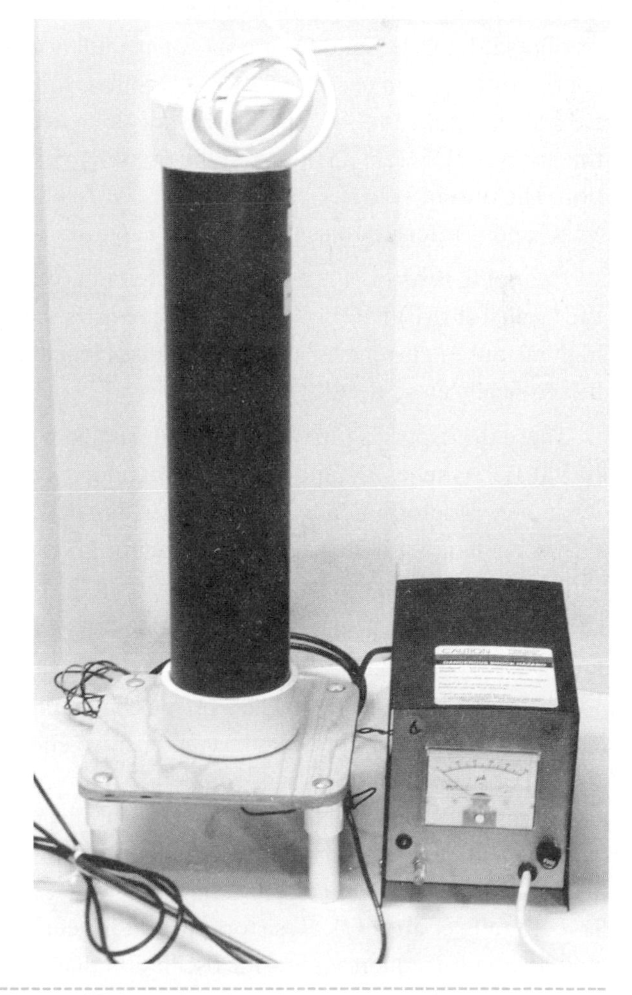

Figure 15-1 *50-Kilovolt power supply*

Basic Description

This project is shown constructed in two sections. The first section is the power conditioning and control circuitry. This is where the 115 VAC level is converted to a variable-frequency, high-voltage current source between 15 and 30 kHz. The second section is the high-voltage multiplier circuitry interconnected via a short umbilical cable from the power control section.

The multiplier section is a Cockcroft-Walton half-wave circuit consisting of 10 capacitors and diodes connected in the ladder configuration, along with a peak-current, high-voltage-limiting resistor.

Circuit Operation

The power cord (CO1) supplies 115/220 VAC from a wall receptacle where the third, green, grounding lead wire must be securely attached to the metal chassis (CHASSIS) for safety (see Figure 15-2).

Switch S1 energizes the primary circuits and fuse FUS1 protects the main circuit from any catastrophic faults. The indicator lamp NEON1 requires current-limiting resistor R12 and lights up when SW1 is turned on. The peak charging current to capacitors C10 and C11 is limited by in-rush resistor Rth. The rectifiers D1, D2, D3, and D4 are shown in a full-bridge configuration when used for 220 VAC input. D1 and D3 are used for 115 VAC operation only along with the jumper, JUMP, in a voltage-doubler configuration. The output voltage across C10 and C11 is 300 VDC and is referred to as the plus and negative rails.

Dropping resistors R2 and R3 provide voltage to the oscillator driver IC1. Capacitor C2 provides the high instant of current for the output pulses and must be physically close to IC1.

These drive pulses turn off and on the main MOSFET switches Q1 and Q2. The switching frequency is determined by the time constant of power control pot R21 and timing capacitor C3. A limit resistor R11 prevents driving the circuit beyond the intended limits. You will note that R21 is part of switch S1.

The switching circuit is in a half-bridge configuration where the MOSFET connected to the positive end (+ rail) must be driven with its source pin referenced at 150 volts above the common end (− rail). This is accomplished by biasing a bootstrap capacitor C4 through ultrafast diode D5, providing the correct DC level to fully control Q1. Resistors R6 and R7 eliminate the high-frequency parasitic oscillation that occurs as a result of rapidly switching the capacitive load of the MOSFET gates. The transition time of the switched pulses across Q1 and Q2 is slowed down by a network consisting of capacitor C7 and resistor R8. The resultant time constant limits the rate of voltage rise dv/dt rate, which could cause premature turn-on at the wrong switch, creating a catastrophic fault mode.

Capacitors C5 and C6 provide a voltage midpoint and produce the necessary storage energy to maintain the voltage level of the individual pulses. The output is taken at the junction of Q1/Q2 and C5/C6 and is a square wave fed to the primary of transformer T1, inducing the high-voltage output in the secondary winding.

A safety shutdown circuit, consisting of a spark breakover switch, fires when the voltage exceeds a certain level and is fed to isolation transformer T2. The output is now rectified by diode D6 and integrated onto capacitor C8. When this voltage exceeds a preset level controlled by trimpot R4, a trigger level now turns on the silicon switch, SCR, and crowbars the voltage to the drive chip IC1, shutting it down and disabling the high voltage until reset by power removal.

The 15,000-volt, high-frequency output is fed to the voltage multiplier stack, consisting of capacitors C14 through C21 and diodes D7 through D14. The resultant 50,000-volt DC output is in series with peak-current-limiting resistors (R17 through R20).

Assembly of the Driver Section

1. Assemble the circuit board as shown in Figure 15-3. Note to allow a length of a half-inch for the pins of Q1 and Q2. This is necessary to allow these parts to be placed flat against the chassis when mounting.

   It is also strongly suggested that you use an eight-pin integrated circuit socket (SO8) because IC1 is very sensitive to static electricity and possible errors during assembly.

   Note that if you are building from a perforated or vector circuit board, it is suggested that you use the indicated traces for the wire runs and insert components starting in the lower left-hand corner. Pay attention to the polarity of the capacitors with polarity signs and all the semiconductors.

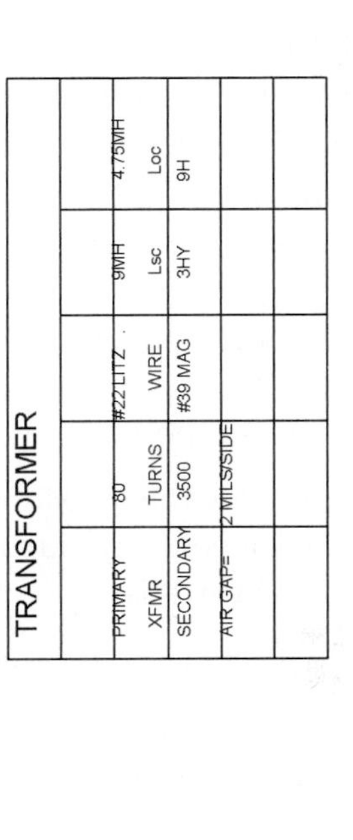

TRANSFORMER				
PRIMARY	80	#22 LITZ	9MH	4.75MH
XFMR	TURNS	WIRE	Lsc	Loc
SECONDARY	3500	#39 MAG	3HY	9H
AIR GAP=	2 MILS/SIDE			

Figure 15-2 *Circuit schematic*

Route the leads of the components as shown and solder as you go, cutting away unused wires. Attempt to use certain leads as the wire runs or use pieces of the #24 bus wire. The heavy foil runs should use the thicker
#20 bus wire as these are the high-current discharge paths.

2. Double-check the accuracy of the wiring and the quality of the solder joints. Avoid wire bridges, shorts, and close proximity to other circuit components. If a wire bridge is necessary, sleeve some insulation onto the lead to avoid any potential shorts.
3. Assemble the transformer as shown in Figure 15-4, or you may purchase this part as noted on the parts list.
4. Connect the external components as shown in Figure 15-5.
5. Fabricate the chassis as shown in Figure 15-6 from a piece of .035-inch 5052 bendable aluminum. Note to trial-fit and measure the diameters of the components before drilling the holes. Pay attention to the holes for mounting Q1 and Q2, as these must be reasonably accurate to align with the layout of the PCB. Also, there must be at least 1/8 inch of clearance between the board and chassis for sandwiching in the PLASTPL plastic plate. This prevents grounding out of the board's wiring traces.
6. Fabricate the plastic base from a 9 × 6-inch piece of 1/16-inch Lexan sheet, as shown in Figure 15-7.
7. Fabricate the parts as shown in Figure 15-8 for the over-voltage shutdown switch.
8. Attach the metal chassis to the plastic base via four screws and nuts, SW12/NU1 (see Figure 15-9).
9. Attach the shutdown switch assembly from Figure 15-8 using sheet metal screws SW1.
10. Mount the assembled PCB, as in Figure 15-3. Note that Q1 and Q2 are mounted using the insulating THERMO1 pads with the nylon screws. Mount the prewired front panel components, as in Figure 15-5.
11. Mount transformer T1 via an 8-inch tie wrap. Note the cores of T1 are wrapped with a single turn of #18 bus wire and this is connected to the ground lug electrically grounding the core.
12. Mount transformer T2 using a piece of double-sided foam tape or glue it in place.
13. Mount the remaining hardware, wiring lugs, and so on.
14. Proceed to wire everything as shown in Figure 15-10. Use twist pairs where noted or tie-wrap to neaten its appearance (see Figure 15-11).
15. Double-check the wiring for any obvious errors. You should have a finished assembly resembling Figure 15-12. The unit is now ready for the basic pretest.

Pretesting the Driver Section

This step will require basic electronic laboratory equipment including a 60 MHz oscilloscope

1. Assembly the test jig as shown in Chapter 14 Figure 14-8. This fixture is a very useful for any 115 VAC project up to 300 watts.

Caution: The circuit commonly noted as −RAIL is electrically above ground by over 150 volts. It is strongly recommended that you obtain a 200-watt isolation transformer before proceeding and that you test this or similar direct-line-powered AC circuits. Experienced people working on dry, wooden floors may choose to unground their test equipment. This still can be a dangerous shock hazard!

2. Connect the scope to the test points as noted in the Figure 15-2 schematic or as shown in Figure 15-4. The lead of R8 is shown connected to the junction of the drain and the source of Q1 and Q2 respectively.
3. Short out the output leads.

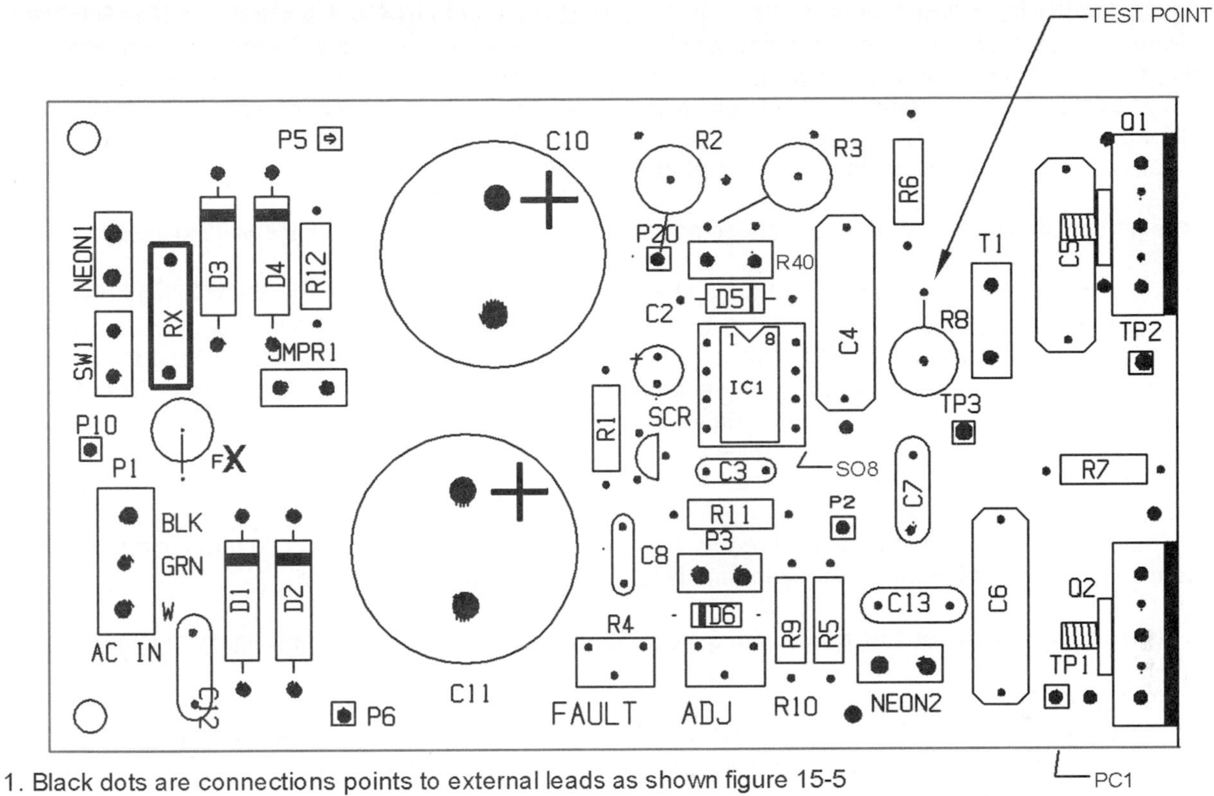

1. Black dots are connections points to external leads as shown figure 15-5

2. The PC board as shown is used in several projects and may not use all the components as indicated.

3. This circuit uses a 8 pin socket for IC1 as it is prone to failing if there are circuit faults

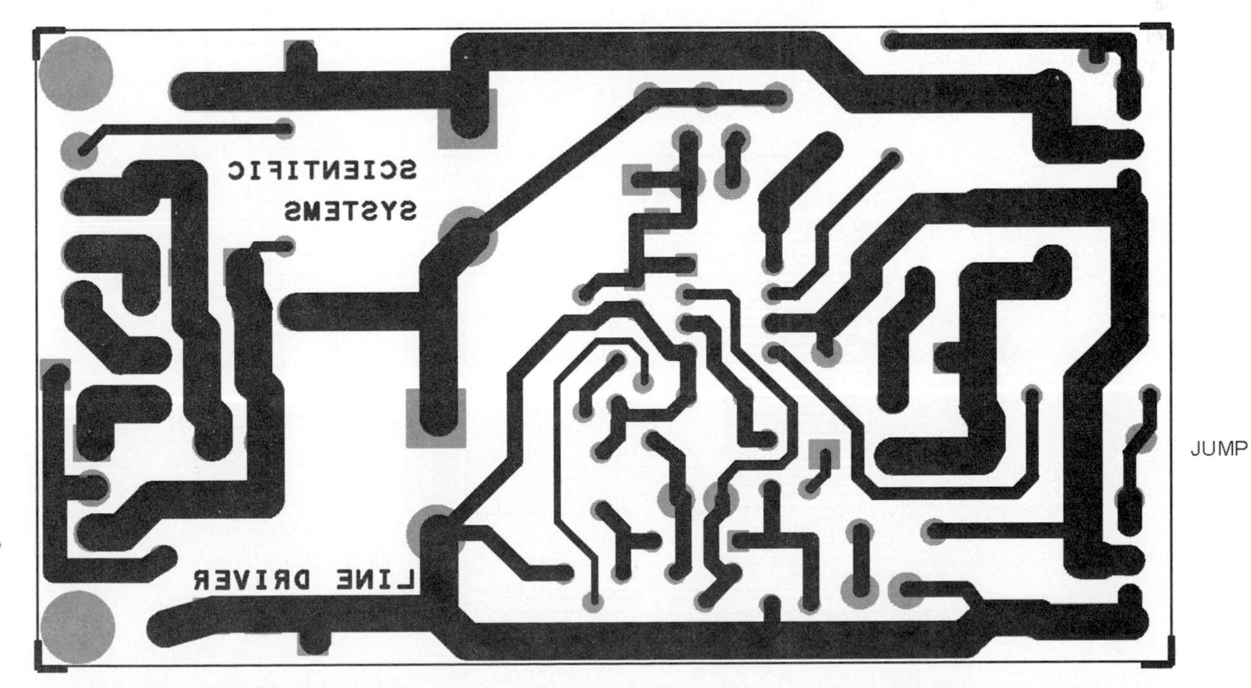

Assembly of the board is shown using a printed circuit board available through www.amazing1.com Builders may use a vector board and use the component leads to duplicate the foil traces as shown.

Figure 15-3 *Assembly of the circuit board*

The special ferrite transformer used in this project is available ready to use and is listed on table 15-1. The transformer is designed with a certain amount of leakage reactance between the primary and secondary winding to allow limited short circuit current. This approach allows charging capacitors without the use of lossy resistors. While this is an advantage in circuit protection is has the disadvantage of requiring transistor switches (FETS) that now must switch the reactive leakage current in addition to the real load current. We show our circuit using high current switches that alleviate this problem.

The design specifications are shown for those advanced builders desiring to attempt winding their own.

1. The secondary is wound with 3500 turns of #39 heavy insulated magnet wire. Start with a bobbin that will fit over the cores and wind about a ¾" layer evenly and without cross overs. The layer should consist of about 200 turns. Note to exit the start of your winding as this is the common connection. Tape the winding using 2 mil polypropylene tape and wind the second layer until you have a total of 3500 turns. This equates out to 18 layers!! (good luck doing this by hand) You now must pot the assembly and this will require a cup to hold the windings and out gassing using a vacuum pump on both the winding assembly and the potting material. We use a two part silicon rubber GE#627.

2. The primary is far easier to do and requires another bobbin now only winding 80 turns of #22 Litz high frequency wire. This will require two layers and is taped together using mylar tape.

3. Assemble the core set and space with 5 mil (.005)" shims. Tape the assembly tightly together.

Note: This type of transformer is often incorrectly referred to as a "flyback transformer". A flyback unit physically looks similar but this is where it ends. A fly back unit stores energy in the core that is mainly across the air gap as this is a very high reluctance compared to the actual ferrite part of the core. This energy builds up as long as the transistor is on and then "flies back" when the transistor switches off.

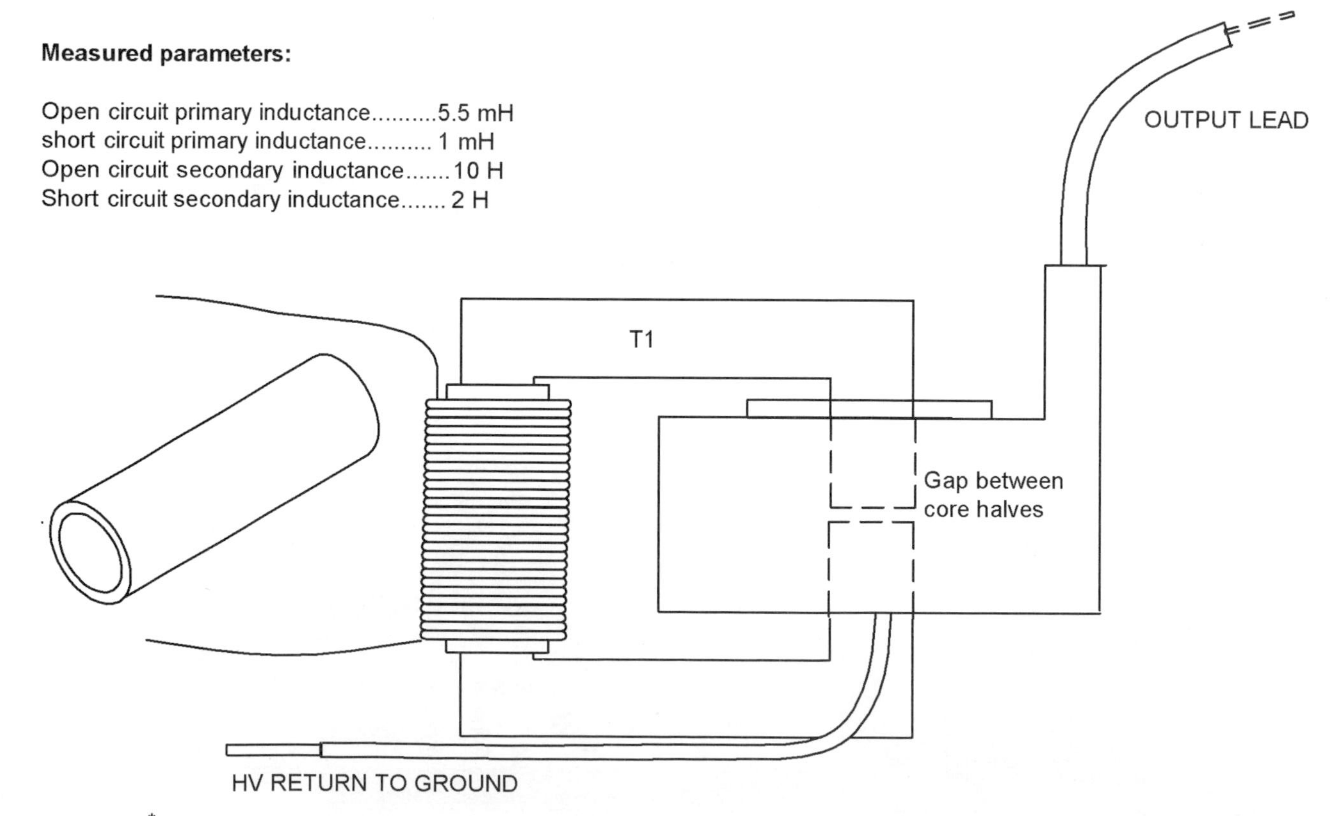

*#18 solid magnet wire can be used however high frequency LITZ wire will give a slight improvement. You can make this wire by obtaining 6 pieces of #26 magnet wire and twist together as a single wire.

Figure 15-4 *Assembly of transformer*

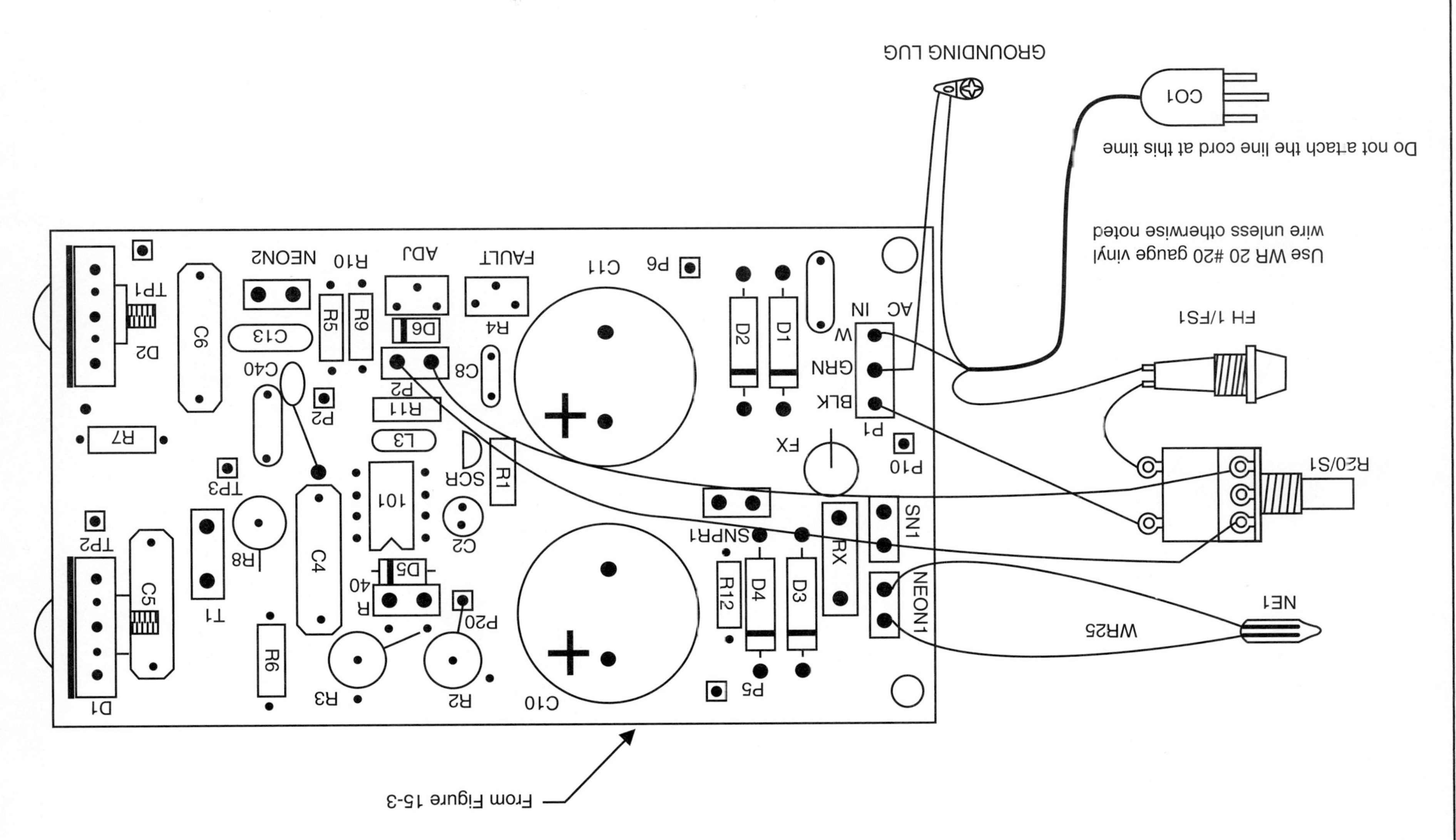

Figure 15-5 *External wiring*

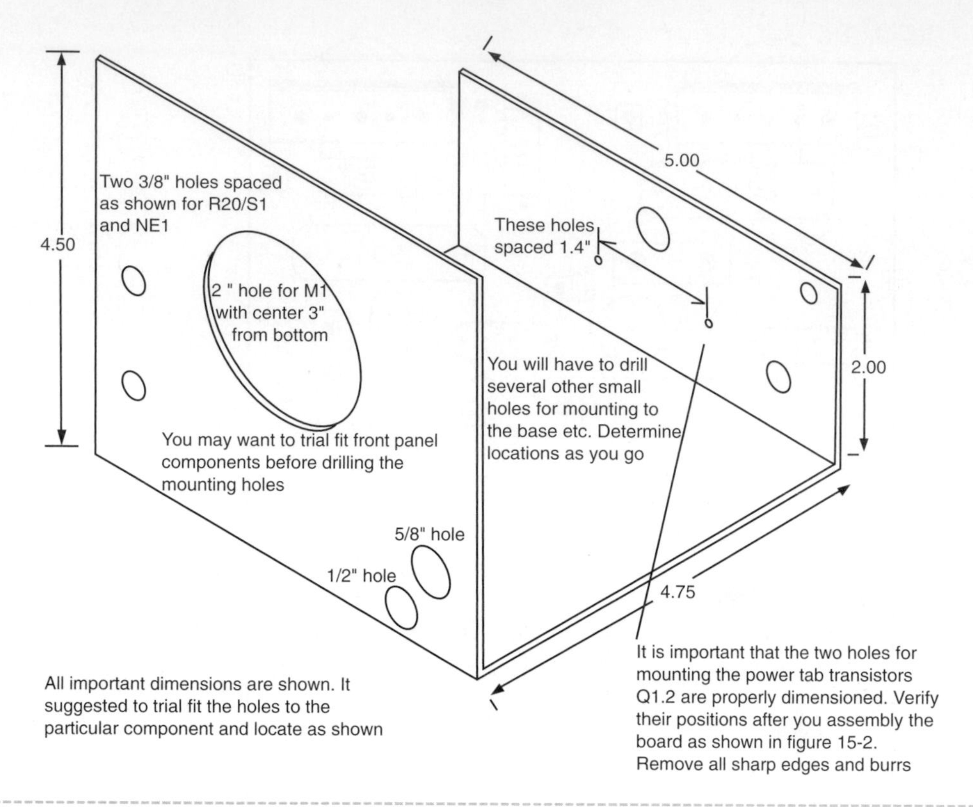

Figure 15-6 *Chassis fabrication*

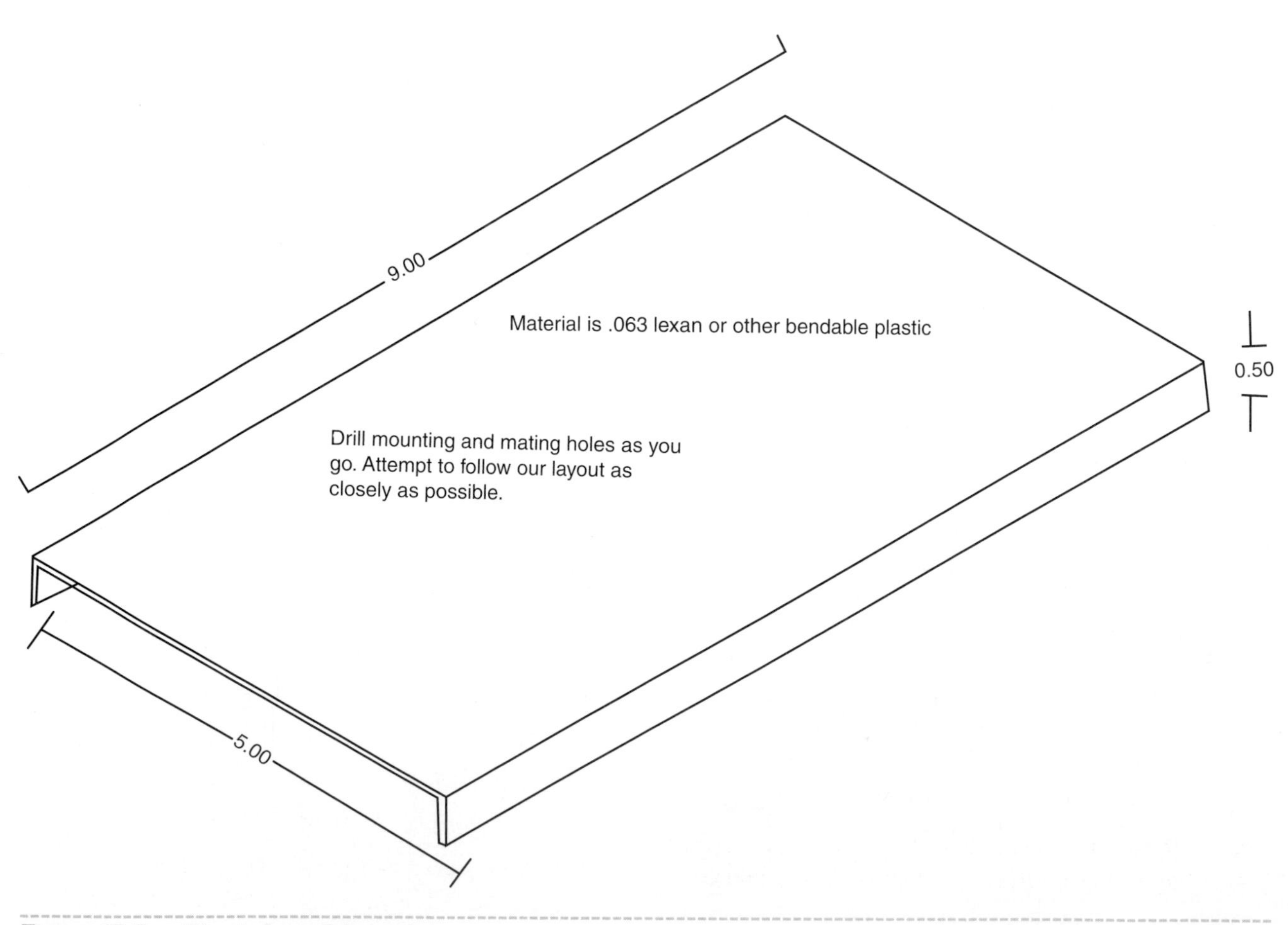

Figure 15-7 *Plastic base fabrication*

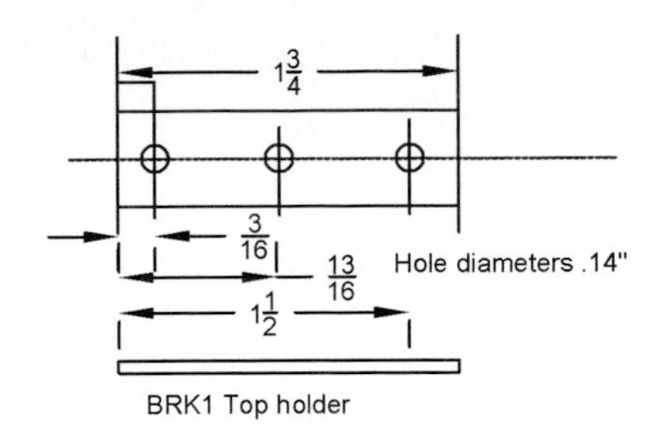

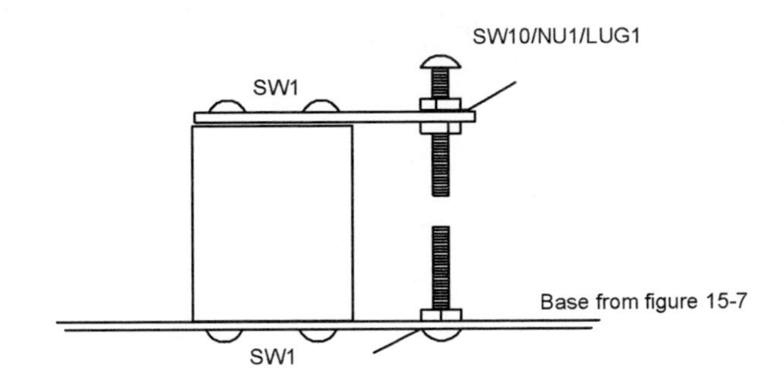

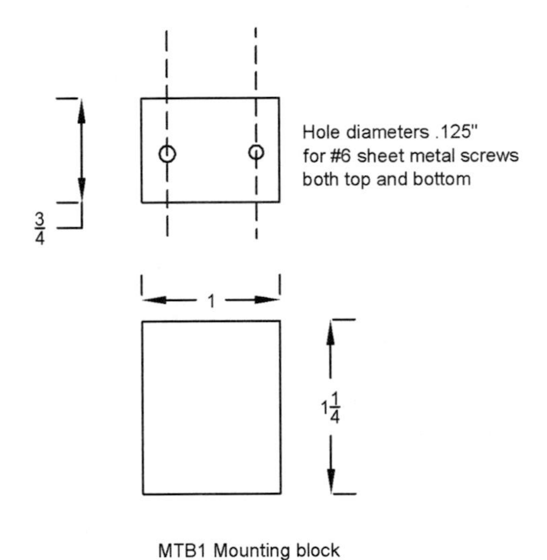

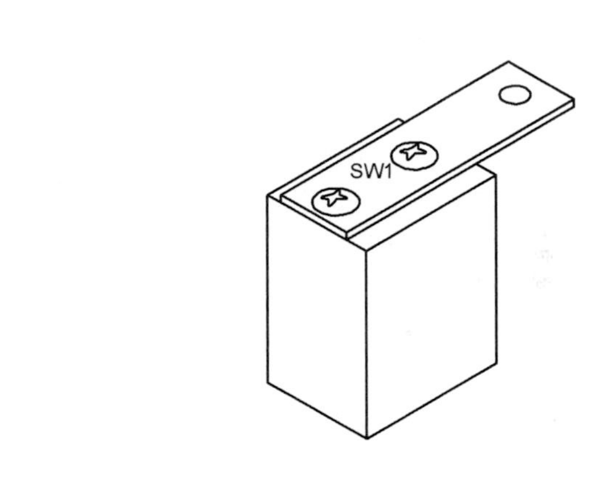

Figure 15-8 *Creating the shutdown switch*

4. Verify the main power switch S1 is off and insert a 2-amp fuse into FH1. Connect CO1 to the female end of the cord on the test jig and adjust both trimpots to midrange. Verify the test jig is not shorting the ballast lamp and plug the jig into the output of the isolation transformer. Quickly turn on S1 and note NEON1 lighting and the ballast lamp not lighting. If it brightly lights up, you must troubleshoot and find your error before proceeding. Otherwise, note a picture-perfect square wave on the scope with the lamp only dimly lighting. Adjust R21 full clockwise and adjust trimpot R10 for an 80-usec pulse period. This corresponds to 12.5 kHz. Experienced builders may wish to monitor the current to T1 using a current probe attachment to the scope. This waveshape should be another picture-perfect saw tooth of 8 amps maximum. This is a reactive current and is not reflected as an input current due to the phase angle.

5. Turn the unit off and unshort the output leads. Adjust the spark switch screws 1/8 of an inch.

6. Turn on R21/S1 and very quickly rotate clockwise, noting the unit shutting down as apparent by the absence of the waveshape at the test point. If this does not occur instantly, you will need to readjust the switch gap or trimpot R4 clockwise until the shutdown activates. If it does not, you must troubleshoot the circuit. This completes the driver section of this system.

Assembly of the Multiplier Section

Assembly of this section requires solder joints to be large, smooth, and globular. This is contrary to normal soldering practices but is necessary to reduce hv leakage.

1. Fabricate the MULTIBOARD board and two BRK2 brackets, as shown in Figure 15-13.

2. Assemble the diodes and components as shown in Figure 15-14, but do not solder at

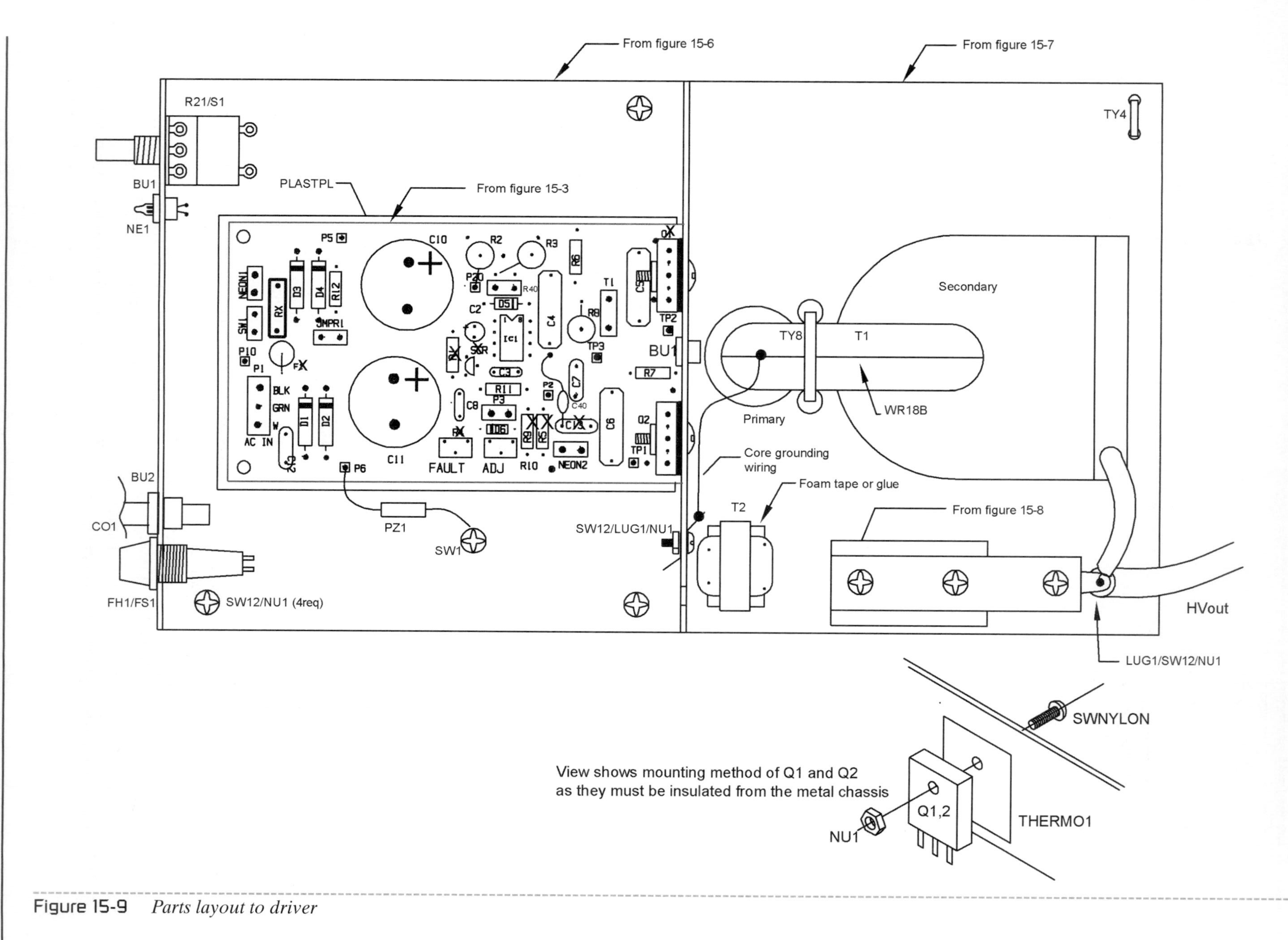

Figure 15-9 *Parts layout to driver*

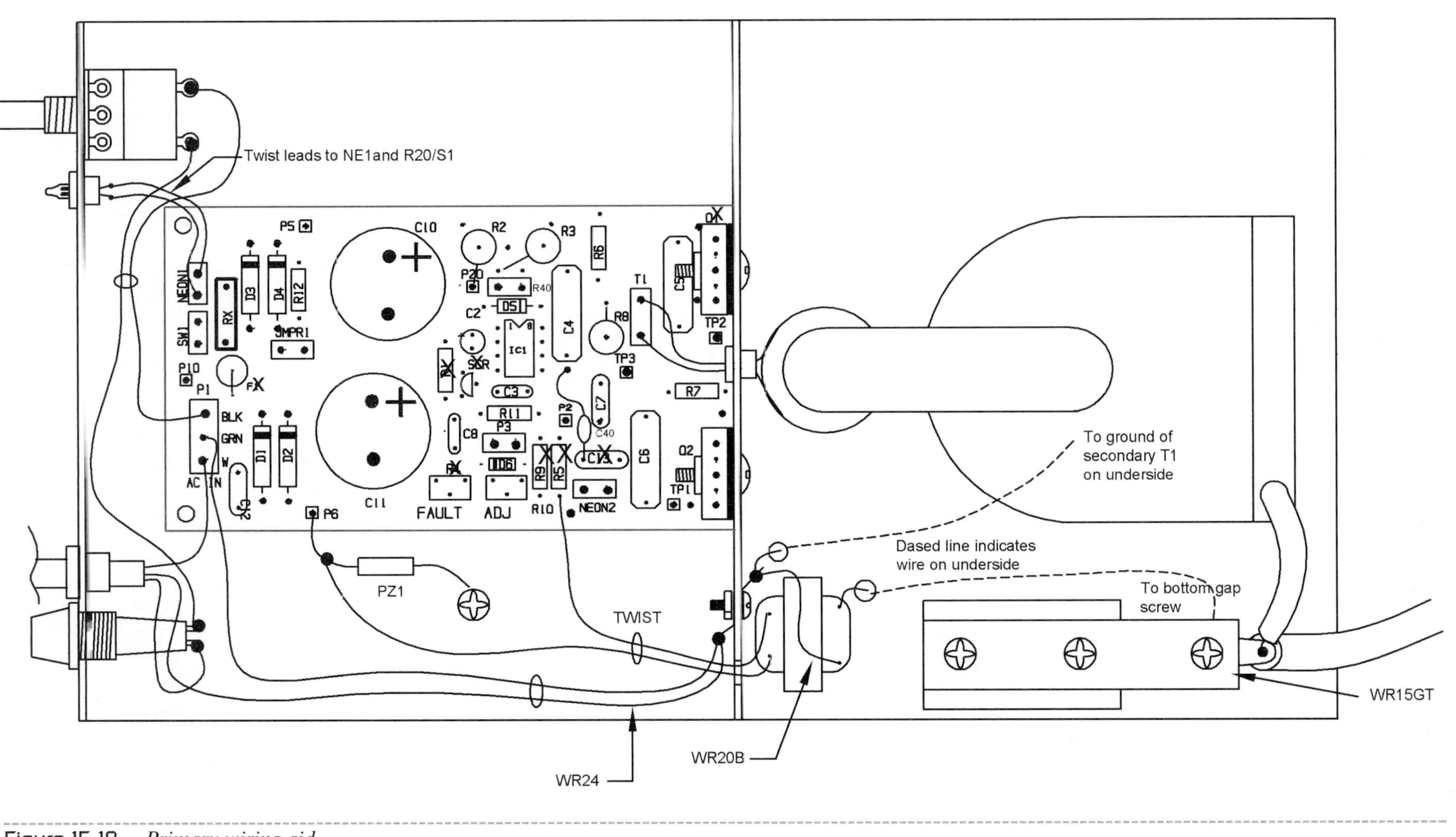

Figure 15-10 *Primary wiring aid*

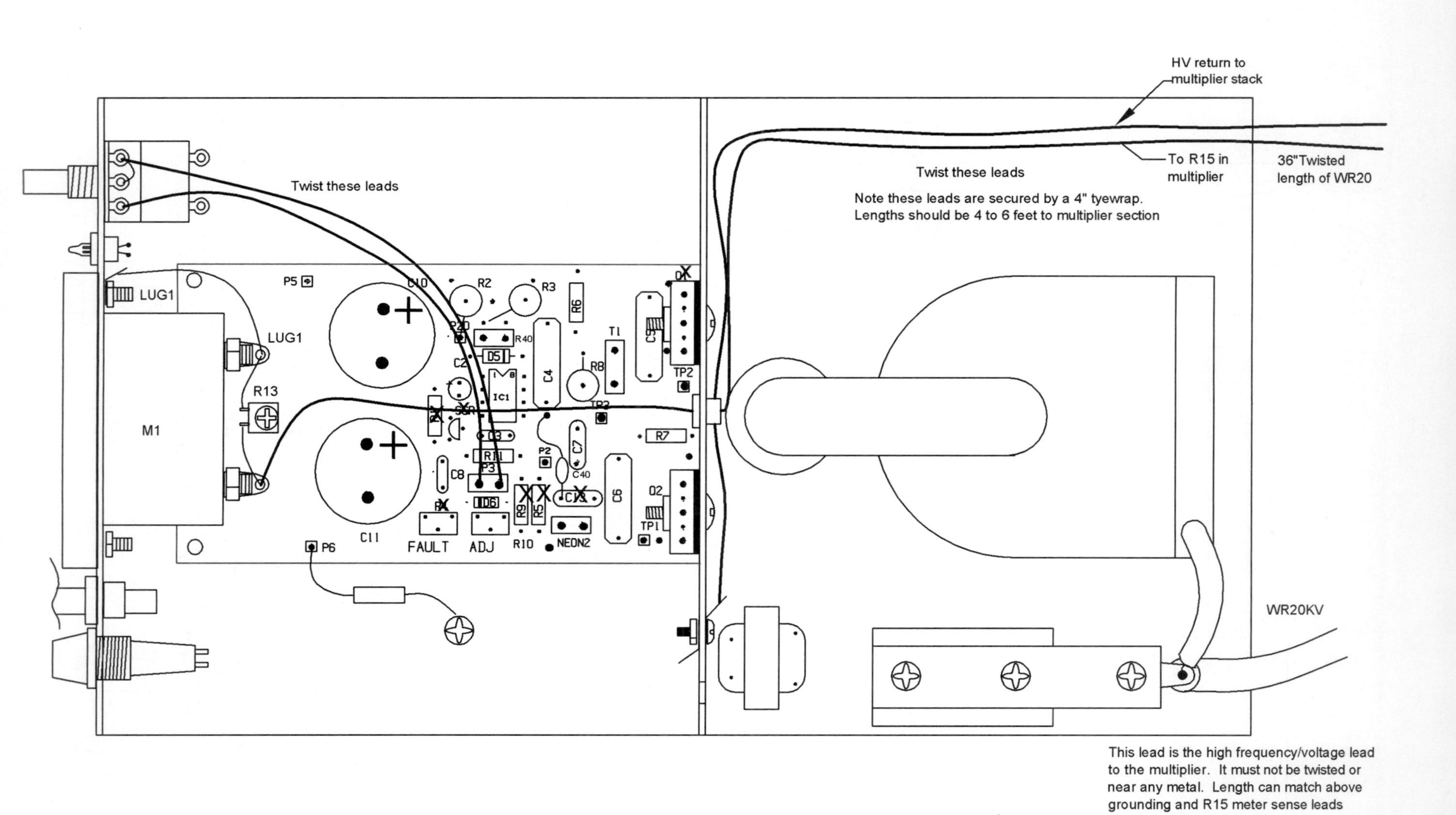

Figure 15-11 *Secondary wiring aid*

this time. Assemble the resistor tube for R17-20 as shown.

3. Mount the capacitors C14 through C21 and the remaining components as shown in Figure 15-15. Insert the BAF1 baffle plate to insulate and separate the left and right rows.

   Proceed to solder all the joints, noting that the capacitor and diode connections should be smooth, round balls of solder 1/8 to 3/16 of an inch wide. This may be tricky with all the leads but should be done to avoid leakage, especially the last four capacitors.

4. Sleeve resistor R14 into a plastic sleeve, as shown in Figure 15-16.

5. The final assembly is shown in Figure 15-17. Note you have a choice of an output lead or a spherical terminal.

Final Testing

This step requires basic electronic laboratory equipment including a 60 MHz oscilloscope along with a high-voltage meter with a 40 kilovolt dc rating. You will need a high-voltage load resistor to verify output power capabilities.

1. Connect the driver and multiplier sections together and separate them by at least 3 feet. Keep the high-voltage wire from the driver away from other conductive objects. Follow Figure 15-11 for the driver and Figure 15-17 for the multiplier interconnecting wiring.

2. Obtain a load resistor of around 10 megohms of at least 100 watts. This resistor must be able to support at least 50,000 volts. Our laboratory load resistor consists of 180 47k, 1-watt resistors connected in series and then inserted into

Figure 15-12 *Driver section*

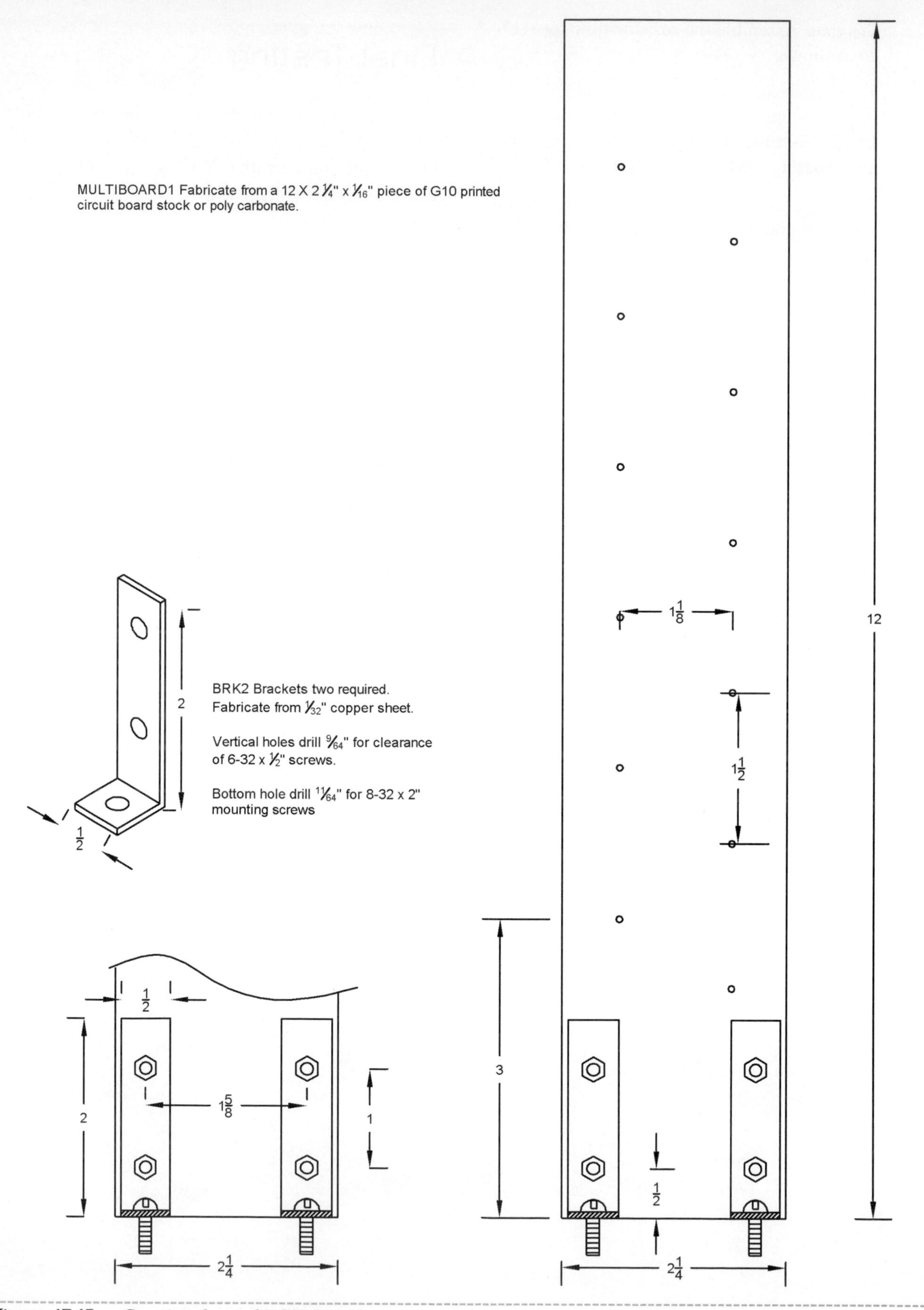

Figure 15-13 *Creating the multiplier board*

some polyethylene tubing, along with the terminating high-voltage leads. The tubing is then coiled, maintaining adequate clearance of the terminating leads. This is a very tedious assembly but works well for powers up to 300 watts at 100 kilovolts. Connect the load to the output and common leads.

3. Assuming that the driver is properly functioning, proceed to apply power and slowly turn the power control pot R21 to full clockwise, noting a line current draw of about 2 amps. Very carefully measure around 35 kilovolts at the output using a 40-kilovolt probe such as an HV44 or similar. The load resistors should start to get warm.

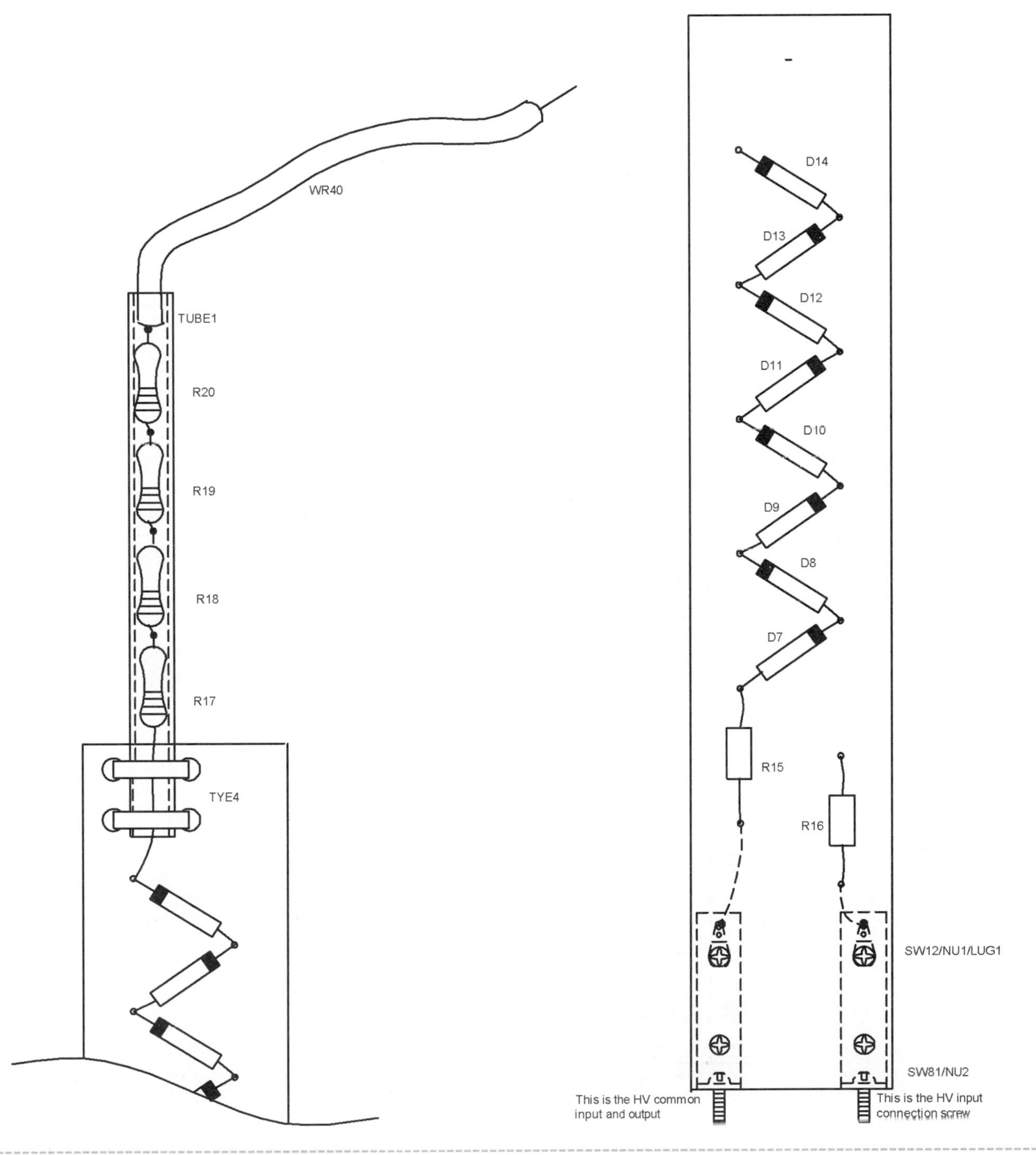

Figure 15-14 *Diode layout and wiring*

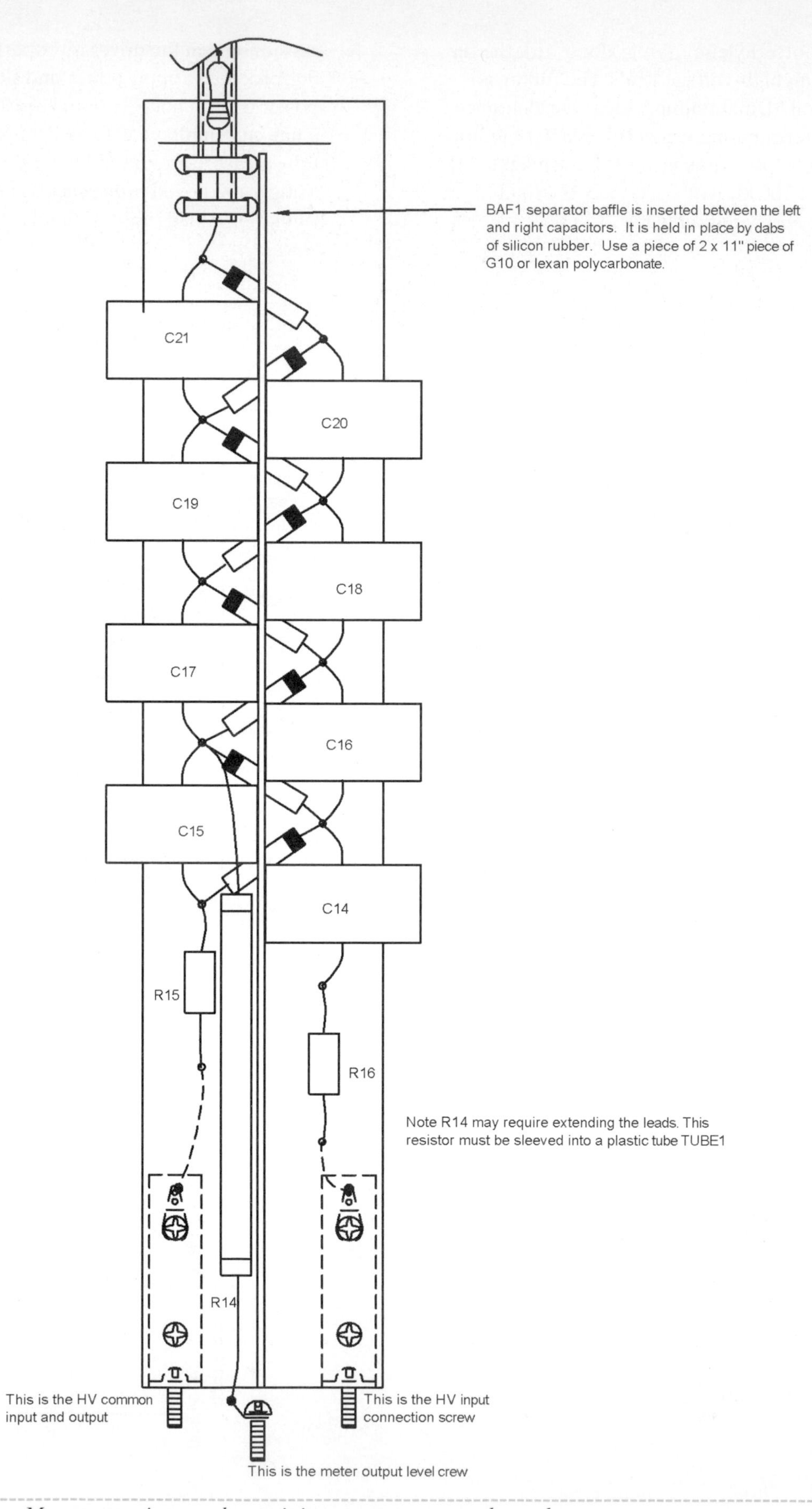

Figure 15-15 *Mount capacitors and remaining components as shown here.*

Figure 15-16 *Multiplier internal photo*

4. Allow to it run for several minutes and note that Q1 and Q2 are only slightly warm to the touch.
5. Reverify the operation of the shutdown circuit to prevent damage to the circuit components in the event of over-volting.

Applications

This unit is excellent to use for powering large anti-gravity lifters because it is not damaged by short-circuiting the output. It can also be used for charging capacitors up to 50,000 volts or as a voltage source with a load resistor of 20 megohms. The charging current is 4 to 5 milliamps and charges as a current source.

DANGER! Improper contact or use of high-energy capacitors can result in death by electrocution or injury due to explosive discharges.

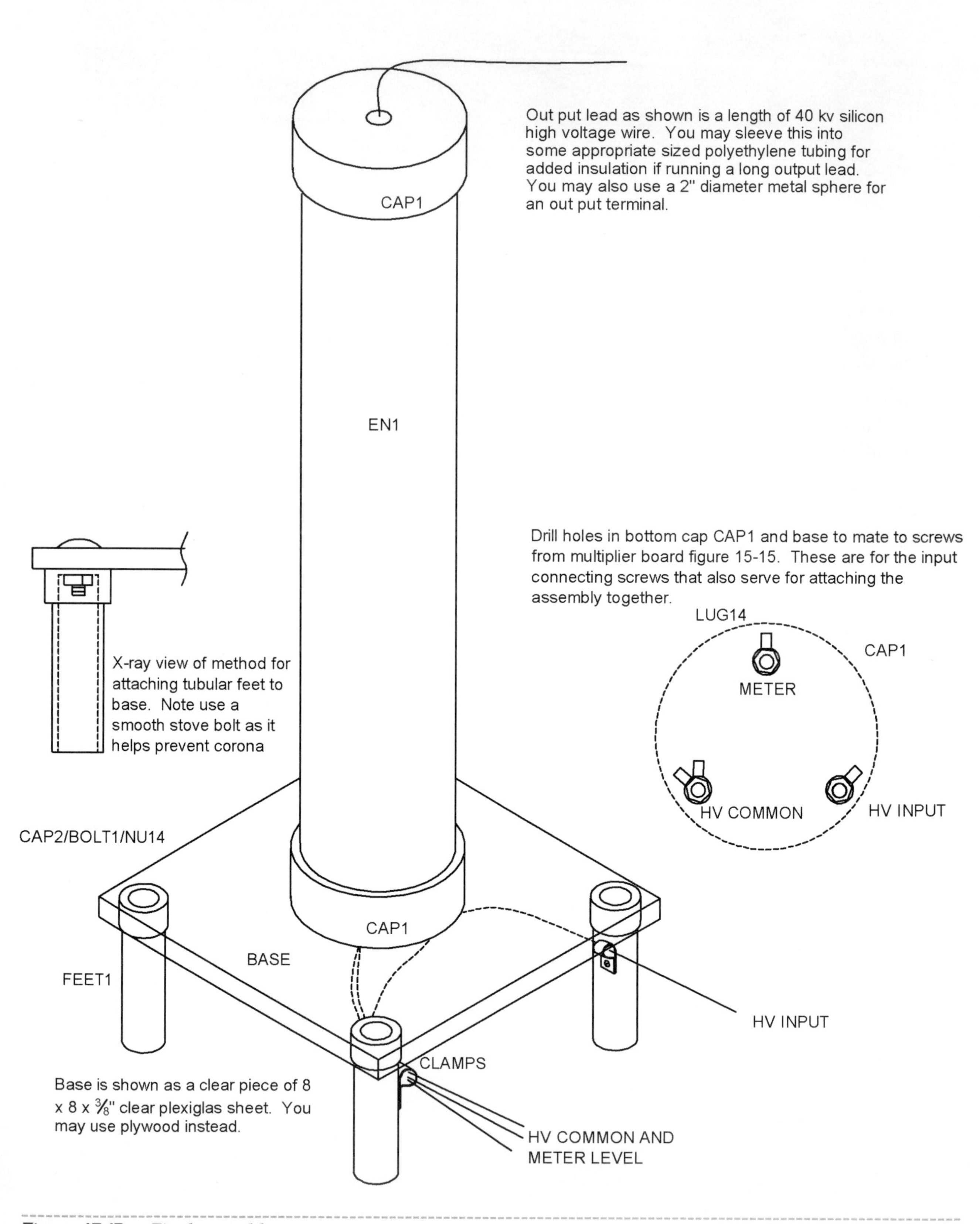

Figure 15-17 *Final assembly*

Table 15-1 50-Kilovolt Power Supply

Ref. #	*Description*	*DB Part #*
R1	100-ohm, 1/4-watt resistor (br-blk-br)	
R2, 3	Two 18K, 3-watt resistors	
R4, 13	Two 2K vertical trimmers	
R6, 7	Two 15 1/4-watt resistors (br-grn-blk)	
R8	10-ohm, 3-watt metal oxide resistor	
R10	5K vertical trimmer	
R5, 11	Two 1K 1/4-watt resistors (br-blk-red)	
R9, 12	Two 100K 1/4-watt resistors (br-blk-yel)	(KC006-LND)
Rth	limiter 5-amp in-rush	
R15, 16	Two 12-ohm, 2-watt carbon resistors (br-red-blk)	
R14	100 meg, 20-kilovolt tiger resistor	
R17–20	4 100K, 2-watt ceramic resistor	DB# 100KCER
R21/S1	10K pot and 115 VAC switch	
C2	10-microfarad, 50-volt vertical electrolytic capacitor	
C3	.01-microfarad, 50-volt plastic capacitor (103)	
C4	.1-microfarad, 600-volt plastic capacitor	
C5, 6	Two 1.5-microfarad, 400-volt metal polyester capacitors	
C7	.0015-microfarad, 600-volt polypropylene capacitor	
C8	.47-microfarad, 50-volt plastic capacitor	
C10, 11	Two 200- to 800-microfarad, 200-volt vertical electrolytic capacitors	
C12, 13	Two .01-microfarad, 2-kilovolt disc capacitor	
C14–21	Eight .0047-microfarad, 20-kilovolt ceramic capacitors	DB#.0047/20 KV
PZ1	420-volt suppressor	
D1, D2, D3, D4	Four 1,000-volt, 3-amp rectifier (1N5408)	
D5	1-kilovolt, 1-amp fast diode (1N4937)	
D6	1N914 small-signal diode	
D7–14	Eight 30-kilovolt, 5-milliamp fast-recovery diodes	DB# 3VG30
SCR	EC103 A, 12 uamp sensitive gate *silicon-controlled rectifier* (SCR)	
Q1, 2	IRFP450/460 *metal-oxide-semiconductor field effect transistor* (MOSFET)	
IC1	IR2153 half-bridge driver	
T1	High-volt transformer assembled as shown in Figure 15-4	DB# FLYPVM
T2	8/500-ohm audio transformer	
SO8	Eight-pin DIP socket	
NE1	Neon indicator bulb with leads	
FH1	Fuse holder panel mount	
FUS1	3-amp slow blow 3AG fuse	
METER	50 uamp, 3-panel meter	DB# METER50L
CO1	Three-wire line cord	
LUG1	Eight #6 solder lugs	
LUG14	Three 1/4-20 lugs	
THERM1	Two thermo mounting pads for Q1 and Q2	
PCLINE	*Printed circuit board* (PCB)	DB# PCLINE
BU38	Three 3/8-inch plastic bushings	
BU58CL	5/8-inch clamp bushing for cord CO1	
WR15GTO	4 feet of 15-kilovolt GTO flexible wire	
WR40KV	12 inches of 40-kilovolt high-voltage silicon wire	DB# WR40KV
WR20	10 feet of #20 vinyl hookup wire	
WR24	12 inches of # 24 vinyl hookup wire	
WR20B	12 inches of # 20 bus wire	
WR18B	24 inches of #18 bus wire	

Table 15-1 *continued*

Fabrication

Ref. #	*Description*	*DB Part #*
CHASSIS	Chassis created from .065-inch aluminum as per Figure 15-6	
COVER	Cover created from plastic sheet to fit chassis	
BASEPLATE	Base plate created from .065-inch Lexan per Figure 15-7	
BRK1	Bracket created from .035-inch copper per Figure 15-8	
BRK2	Two brackets created from .035-inch copper per Figure 15-13	
MTB1	Mounting block created from .75-inch PVC plastic	
PLASTPL	4.5 × 2.5 piece of insulating plastic, shown in Figure 15-9	
MULTIBOARD1	Multiplier board created from .065-inch Lexan per Figure 15-13	
BASE	Base created from 8 × 8 × 3/8-inch Plexiglas per Figure 15-17	
BAF1	Baffle separator created from .065-inch Lexan per Figure 15-15, available from a hardware store	
SW1	Five #6 1/4-inch sheet metal screws for cover and MTB1	
SWNYLON	Two 6-32 1/2-inch nylon screws	
SW12	Ten 6-32 1/2-inch screws	
SW10	Two 6-32 × 1-inch screws	
SW81	Three 8-32 × 2-inch screws	
BOLT1	Four 1/4-20 × 1 1/2-inch stove bolts	
NU1	Sixteen 6-32 keep nuts	
NU2	Six 8-32 keep nuts	
NU14	Four 14-20 hex nuts	
CAP1	Two 3-inch PVC flat caps (GENOVA# 70153)	
CAP2	Four 1/2-inch PVC flat caps (GENOVA# 30155)	
FEET1	Four pieces of 1/2 × 3-inch PVC tubing	
EN1	3 × 15-inch PVC tubing	
TUBE1	10 inches of 3/8 ID × 1/2 OD flexible plastic tubing	

Chapter Sixteen

Magnetic High-Impact Cannon

This very advanced project (shown in Figure 16-1) is intended for the experienced experimenter and researcher in the field of high-voltage and high-energy circuitry. The magnetic cannon accelerates a projectile at high velocities with considerable kinetic energy; this is accomplished strictly from a magnetic pulse. Not only is this project a dangerous electrical device, it is also kinetically hazardous, as the accelerated projectile can cause serious injury and even death. The same consideration and respect given to a firearm must be given to this project. A video demonstration of this device blasting large holes through a wall can be seen at www.amazing1.com.

Figure 16-1 *The magnetic cannon*

Under the correct supervision, the magnetic cannon can provide a high-action science project demonstrating several important electrical laws that involve electromagnetic reactions. Lenz's law and the Lorentz JXB forces are clearly utilized in this device. The system as shown can provide the experienced hobbyist hours of fun and entertainment by experimenting with its effect (its impact) on various objects.

This is an advanced level project requiring electronic skills and high-voltage experience. Expect to spend $300 to $500 unless you have access to the surplus market. All parts are readily available and any specialized parts are available through Information Unlimited (www.amazing1.com) and are listed in the parts list at the end of the chapter.

Theory of Operation

A nonmagnetic conductor such as aluminum is placed in a time-variant magnetic field. Induced currents in the aluminum produce currents that, in turn, produce opposing magnetic fields thereby causing a moment of acceleration of the aluminum piece. In this project the aluminum piece is in the form of a large flat washer that is the projectile. The aluminum projectile has a hole in its center with a mandrel guide to keep it traveling in a straight line.

The heart of this project is the accelerating coil. It is what couples the energy (the electrical current) to the aluminum ring (the projectile). Optimum efficiency is dependent on the coupling to achieve maximum projectile kinetic energy. This requires a minimal proximity of the projectile to the coil along with minimal air gaps (such as the spacing between the wires of the coil and the geometry of the coil as related to the dimensional format of the projectile). Also the inductive value of the coil is related to the storage capacity to provide the current rise over a time period that is dependent on the physical parameters of the projectile. The associated second-order differential equations that mathematically determine these parameters are beyond the scope of this material. Although the mathematical purist will find a deviation from maximum efficiency by using different materials, practically speaking, there are materials of different sizes and materials that offer a cost-effective compromise. An example is the use of square magnetic wire in place of conventional round stock wire. Using square magnetic wire for the coil, when properly wound, will provide more kinetic energy to the projectile due to reduced air space (reluctance); however it is quite difficult to wind and usually requires purchasing a significant amount at a healthy price.

The shape and timing parameters of the coil-magnetizing pulse must be related to the projectile to achieve optimum efficiency. A current rise that is too fast will cause slippage (magnetic cavitation). Note that one definite disadvantage of this method of acceleration is that the projectile is influenced over a very short distance. Velocity, now being the square root of the acceleration times the distance, is limited. To achieve high resultant velocities requires a very high moment of acceleration. Projectiles using conventional explosives current detonators and boosters could be prone to sympathetic initiation by these high g accelerating forces at the time of launching.

Circuit Theory

This project as shown in Figure 16-2 is constructed operating from 115 vac household current. It also can be built to operate from 12 volts or built-in batteries.

A high-voltage current-limited 60 Hz transformer (T30) steps up the 115 vac to 6,500 vac and is rectified by high-voltage diodes (D31 through D34). DC current now charges the energy storage capacitors (C30) through the isolation resistor (R30) to a programmable value, as selected by the operator. It is this stored energy that is discharged into the accelerator coil (L1) as it is switched by the spark gap (GAP1). Once switched into the coil, the now rapidly rising current wave induces a current into the aluminum ring projectile (PROJ1). It is this induced current that now generates a very high magnetic moment and repels the initial field in the accelerator coil, causing a moment of intense accelerative forces. Those not familiar with this concept often ask why an aluminum ring? The answer is that a magnetic material would now be attracted and thus would neutralize the repulsion.

The initializing of the circuit commences by turning on the key switch (S2). This switch is intended to keep unauthorized personnel from powering up the system. The key switch controls 12 volts of DC power necessary to energize the relay (RE1), with normally open contacts controlling power to the high-voltage transformer (T30).

The controlling system, as shown, consists of momentary pushbutton switches that start the charging action (S4) and can stop this action via switch (S5). Triggering the momentary pushbutton switch (S3) supplies power to the trigger module (TRIG10) and firing trigger gap (GAP2), thereby initiating the main gap GAP2 and switching the energy from the storage capacitors C30 into the accelerator coil L1. Charging voltage to C30 is controlled by the potentiometer R14. Once set, the voltage will maintain its preset level until triggered or readjusted. Meter M1 indicates the charge voltage and is calibrated by the trimpot (R16).

The transformer (T1) supplies 12 volts of AC that is rectified by diodes (D1 through D4), filtered by capacitor C4, and regulated by zener diodes (Z1 and Z2) to 12 volts dc in order to power the control circuits. The indicator *light-emitting diode* (LED1) illuminates when the key switch is energized. The second indicator LED (LED2) illuminates when charging of C30 is taking place. A buzzer (BUZ1) sounds whenever there is a charge voltage on C30. This is a safety

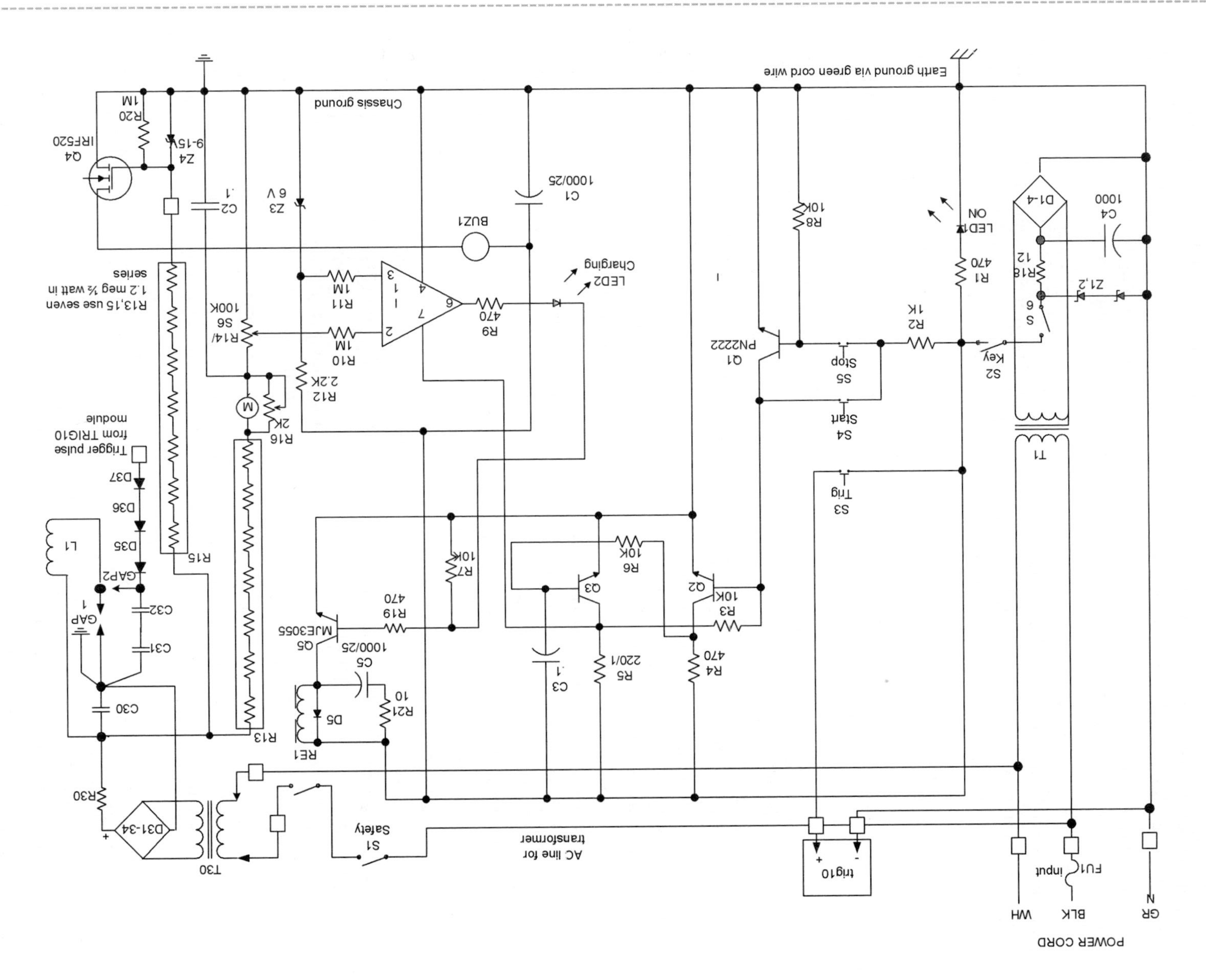

Figure 16-2 *Schematic*

device to warn the operator that a voltage exists on the storage capacitors. Sense voltage is obtained for the meter and charge voltage control circuit via string resistor R13. The safety level voltage is obtained through the resistor string (R15).

Assembly Steps

1. Cut a piece of .1 × .1 grid perforated circuit board (PB1) 6 × 4½ inch. You will have to drill holes for transformer T1 and relay RE1.
2. Insert the components as shown in Figure 16-3, starting from the lower left-hand corner. Count the perforated holes as a guide. Note the polarity marks on capacitors and diodes. It is suggested that you use a socket (SO8) for the LM741 integrated circuit (I1).
3. Wire the components as shown in Figure 16-3 using the leads of the actual components as the connection runs. These are indicated by the dashed lines. Always avoid bare-wire bridges and globby solder joints. Check for cold or loose solder joints. Note the symbols indicating wires to external components and to solder junctions beneath board.
4. Assemble the resistor dividers (R13 and R15) as shown in Figure 16-4 from seven 1.2 meg, ½-watt resistors that are all connected in series and sleeved into a ¼-inch ID flexible plastic tubing. Place these along with the connecting leads. The leads connected to R30 must be rated for 10 kilovolts. Connect the external components using #20 vinyl-jacketed, stranded wire. Attach and solder shunt resistor R16 across meter M1. Sleeve in the connection.
5. Connect the three high-voltage diodes (D35, D36, and D37) in series as shown in Figure 16-5. Make sure there are no sharp edges on solder joints and then sleeve into a plastic tube with a short piece of high-voltage wire lead.

   Connect the remaining components, noting that some may have to be unsoldered when routing through the panel, as shown. Verify proper lengths of connecting leads with other figures in the project.
6. Fabricate the chassis from a sheet of .063 aluminum as shown in Figure 16-6. The front panel holes are shown to approximate layout and size. These should be verified with all parts for hole size and location.
7. Fabricate a 6 × 7½ inch sheet of plastic (PLATE1) to insulate the assembly board connections as shown in Figure 16-7. Assemble T30 to the chassis using screws and nuts. Include solder lugs for grounding the power cord and assembly board.
8. Fabricate a 2¼ × 7 inch piece of plastic (PLATE2) for mounting the four rectifiers (D31 through D34) and resistor R30. Use two-sided sticky tape or silicon rubber to secure. Use a long screw or piece of threaded rod with an insulating washer on top side of resistor.
9. Assemble components to the front panel, as shown in Figure 16-8, and then finalize the wiring to complete assembly. Use pieces of 10 kilovolt-rated wire for those that are shown as heavy traces. Other leads are made from #20 vinyl-jacketed, stranded wire. The grounding leads shown are from the power cord, assembly board, and D33/D34 to the lug on T30. Output leads are shown as the TRIGGER, COMMON GROUND, and HV OUTPUT. Observe all notes on this figure.

Circuit Testing

10. Obtain a ballasted 115 vac power source. You can make this simply by placing a 60-watt lightbulb in series with the hot side of the power line, which is usually designated by the black lead. Do not eliminate the green earth ground connection.
11. Verify all switches are off and then insert a 3-amp fuse into FU1. Temporarily connect the high-voltage output to the chassis ground. Output is current limited by the leakage inductance of T30.

12. Plug into the ballasted 115 vac source. Turn on S6 and the key switch. Note LED1 lighting. Check control of both these switches as they are connected in series.
13. Push S3 and note a high-voltage spark at the trigger output lead.
14. Push S4 and note the relay latching and 60-watt lamp starting to light. The unit may chatter in this mode. LED2 should light in coincidence with the relay.

This concludes the basic test of this circuit but does not verify the programming voltage function, meter reading and calibration, or the safety buzzer.

Assembly of the Completed System

15. Fabricate the upper and lower deck pieces, as shown in Figure 16-9. Fabricate four pieces of 5½-inch PVC tubing for the pillar spacers located at each corner.

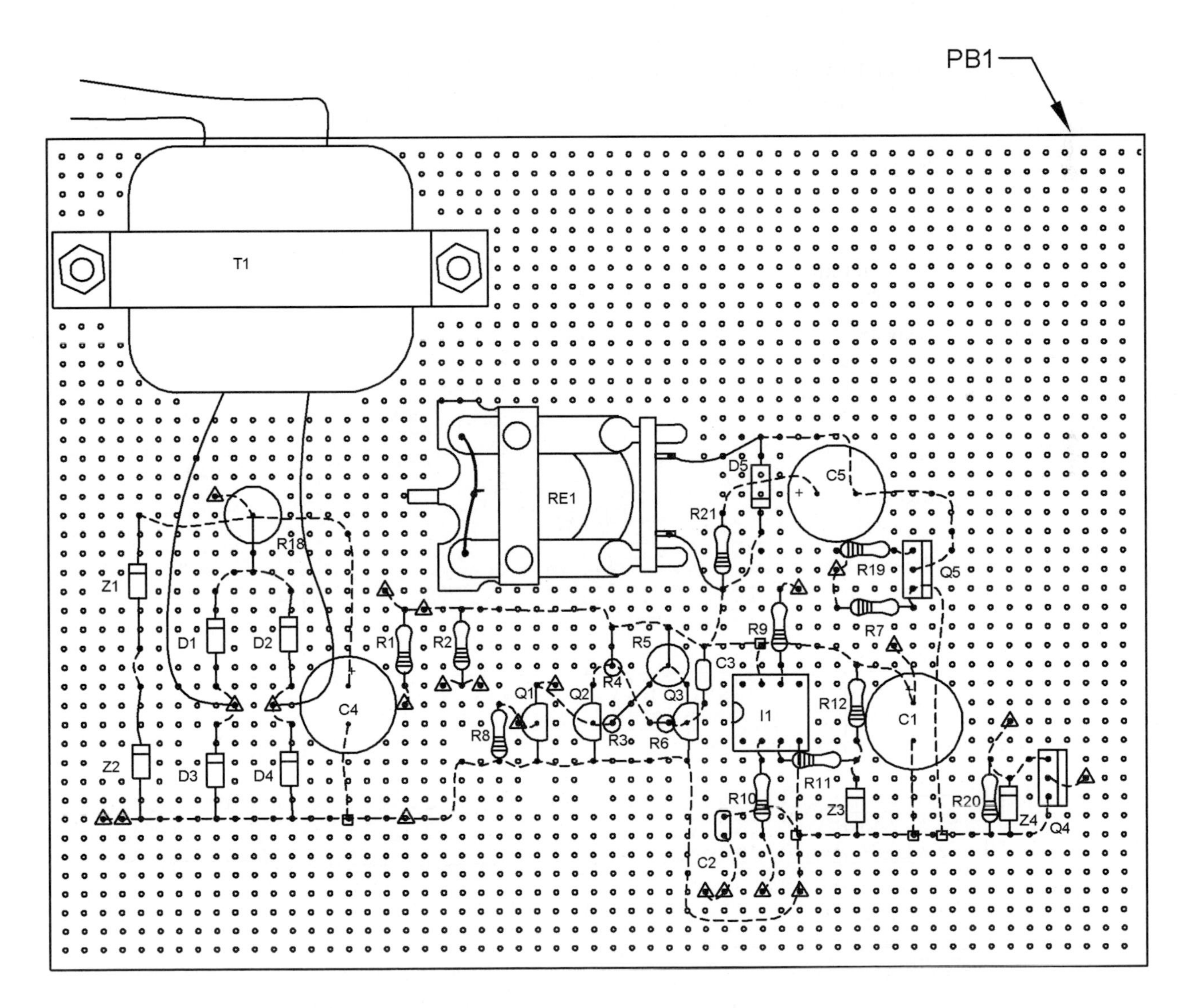

This symbol indicates points for wires to external components
This symbol indicates solder junction of leads
This symbol indicates wiring on underside of board

Figure 16-3 *Assembly board parts layout*

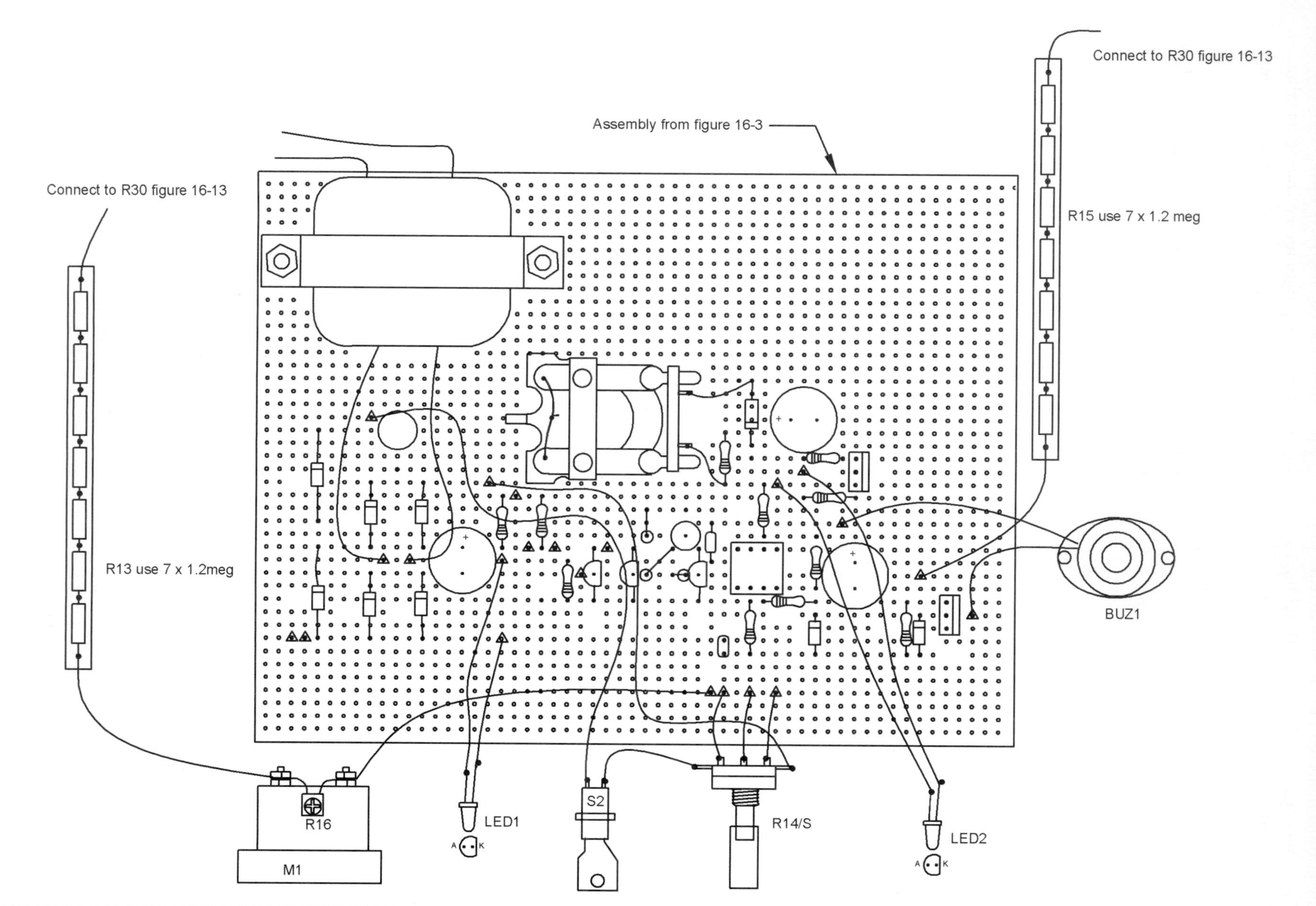

Figure 16-4 *Assembly board external wiring first level*

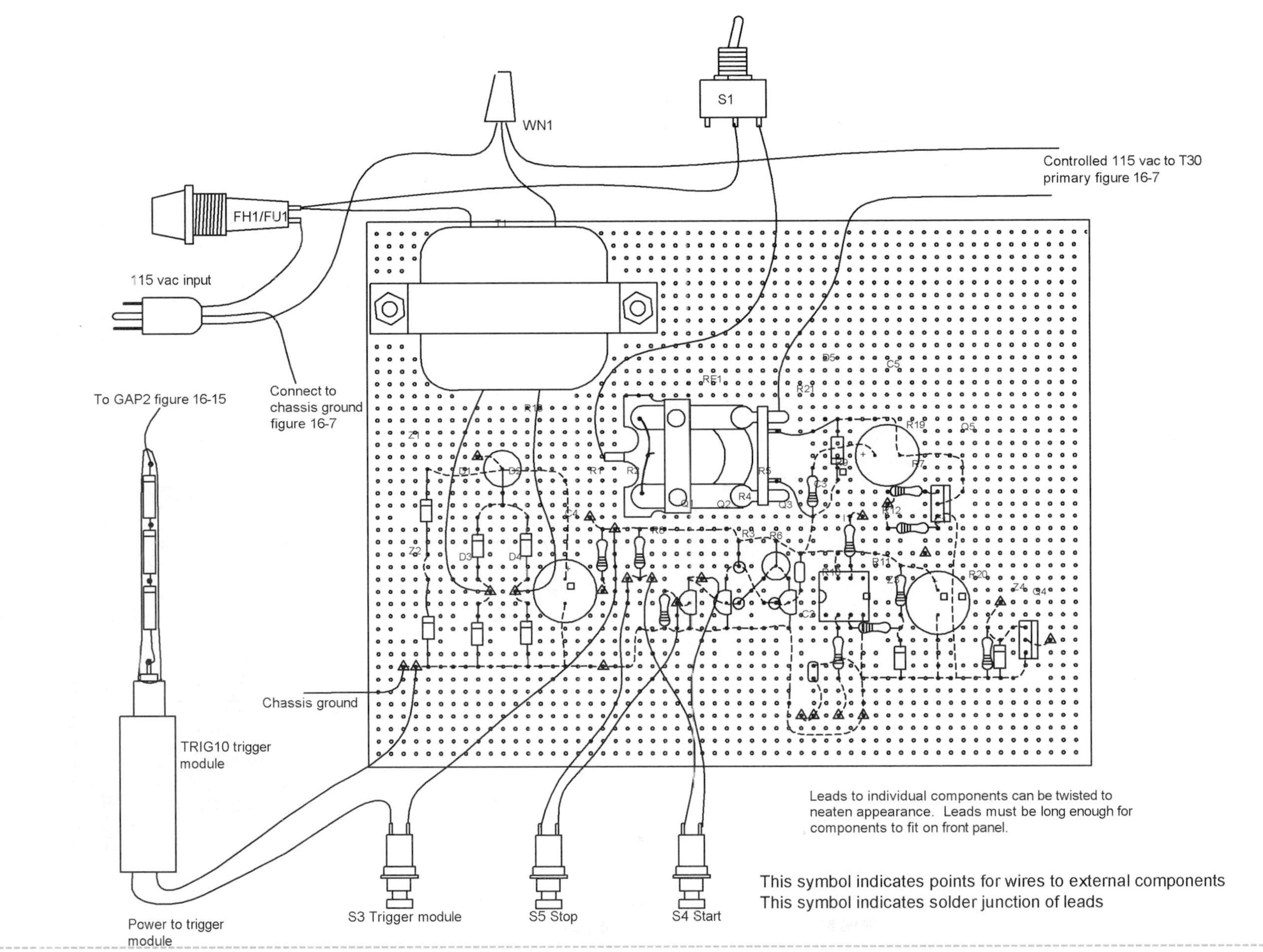

Figure 16-5 *Assembly board external wiring second level*

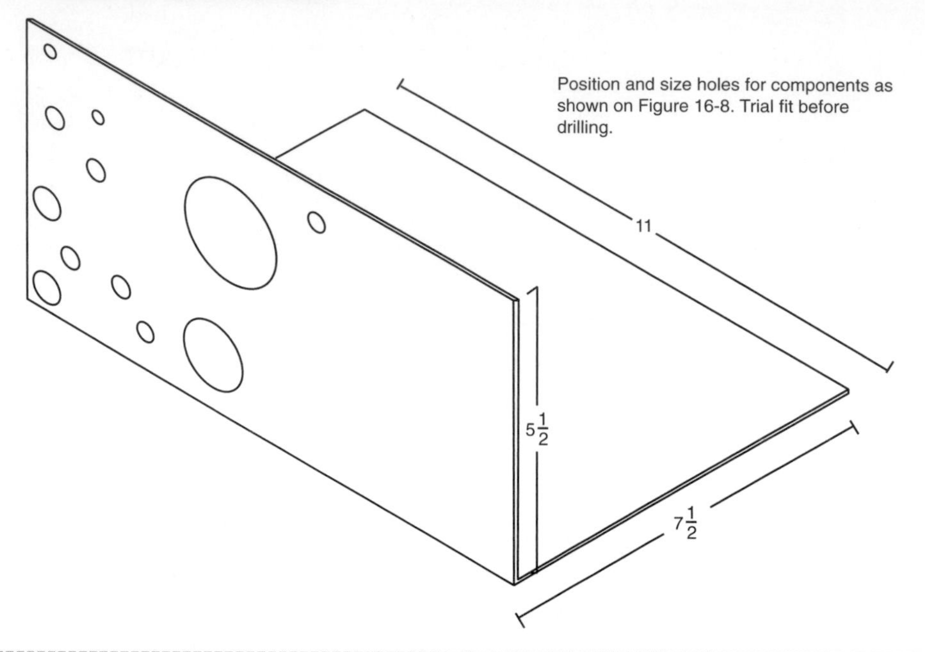

Figure 16-6 *Fabrication of chassis*

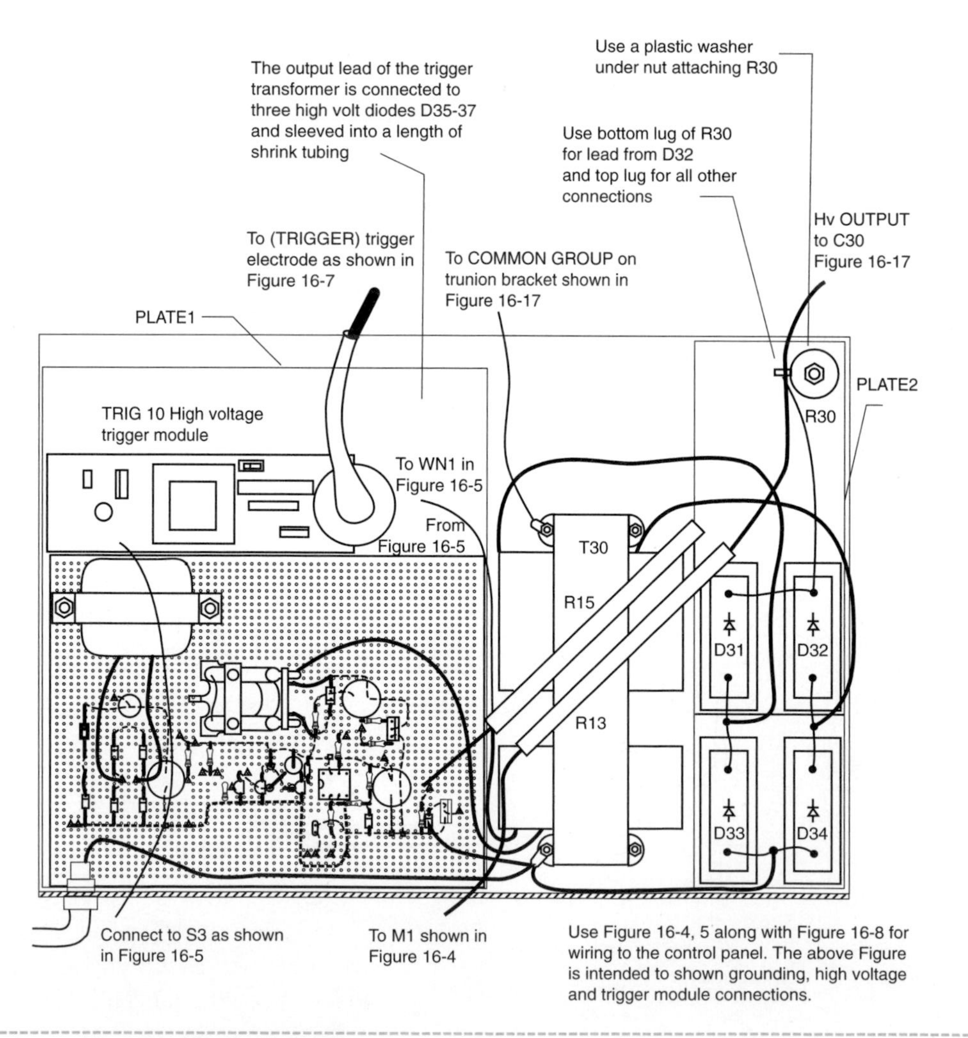

Figure 16-7 *Assembly and wiring of chassis*

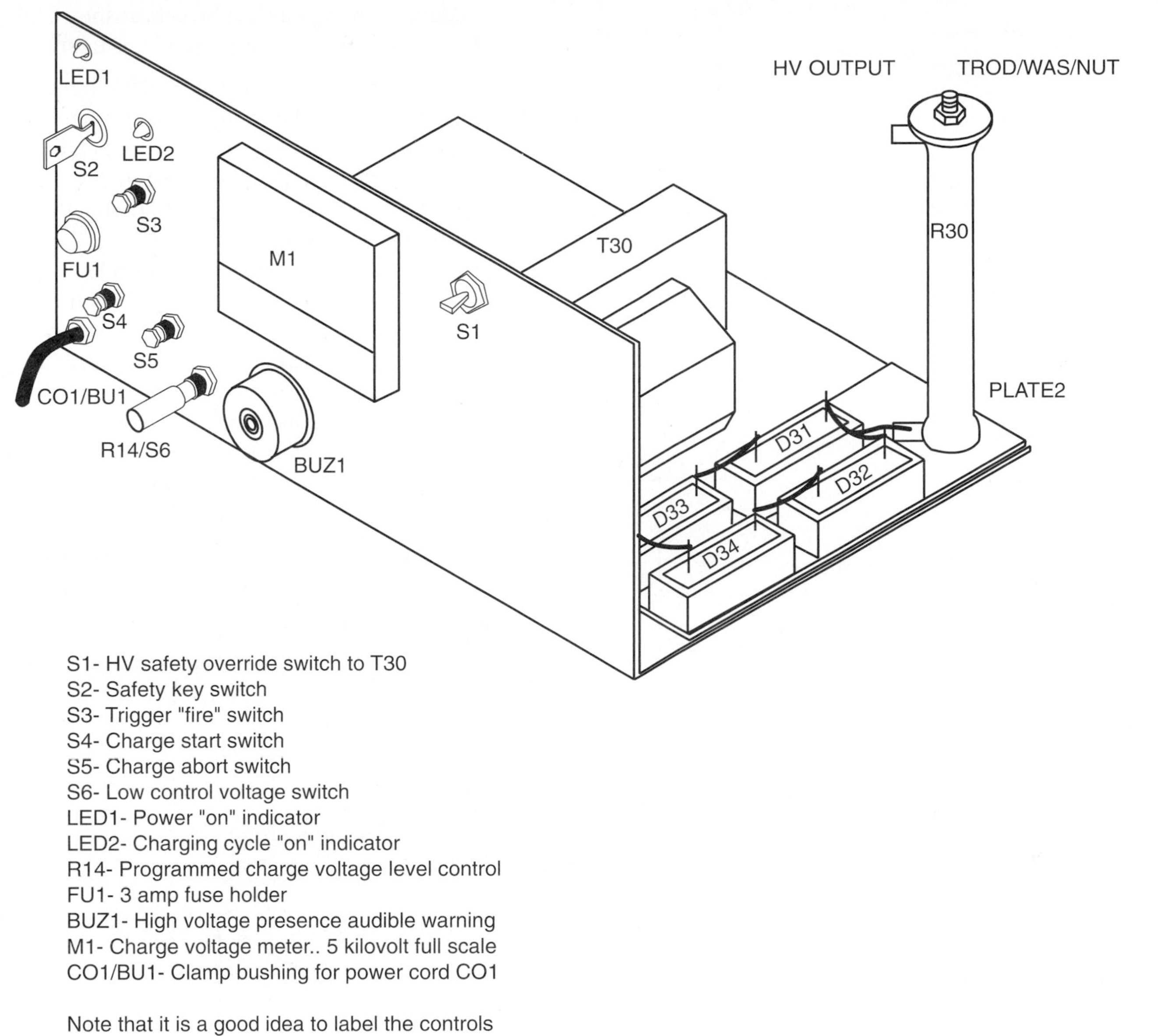

Figure 16-8 *Front-panel controls and wiring*

16. Fabricate the capacitor bracket, as shown in Figure 16-10. This piece secures the three energy storage capacitors C10A, C10B, and C10C on the bottom deck.

17. Fabricate the trunion bracket, as shown in Figure 16-11. This piece secures the actual flyway and accelerating coil and therefore must be able to withstand the kinetic recoil from the accelerating reaction of the projectile. This bracket is the central electrical grounding point of the discharge circuit and must also be actively grounded to earth ground via the power cord green wire. There is also a 3/8-inch hole for the electrode holding collar (COLLAR1) for the ground electrode (TUNG38) of the main spark switch (GAP1). This part must be centered and carefully soldered in place using a propane torch.

18. Fabricate the front and rear disk section, as shown in Figure 16-12. You may also fabricate the projectile (PROJ1) from a piece of aluminum, as shown.

19. Wind the coils, as shown in Figure 16-13. It is a good idea to study Figures 16-12, 16-13, and 16-14 before attempting this step. Make sure to leave connection leads that are a minimum of 10 inches long.

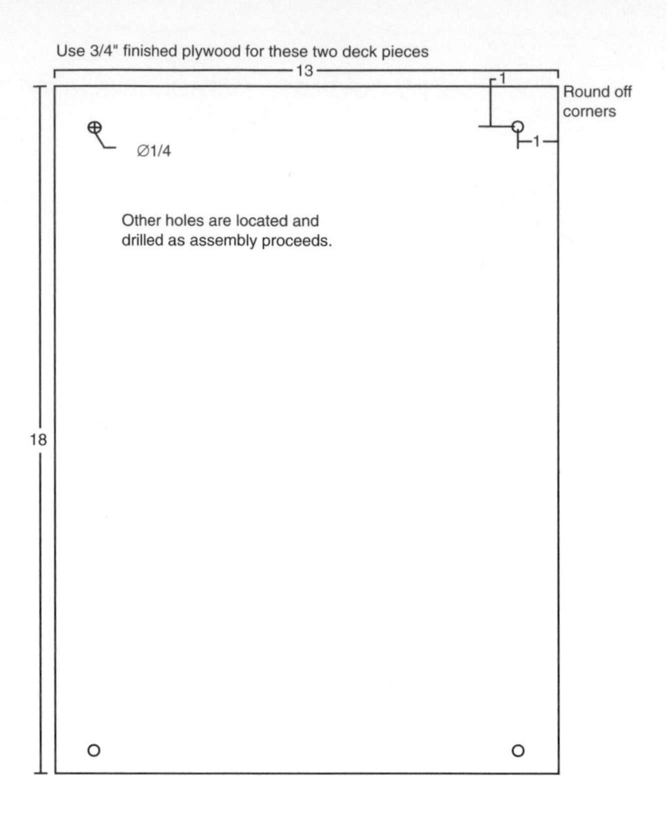

Figure 16-9 *Fabrication of deck sections*

20. Assemble the flyway and breech, as shown in Figure 16-14. Study the notes on this figure.

21. Fabricate the two plastic blocks (PVCBLK and TFBLK), as shown in Figure 16-15, to the suggested dimensions. Note that the heights of these pieces must allow close alignment of the electrodes as dictated by the position of the soldered collar on the trunion bracket. C31 and C32 are wired in place and must be spread apart to prevent sparking. The blocks are secured to the top deck using sheet metal screws. Use two screws for the PVC block. Do not allow these screws to penetrate more than 1/2 inch into the plastic material. Note the clearance hole in the deck for the high-voltage lead from the trigger module, shown in Figure 16-7.

22. Assemble the three energy-discharge capacitors (shown in Figure 16-16) along with the charger and control module.

23. Make the final assembly steps and wire, as shown in Figure 16-17. Note the direct and

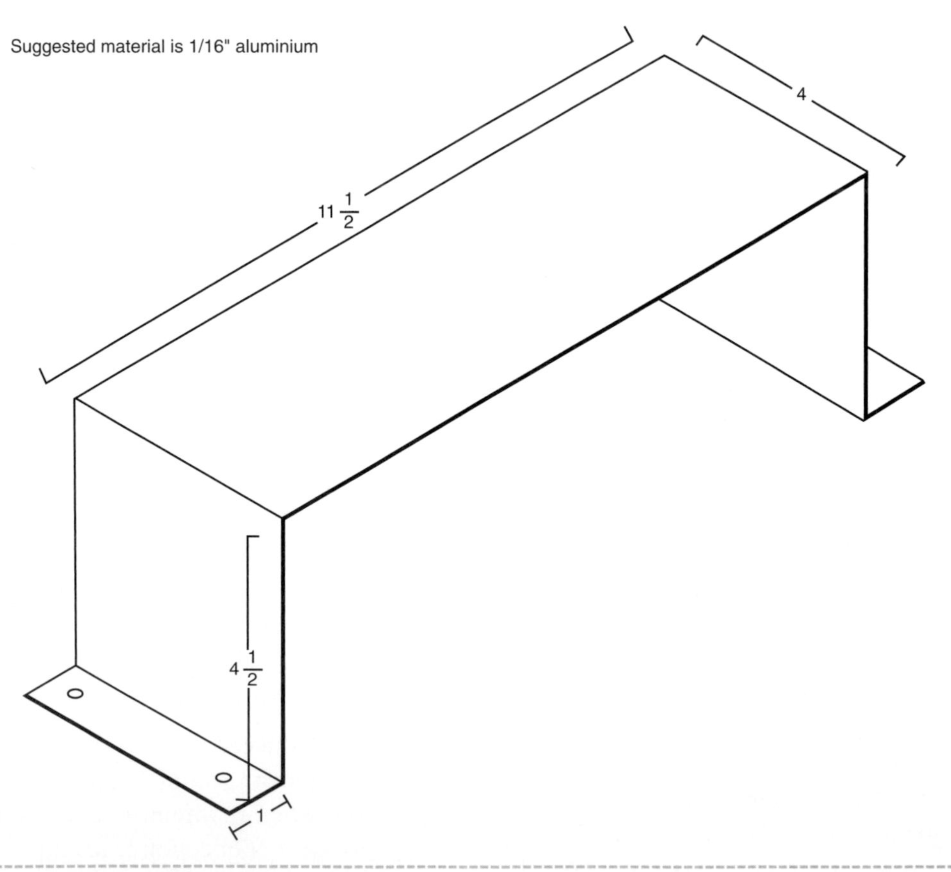

Figure 16-10 *Fabrication of the capacitor bracket*

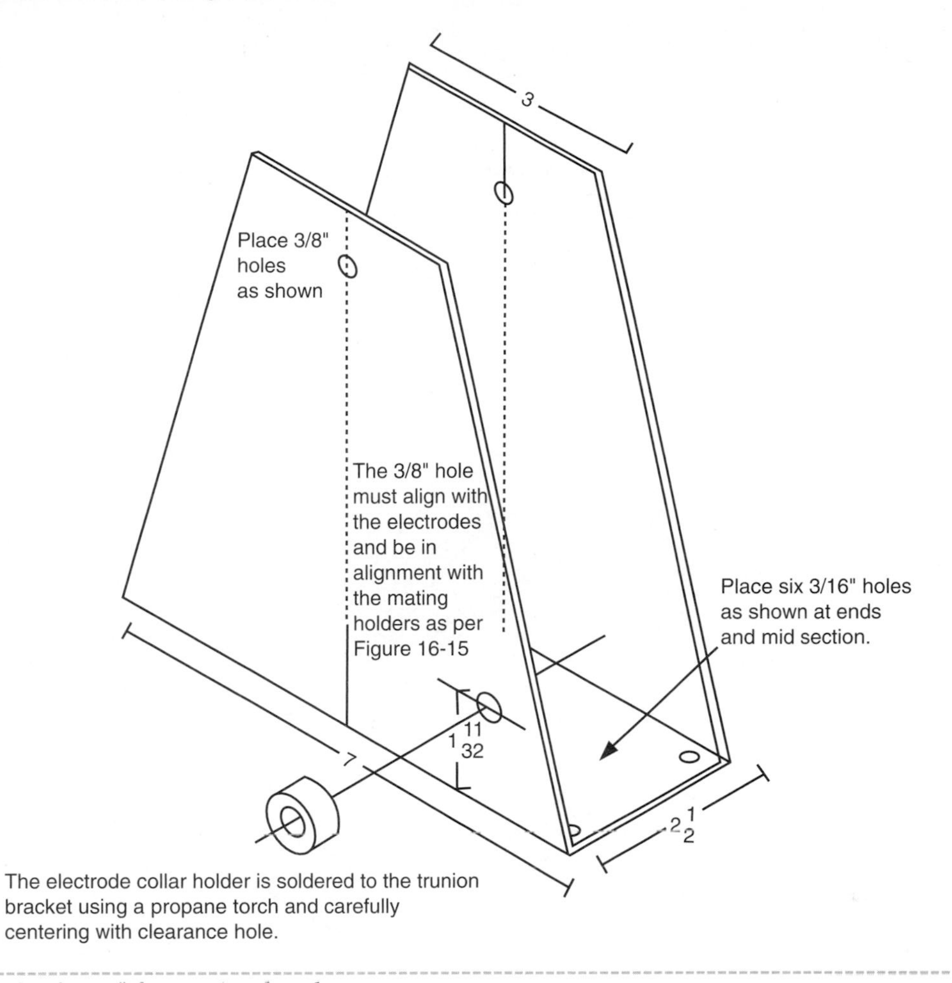

Figure 16-11 *Fabrication of the trunion bracket*

heavy leads for the discharge path. You may want to sleeve some plastic tubing over these leads, as they are rated only for 600 volts.

Note: The trunion bracket is the common grounding point for the system. It is very important that the earth ground green lead of the power cord be firmly attached at this point to ensure system operating safety.

24. Verify and use the diagram shown in Figure 16-18 as a wiring aid and reconfirm grounding and general circuitry and integrity.

Testing and Operation of the Cannon

25. Locate a test site with a backstop capable of stopping the projectile. Obtain a 5,000 volt DC reading meter to set M1. It is assumed that the circuitry is correctly assembled and the basic electrical pretest as described in steps 10 through 14 was successively performed. Preset the main gap to 1/16 inch and the trigger gap to 1/8 inch.

26. Preset all switches shown back in Figure 16-8 to the off position. Familiarize yourself with these controls

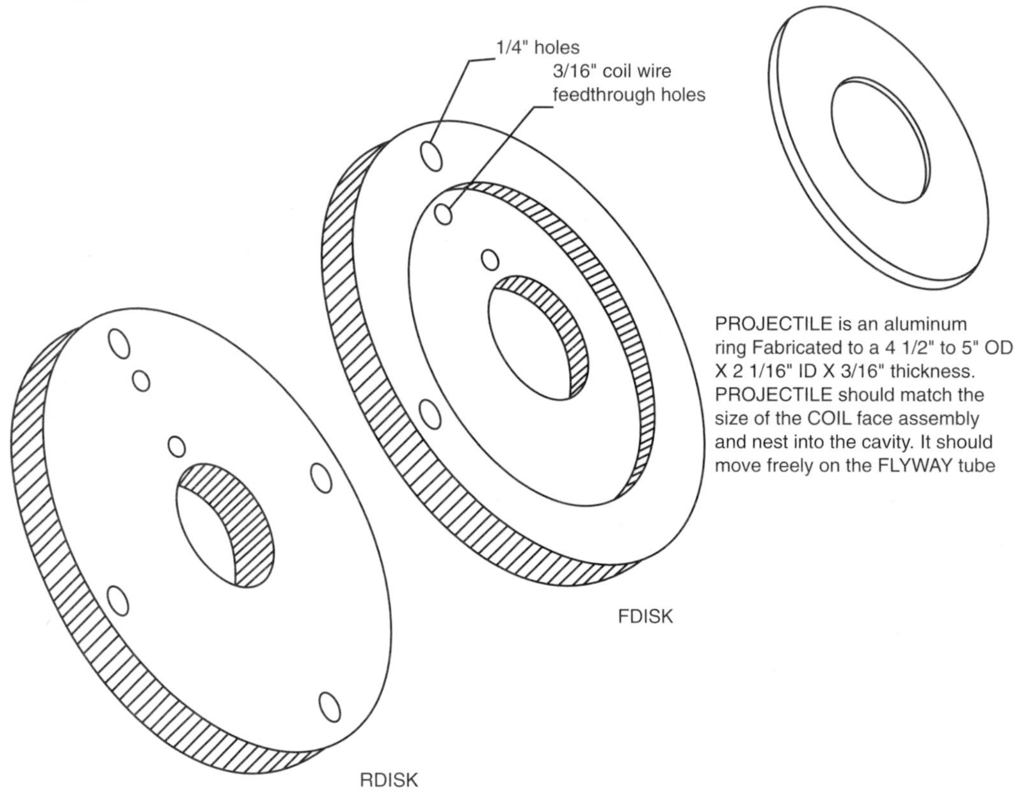

Figure 16-12 *Fabrication of the accelerator coil holder*

Danger: The next steps involve exposure to dangerous high voltages. Failure to take any of the following described steps will require a complete shutdown and power removal, along with a shorting of C30 before performing any trouble shooting.

27. Connect the test meter from chassis ground to the output end of resistor R30. Plug the unit into 115 vac and turn on S6. Turn on the key switch S2 and note LED1 lighting. Place the projectile over the accelerator coil.
28. Depress pushbutton switch S3 and note a spark occurring in both gaps.
29. Verify S1 safety switch is off. Depress pushbutton switch S4 and note the relay RE1 energizing and LED2 lighting. Depress pushbutton switch S5 and note RE1 de-energizing and LED2 turning off. Verify these on and off functions of S4 and S5 several times before the next steps.

    S5 is important as it must stop the charging action in the event that you choose to quickly abort the cycle.
30. Turn on toggle switch S1. Restart the system by again depressing R4, noting a voltage reading on both the panel meter M1 and the external test meter. This voltage will build up to a level as set by the charge-voltage level control R14 and then shut down, only to restart the charge cycle when the voltage charge level

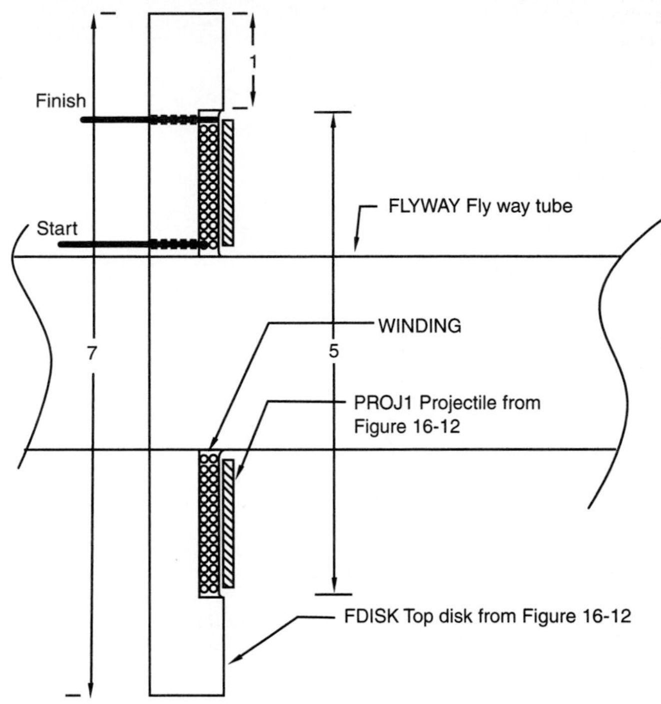

1. Fabricate the FLYWAY from a 18" piece of 2" PVC tubing
2. Wind a coil of 30 turns of #12 in fifteen layers of 2 turns each. Finished coil must be in a tight pancake configuration and have 10 inch connection leads. *You may have to make up a temporary bobbin jig for this step.* Hold together with tape etc.
3. Place COIL section into cavity routing the leads as shown. COIL section should nest into cavity and be positioned flat.
4. Seal all points that can leak as epoxy must be carefully poured over the COIL forming a 1/16" layer on top.

Figure 16-13 *Side view showing winding*

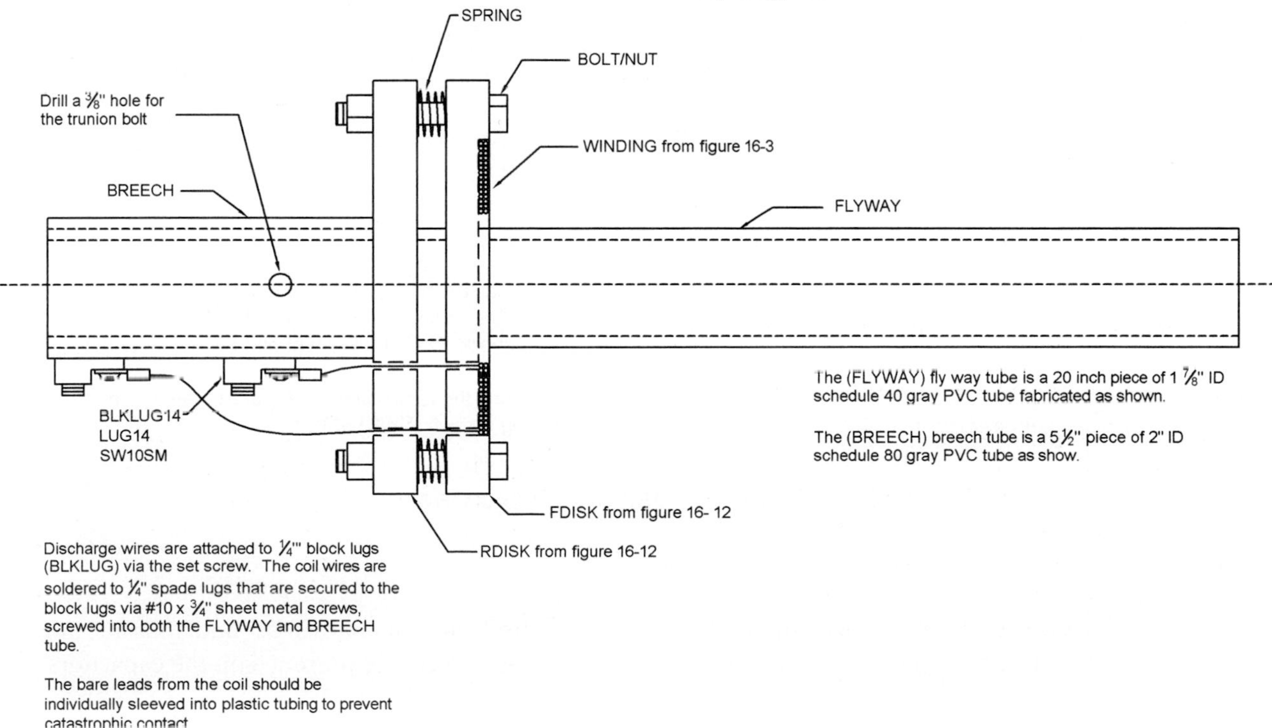

Figure 16-14 *Assembly of the breech and flyway*

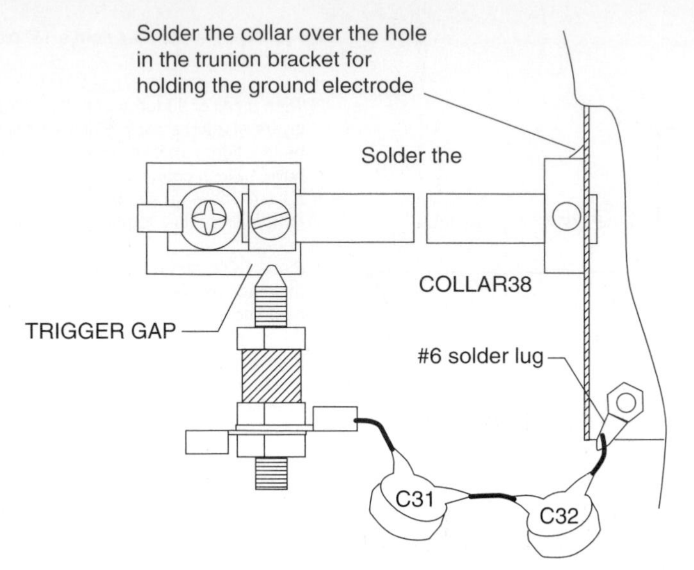

Solder gap switch assembly top view

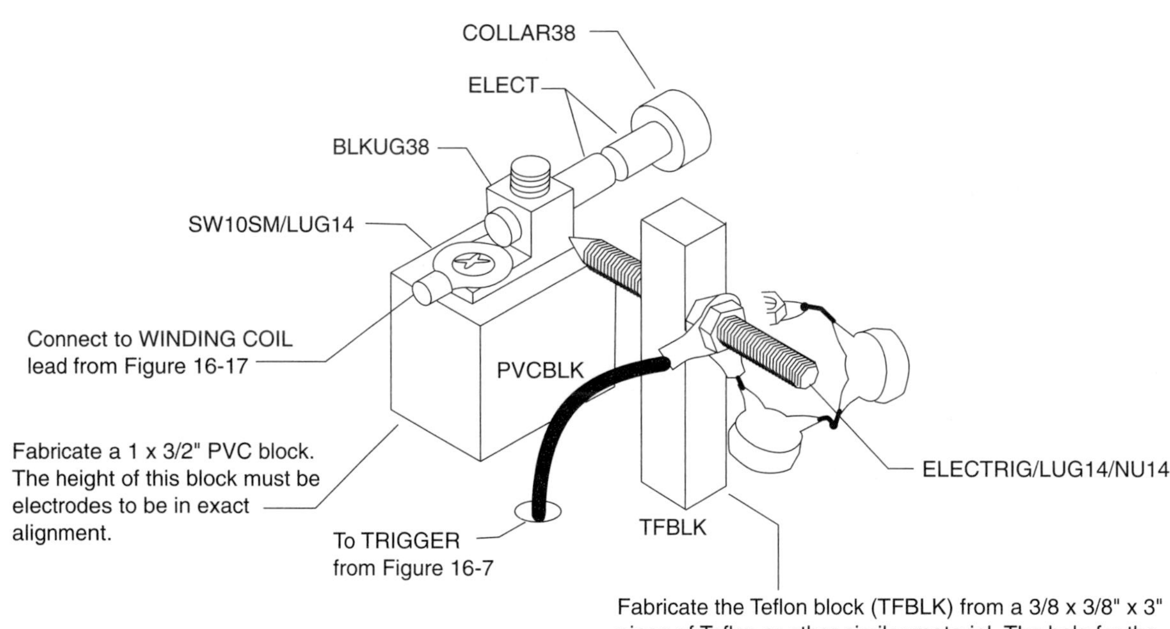

Spark gap switch is o view

Figure 16-15 *Details of the spark and trigger switches*

falls below the set level. Obviously there is an amount of hysteresis that prevents off-and-on chatter. With R14 set to full counterclockwise (CCW), this voltage should reset at less than 1,000 volts. The buzzer BUZ1 should be producing an annoying sound, thus alerting you to a charge presence on the capacitors.

31. Slightly turn R14 clockwise and allow it to cycle, repeating up to 1,500 volts as read on the external test meter. Depress the trigger

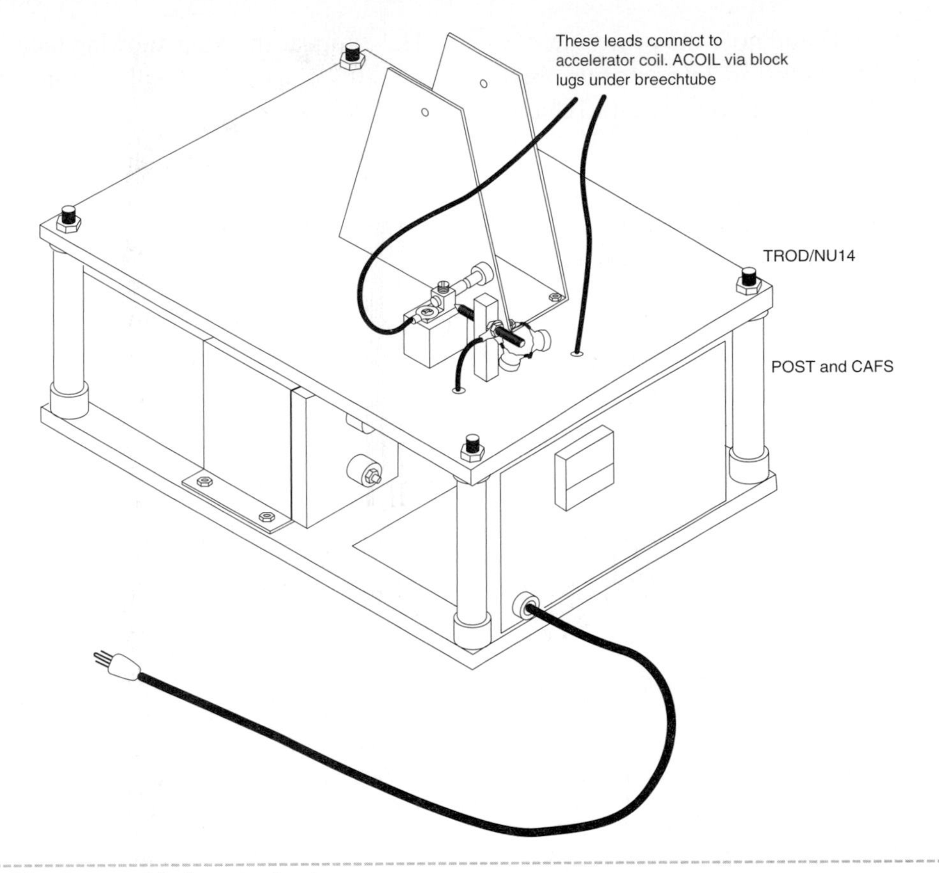

Figure 16-16 *Final assembly isometric view*

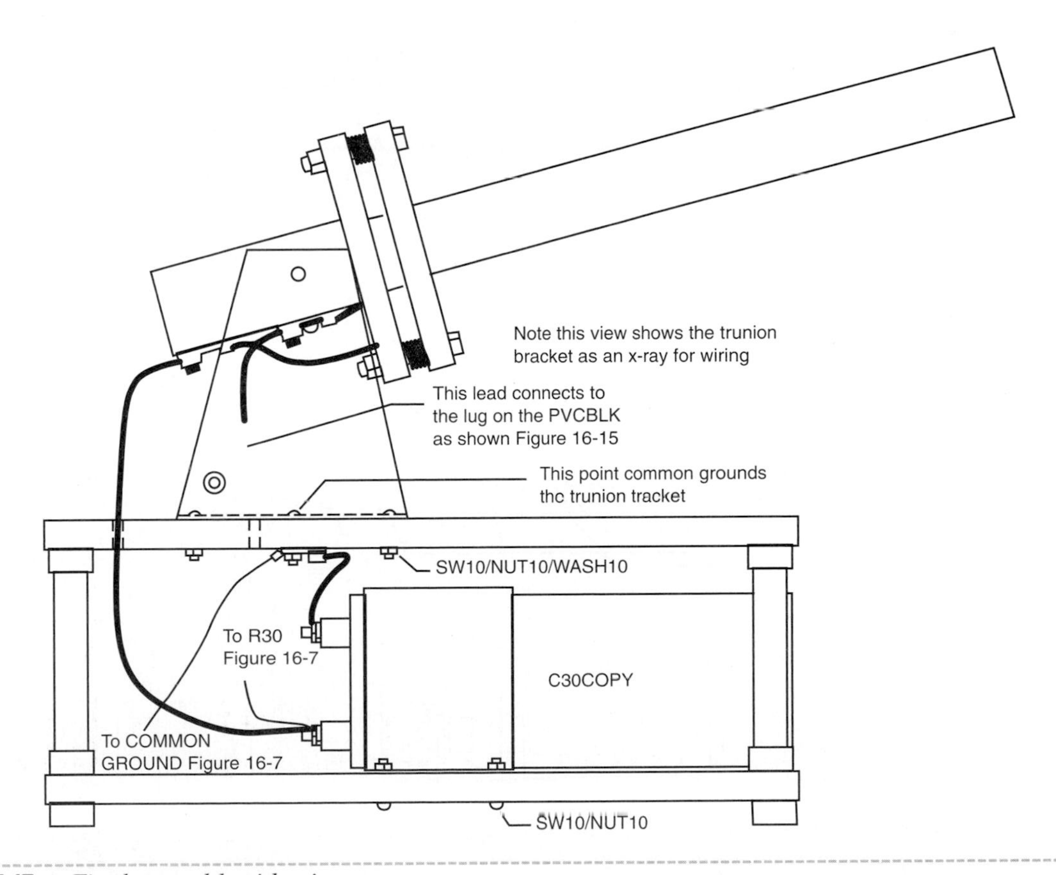

Figure 16-17 *Final assembly side view*

fire pushbutton S3 and note the projectile shooting into the backstop. Immediately press the stop switch S5 unless you want it to charge up for the next shot.

Note: The system will automatically recharge to the 1,500-volt present level unless the stop switch is activated.

32. Repeat the cycle, increasing the charge voltage up to 2,500 volts, and then adjust the trimpot R16 on the meter M1 to agree with the test meter.

33. Repeat this step allowing the charge voltage to go up to 5,000 volts, noting the impact of the projectile.

34. Verify the unit is not breaking apart as a result of recoil. Rethink the recoil springs if a problem exists.

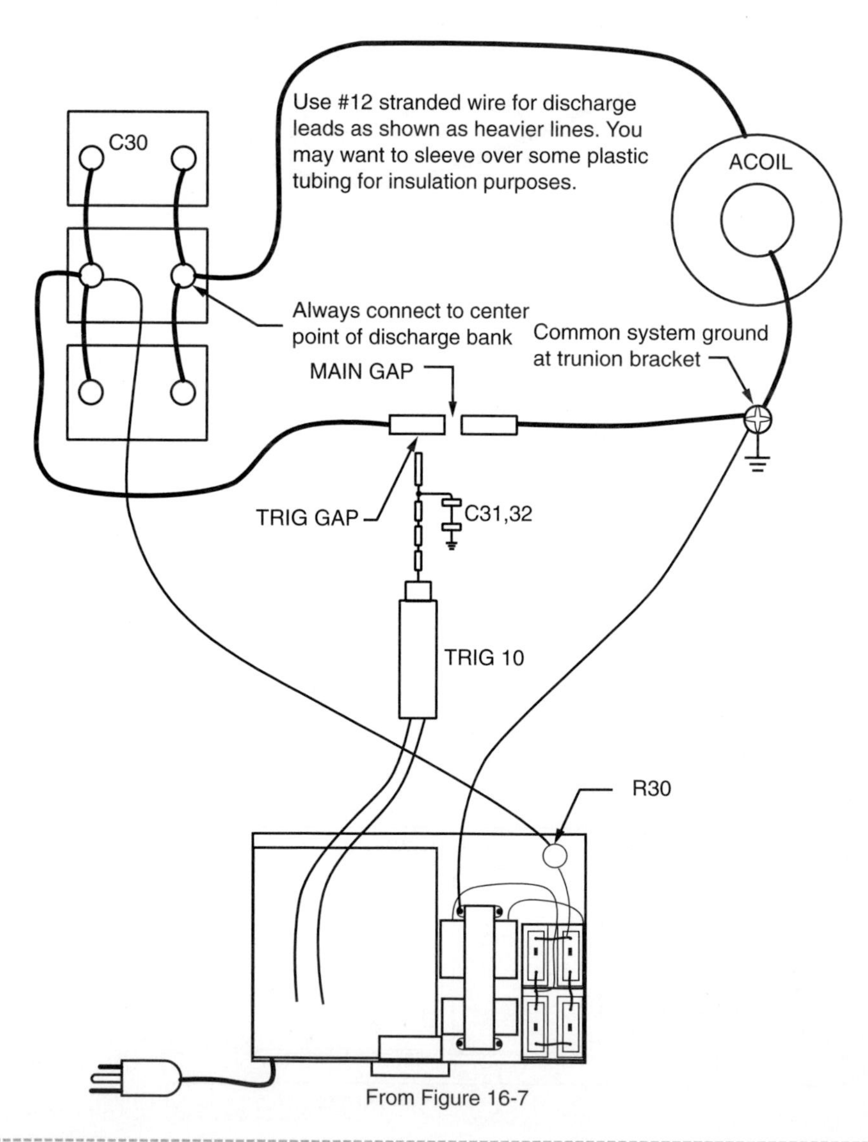

Figure 16-18 *Overall system wiring diagram*

Table 16.1 Magnetic Cannon Part List

Ref. #	Description	DB Part #
R1, 4, 9, 19	Four 470, 1/4-watt resistors (yel-pur-br)	
R2	1K, 1/4-watt resistor (br-blk-red)	
R3, 6, 7, 8	Four 10 kilohm, 1/4-watt resistors (br-blk-or)	
R5	220, 1/4-watt resistor (red-red-br)	
R10, 11	Two 10M, 1/4-watt resistors (br-blk-bl)	
R12	2.2K, 1/4-watt resistor (red-red-red)	
R13, 15	Fourteen 1.2M, 1/4-watt resistors (br-red-gr) (see schematic in Figure 16-2)	
R14/S6	100K potentiometer, 17 mm with low-volt switch	
R16	2K trimpot	
R20	1M, 1/4-watt resistor (br-blk-gr)	
R21	10-ohm, 1/4-watt resistors (br-blk-blk)	
R18	12-ohm, 3-watt resistor	
R30	16K, 50-watt power resistor	DB# 16K50W
C1, 4, 5	Three 1,000 mfd, 25-volt vertical electrolytic capacitors	
C2, 3	Two .1 mfd, 50-volt plastic capacitors	
*C30A	Three 32 mfd, 4,500-volt oil-filled cans in parallel for 96 mfd/4.5 kv	DB# 32m/4500
*C30B	Four 24 mfd, 5,200-volt oil-filled cans in parallel for 96 mfd/5 kv	DB# 24m/5200
I1	741 op-amp DIP IC	
Q1, 2, 3	Three PN2222 GP NPN transistors	
Q4	IRF520 Mosfet	
Q5	MJE3055 NPN power transistor	
D1-4	Four 1N4001 50-volt 1-amp diodes	
D5	1N4007 1,000-volt, 1-amp diode	
D31-34	Four 8-kilovolt, 100 ma standard recovery diodes	DB# H407
D35-37	Three 30 kilovolt, 10 ma diodes	DB# VG30
Z1, 2, 3	Three 1N4733 6-volt zener diodes	
Z4	1N4745 15-volt zener diode	
LED1	Red LED indicator	
LED2	Green LED indicator	
S1	SPST toggle switch	
S2	Key switch	
S3, 4, 5	Three pushbutton NO switches	
RE1	12-volt, 5-amp relay	
BUZ1	12-volt buzzer	DB# BUZROUND
Q1	100-volt IRF540 MOSFET	
T1	12-volt, 100 ma transformer	DB# 12/.1
T2	6,500-volt, 20 ma current-limited transformer	DB# 6kv/20
*L10	Accelerator coil assembly, as shown in Figures 16-12 and 16-13	DB# ACCOIL
*GAP1	Main gap assembly, as shown in Figure 16-15	DB# GAPMAIN
*GAP2	Trigger gap assembly, as shown in Figure 16-15	DB# GAPTRIG
CHASSIS1	Metal chassis fabricated as shown in Figure 16-6	
PLATE1	6 × 7 1/2 × 1/16 inch plastic insulating plate	
PLATE2	2 1/4 × 7 1/2 × 1/16 inch plastic insulating plate	
TAPE1	12 inches of 1-inch wide, thin sticky tape	
PLTUBE	18 inches of 1/4-inch ID vinyl flexible tubing for R13, R15, and trigger diodes	
CO1	6 feet of 3-wire #18 power cord	
HVWIRE	12 inches 20 kv silicon high-voltage wire	DB# WIRE20KV
BUSSWIRE	24 inches #20 bare busswire	

WR20	10 feet #20 vinyl stranded hook-up wire	
WN1	Medium wire nut	
BU2, 3	Two bushings for LEDS	
BU1	Power cord clamp bushing	
SW6	Six 6-32 × 3/4-inch screws	
SW1	6-32 × 1/4-inch screw for mounting relay	
NU6	Six #6 hex nuts	
LUG6	#6 solder lug	
TROD	5-inch length of 1/4-20 threaded rod, 2 nuts, and plastic washers	
TRIG10	Assembled trigger pulse module	DB# TRIG10
TRIG1K	Kit of trigger pulse module	DB# TRIG1K

Parts List to Final System

DECKS	Two fabricated pieces of 7/8-inch birch plywood 13 × 17 inches
POSTS	Four fabricated pieces of 5 3/4 × 7/8 inch PVC tubing
PLATE1	Fabricate a piece of 1/16-inch plastic sheet 6 × 7 inches
PLATE2	Fabricate a piece of 1/16-inch plastic sheet 2 1/4 × 7 inches
BRKCAP1	Fabricate the capacitor bracket seen in Figure 16-10
BRKTRUN1	Fabricate the trunion bracket seen in Figure 16-11
RDISK	Fabricate the rear disk seen in Figure 16-12
FDISK	Fabricate the front disk seen in Figure 16-12
PROJ1	Fabricate the projectile seen in Figure 16-12
BLKELECT	Fabricate the electrode holding block seen in Figure 16-15
BLKTRIG	Fabricate the trigger electrode holding block seen in Figure 16-15
FLYWAY	Fabricate the flyway tube seen in Figure 16-14
BREECH	Fabricate the breech tube seen in Figure 16-14
CAPS	Twelve 7/8-inch PVC caps for posts and feet seen in Figure 16-1
SPRINGS	Four 1/2-inch diameter × 3/4-inch stiff compression springs
COLLAR	3/8-inch shaft collar
BLKLUG25	Three 1/4-inch block lugs for breech connections
BLKLUG38	3/8-inch block lug for electrode
LUG14	Five 1/4-inch wide solder lugs
ELECTRGAP	Two 3/8 × 2-inch tungsten or tool steel electrodes
ELECTRTRIG	1/4-20 × 3-inch threaded rod with point
TROD	Four 1/4-20 × 8-inch threaded rod
BOLT/NUT	Five 3 1/2 × 3/8-inch bolt and nut
NUT14	Eleven 1/4 × 20 nuts for trigger gap and posts
SW10M	Five #10 × 1/2-inch sheet metal screws
SW10	Ten #10-32 × 1.5-inch machine screw
NU10	Ten #10-32 nuts
WASH10	Twenty #10 wide shoulder washers
WIRE12	4 feet #12 stranded, vinyl-covered wire
WIRE12	30 feet #12 heavy-coated, magnetic wire for winding ACOIL

Chapter Seventeen

Vacuum Tube Tesla Coil Project

This advanced project, as shown in Figure 17-1, is similar to a coil made by John Weisner some years ago. The geometry of the secondary not being the most efficient is a good compromise for wireless energy transmission and also provides a reasonable output display. The project is intended for the experienced experimenter and researcher in the field of high-voltage and high-frequency circuitry. Not only is this project a dangerous electrical device, but it can also be a bothersome source of radio interference, especially in the standard AM broadcast band and should be used in a shielded enclosure or a Faraday cage if operated on a continuous basis. A gas discharge lamp such as a standard household fluorescent lamp will glow up to 10 feet away when this device is used. Annoying burns are experienced when contact is made with conductive objects up to similar distances.

Figure 17-1 *The vacuum tube Tesla coil*

We also have built coils using 304TH, 450TH, and parallel 2500THs with a full 5,000 watts of plate dissipation. This coil produced spectacular flaming arcs up to 8 feet long with 8 kilowatts of input. A video showing the operation is on our web site. The coil shown here will easily produce 1 to 2 feet of flaming arcs.

This system can provide the experienced hobbyist with hours of fun and entertainment experimenting with terminals and settings to generate wireless energy as well as output arcs and sparks. This is an advanced-level and potentially expensive project requiring electronic skills and high-voltage experience. Expect to spend $300 to $750. All the parts are readily available, with specialized parts obtainable through Information Unlimited (www.amazing1.com), and they are listed in Table 17-1.

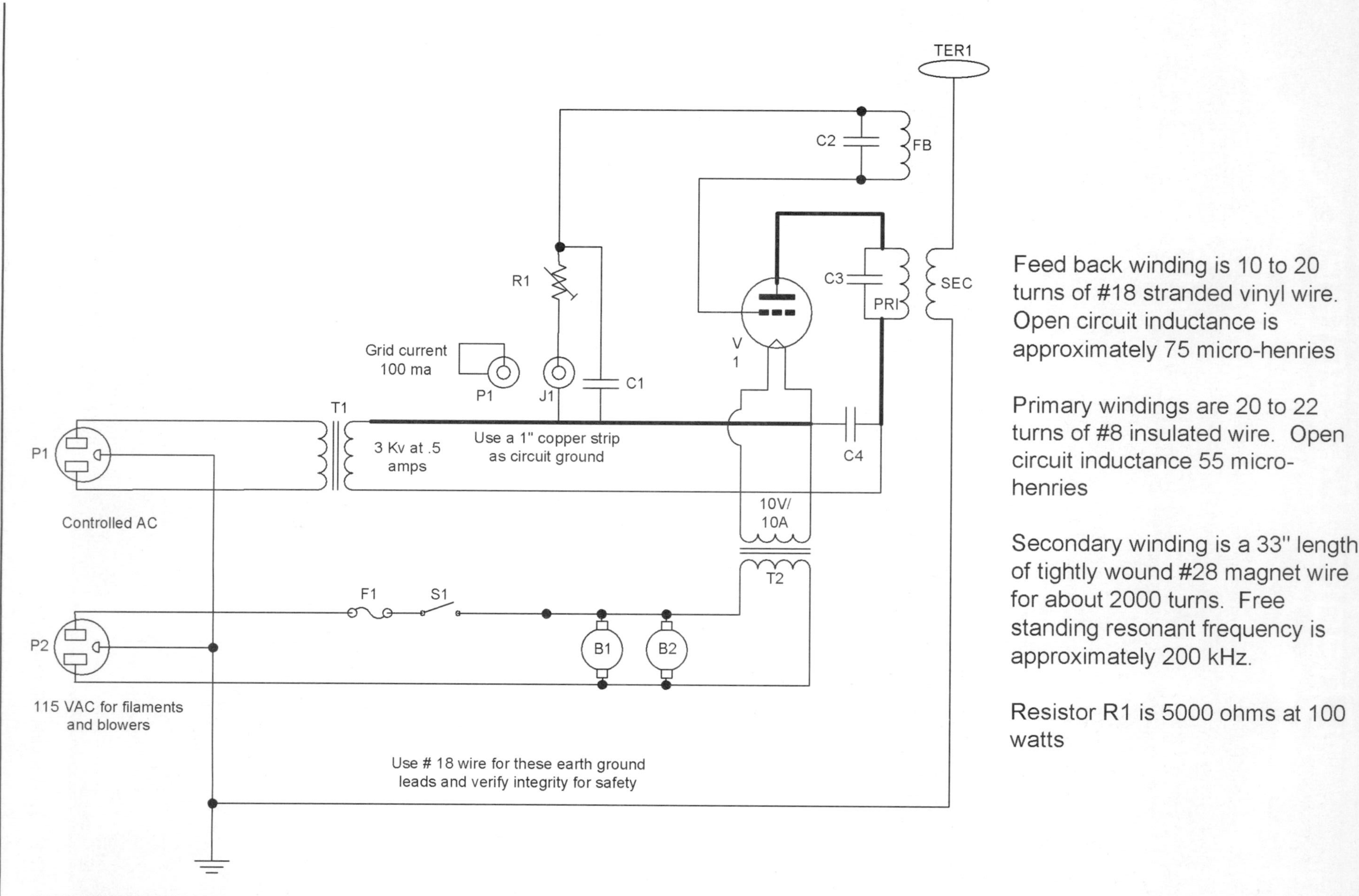

Figure 17-2 *Circuit schematic*

Basic Description

These spectacular display devices produce arcs and sparks quite unlike the undamped spark gap-driven Tesla coils. Operation does not require a noisy spark gap that produces copious amounts of *radio frequency interference* (RFI) but operates efficiently at the quarter-wave frequency of the secondary coil. Vacuum tubes, while large and requiring filament power, do offer robust and forgiving operation and are intended as a good starter approach to solid-state Tesla coils.

This approach is far less temperamental than the solid-state *metal-oxide-semiconductor field effect transistors* (MOSFET) or *insulated gate bipolar transistor* (IGBT) drivers. However, once these FET devices are debugged and tweaked, one can obtain some excellent results. Our lab FET coil generates over 4-foot sparks using only 3 kilowatts. The output frequency tracks within 5 percent of the resonant point using one of our proprietary circuits that will be on our site in the late spring of 2006 with the complete system scheduled for publishing in the third edition of this book series.

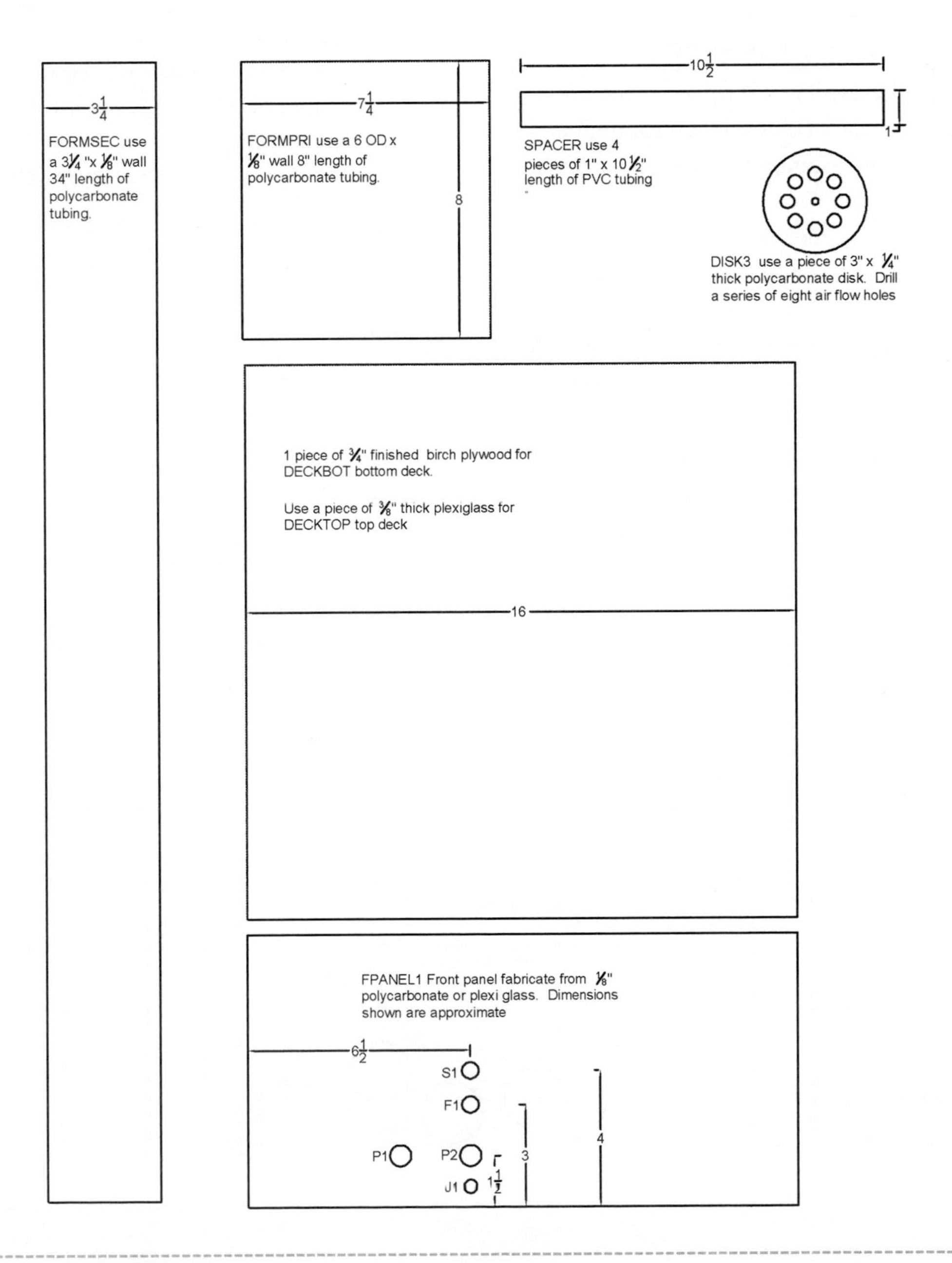

Figure 17-3 *Creating the major parts*

The circuit as shown on the schematic in Figure 17-2 of the vacuum tube device is nothing more than a medium-powered Hartley radio frequency oscillator tuned to the resonant frequency of the secondary coil. The circuit uses a medium-powered, readily available 833A triode transmitting tube that inherently has a high grid to plate capacitance necessary for self-oscillation. The plate load is a resonant transformer with both the primary and secondary tuned to the approximate identical frequencies. The output potential is now a function of the secondary-to-primary coil inductances times the peak primary plate voltage. This system generates voltages at frequencies sufficient to cause air breakdown, thus producing the sparks and arc jumping into open air.

The output of the oscillator is relatively closely coupled to the secondary coil designed for high Q performance and self-resonant to the quarter-wave of the oscillator frequency. The voltage distribution is now that of a quarter section with a current node at the base and voltage node at the top. The input power to the coil is raw, unrectified AC at 3,000 volts *root mean square* (rms) with a current of .5 amps. The peak voltage is over 4,000 volts and is supplied by a conventional plate transformer being fed by a voltage-adjustable variac, which allows you to adjust the output voltage of the system. A pulse signal is also shown that controls the grid of the tube, allowing a wide range of spark texture variation by changing the duty cycle and frequency.

Assembly Steps

1. Fabricate the secondary coil form (FORMSEC) from a 3¼ × ⅛-inch wall polycarbonate tubing, as shown in Figure 17-3. It is not suggested that you use PVC tubing due to moisture retention and breakdown.

   Fabricate the primary coil form (FORMPR1)as shown in Figure 17-3 from a length of 7½ × ¼-inch wall plastic tubing. Polycarbonate tubing is preferred.

   Fabricate two 3-inch diameter discs (DISK3) as shown in Figure 17-3 from ¼-inch polycarbonate. These pieces are for the ends of the secondary coil and are secured via three #6 × ⅜-inch brass wood screws. Drill air passage holes as shown for the blower (B2).

   Fabricate four 10.5-inch spacers (SPACER) as shown in Figure 17-3 from lengths of 1-inch OD PVC tubing.

   Fabricate the 16 × 16-inch square bottom deck (DECKBOT) as shown in Figure 17-3 from a piece of ¾-inch finished plywood and paint it black.

   Fabricate the 16 × 16-inch deck top (DECKTOP) as shown in Figure 17-3 from a ⅜-inch-thick Plexiglas or you may use plywood.

2. Fabricate the parts as shown in Figure 17-4. Note to use 1/16-inch G10 or another high-temperature plastic for the actual tube

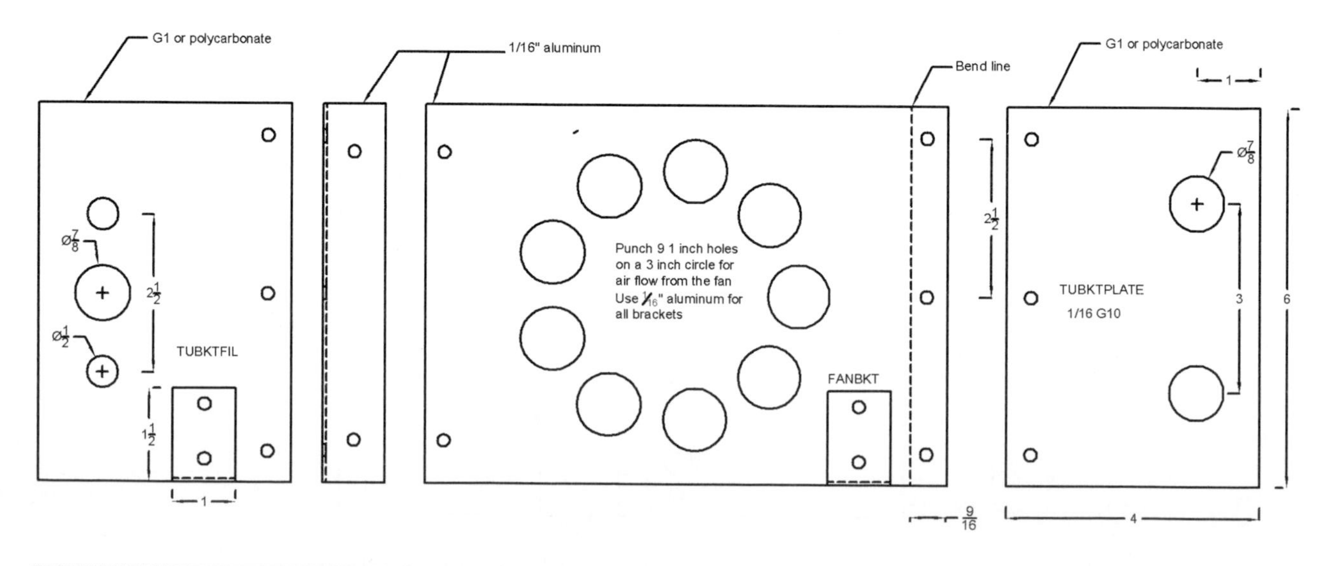

Figure 17-4 *Assembly of the tube and mounting bracket*

holders (TUBKTFIL and TUBKTPLATE). Use 1/16-inch aluminum for the other parts. Note the mating holes for assembling them together, as shown in Figure 17-4.

3. Assemble the fan and tube-mounting bracket from the parts in Figure 17-5. Use thirteen 6-32 × 1/2-inch screws and nuts. Foot brackets secure the assembly to the wooden deck via four #8 × 1/2-inch wood screws. TUBKTPLATE must be disassembled to insert the tube V1 in place. The blower fan B1 is centered to maximize air flow across the plate section of V1. It is mounted using four 8-32 × 1/2-inch screws and nuts. The entire assembly

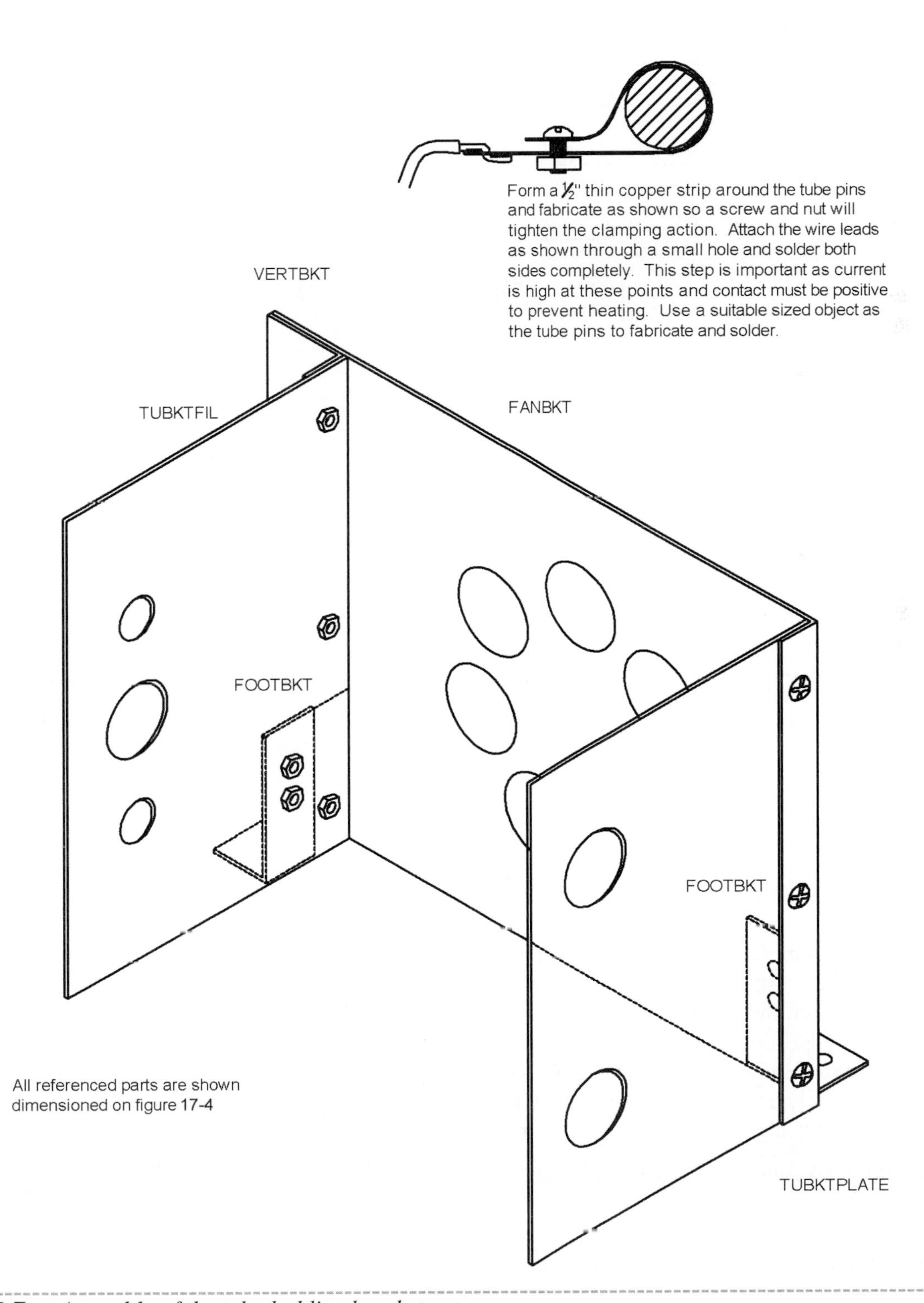

Figure 17-5 *Assembly of the tube-holding bracket*

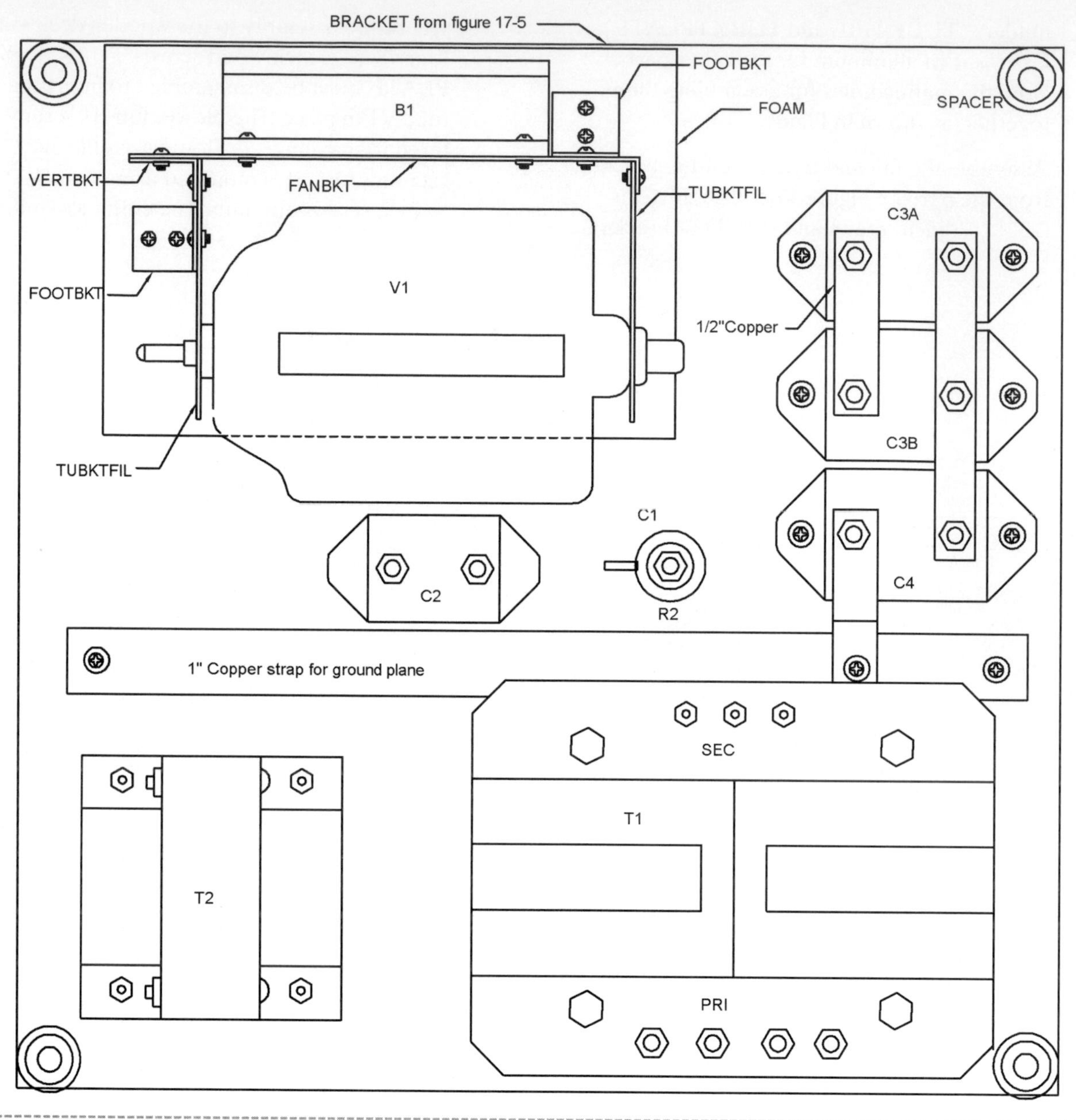

Figure 17-6 *Bottom deck assembly*

must set on a 1/4-inch foam rubber pad to avoid any vibration. Fabricate the tube pin connections from 1/2-inch strips of thin copper in the form of clamping connectors. Use screws to tighten the clamps securely around the tube pins. Crimp the connecting wires in place and carefully solder the connecting leads completely. This is important as the tube pins must pass a high current.

4. Lay out the bottom deck DECKBOT and trial-fit the components, as shown in Figure 17-6. (You may want to consider castors at the corners for system mobility.) Cut a 14-inch length of a 1-inch-wide copper strip for the main ground plane. Secure it via two wood screws at the end points. Secure the plate transformer (T1) using four 1/4-20 × 1 1/2-inch bolts with flat washers on the bottom side of the wooden deck. Mount the three mica capacitors (C3A, C3B, and C4) using 6-32 × 1 1/2-inch screws, nuts, and washers. Cut 1/2-inch lengths of copper straps and attach them as shown. Note the grounding piece on C4 must be shaped and formed to have a flat surface abutting the ground plane strap and fitting as shown.

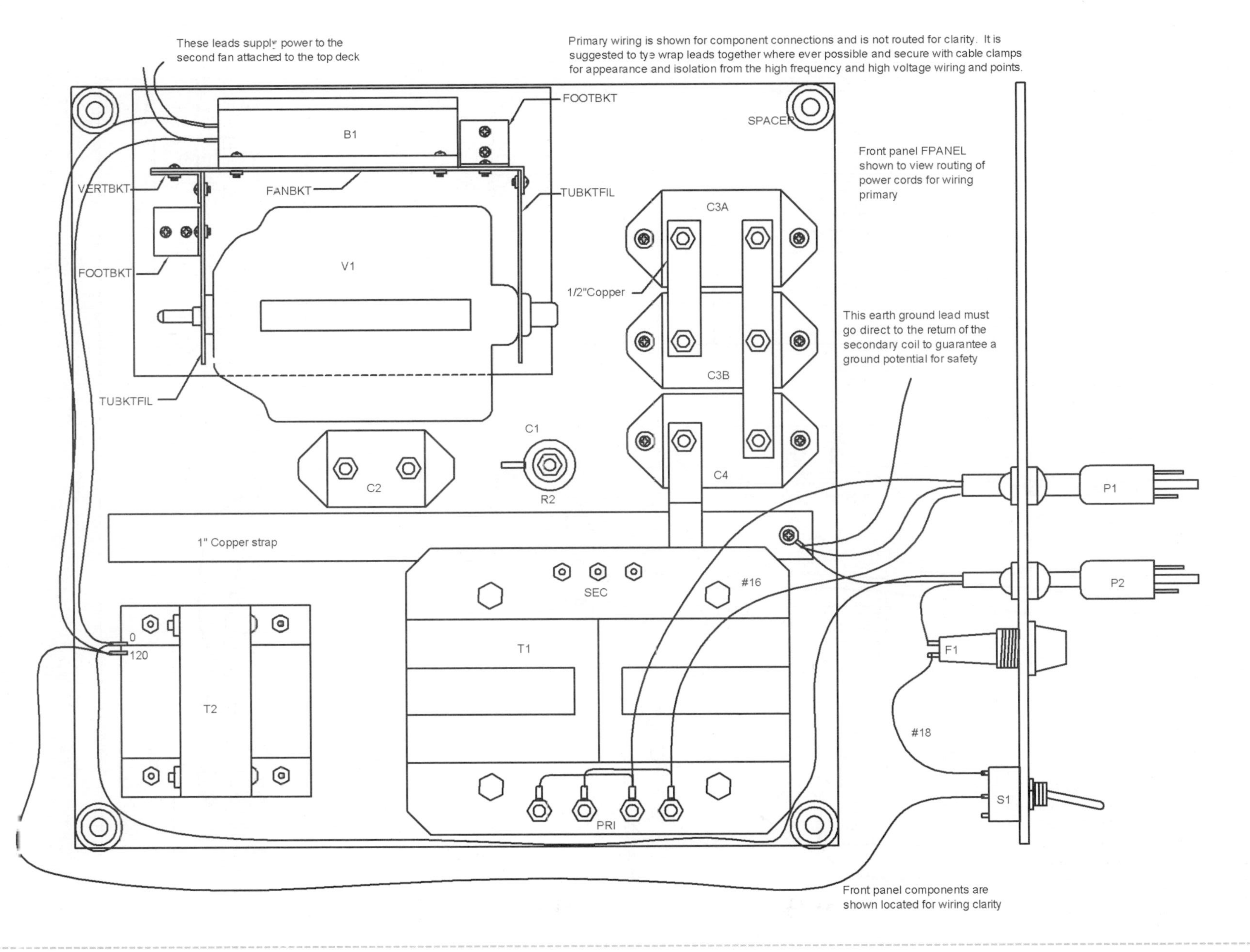

Figure 17-7 *Bottom deck primary wiring*

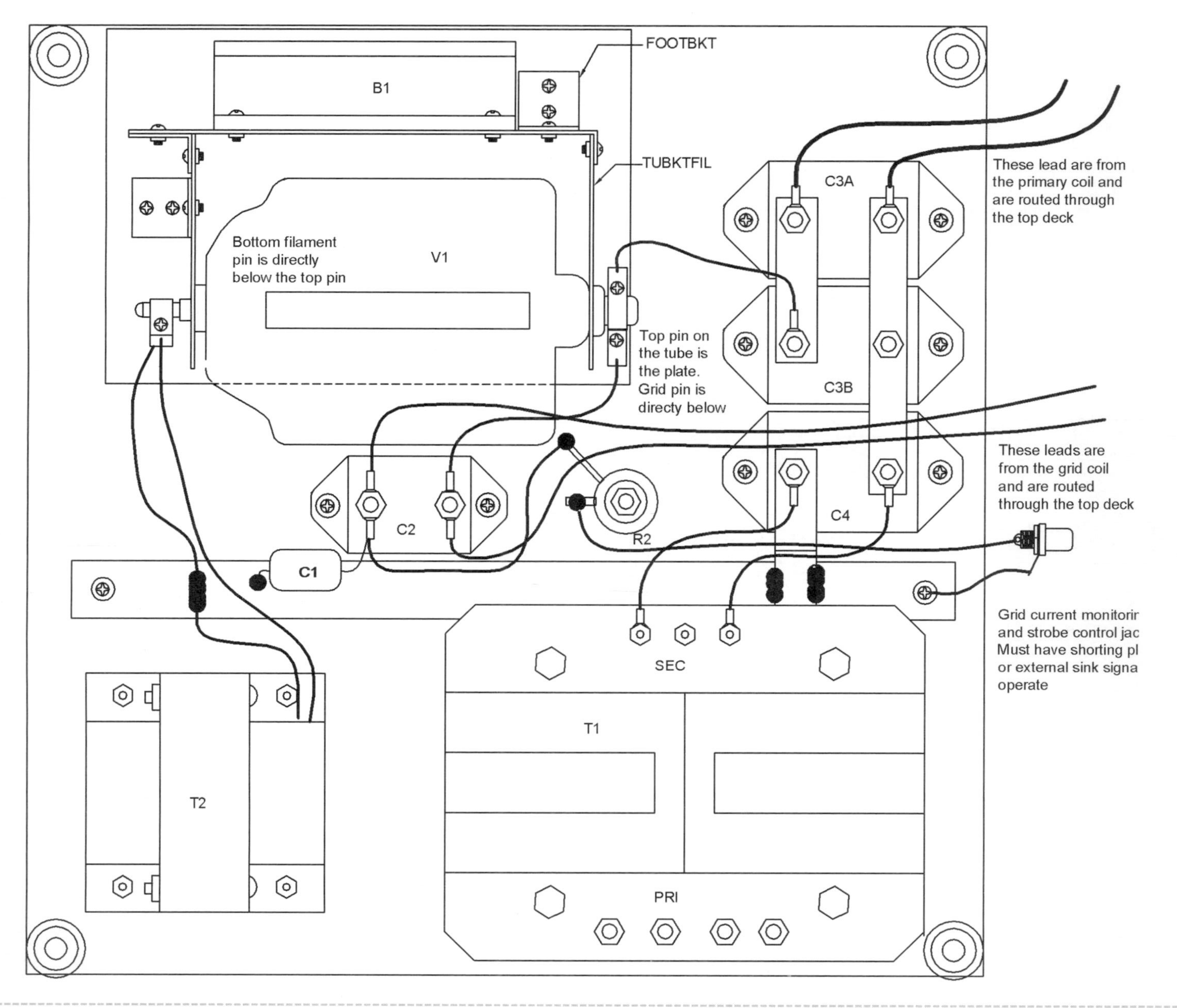

Figure 17-8 *Bottom deck high-voltage wiring*

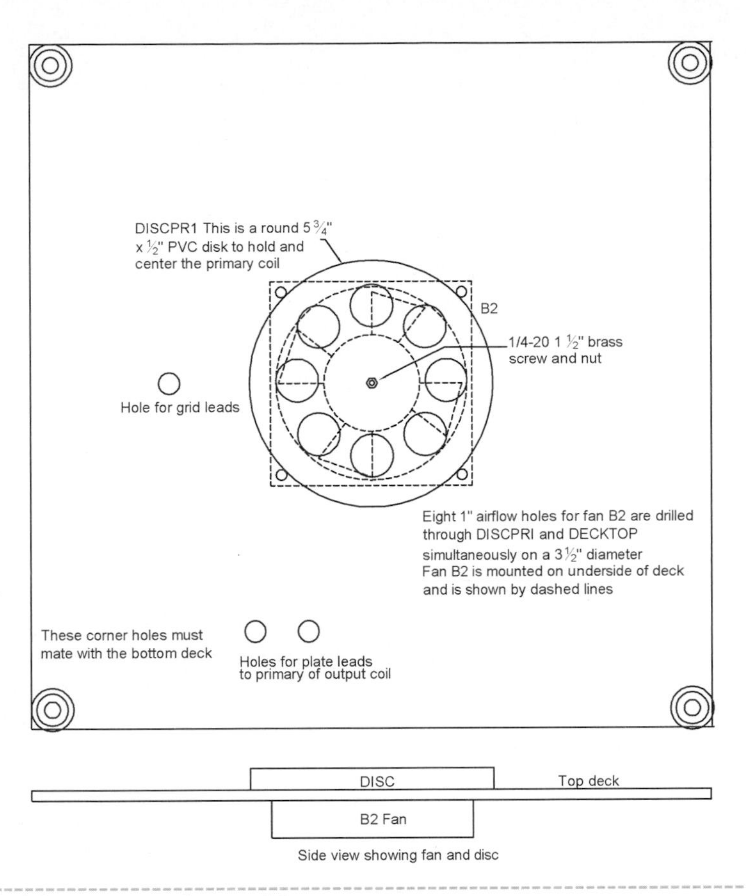

Figure 17-9 *Top deck fabrication and assembly*

Do not tighten the nuts at this time. Mount mica capacitor C2 using similar hardware and grid resistor R2 vertically using a long 1/4-inch bolt or a threaded rod. Do not overtighten. Secure the filament transformer T2 using four 8-32 × 1 1/2-inch screws and nuts. Secure the tube and fan bracket via four #8 × 1/2-inch wood screws through the two foot brackets FOOBKT. Also note the foam rubber mount FOAM. Drill holes for the 1/4-inch −20 threaded rods for the deck spacer.

5. Thread the power cords (P1, P2) through the clamp bushings (BU1, BU2) as shown in Figure 17-7 on the front-panel section (FPANEL). The actual location on the panel section is shown in Figure 17-3. Figure 17-7 is shown to clarify the wiring. Power cord P1 must be a heavy-duty, #16, three-wire cord, with P2 being a #18, three-wire cord. Connect the two green ground leads to a crimp lug along with a 24-inch piece of #18 stranded wire that eventually connects to the center ground lug of the secondary coil. Solder this lug and verify the integrity as this is mandatory for electrical safety. Proceed to wire as shown using #18 stranded wires. Note that the two leads eventually connect to the second blower (B2) mounted on the top deck. Use crimp lugs and solder wherever shown. Route the wires, tie wrap, and clamp into place for appearance and isolation from other components.

6. Wire the filament clips to the 10-volt output leads of T2 as shown in Figure 17-8. These clips are shown in Figure 17-5. You will note that insulation is removed from one of these leads, and the uncovered section is soldered to the 1-inch copper strip. Solder in the bypass resistor (C1) to the copper strip and to a lug with the other leads to R2. High-voltage wire should be used from transformer T2 to capacitor C4. Proceed to complete the wiring using crimp lugs and soldering. Two pairs of wires are shown from the primary and grid coil. These are not actually connected at this time as they are part of the top deck assembly (DECKTOP). Jack (J1) is attached to the front panel shown located in Figure 17-3.

7. Fabricate the DECKTOP, as shown in Figure 17-9. This piece was originally shown as being

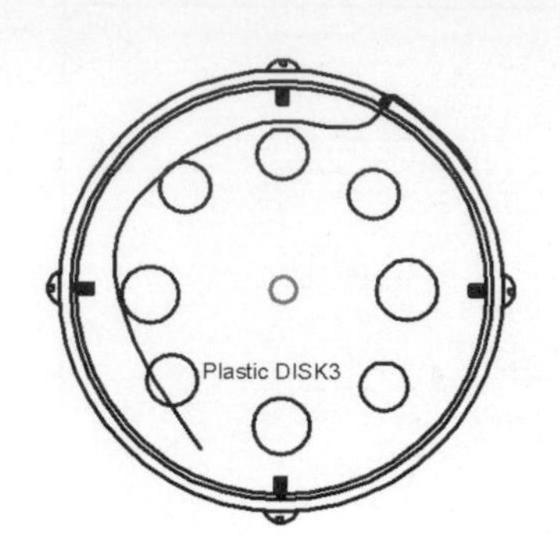

The secondary is wound on the form as shown for FORMSEC in figure 17-3.
Special note: The coil form must be totally clean both inside and out. Use isopropyl alcohol and allow the form to completely dry. Coat with a layer of orange shellac should do this on a dry day or in a dehumidified area.

Start by inserting the polycarbonate disk DISK3 into the output end of the coil form flush with the end. Secure using 3 or 4 small plastic screws. Do not use metal on the coil output end as arcing will occur.

Insert a 1" width aluminum bracket into the base end with a 1/4-20 tapped hole for screwing on to the brass screw from figure 17-9. Bend down ½" 90 degree ends to attach to inner diameter of coil via brass screws and nuts.

Position roll of #28 heavy magnet wire and thread into holes on base end of form. Attach free end to bracket via one of the screws. Drill small holes at each end of the coil form ½" from ends for threading the "start" and "finish" leads of the windings.

Start to wind the turns and continue for 33" in length being careful to keep the wire tight, free of kinks and avoiding any overlaps. Do approximately several inches at a time and shellac in place using orange shellac. Always secure lead with a piece of good adhesive sticky tape as an "unwind" will be disastrous. Winding this coil with two people is much easier.

Completed coil should contain approximately 2400 turns and tune to around 200 khz when free standing. Note winding should be in the same direction as primary coil LP1

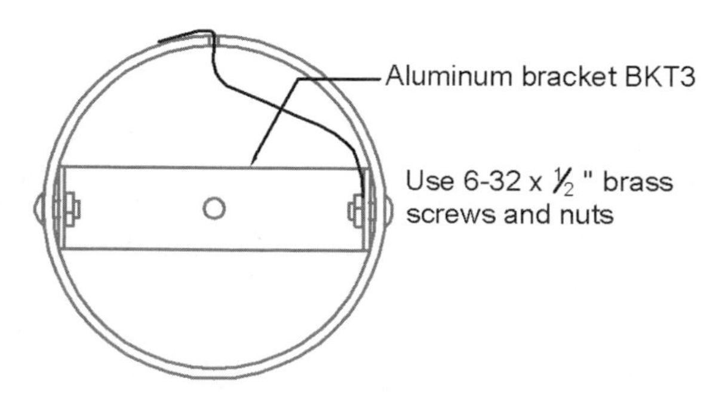

Figure 17-10 *Assembly of the secondary coil*

a 3/8-inch Plexiglas square from Figure 17-3. Fabricate a 5 3/4-inch PVC disc (DISCPRI) from 1/2-inch-thick sheet. This piece is for securing and centering the primary coil assembly, shown later in the plans. Secure it in place on the deck via a 1/4-20 × 1-inch brass screw and nut. This will be the grounding point for the bottom of the secondary coil and

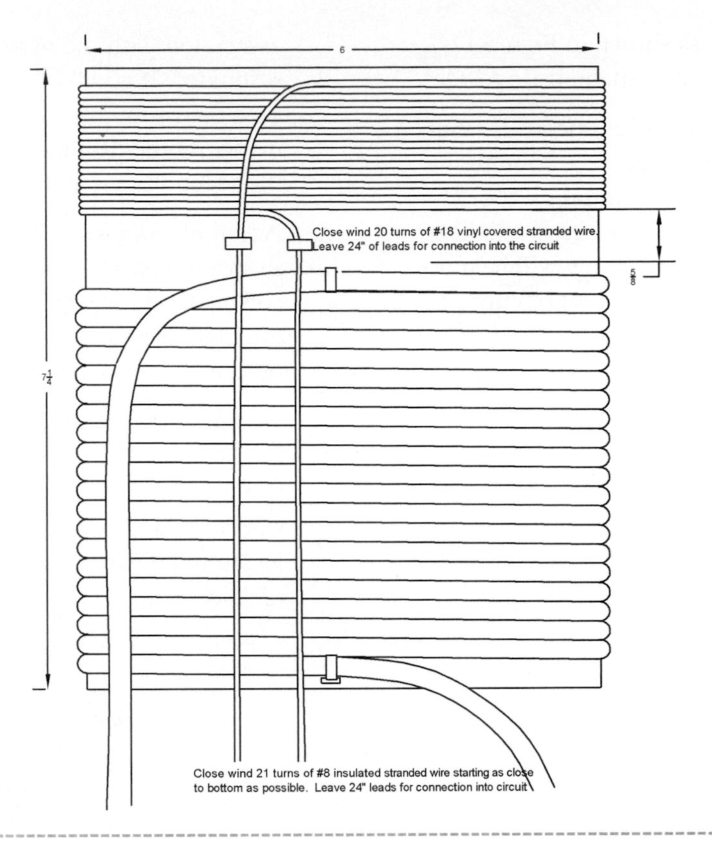

Figure 17-11 *Assembly of the primary coil*

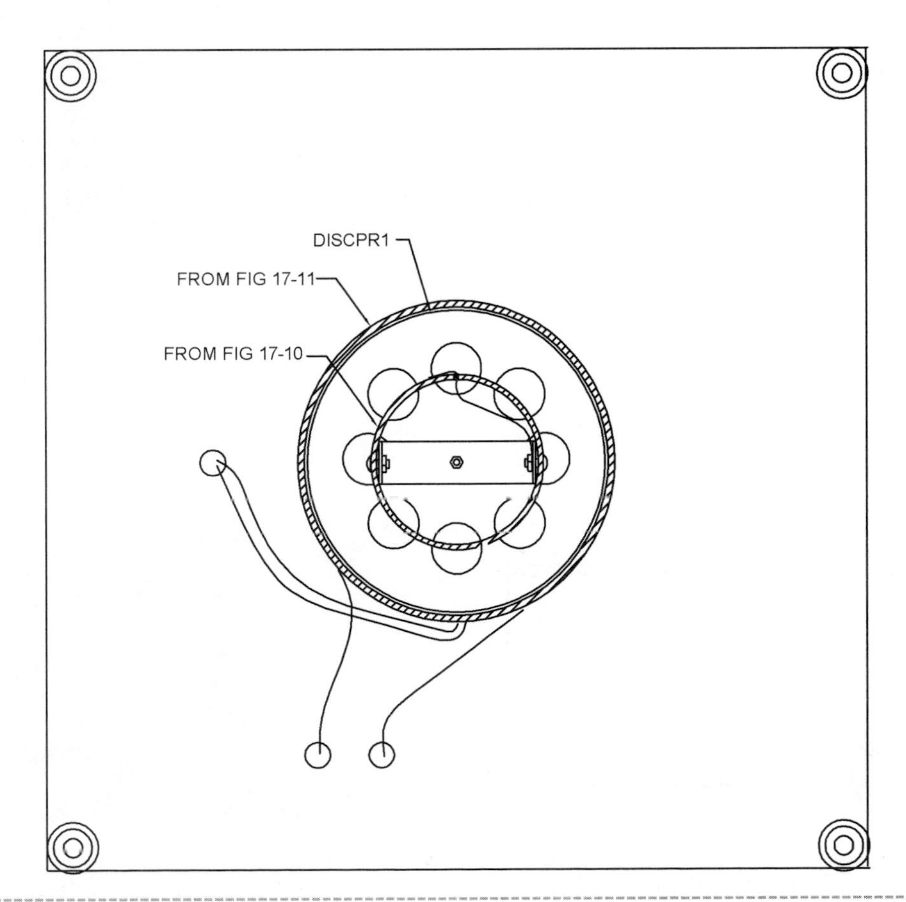

Figure 17-12 *Final assembly of top deck*

goes to earth ground, as shown in Figure 17-7. Proceed to drill the eight 1-inch air flow holes.

At this time, the builder should be aware that if his or her coil is to be used on a continuous basis, the second blower mounted on the underside of the top deck is necessary to keep the bottom of the coils cool. They will heat up over time due to the resonant rise of current associated with the base of a quarter-wave resonator. If your coil is to be used for a short-time operation of several minutes, the part involving the drilling of the air flow holes and the second fan may be omitted.

8. Assemble and wind the secondary coil as shown in Figure 17-10. Proper winding of the coil is mandatory for achieving optimal operation. Kinks and overlaps will seriously affect output.

9. Assemble the primary coil assembly, as shown in Figure 17-11. Start at the bottom and tightly wind the turns. Use tie wraps to secure the start and finish of this heavier 21 turns of the #8 stranded wire. You may want to place dabs of silicon rubber to secure them in place. Start

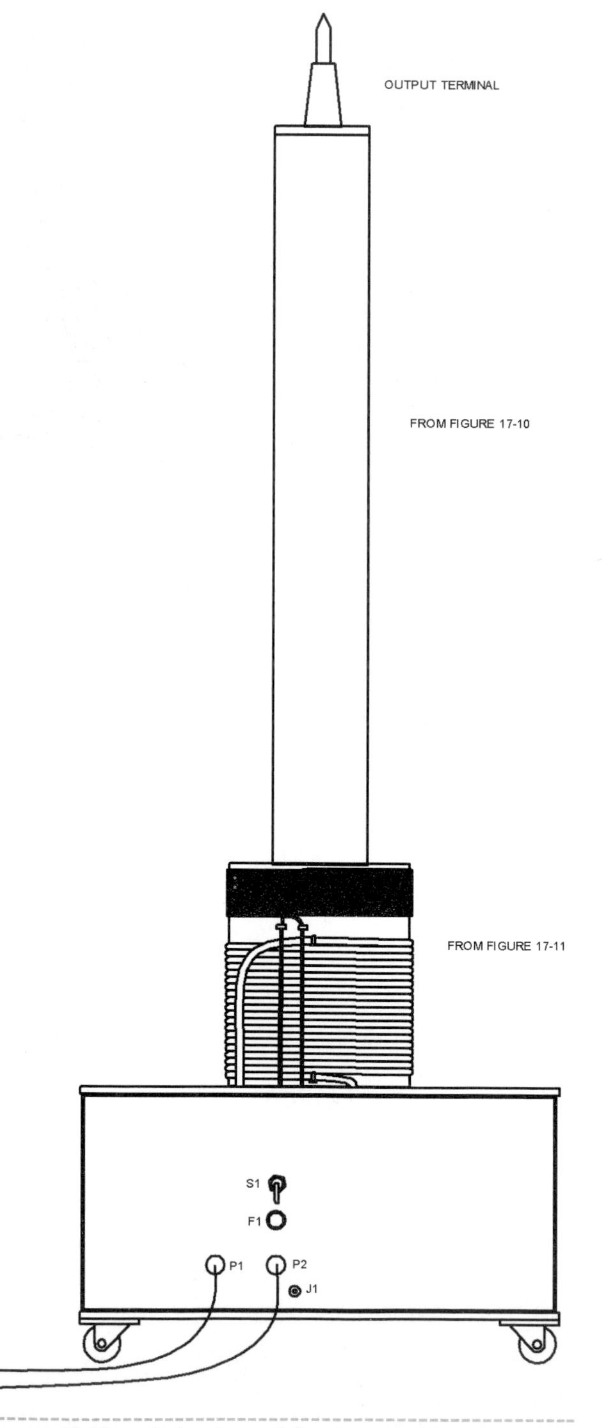

Figure 17-13 *Side view of final assembly*

Figure 17-14 *The final assembly*

Figure 17-15 *The right-side view*

from the top and wind 20 turns of #18 vinyl-covered stranded wire and tie wrap it in place. at there must be at least ½-inch the closest proximities of these two a flashover (or spark break) could the vacuum tube. Leave at least 2 feet leads for connections, as shown in 7-8.

assemble the top deck, as shown in 7-12. The primary coil section slips over the disc DISCPRI, as shown in Figure 17-8. The tapped hole in the bracket of the secondary coil screws onto the center brass screw shown in Figure 17-8. It is this screw that grounds and secures the secondary coil.

Route the leads from the coils through their respective holes for connection into the circuit. You will note back in Figure 17-8 that the heavy plate leads are shown connected to C3A via heavy crimp lugs and are also soldered. The same applies for the smaller grid leads that connect across C2. These leads may require switching during the test steps. Both sets of these leads carry high-frequency energy with the heavier leads carrying high-radio-frequency voltages and currents. These heavier leads must be as short and direct as possible and not be in close proximity to other objects.

Figure 17-13 shows the completed graphic of the coil assembly with an insulator and a pointed discharge pin. Any suitable feed through the insulator around 2 to 3 inches in height can be used. You may also use a piece of polycarbonate or Teflon rod for this part. Other plastics may carbonize and burn.

The following figures are actual photos showing different views of the assembly apart and intact. They are intended as assembly aids.

Figure 17-14 shows a 12 × 3-inch toroid as the output terminal. You will note the two power cords exiting through the front panel.

Figure 17-15 shows the right side with a view of the blower B1 and tube-mounting bracket.

Figure 17-16 shows the left side with a view of the filament and plate transformers.

Figure 17-16 *The left-side view*

Figure 17-17 *The front view*

Figure 17-18 *The top view*

Figure 17-19 *Another top view*

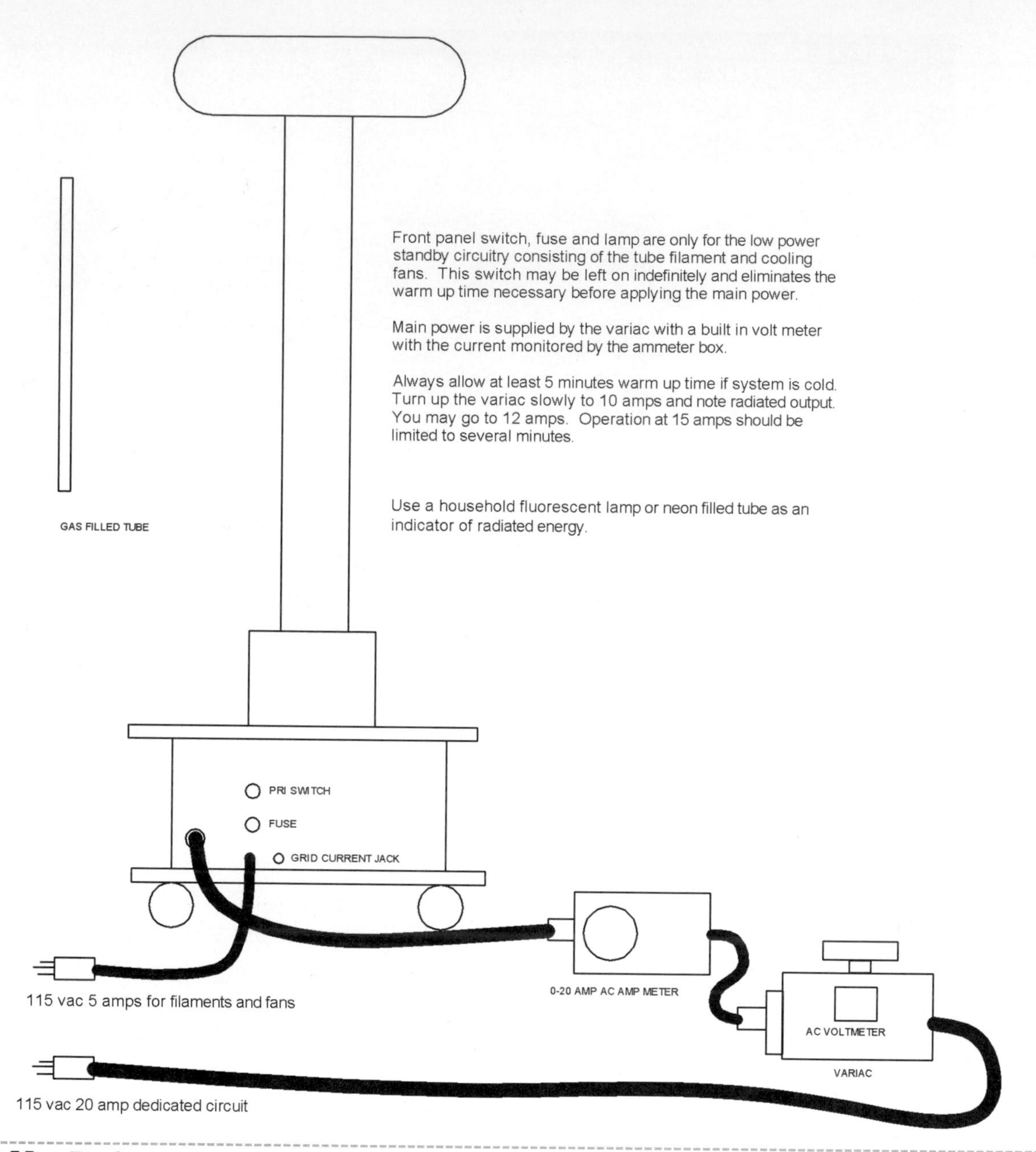

Figure 17-20 *Final system setup*

Figure 17-17 shows the front view through the clear plastic panel. Note the three mica capacitors C3A, C3B, and C4.

Figure 17-18 shows the top view with the clear plastic top deck removed. We used a piece of shielded strap to connect the plate terminal of the tube to C3A and C3B. This is not necessary and may be heavy flexible wire.

Figure 17-19 also shows the top view with the clear plastic top deck removed but at a different angle.

Testing and Operating

1. You will need a dedicated 20-amp line to obtain full output. Set up the system as shown in Figure 17-20. Place the unit away from metal objects, insert a 7-amp fuse in the holder, and insert a shorting into the grid current jack (J1). Plug in the 115 VAC plug for the filament and blower fans, and wait for the tube to glow and the fans to blow. Always allow at least 3 minutes for the tube filament to reach the proper emission temperature.

2. Verify that the variac is properly wired and turned to zero volts, as indicated on the voltmeter. Plug it into a 20-amp circuit. Slowly turn up the variac to 30 volts and verify some output as indicated by the bulb glowing or some corona discharge occurring on the output pin.

   If you do not observe some output, it will be necessary to completely shut the system down by unplugging both plugs, reversing the grid coil wires that attach to C2, and repeating the previous step. This is referred to as phasing the system. You should get substantial output and easily light the tube when properly phased.

3. Once you verify that the coil is properly phased, you may start to increase the voltage, noting the output discharge, the input current, and the applied AC voltage as indicated on the variac meter. You must also observe that the plate of the tube does not start to glow, as this is an indication of mistuning. The tube may have a very soft glow at full 120 VAC where the output should be 15 inches or more. The input current will approach 20 amps.

Notes

Tuning the coil for maximum output is tricky and may take some effort on the part of the builder. The tuning values will vary as the discharge increases due to the virtual capacitance of the ionization cloud at the top of the coil. The output terminal will also greatly effect the output tuning. Try adding or subtracting a turn or two on the primary plate coil, as this will change the resonant drive signal. You will need more inductance as the output increases, and the trick is to select the optimum value at the desired input power.

A 12-inch toroid or another large capacitive terminal will cause larger circulating currents within the coil and greatly enhance the wireless transmission of energy, as can be demonstrated by observing the distance that is possible for lighting the gas discharge tube without a wire. This will eliminate the air breakdown discharge when operating in this mode as the terminal voltage drops unless you increase the input power to overcome this loading effect.

The geometry of the secondary coil as shown is not the optimum for maximum length discharges. A squatter size would be more efficient but would radiate less.

Table 17-1 Vacuum Tube Tesla Coil Parts List

Ref. #	Description	DB Part #
R2	5K, 100-watt wire-wound resistor	
C1	.004-microfarad, 1,000-volt polypropylene capacitor	DB# .0039/400
C2	.003-microfarad, 3-kilovolt mica capacitor	DB# .003/3 KVM
C3A, B, C4	Three .01-microfarad, 10-kilovolt mica capacitors	DB# .01/10 KVM
V1	833A Triode vacuum tube	DB# 833A
T1	3,000-volt, .5-amp plate transformer	DB# 3KV/.5A
T2	10-volt, 10-amp filament transformer	DB# 10V/10A
B1, 2	5 × 5-inch muffin fan and 5 × 5-inch high-output muffin fan	DB # FAN1
S1	*Single pole, single throw* (SPST) 5-amp toggle switch	
F1/FS1	Panel-mount fuse holder and 5-amp fuse	
P1	8-foot #16 three-wire power cord	
P2	8-foot #18 three-wire power cord	
AMPMETER	20-amp AC panel mount ammeter	
VARIAC	18-amp variac with built-in meter	DB# VARIAC18A
TUBE	15-inch neon tube	DB# NE15
TER1	12 × 3-inch toroidal terminal	DB# TO12

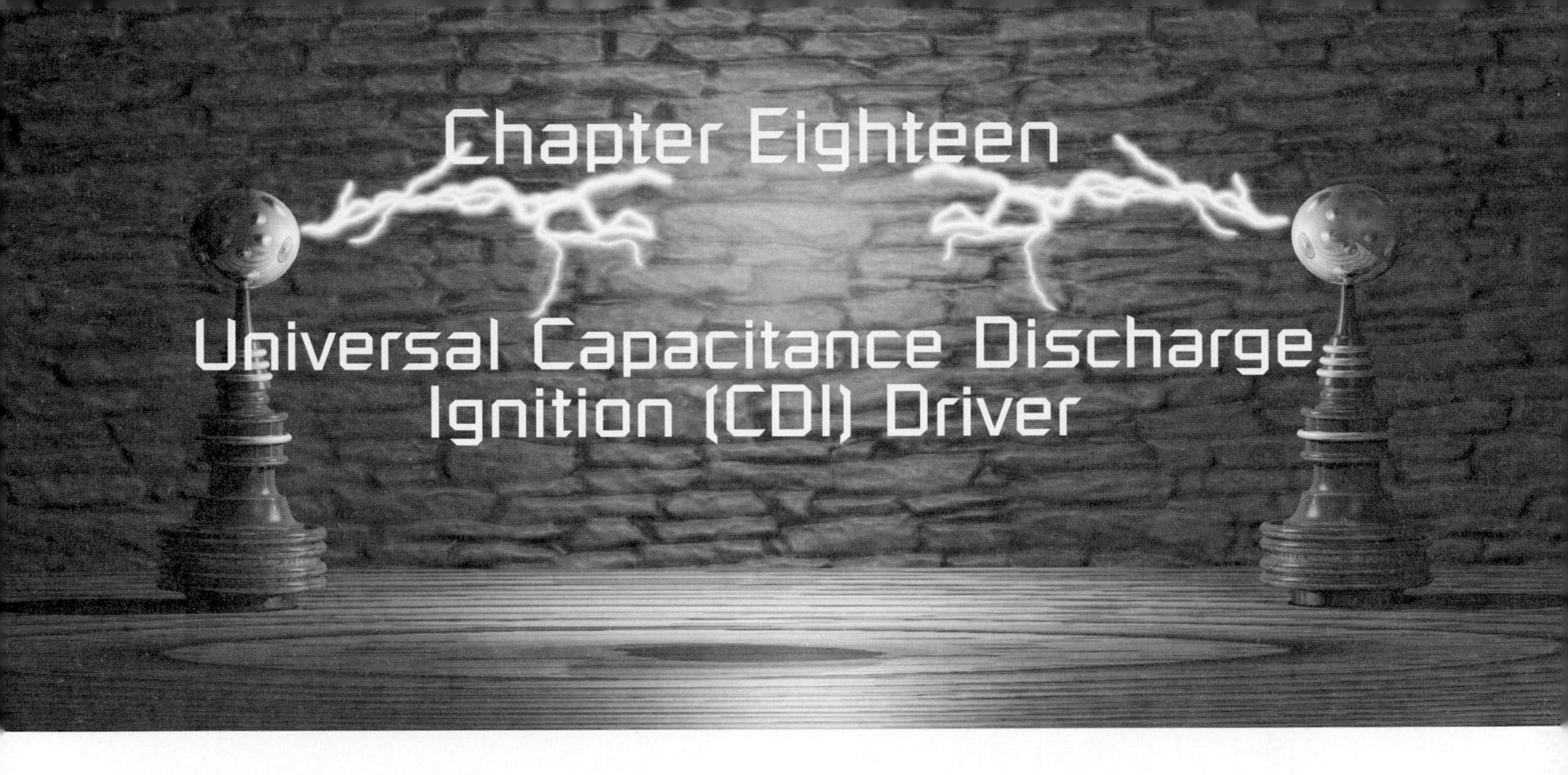

Chapter Eighteen

Universal Capacitance Discharge Ignition (CDI) Driver

Figure 18-1 shows how to construct a very useful driver circuit that can be used to pulse most induction and ignition coils in the forward direction. High-quality automotive ignition coils can produce 50- to 75,000-volt pulses capable of jumping several inches. The system allows control of the pulse repetition rate up to 100 reps/second and individual pulse energy up to .3 joules. A maximum power output of over 30 watts is possible. Our laboratory model has generated discharges of up to 10 inches from some high-performance ignition coils.

The advantage of a *capacitance discharge ignition* (CD) system is its ability to produce high peak currents and voltages simultaneously. This property is well understood and appreciated in the automotive industry where fouled spark plugs are prevented by the brisance of the discharge. Electric fences burn away vegetation growth from contacting the wire by using high current pulses. Stun guns use these high peak current pulses to thwart the neuromuscular system. All this peak power with very low resultant energy allows many useful applications where battery-operated systems can be efficiently utilized.

Figure 18-1 *The capacitor discharge driver*

This is an intermediate-level project requiring basic electronic skills and one should expect to spend $25 to $50. All the parts are readily available, with specialized parts obtainable through Information Unlimited (www.amazing1.com), and they are listed in Table 18-1.

Circuit Description

Figure 18-2 shows the complete circuit schematic of the adjustable energy and repetition rate capacitor discharge driver. The adjustable voltage regulator I2 controls the operating voltage to the system via control pot R9/S1 from 6 to 12 volts. Input to I2 is a 12-volt DC wall adapter transformer with a current rating of 1 to 2 amps being controlled by switch S1. This adjustable voltage is fed to the switching transistors Q1 and Q2 via the center tap of stepup transformer T1. The switching action of these transistors is controlled by a complementary pulse driver circuit I1 with its frequency controlled by trimpot R3. This circuit generates a high-frequency signal that alternates between the two primary windings, in turn inducing a higher voltage in the secondary winding.

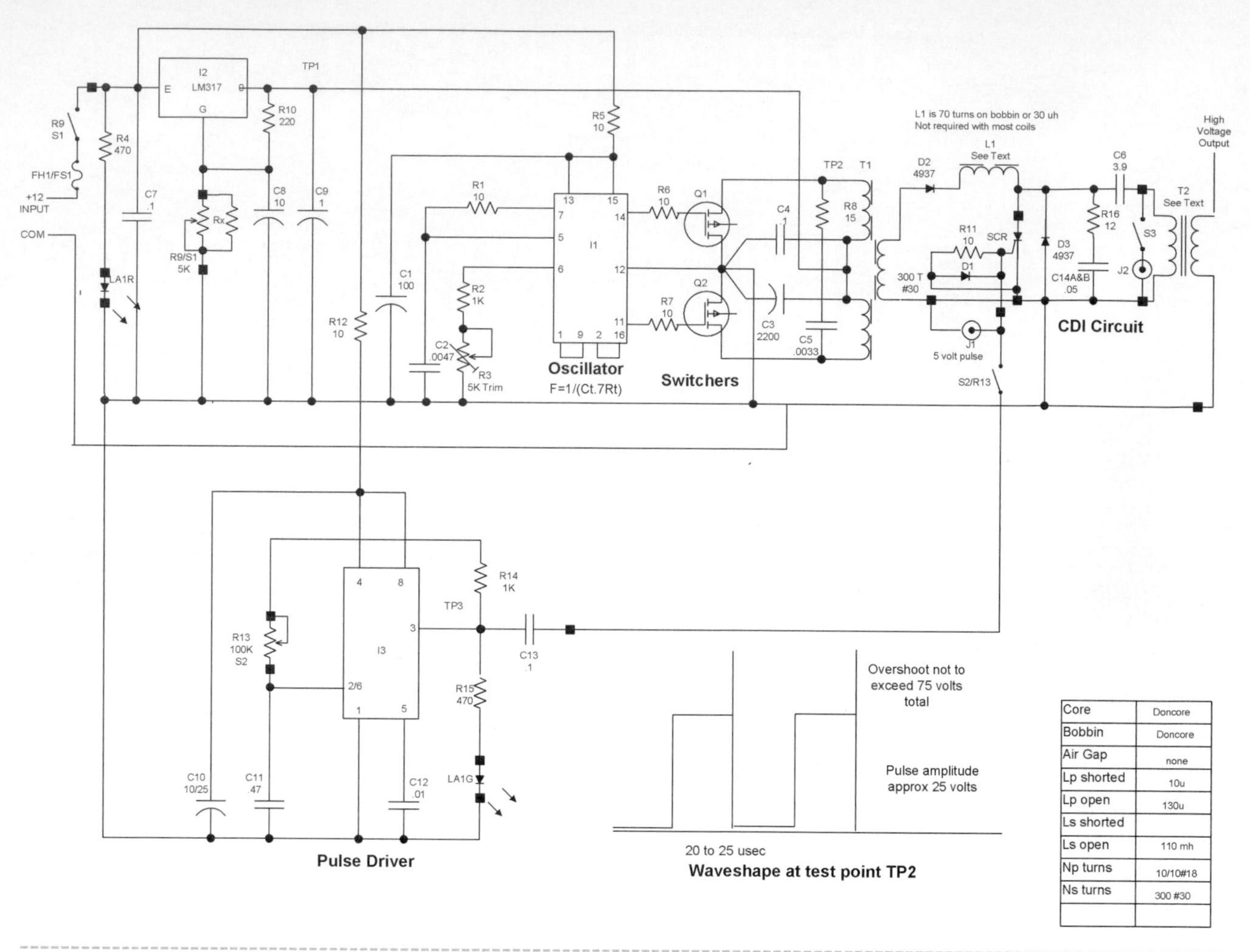

Figure 18-2 *Circuit schematic*

T1 is a transformer with a preset, built-in leakage reactance that controls the short-circuit-charging current to discharge capacitor C6 through rectifying diode D2. Once C6 is charged, it can be discharged by the silicon controlled rectifier switch (SCR) through the primary high-voltage pulse transformer T2. The SCR is controlled by timer chip I3, connected as a variable pulse generator via pulse rate adjust pot R13/S2. It is these pulses that trigger the SCR to switch energy from C6 into T2, generating the high-voltage output pulses. You may also externally switch the SCR via a pulse signal at input jack J1. R13/S2 must be off for this external function. The pulse transformer is shown as a low-power unit in the parts list.

The real intention of the circuit is the ability to pulse-drive many different ignition coils at different energy levels as selected by R9 and pulse repetition rates as selected by R13. We have obtained sparks up to 4 inches with conventional automotive coils.

Assembly Steps

1. Assemble the switching transformer as shown in Figure 18-3.
2. Fabricate the heatsink bracket (HS1) and transistor-mounting bracket (BR1), as shown in Figure 18-4. Make sure you've removed any sharp edges or burrs in the mounting holes that could break through the insulating pads.
3. Attach Q1 and Q2 as shown, noting that the mounting tabs are insulated from the actual metal heatsink bracket by the nylon screws.

1. Bifilar (parallel) wind two different color #18 magnet wire for 10 turns on the primary bobbin. Tape in place leaving 8" leads. Note that different colors will help identifying the lead ends.
   #18 solid magnet wire can be used however high frequency LITZ wire will give a slight improvement. You can make this wire by obtaining 6 pieces of #26 magnet wire and twist together as a single wire.

2. Wind 300 turns of #30 magnet wire in the 10 segments of the secondary bobbin. Wind 30 turns per segment and attach the start and finish leads to the connection lugs

3. Place the wound coils onto the cores as shown and tape tightly into place.

4. You may verify the inductance as follows secondary open:
   A to B&C are around 130 micro henries.
   D to B&C are around 130 micro henries
   Secondary winding 110 millihenrys

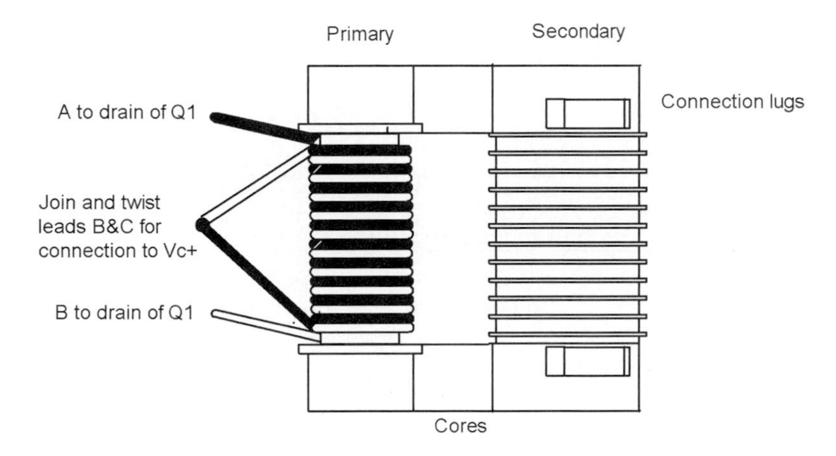

*#18 solid magnet wire can be used however high frequency LITZ wire will give a slight improvement. You can make this wire by obtaining 6 pieces of #26 magnet wire and twist together as a single wire.

Figure 18-3 *T1 Transformer construction*

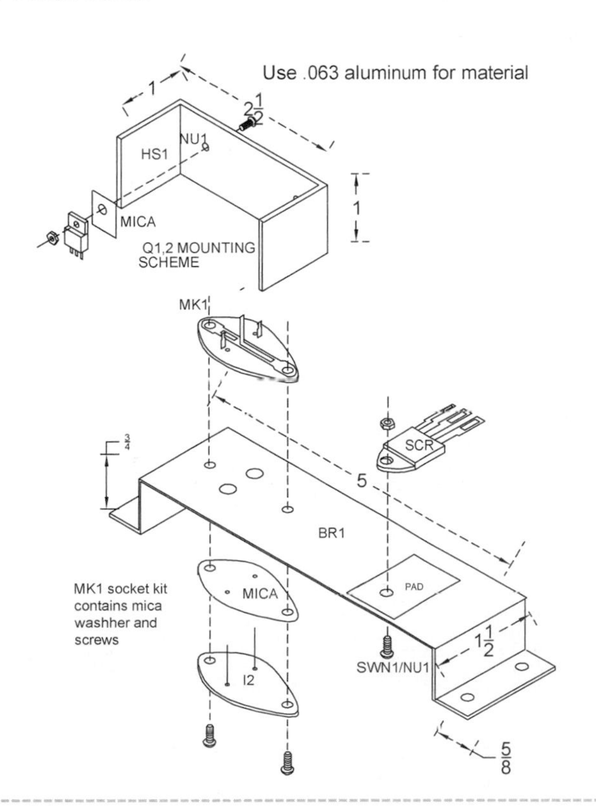

Figure 18-4 *Assembly of brackets*

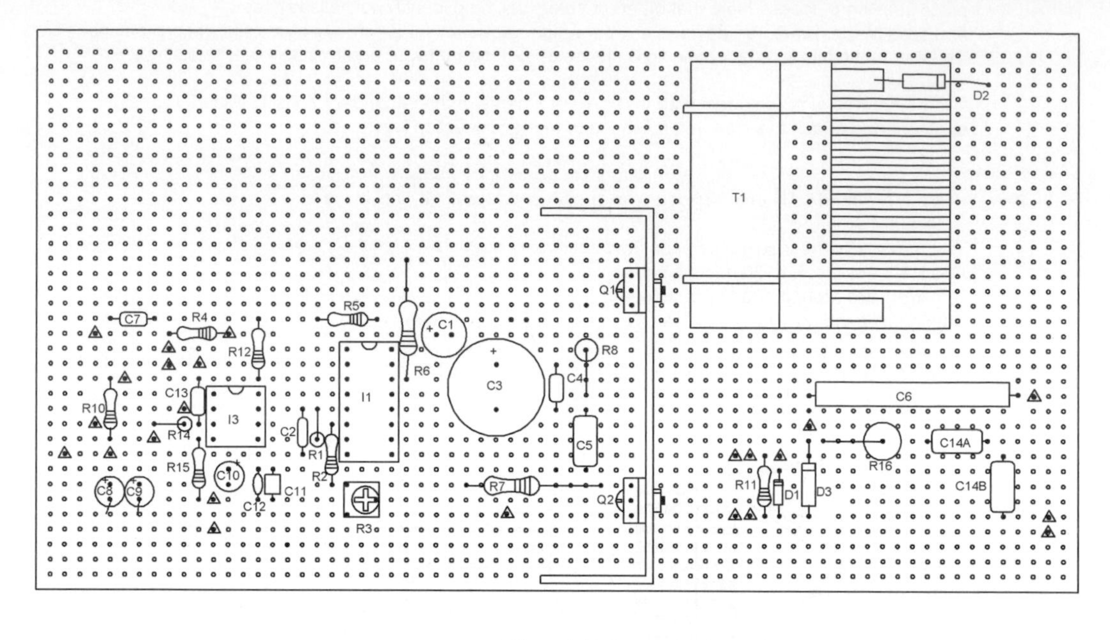

Figure 18-5 *Assembly board and parts location*

Attach I2 using the socket assembly kit (MK1) and attach the SCR using a nylon screw and an insulating pad (PAD).

4. Cut a piece of .1 × .1 grid perforated circuit board (PB1) at 7 × 3.8 inches.

5. Insert the components as shown in Figure 18-5, starting from the lower left-hand corner. Count the perforated holes for a guide. Note the polarity marks on C1, C3, C8, C9, and C10 and diodes D1, D2, and D3. It is suggested that you use sockets SO16 for I1 and SO8 for I3.

6. Wire the components as shown in Figure 18-6 using the leads of the actual components as the connection runs, which are indicated by the dashed lines. Always avoid bare-wire bridges and messy solder joints. Check for cold or loose solder joints and secure T1 using double-sided sticky tape or plastic tie wraps.

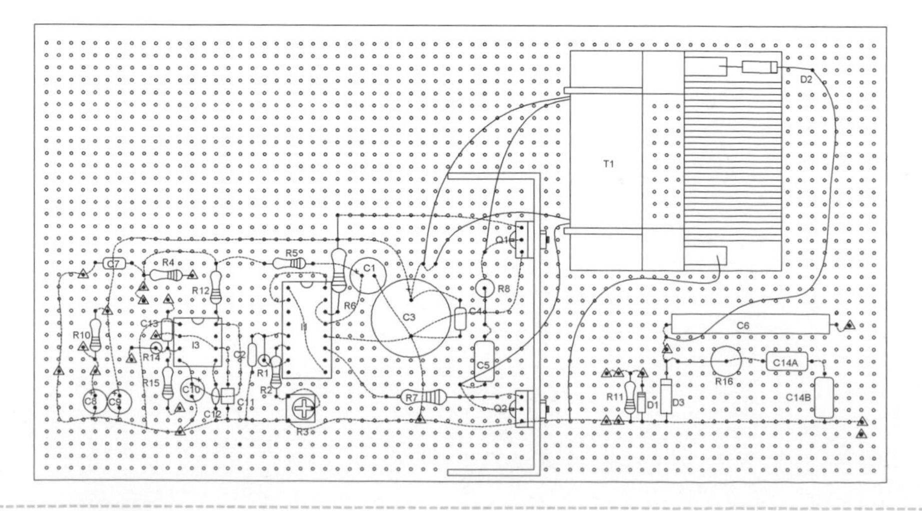

Figure 18-6 *Internal onboard wiring*

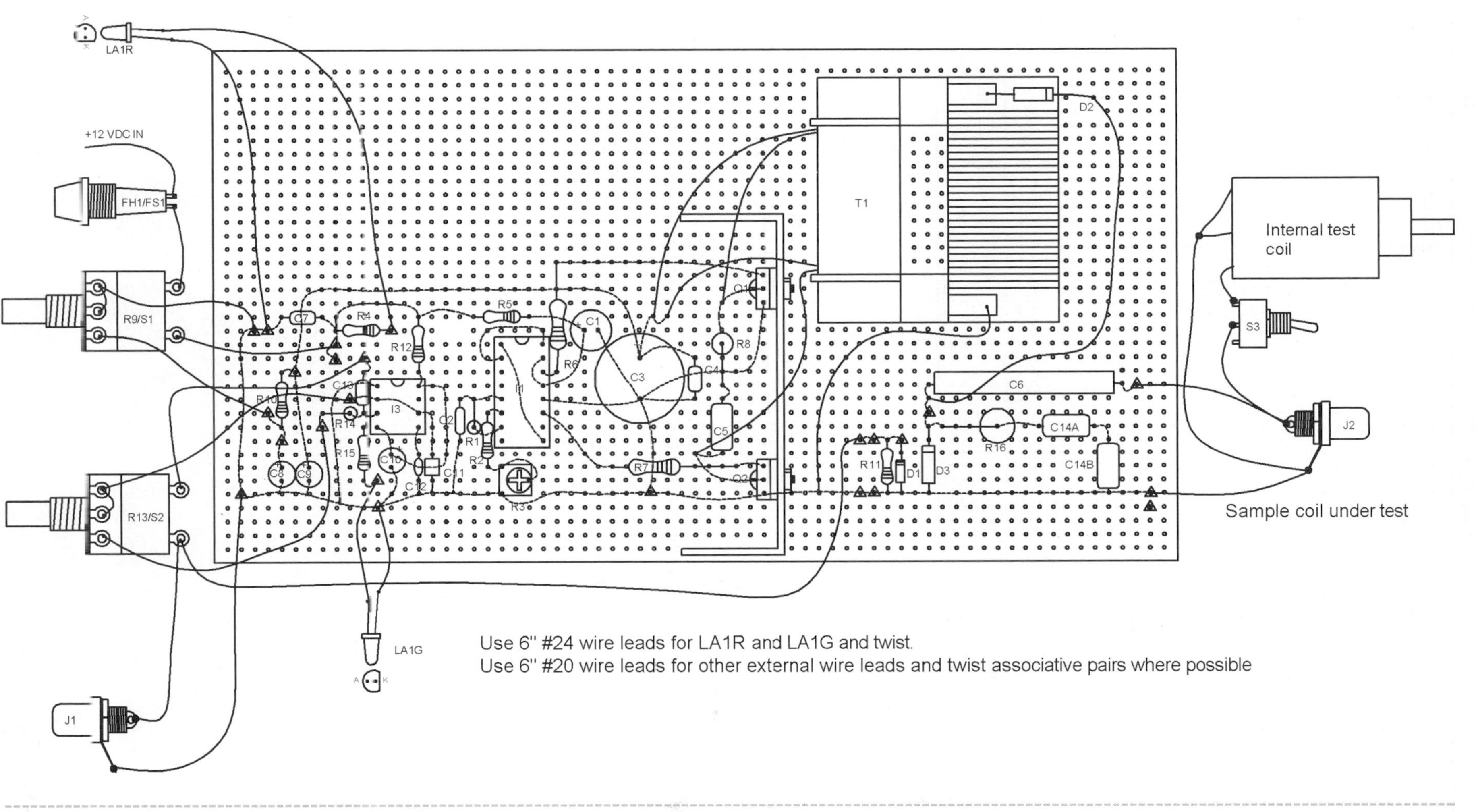

Figure 18-7 *External wiring to controls and input*

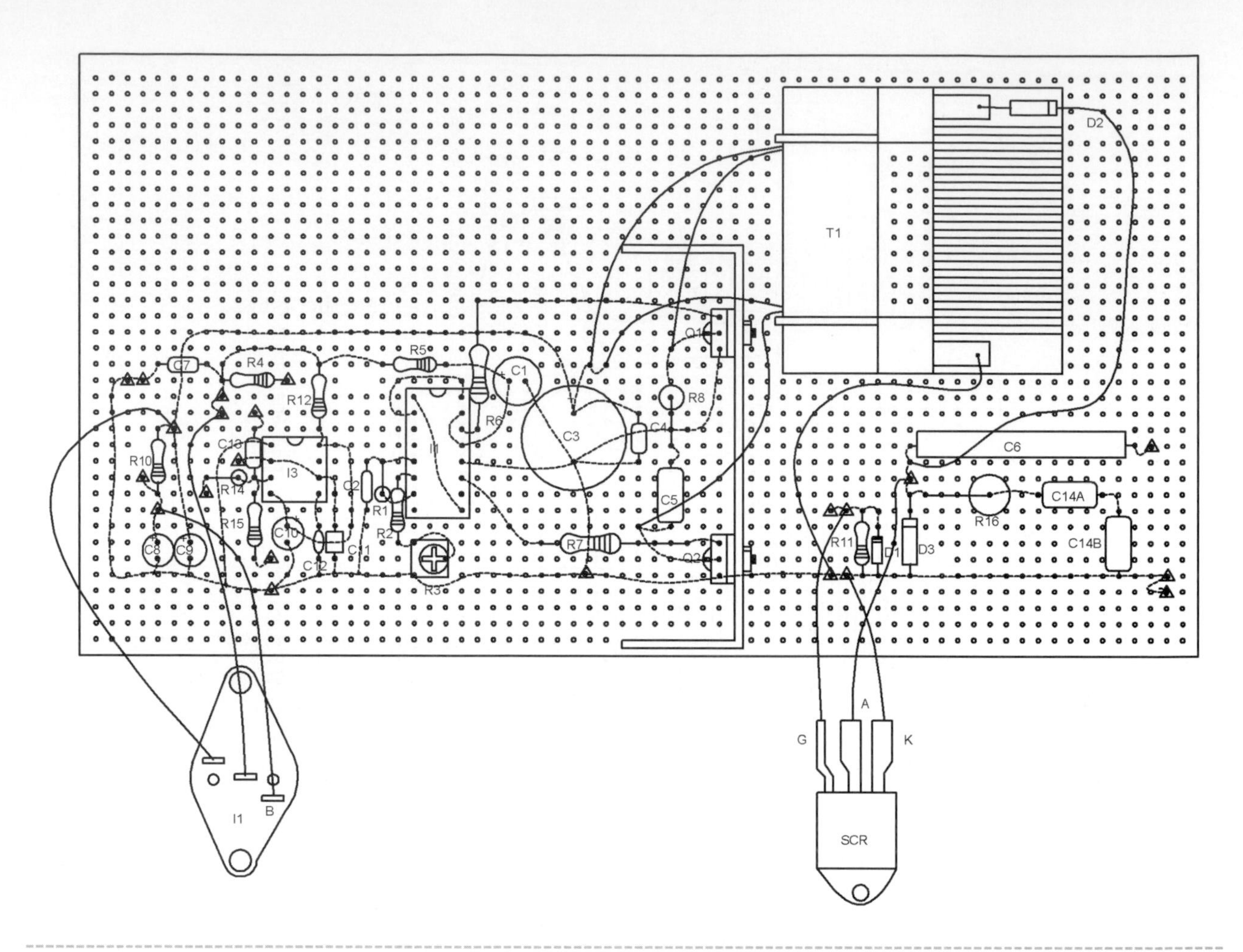

Figure 18-8 *External wiring to transistor modules*

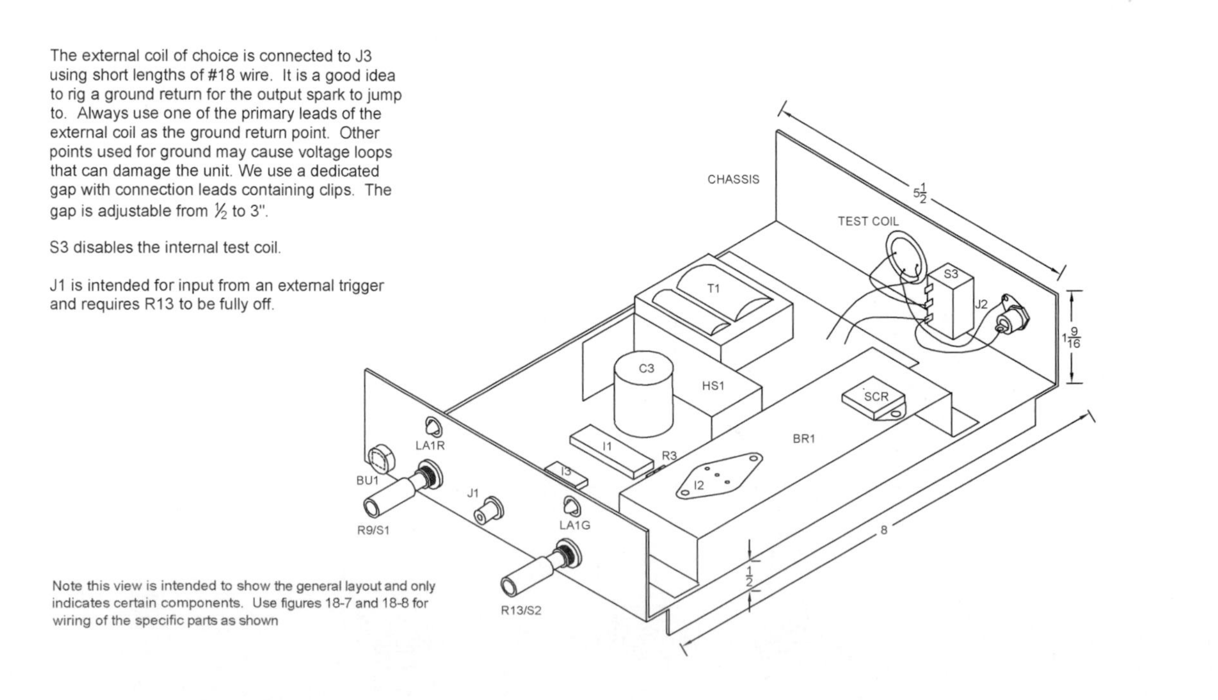

Figure 18-9 *Final assembly*

7. Wire the external components as shown in Figure 18-7.

8. Wire the transistors from the assembled bracket, as shown in Figure 18-8 and detailed in Figure 18-4.

Fabricate the mounting chassis at the overall dimensions shown in Figure 18-9. Trial-fit the remaining components, proceeding to drill and locate with the proper sized holes. Attempt to follow the figure and note the position of the BR1 bracket assembly and front panel controls. Route and twist all lead pair to neaten appearance. You may want to splice in some output leads to replace the connections to the test pulse transformer T2. These should be #18 wire and as short as physically possible to the external ignition coil. Also create a suitable cover if required and ventilate it with holes for air cooling.

Test Steps

This step will require basic electronic laboratory equipment, including a 60 MHz oscilloscope.

1. Turn off both controls R9 and R13. Set the R3 trimpot at midrange.
2. Connect a scope to TP2 and a voltmeter to TP1.
3. Apply 12 VDC to the input leads and turn on R9/S1. Adjust the meter for 5 volts and adjust the period of the waveshape at TP2 to a 25 us period. The input current should be less than .2 amps on a 12-volt power supply.
4. Note that the LED indicators are on with LA1G flashing at the pulse repetition rate determined by the R13 setting. This occurs even with R13/S2 fully off.
5. Turn on R13 and adjust it for 10 volts. Note the test coil sparking over its outer surface. Rotate R13 and notice the pulse rate increasing. Only operate the test coil to test circuit performance.

Notes on Operation

The unit is intended to drive larger ignition and induction coils. Jack J2 is intended as the output to the external coil under operation.

Note that the test coil must be disabled for external operation and may be disconnected by unsoldering or using a suitable high-current switch that may be mounted on the rear panel adjacent to J2.

When using unknown coils, it is advised that you monitor the DC voltage across the SCR. It should never exceed 400 volts, as the storage capacitor will be overrated. The SCR can handle up to 800 volts.

A change occurs in the charging voltage as the pulse repetition rate is changed. It decreases as the repetition rate increases. You can compensate for this by adjusting R9, but you must be careful to turn it down before lowering the repetition rate. Always start with the voltage control R9 at the lowest setting and increase as necessary. Additionally, jack J1 is intended for external triggering and must be used with R13/S2 in the off position.

Table 18-1 Parts List for CDI Driver

Ref. #	Description	DB Part #
R1, 5, 6, 7, 11, 12	Six 10-ohm 1/4-watt resistors (br-blk-blk)	
R2, 14	Two 1K, 1/4-watt resistors (br-blk-red)	
R3	5K horizontal trimmer	
R4, 15	Two 470 ohm, 1/4-watt resistors (yel-pur- br)	
R8	15-ohm, 3-watt metal oxide resistor	
R9/S1, R13/S2	Two 10K pots and switches	
R10	220-ohm 1/2-watt resistor (red-red-br)	
R16	12-ohm, 2-watt metal oxide resistor	
Rx	Select a ¼-watt resistor to shunt R9 to improve range	
C1	100-microfarad, 25-volt vertical electrolytic capacitor	
C2	4,700-picofarad, 50-volt polyester capacitor	
C3	2,200-microfarad, 25-volt vertical electrolytic capacitor	
C4, 7, 13	Three .1-microfarad, 50-volt polyester capacitors	
C5	.0033-microfarad, 250-volt polypropylene capacitor	
C6	3- to 5-microfarad, 350-volt polyester capacitor	DB# 3.9M/350
C9	1-microfarad, 25-volt vertical electrolytic capacitor	
C8, 10	Two 10-microfarad, 25-volt vertical electrolytic capacitors	
C11	.47-microfarad, 50-volt polyester capacitor	
C12	.01-microfarad, 50-volt disc capacitor	
C14	.05-microfarad, 400-volt metalized polypropylene	
I1	*Integrated circuit* (IC) driver and pulse-width modulator (LM3525)	
I2	1.5-amp adjustable voltage regulator (LM317)	
I3	Timer *dual inline package* (DIP) IC (LM555)	
Q1, 2	Two 100-volt *metal-oxide-semiconductor field effect transistors* (MOSFETS) (IRF540)	
SCR1	70-amp, 700-volt *silicon-controlled rectifier* (SCR)	DB# S8070W
D1	IN914 small-signal diode	
D2, 3	Two IN4937 fast-recovery, 1,000-volt, 1-amp diodes	
LA1R	Red *light-emitting diode* (LED)	
LA1G	Green LED	
T1	Inverter high-voltage transformer, as shown in Figure 18-3	DB# TRANCDI
T2	20-kilovolt pulse transformer (see text)	DB# CD25B
FH1	Panel-mount fuse holder	
FS1	3-amp fuse	
J1, 2	RCA phono jack	
S3	*Single pole, single throw* (SPST) 5-amp or more toggle switch	
PB1	7 × 3.8 × .1-inch perforated circuit vector board	
SOCK8	Eight-pin IC socket	
SOCK16	16-pin IC socket	
MK1	TO3 transistor mounting kit	
PAD	Thermo pad insulator	
CHASSIS	#20-gauge aluminum chassis, as shown in Figure 18-9	
HS1	Heatsink, as shown in Figure 18-4	
BR1	Transistor-mounting bracket, as shown in Figure 18-4	
MICA	Two Mica insulating pads for Q1 and Q2	
SWN1	Two 6-32 × 1/2-inch nylon screws	
SW1	6-32 × 1/2-inch screws	
NU1	Seven 6-32 nuts	
WIRE20B	12 inches of #20 bus wire	
WR20	Six feet of #20 vinyl hookup wire	
BU1	Small clamp bushing	
BU2, 3	Two 3/8-inch plastic bushings for LED	

Chapter Nineteen

Long-range Telephone Conversation Transmitter

Figure 19-1 shows a device that allows the user to monitor a phone's incoming and outgoing calls without any connecting wires. The system is intended to transmit both sides of a telephone conversation to any FM radio or headset-type receiver, such as a Walkman radio. It allows the user to perform outside activities such as mowing the lawn or working outside the house while completely monitoring the telephone over the radio.

The unique feature of this device is that activation occurs only when the phone is off the hook. This eliminates transmitting a continuous signal, thus preventing potential interference and unnecessarily wearing down the batteries. It also eliminates dead-tape listening time, as now the tape only records voices when the telephone is used. You may also listen to a radio station while monitoring the telephone. When the phone is used, the radio station is replaced by the transmission of the unit.

Figure 19-1 *Wireless circuit*

The device is shown assembled on a perforated circuit board and may be housed in a small plastic enclosure. Potting (or encapsulating with epoxy) and sealing is at the discretion of the builder.

Circuit Description

This chapter shows how to build a mini-powered FM transmitter that transmits both sides of a telephone conversation to a nearby FM radio. The circuit consists of a *radio-frequency* (RF) oscillator section combined with a telephone-activated switch section that controls the oscillator power (see Table 19-1). Transistor Q1 forms a relatively stable RF oscillator whose frequency is determined by the values of coil LI and tuning capacitor C4. The C4 setting determines the desired operating frequency and is in the standard FM broadcast band with tuned circuit design, favoring the high end up to 110 MHz. Capacitor C2 supplies the necessary feedback voltage developed across resistor R3 in the emitter circuit of Q1, sustaining an oscillating condition. Resistors R1 and R2 provide the necessary bias of the base emitter junction for proper operation, while capacitor C1 bypasses any RF to ground fed to the base circuit. Capacitor C3 provides an RF return path for the tank circuit of L1 and C4 while blocking the DC supply voltage fed to the collector of Q1 (see Figure 19-2).

You will note that the junction of the base bias resistors R1 and R2 is a feed point consisting of capacitor C5 and resistor R4. This point is where the speech voltage from the telephone is applied. Because of the nature of the oscillator frequency being subject to change by varying the base bias condition, a varying AC voltage superimposed at this point causes a corresponding frequency shift (FM) along with an amplitude-modulated (AM) condition. It is this property that allows the circuit to be FM modulated by the speech occurring on the telephone lines. The signal is clearly detected by any FM receiver when properly tuned.

Be aware that before the phone receiver is lifted off the hook, a DC voltage of approximately 50 volts along with some AC hum is measured across the green and red wires of the phone line. This 50 volts is used to keep transistors Q2 and Q3 "off" by preventing forward biasing of the Q2 junction. The zener diode Z1 reverse voltage is exceeded as long as the voltage on the telephone lines remains above the zener voltage. When the phone is off the hook, the voltage is now below the zener voltage, and resistor R5 now biases Q2 "on," which turns Q3 on and clamps the common return of the oscillator to the negative of the battery, commencing operation by turning on.

Note that Q2 and Q3 make a simple DC switch that produces a minimal battery current during the on-hook conditions. When the receiver is lifted, commencing normal conversation, this DC voltage drops to less than 10 volts, and an existing AC voltage corresponding to the speech conversation now modulates the oscillator as described. You will also note Q2 and Q3 being a simple DC switch that only allows a minimal battery current during the on-hook condition.

Capacitor C5 is necessary to block the DC component of the phone line when an on-hook condition exists and allows little attenuation of the varying AC speech voltage. Resistor R4 is necessary for attenuation of this speech voltage that could cause overmodulation and unnecessary serious sidebands if not properly selected.

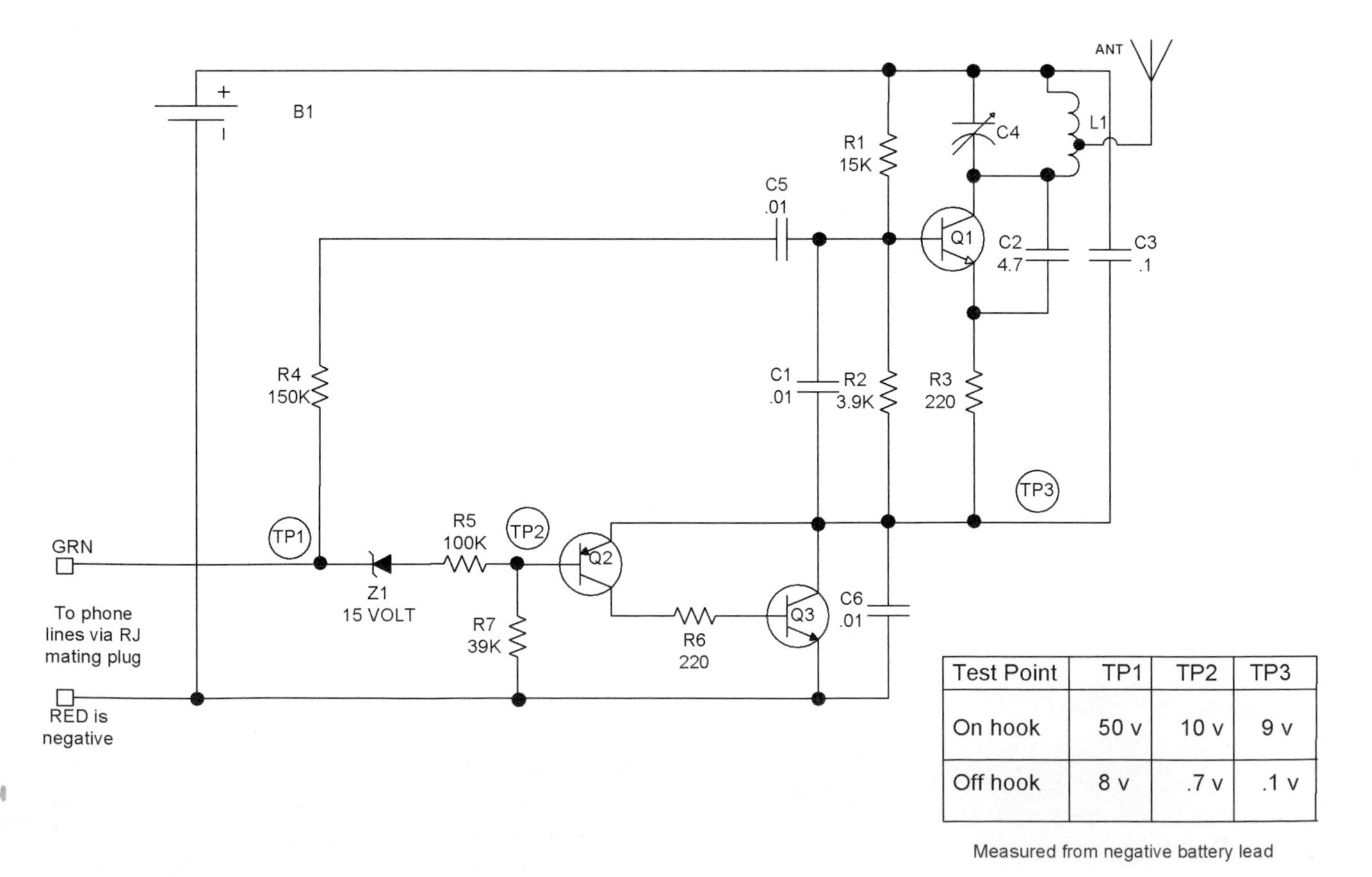

Test Point	TP1	TP2	TP3
On hook	50 v	10 v	9 v
Off hook	8 v	.7 v	.1 v

Figure 19-2 *Circuit schematic*

Assembly of the Circuit Board

Assemble the circuit as shown in Figure 19-3.

1. Fabricate a piece of .1-inch grid perforated board to a size of 2 × 1 inches.
2. Form L1 by tightly wrapping eight turns of #16 bus wire on a #8 wood screw. This produces an eight-turn coil with an inner diameter of approximately .135 inches and is about .625 inches in length. Insert it in the proper hole and solder as shown.
3. Insert trimmer capacitor C4 into the holes as shown. You may put this component on either side of the PC board. This choice is up to you and should be determined by the final packaging scheme you require. Three holes are available for the part, with two of them electrically the same point. This is necessary as some trimmers may have three pins. Make sure that the common pins are connected to the same electrical points with the odd pin to the other point.
4. If you are building from a perforated board, it is suggested that you insert components starting in the lower left-hand corner. Pay attention to the polarity of the capacitors with polarity signs and all the semiconductors. Route the leads of the components as shown and solder as you go, cutting away unused wires. Attempt to use certain leads as the wire runs or use pieces of the #24 bus wire. Follow the dashed lines on the assembly drawing as these indicate connection runs on the underside of the assembly board.
5. Attach the external leads of battery clip CL1 and the RJ cable and plug them into the phone connection jack.
6. Double-check the accuracy of the wiring and the quality of the solder joints. Avoid wire bridges, shorts, and close proximity to other circuit components. If a wire bridge is necessary, sleeve some insulation onto the lead to avoid any potential shorts.
7. Tune an FM receiver to a fairly strong station at the high end of the band (108 MHz or higher). Turn up the volume and position it at a distance of 25 to 50 feet.
8. If a multimeter or 50-milliamp meter is available, connect it in series with the battery lead.

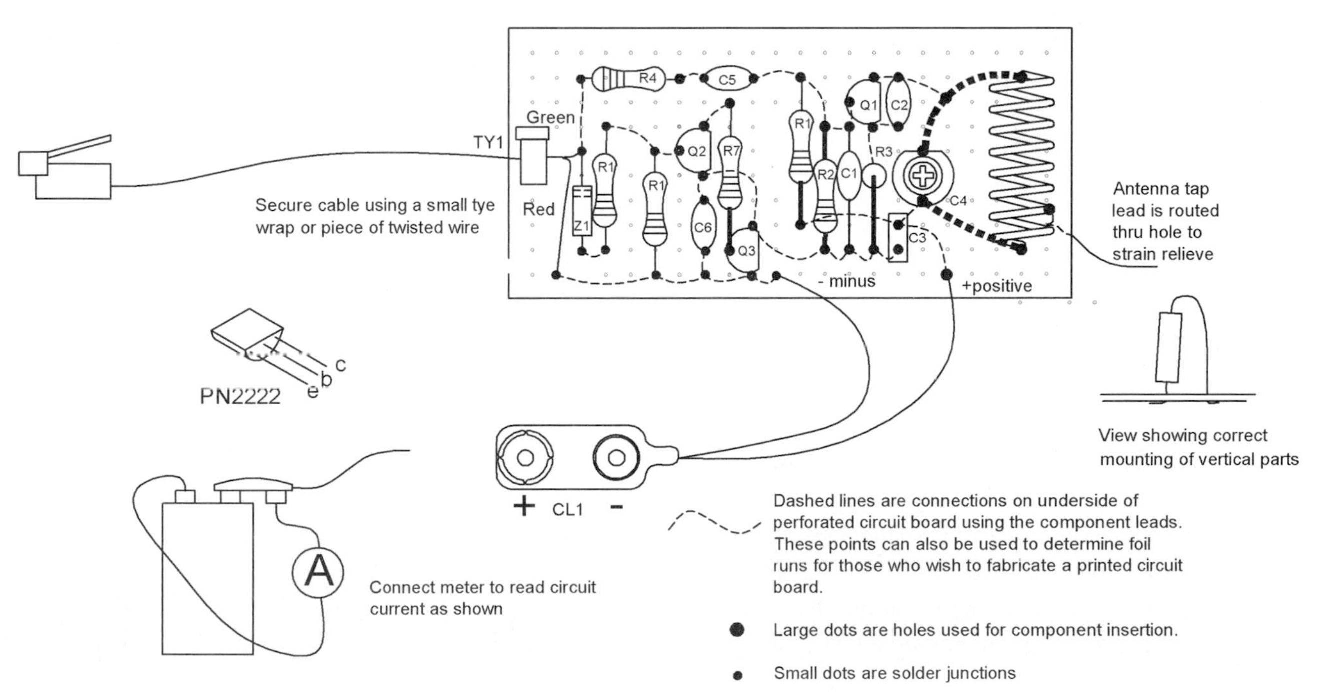

Figure 19-3 *Assembly of circuit board*

This can be done by removing one of the clip fasteners and connecting the meter to the free contacts, as shown in Figure 19-3. The meter should read 5 to 7 milliamps. Use a short piece of bare wire to touch the coil L1 a turn at a time, starting from the C3 end. Note that as you progress turn by turn away from C3 that the indicated meter current will drop or change.

9. Connect the operating unit to the phone line with the phone on hook and notice the meter reading dropping to almost nothing. You may have to reverse the connections to the phone lines. Check the following voltage points indicated on the schematic to verify proper circuit operation, as shown in Figure 19-2.
10. Take the phone off the hook and tune C4 with a tuning wand until you hear the telephone tone. It may be difficult at first to spot the signal as this adjustment is very touchy. It is important to know that several spots in the adjustment may be erroneous and will be weak and unstable. The correct signal will be strong, stable, and clear, and it can be verified by distance-of-transmission testing.
11. Once the desired setting of C4 is found, the position should be marked on the adjustment capacitor with the frequency noted.
12. Conduct the final test by making an actual call and verifying a clear transmission of both parties to the FM radio.
13. When operation is verified, you may want to enclose the circuit board in a plastic box along with the battery. This allows easier tuning and handling.

Notes

One of the things to watch for when using a device of this type is proper tuning. The adjustment capacitor, C4, is quite sensitive and requires only a slight movement to change the frequency, so always use a tuning wand. It is very easy when one is not familiar with this unit to tune to an erroneous signal. This phenomenon is likely to occur when the unit is close to the monitoring receiver. As stated previously, an erroneous signal will be weak, distorted, and unstable (often it is mistaken for the main signal and the unit is blamed for poor performance), and the main signal will be strong, stable, and undistorted, if modulated.

Several experiments in tuning the unit should be done before attempting to use it for the desired application. Also, whenever possible, the unit should be used around 108 to 109 MHz, which is at the boundary of the aircraft band and upper FM broadcast band. When the approximate desired spot is found, final touchup tuning should be done at the receiver end for clarity. In most areas, these upper frequencies are clear and allow uninterrupted use in contrast to the lower ones in the FM band where a slight change in frequency from a clear spot results in the unit being drowned out by a strong broadcast station. Do not go above 108 MHz if near an airport or air traffic lane.

Most available FM radios can easily be detuned slightly to shift the dial readings down to where the dial reading of 108 MHz is actually 109 MHz. This is accomplished by carefully adjusting the "oscillator," the padding trimmer located on the main tuning capacitor, and "walking" a known station down the necessary megahertz or two. An antenna-peaking trimmer should now be adjusted for a maximum signal at the high end.

Optimum performance will require a good-quality receiver with an analog slide-rule-type tuning dial. Digital tuned receivers will not work that well.

Table 19-1 Long-Range Phone Conversation Transmitter Parts List

Ref. #	*Description*	*DB Part #*
R1	15K, 1/4-watt resistor (br-gr-or)	
R2	3.9K, 1/4-watt resistor (or-wh-red)	
R3, 6	Two 220 ohm, 1/4-watt resistor (red-red-br)	
R4	150K, 1/4-watt resistor (br-gr-yel)	
R5	100K, 1/4-watt resistor (br-blk-yel)	
R7	39K, 1/4-watt resistor (or-wh-or)	
C1, 5, 6	Three .01-microfarad, 50-volt disc capacitors	
C2	4- to 6-picofarad small silver mica disk (4.7)	DB# 4.7P
C3	.1-microfarad, 50-volt disc capacitor	
C4	6- to 35-picofarad trimmer capacitor	DB# 635P
Z1	15-volt zener diode (1N5245)	
Q1, 3	Two PN2222 NPN general-purpose transistors	
Q2	PN2907 PNP general-purpose transistor	
L1	Coil wound as described in assembly steps	DB# COIL8T
CL1	Battery snap clip	
PBOARD	2 × 1 × .1-inch grid perforated board	
WR24BUSS	12 inches of #24 bus wire for wiring and antenna tap	
WR16BUSS	6 inches of #16 bus wire for making L1 coil	
WR20	24 inches of #20 vinyl hookup wire	
RJCABLE	Telephone cable with RJ11 plug	

Chapter Twenty

Line-Powered Telephone Conversation Transmitter

This project shown in Figure 20-1 is similar to that described in Chapter 19, but this one can be permanently implanted in a telephone and can operate indefinitely, as it does not require batteries. The circuit as shown derives its power from the phone lines once the receiver is lifted off of the hook. Operating range is limited to several hundred feet, far less than the battery-powered version.

The unique features of this device is that it does not require a battery, and activation occurs only when the phone is off the hook. This eliminates transmitting a continuous signal, thus preventing potential interference and eliminates dead-tape listening time. You may also listen to a radio station while monitoring the telephone. When the phone is used, the radio station is replaced by the transmission of the unit. The device is shown assembled on a perforated circuit board and may be housed in a small plastic enclosure. Potting and sealing is at the discretion of the builder.

Figure 20-1 *Photograph of circuit*

Circuit Description

This chapter shows how to build a mini-powered FM transmitter that transmits both sides of a telephone conversation to a nearby FM radio. The circuit consists of a *radio-frequency* (RF) oscillator section combined with a telephone power conditioner that powers the oscillator (see Table 20-1). Transistor Q1 forms a relatively stable RF oscillator whose frequency is determined by the value of coil LI and tuning capacitor C3. The setting of C3 determines the desired operating frequency and is in the standard FM broadcast band with tuned circuit design favoring the high end up to 110 MHz. Capacitor C2 supplies the necessary feedback voltage necessary to sustain oscillation. Resistor R2 provides the necessary bias of the base emitter junction for proper operation, while

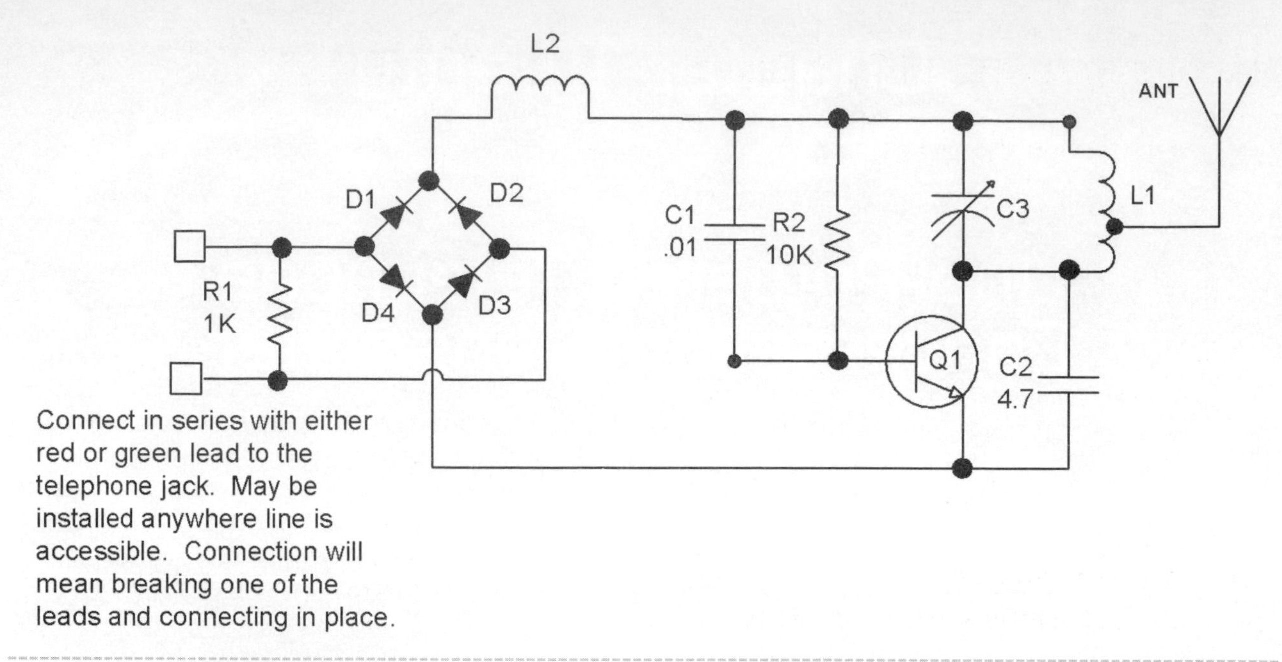

Figure 20-2 *Circuit schematic*

capacitor C1 bypasses any RF to ground. Choke L2 blocks the signal energy from going back to the phone lines yet allows DC power to Q1 (see Figure 20-2).

Connecting the circuit requires breaking one of the lines and connecting the input leads in series with the telephone. It is this method that leaches voltage from the series load resistor R1 when the telephone is in use, as current now flows across R1. The bridge diodes consisting of D1 through D4 polarize the audio voltages necessary to modulate the oscillator.

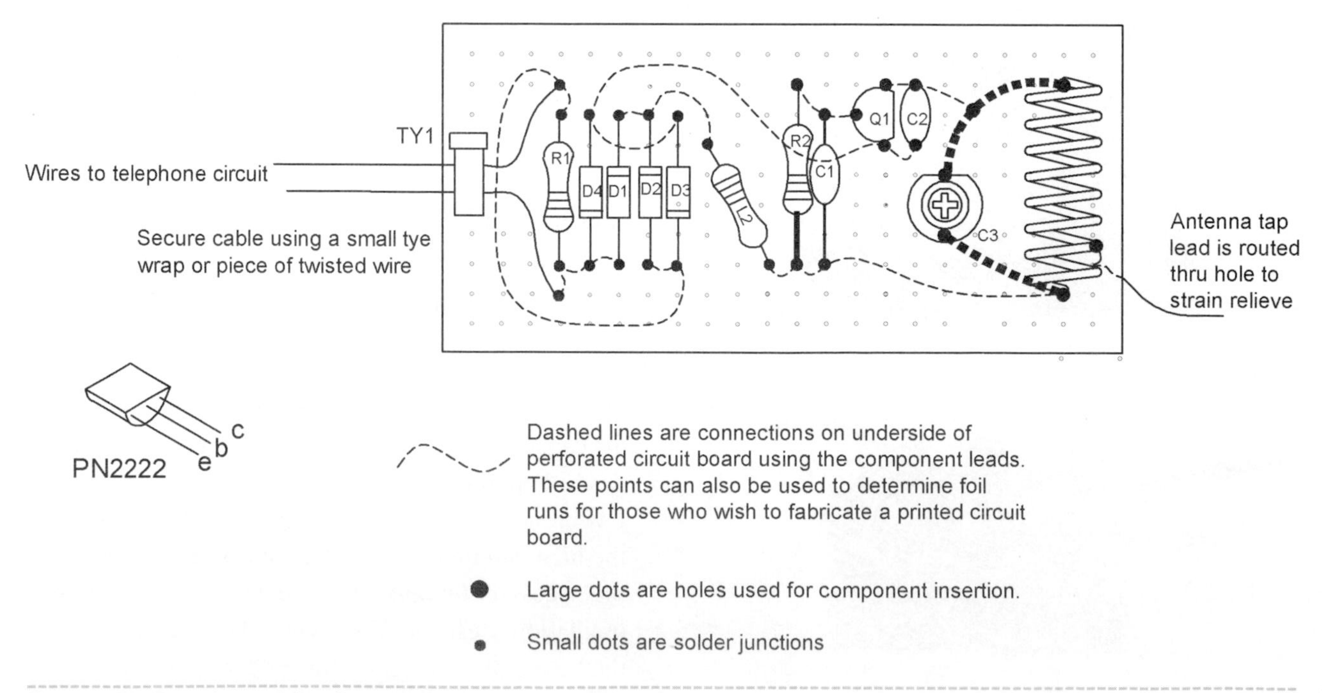

Figure 20-3 *Assembly of circuit board*

Assembly of the Circuit Board

Assemble the circuit as shown in Figure 20-3:

1. Cut a piece of .1-inch grid perforated board to 2 × 1 inches.
2. Form L1 by tightly wrapping eight turns of #16 bus wire on a #8 wood screw. This produces an eight-turn coil with an inner diameter of approximately .135 inches and is about .625 inches in length. Insert it in the proper hole and solder as shown.
3. Insert trimmer capacitor C4 into the holes as shown. You may put this component on either side of the PC board. The choice is up to you and should be determined by the final packaging scheme you require. Three holes are available for the part, with two of them electrically the same point. This is necessary, as some trimmers may have three pins. Make sure that the common pins are connected to the same electrical points, with the odd pin to the other point.
4. If you are building from a perforated board, it is suggested that you insert the components starting in the lower left-hand corner. Pay attention to the polarity of the capacitors that have polarity signs and all the semiconductors. Route the leads of the components as shown and solder as you go, cutting away unused wires. Attempt to use certain leads as the wire runs or use pieces of the #24 bus wire. Follow the dashed lines on the assembly drawing as these indicate connection runs on the underside of the assembly board.
5. Attach the external leads that are connected to the telephone lines.
6. Double-check the accuracy of the wiring and the quality of the solder joints. Avoid wire bridges, shorts, and close proximity to other circuit components. If a wire bridge is necessary, sleeve some insulation onto the lead to avoid any potential shorts.
7. Tune an FM receiver to a fairly strong station at the high end of the band (108 MHz or higher). Turn up the volume and position it 25 to 50 feet away from the transmitter.
8. Connect the operating unit to the phone line with the phone on the hook. Remove the phone and adjust C3 until you hear the dial tone. Tuning the trimmer must be done with an insulating tuning wand. It is best to tune the unit by placing it on a nonconductive surface and securing it with a sponge or a piece of rubber.

   It may be difficult at first to spot the signal as this adjustment is very touchy. Also note that several spots in the adjustment may be erroneous and will be weak and unstable. The correct signal will be strong, stable, and clear and can be verified through distance-of-transmission testing.
9. Once the desired setting of C4 is found, it should be marked, with the frequency noted.
10. Conduct a final test by making an actual call and verifying the clear transmission of the FM radio to both parties.
11. When operation is verified, it may be a good idea to enclose the circuit board in a plastic box. This allows easier tuning and handling.

Notes

Please see notes in Chapter 19.

Table 20-1 Line-Powered Telephone Conversation Transmitter Parts List

Ref. #	Description	DB Part #
R1	1K 1/4-watt resistor (br-blk-red)	
R2	10K 1/4-watt resistor (br-blk-or)	
C1	.01-microfarad, 50-volt disc capacitor	
C2	4- to 6-picofarad small silver mica disk (4.7)	DB# 4.7P
C3	6- to 35-picofarad trimmer capacitor	DB# 635P
D1–4	100-volt diodes IN4002	
Q1	PN2222 NPN general-purpose transistor	
L1	Coil wound as described in assembly steps	DB# COIL8T
L2	3.6 microhenry inductor	DB# 3.6U
PBOARD	2 × 1 × .1-inch grid perforated board	
WR24BUSS	12 inches of #24 bus wire for wiring and antenna tap	
WR16BUSS	6 inches of #16 bus wire for making L1 coil	
WR20	24 inches of #20 vinyl hookup wire	

Chapter Twenty-One

Remote Wireless FM Repeater

Figure 21-1 shows a wireless FM transmitter designed to transmit the output of a tape recorder, VCR, DVD, or any device having composite audio and video outputs to a nearby radio or TV tuned to the unit's frequency. Assembly is built on a piece of circuit board and may be placed in any suitable plastic container along with a 9-volt battery.

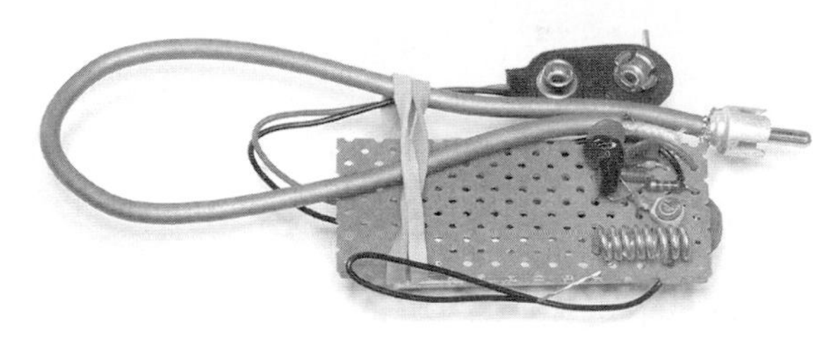

Figure 21-1 *The completed device*

Circuit Description

The parts for this project are listed in Table 21-1. Transistor Q1 forms a relatively stable *radio-frequency* (RF) oscillator whose frequency is determined by the value of coil L1 and tuning capacitor C4. The C4 setting determines the desired operating frequency and is in the standard FM broadcast band with tuned circuit design favoring the high end up to 110 MHz. Capacitor C2 supplies the necessary feedback voltage developed across emitter resistor R3 in the emitter circuit of Q1, sustaining an oscillating condition. Resistor R1 and R2 provide the necessary bias of the base emitter junction for proper operation, while capacitor C1 bypasses any RF to ground fed to the base circuit. Capacitor C9 provides an RF return path for the tank circuit of L1 and C4 while blocking the DC supply voltage fed to the collector of Q1 (see Figure 21-2). R4 is connected in series with DC-blocking capacitor C5 to the input cable and plug.

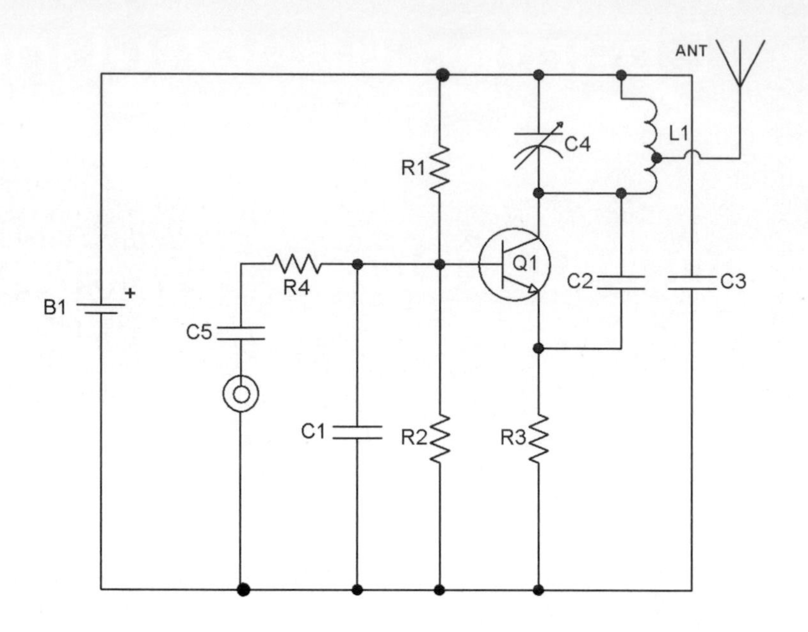

Figure 21-2 *Circuit schematic*

Assembly

Assemble the circuit as shown Figure 21-3.

1. Cut a piece of .1-inch grid perforated board to a size of 2 × 1 inches.
2. Form L1 by tightly wrapping eight turns of #16 bus wire on a #8 wood screw. This produces an eight-turn coil with an inner diameter of approximately .135 inches and a length of .625 inches. Insert it in the proper hole and solder as shown.
3. Insert trimmer capacitor C4 into the holes as shown. You may put this component on either side of the PC board. This choice is up to you and should be determined by the final packaging scheme you require. Three holes are available for the part with two of them electrically the same point. This is necessary, as some trimmers may have three pins. Make sure that the common pins are connected to the same electrical points with the odd pin to the other point.
4. If you are building from a perforated board, it is suggested that you insert components starting in the lower left-hand corner. Pay attention to the polarity of the capacitors that have polarity signs and all the semiconductors. Route the leads of the components as shown and solder as you go, cutting away unused wires. Attempt to use certain leads as the wire runs or use pieces of the #24 bus wire. Follow the dashed lines on the assembly drawing as these indicate the connection runs on the underside of the assembly board.
5. Attach the external leads of the battery clip (CL1) and shielded cable (WR10).
6. Double-check the accuracy of the wiring and the quality of the solder joints. Avoid wire bridges, shorts, and close proximity to other circuit components. If a wire bridge is necessary, sleeve some insulation onto the lead to avoid any potential shorts.
7. Tune an FM receiver to a fairly strong station at the high end of the band (108 MHz or higher). Turn up the volume and position the receiver about 25 to 50 feet from the transmitter circuit.
8. If a multimeter or 50-milliamp meter is available, connect it in series with a battery lead. This can be done by removing one of the clip fasteners and connecting the meter to one of the free contacts, as shown in Figure 21-2. The meter should read 5 to 7 milliamps. Pick up a short piece of bare wire and touch coil L1. Touch it a turn at a time, starting from the C3

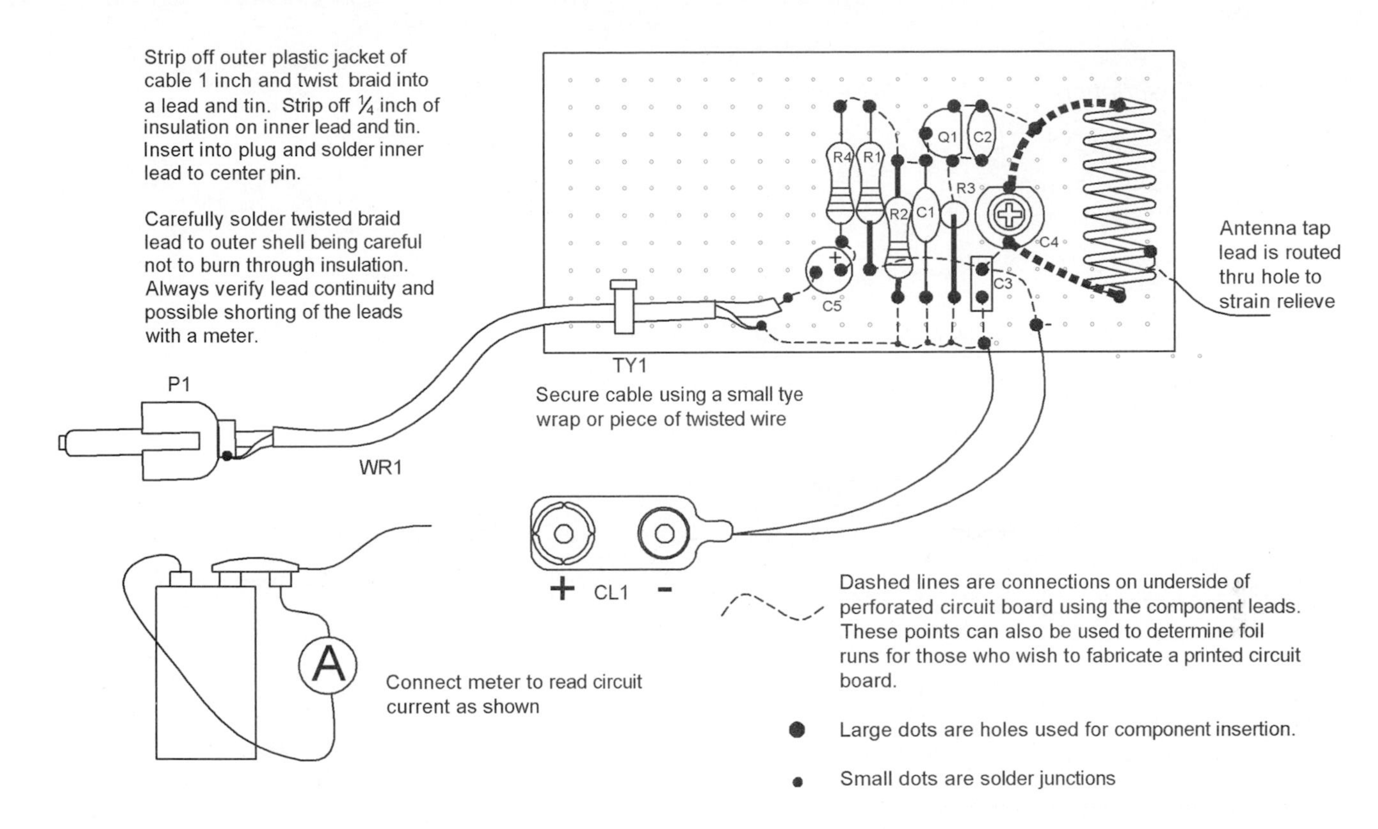

Figure 21-3 *Assembly of board*

end. Note that as you progress turn by turn away from C3 that the indicated meter current will drop or change.

9. Connect the cable into an audio source such as a recorder or another radio. Slowly rotate C4 with an insulated tuning wand until the station being received by the radio is at approximately 108 MHz. It may be difficult at first to spot the signal, as this adjustment is very touchy. Also note that several spots in the adjustment may be erroneous and will be weak and unstable. The correct signal will be strong, stable, and clear and can be verified by distance-of-transmission testing.
10. Once the desired setting of C4 *is* found, it should be marked with the frequency noted.
11. When the operation is verified, it may be desired to pot the assembly as shown. This allows easier tuning and handling.

Notes

Please refer to the notes in Chapter 19 on operation of wireless circuits.

Table 21-1 Remote Wireless FM Repeater Parts List

Ref. #	*Description*	*DB Part #*
R1	15K, 1/4-watt resistor (br-gr-or)	
R2	3.9K, 1/4-watt resistor (or-wh-red)	
R3	220-ohm, 1/4-watt resistor (red-red-br)	
R4	1K, 1/4-watt resistor (br-blk-red)	
C1, 5	Two .01-microfarad, 50-volt disc capacitors	
C2	Small 4- to 6-picofarad silver mica disk (4.7)	DB #4.7P
C3	.1-microfarad, 50-volt plastic capacitor	
C4	6- to 35-picofarad trimmer capacitor	DB #635P
CL1	Battery snap clip and leads	
Q1	PN2222 NPN general-purpose transistor	
L1	Coil wound as described in assembly steps	DB #Coil8T
PBOARD	2 × 1 × .1-inch grid perforated board	
WR24BUSS	12 inches of #24 bus wire for wiring and antenna tap	
WR16BUSS	6 inches of #16 bus wire for making L1 coil	
WR1	12 inches of shielded cable	
P1	RCA phono plug	

Chapter Twenty-Two

Tracking and Homing Transmitter

This project shows how to construct a beeping transmitter, as shown in Figure 22-1, for use as a tracking or homing device. It transmits audio pulses (beeps) to a nearby FM radio tuned to the desired frequency. The unit is intended to be used in transmitter hunt games or as a homing beacon for hunters or hikers. It is suggested that you mark a bearing heading on a small but good-quality FM radio and use it as a low-cost receiver for determining the direction of the signal pickup. The range of the device, when used as a homing device with a quarter-wave vertical placed on top of a metal roof, such as a vehicle, can be in excess of 3 miles when using a quality analog FM receiver.

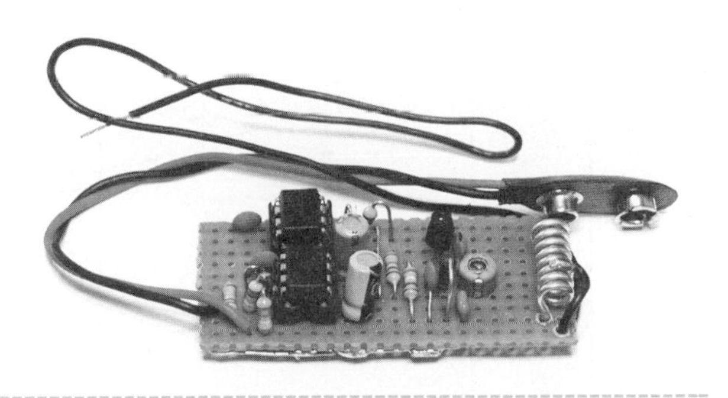

Figure 22-1 *Transmitter circuit*

Circuit Description

This chapter shows how to build a mini-powered FM transmitter that transmits audio pulses for tracking and homing applications (see Figure 22-2). The circuit consists of a *radio-frequency* (RF) oscillator section interfaced with a tone modulator and a pulse-switching section that controls the oscillator power. Transistor Q1 forms a relatively stable RF oscillator whose frequency is determined by the value of coil LI and tuning capacitor C4. The C4 setting determines the desired operating frequency and is in the standard FM broadcast band with the tuned circuit design favoring the high end up to 110 MHz. Capacitor C2 supplies the necessary feedback voltage developed across resistor R3 in the emitter circuit of Q1, sustaining an oscillating condition. Resistors R1 and R2 provide the necessary bias of the base emitter junction for proper operation, while capacitor C1 bypasses any RF to ground fed to the base circuit. Capacitor C3 provides an RF return path for the tank circuit of LI and C4 while blocking the DC supply.

You will note that the junction of the base bias resistors R1 and R2 is a feed point consisting of resistor R4. This point is where the beeping tone from the tone generator chip I2 is applied. Because of the nature of the oscillator frequency being subject to change by varying the base bias condition, a pulsed voltage superimposed at this point causes a corresponding frequency shift (FM) along with an amplitude modulated shift (AM) condition. It is this property that allows the circuit to be FM modulated, producing the tone frequency. The signal is clearly detected on any FM receiver when the receiver is properly tuned. The tone frequency is the pulse repetition rate of I1 wired as an astable pulse generator. This repetition rate is determined by timing resistor R5 and capacitor C6.

The ratio of beeping on to off time is controlled by chip I1, wired as a free-running astable device with duty cycle control. This function is determined by the ratio of resistors R7 and R8 and timing capacitor C8. The output of I1 controls the bias voltage of oscillator Q1 through resistor R1, turning the oscillator on and off (see Figure 22-2).

Assembly of the Circuit Board

Assemble the circuit as shown in Figure 22-3.

1. Cut a piece of .1-inch grid perforated board to a size of 2 × 1 inches.
2. Form L1 by tightly wrapping eight turns of #16 bus wire on a #8 wood screw. This produces an eight-turn coil with an inner diameter of .135 inches and a length of .625 inches. Insert it in the proper hole and solder as shown.
3. Insert trimmer capacitor C4 into the holes as shown. You may put this component on either side of the PC board. The choice is up to you and should be determined by the final packaging scheme you require. Three holes are available for the part, with two of them electrically the same point. This is necessary, as some trimmers may have three pins. Make sure that the common pins are connected to the same electrical points with the odd pin to the other point.
4. If you are building from a perforated board, it is suggested that you insert components, starting in the lower left-hand corner. Pay atten-

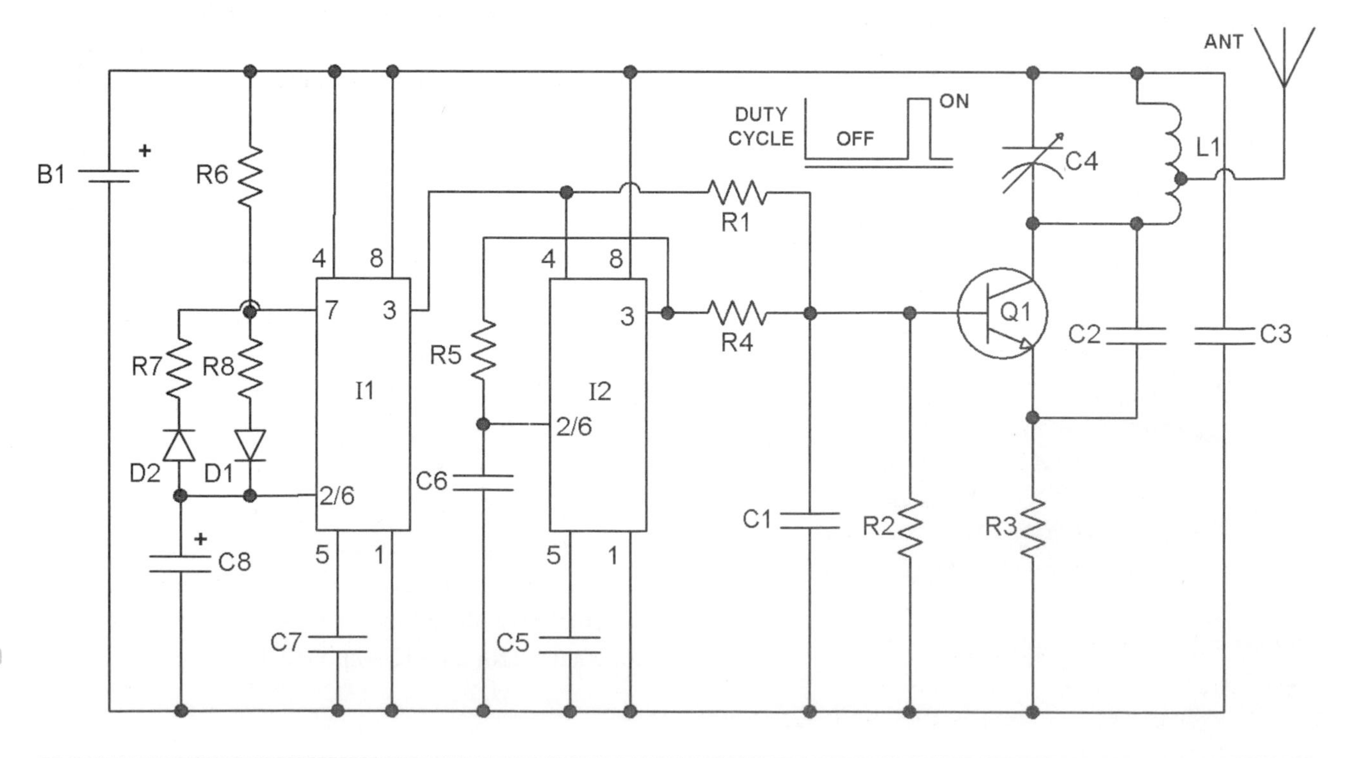

Figure 22-2 *Circuit schematic*

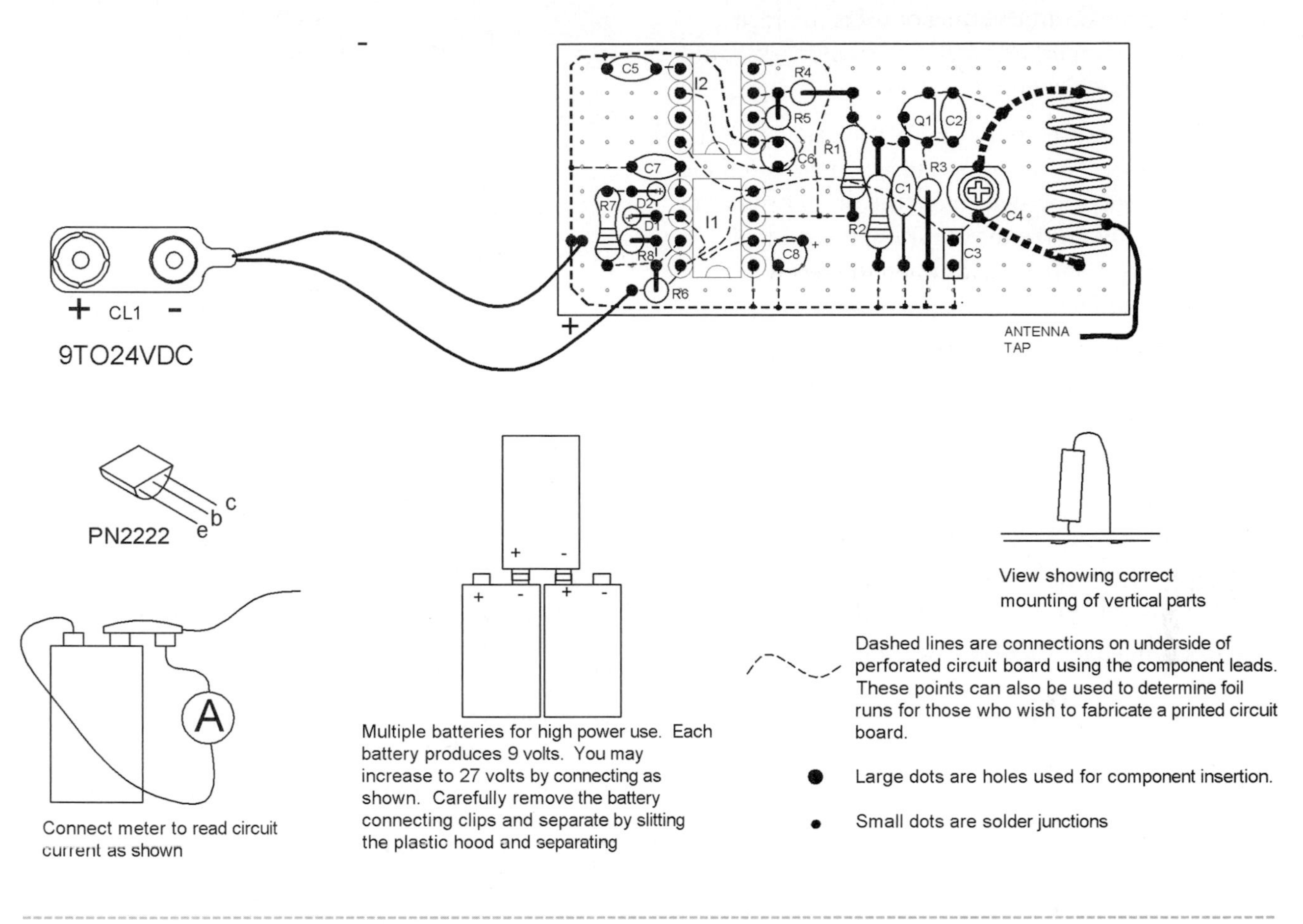

Figure 22-3 *Assembly of the circuit board*

tion to the polarity of the capacitors with polarity signs and all the semiconductors. Route the leads of the components as shown and solder as you go, cutting away the unused wires. Attempt to use certain leads as the wire runs or use pieces of the #24 bus wire. Follow the dashed lines on the assembly drawing as these indicate the connection runs on the underside of the assembly board.

5. Attach the external leads of battery clip CL1.
6. Double-check the accuracy of the wiring and the quality of the solder joints. Avoid wire bridges, shorts, and close proximity to other circuit components. If a wire bridge is necessary, sleeve some insulation onto the lead to avoid any potential shorts.
7. Tune an FM receiver to a fairly strong station at the high end of the band (108 MHz or higher). Turn up the volume and position the receiver 25 to 50 feet from the transmitter circuit.
8. If a multimeter or 50-milliamp meter is available, connect it in a series with a battery lead. This can be done by removing one of the clip fasteners and connect the meter to the free contacts, as shown in Figure 22-2. The meter should read a pulsing of 5 to 7 milliamps. Use a short piece of bare wire to touch the coil L1 a turn at a time, starting from the C3 end. Note that as you progress turn by turn away from C3 that the indicated meter current will drop or change.
9. Tune C4 with an insulated tuning tool until you hear the beeping tone. It may be difficult at first to spot the signal, as this adjustment is very touchy. Also note that several spots in the adjustment may be erroneous and will be weak and unstable. The correct signal will be strong, stable, and clear and can be verified by distance-of-transmission testing.

10. Once the desired setting of C4 is found, it should be marked with alignment marks, with the frequency noted.

11. Finally, test the unit using a good-quality FM analog receiver. Note to place an arrow on the radio once you determine the proper positioning to obtain the maximum signal strength. Do this step from a distance of at least several hundred feet.

12. When operation is verified, it may be desired to enclose the circuit board into a plastic box along with the battery. This allows easier tuning and handling.

Notes

Please see the notes in Chapter 19 on operating.

Table 22-1 Tracking and Homing Transmitter Parts List

Ref. #	*Description*	*DB Part #*
R1	Two 10K, 1/4-watt resistor (br-gr-or)	
R2	3.9K, 1/4-watt resistor (or-wh-red)	
R3	220-ohm, 1/4-watt resistor (red-red-br)	
R5	1K, 1/4-watt resistor (br-blk-red)	
R6	39K, 1/4-watt resistor (or-wh-or)	
R7	100K, 1/4-watt resistor (br-blk-yel)	
R8	22K, 1/4-watt resistor (red-red-or)	
C1, 5, 7	Three .01-microfarad, 50-volt disc capacitors	
C2	Small, 4- to 6-picofarad, silver mica disk (4.7)	DB# 4.7P
C3	.1-microfarad, 50-volt plastic capacitor	
C4	6- to 35-picofarad trimmer capacitor	DB# 635P
C6	.47-microfarad, 25-volt vertical electrolytic capacitor	
C8	4.7-microfarad, 25-volt vertical electrolytic capacitor	
D1, 2	Two 1N914 small-signal diodes	
I1, 2	Two 555 timers in a *dual inline package* (DIP)	
CL1	Battery snap clip and leads	
Q1	PN2222 NPN general-purpose transistor	
L1	Coil wound as described in assembly steps	DB# COIL8T
PBOARD	2 × 1 × .1-inch grid perforated board	
WR24BUSS	12 inches of #24 bus wire for wiring and antenna tap	
WR16BUSS	6 inches of #16 bus wire for making L1 coil	

Chapter Twenty-Three

Snooper Phone Room-Listening Device

Figure 23-1 shows a simple electronic device intended to be an intrusion detector and listening device for checking a home or office while away or on vacation. It may also be used to trigger other electrical devices. The circuitry does not require a tone encoder, so it does not provide selected activation by a particular caller. Thus, it limits the device from being used for illegal purposes and allows the sale of the kit or fully assembled unit without the requirement of special authorization.

This is an intermediate-level project requiring basic electronic skills. Expect to spend $25 to $40. All parts are readily available, with specialized parts obtainable through Information Unlimited (www.amazing1.com), and they are listed in Table 23-1.

Figure 23-1 *The snooper phone*

Unlike the controversial Infinity transmitter, our system does not utilize decoder and encoder circuitry. Once connected to the phone lines, it will open a microphone to anyone who happens to dial the number. This feature obviously reduces the potential use of the device for illegal interception of oral communication. Even though the phone does not ring, it would not be long before the potential victim of an illegal surveillance would be aware of the setup. It is therefore quite obvious that the applications are strictly limited to use as a security device that checks for unauthorized intrusion or checks household appliances by placing a pickup device in certain favorable locations to overhear these desired sounds and noises. I am sure that anyone with this capability would breathe a sigh of relief to be able to dial his or her home or office phone while being away and hear the familiar sounds of appliances and other systems properly performing their duties and functions.

Circuit Operation

Your circuit consists of a high-gain amplifier (I1) fed into the telephone lines via transformer T1. The circuit is initiated by the action of a voltage transient pulse occurring across the phone line at the instant of a telephone circuit connection being made. This transient is a voltage change of about 48 to 5 VDC and usually occurs before the ring signal. It is this change that immediately forward-biases diode D1 and triggers pin 2 of timer I2, whose output pin 3 goes positive, turning on transistors Q1 and Q2. Timer I2 now remains in this state for a period depending on the values of timing resistor R12 and the selected value of the capacity determined by the activated time of switch S2. This four-position switch allows time selections of 10, 30, 100, and 1,000 seconds by the selection of the respective capacitors C12 through C15 (see Figure 23-2).

You will note that when Q2 is turned on by timer I2 that a simulated off-hook condition exists by the switching action connecting the 500-ohm winding of the transformer directly across the phone lines. Simultaneously, Q1 clamps the ground of the I1 amplifier to the negative return of battery Bl, enabling this amplifier section. System power is controlled by switch S1 being part of the pickup sensitivity control R1.

Electret microphone M1 picks up the sounds fed to operational amplifier I1 with the output coupled by capacitor C6 to the 1,500-ohm winding of Tl, and R1 controls the pickup sensitivity of the system. The system described will operate when any incoming call is received without the phone ever ringing.

Construction Steps

If you are a beginner, it is suggested to obtain our *GCAT1 General Construction Practices and Techniques*. This informative literature explains basic practices that are necessary in proper construction of electromechanical kits and is listed in Table 23-1.

1. Lay out and identify all the parts and pieces. Verify them with the parts list, and separate the resistors as they have a color code to determine their value. Colors are noted on the parts list.
2. Cut a piece of .1-inch grid perforated board to a size of 4.2 × 1.5. Locate and drill the holes as shown in Figure 23-3.

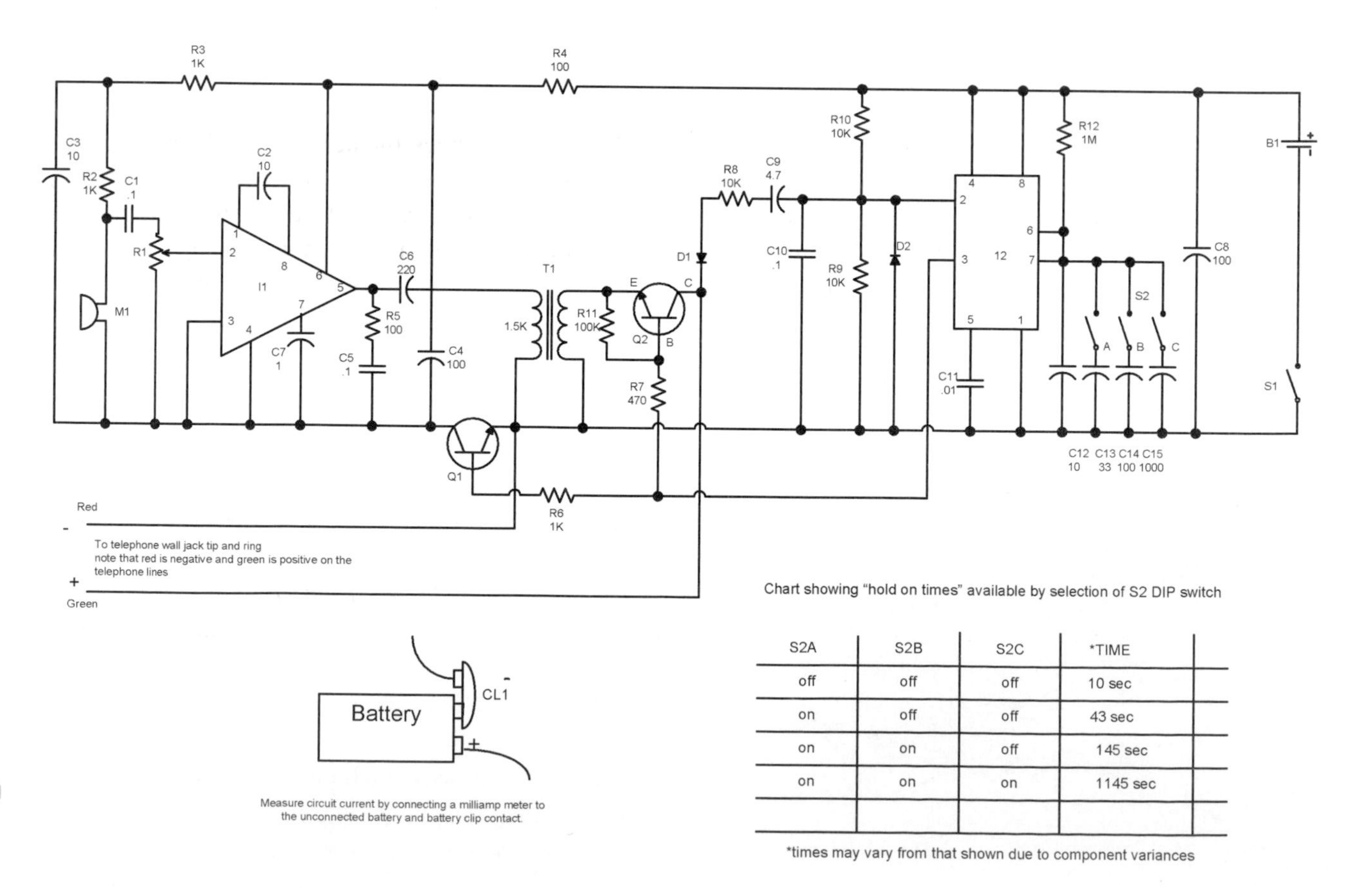

S2A	S2B	S2C	*TIME
off	off	off	10 sec
on	off	off	43 sec
on	on	off	145 sec
on	on	on	1145 sec

*times may vary from that shown due to component variances

Figure 23-2 *Circuit schematic*

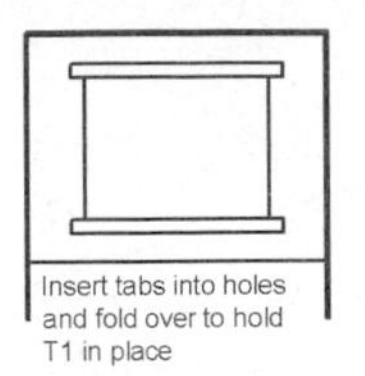

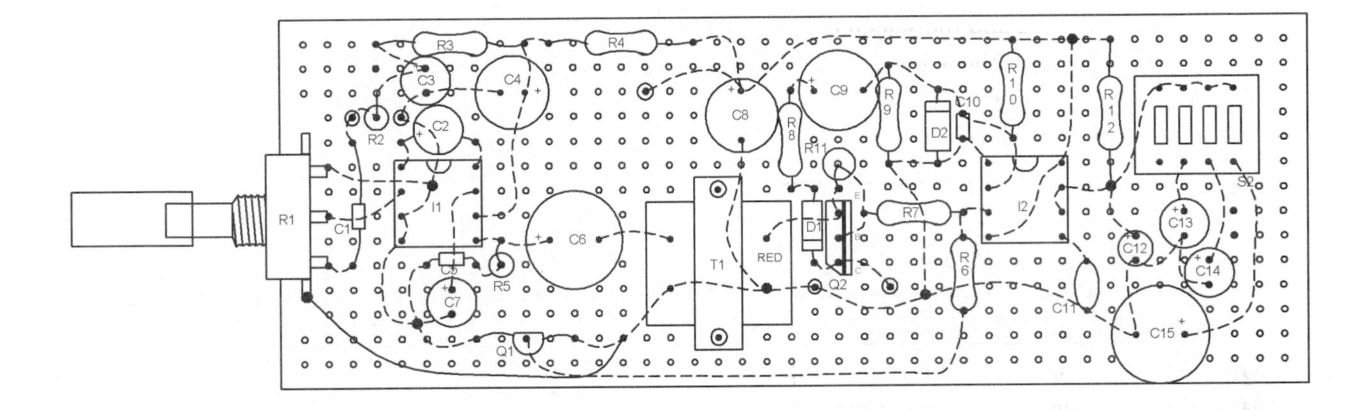

Dashed lines are connections on underside of perforated circuit board using the component leads. These points can also be used to determine foil runs for those who wish to fabricate a printed circuit board.

- Small dots are holes used for component insertion.
- Large dots are solder junctions
- Circles with dots are holes that may require drilling for component mounting and points for external connecting wires indicating strain relief points. It is suggested to drill clearance holes for the leads and solder to points beneath board.

Figure 23-3 *Assembly board layout*

3. If you are building from a perforated board, it is suggested that you insert the components starting in the lower left-hand corner (see Figure 23-3). Pay attention to the polarity of the capacitors with polarity signs and all the semiconductors.

   Route the leads of the components as shown and solder as you go, cutting away unused wires. Attempt to use certain leads as the wire runs or use pieces of the #24 bus wire. Follow the dashed lines on the assembly drawing as these indicate connection runs on the underside of the assembly board.

4. Attach the external leads as shown in Figure 23-4. You may relieve any strain they might have by drilling holes to allow the insertion of the entire wire. Note that the short pieces of bus wire are soldered to R1 for connection to the circuit board.

5. Double-check the accuracy of the wiring and the quality of the solder joints. Avoid wire bridges, shorts, and close proximity to other circuit components. If a wire bridge is necessary, sleeve some insulation onto the lead to avoid any potential shorts.

Final Assembly

Create a suitable housing from a piece of aluminum or purchase a ready-made aluminum box approximately 5 × 3 × 2 inches. Drill the holes as required for the microphone grommet GR1, gain control R1, and strain relief bushing for the phone line cable assembly RJ11. The metal box should be grounded to the circuit common line to reduce hum pickup. Note that a piece of Velcro sticky tape can be used to secure the battery or use your own ingenuity.

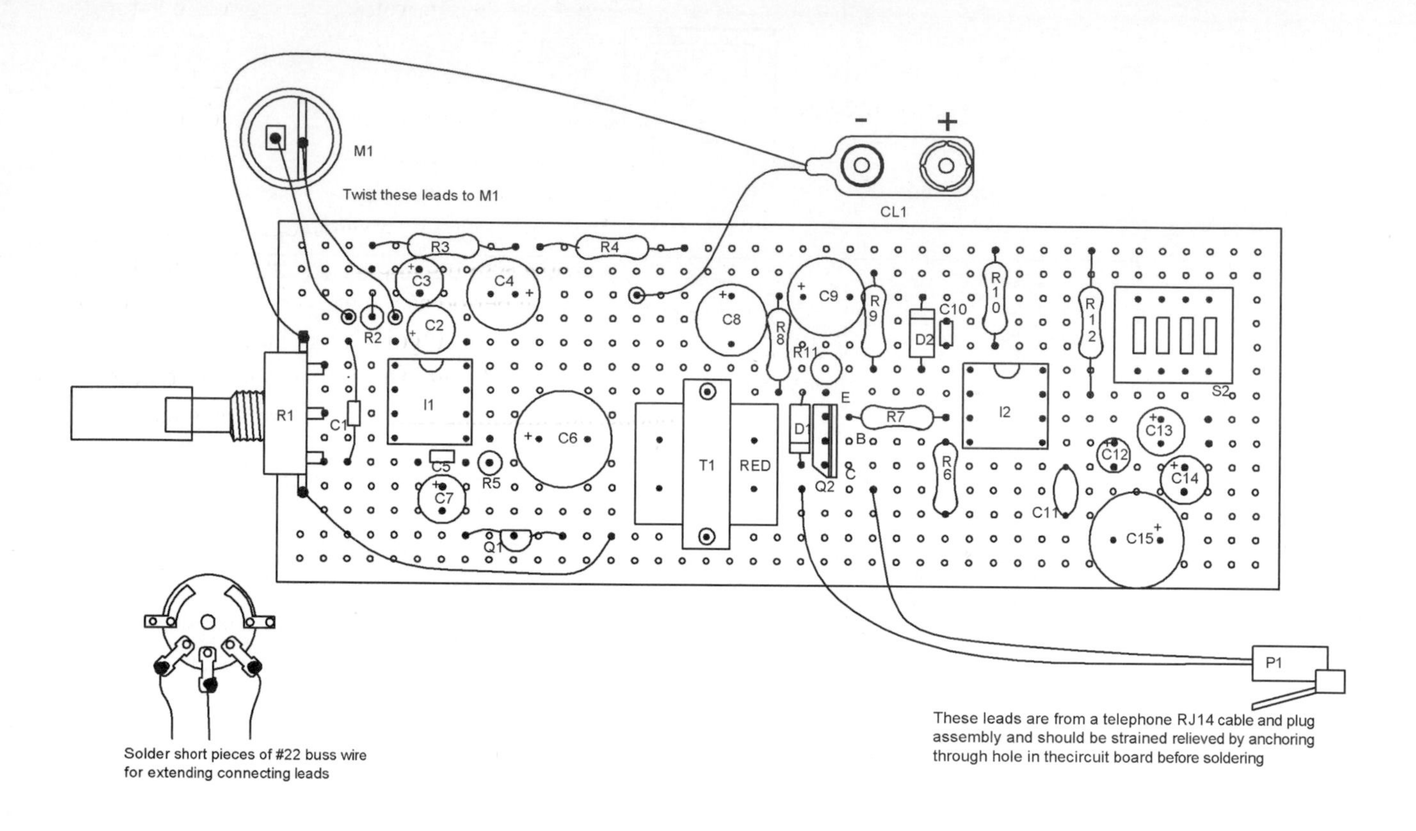

Figure 23-4 *External connection and parts identification*

Testing Steps

This step will require basic electronic laboratory equipment including a 60 MHz oscilloscope if trouble shooting the circuit is necessary.

1. Turn the unit off by rotating R1 full counter-clockwise, noting the click of the switch. Verify that all the DIP switch levers of S2 are down (off). Connect the battery as shown in series with a current meter and note zero current flow. Turn the unit on and note a current of 5 milliamps, which may momentarily jump to 30 milliamps as the unit times out.

2. Locate a phone jack so that you can monitor the DC voltage. Connect a meter across the tip and ring leads (green and red), and note the 40 to 50 volts with the phone on the hook. Also note the green lead is positive and the red is negative. This is the opposite of conventional wiring. Lift the phone and notice the 40 to 50 volts dropping down to 8 to 12 volts. This step verifies a properly working phone line.

3. Turn the unit off and connect it to the phone line. Rotate the R1 so that it just clicks on, and note the voltage remaining as or timing out to read the on-hook 40 to 50 volts. Preset R2 to midway.

4. Call the target phone using a cell or another telephone. The unit should connect at the beginning of the ring signal before the phone has a chance to ring. You should clearly hear sounds until the unit times out, at which time the device disconnects and restores the phone to normal operation. Adjust R1 for desired pickup sensitivity. The timeout time should be only about 10 seconds for the presetting of S2, as instructed in step 1. You may lengthen these times as shown in Figure 23-2.

Notes

The unit can also operate with extension phones without dialing the number from another place. Simply connect the unit to a phone jack and lift the extension phone located elsewhere, activating the circuit. Applications for monitoring other rooms and floors via the telephone lines now is possible.

Connections can be made in simple ways across the phone lines at any convenient point, such as the phone jack, lead-in box, or inside the telephone itself via clips or lugs. They can also be made via a simple, standard four-pin phone plug mated to any standard phone jack. The latter method is usually recommended when a home or office is equipped with these convenient jacks, which allow simple placement of the telephone or device.

When the phone activates, allowing you to listen to the target premises, it remains on for a time determined by the setting of the DIP switch S2. This time can be selected from 10 seconds to 20 minutes by proper selection of these levers. Even when you hang up, the system will keep the line connected so another call will get a busy signal until the unit times out. One now sees the importance of not overselecting the timing period as the phone cannot be used for either incoming or outgoing calls. Outgoing calls can normally be made with the unit installed to the target phone, but incoming calls will activate the microphone.

Table 23-1 Snooper Phone Room-Listening Device Parts List

Ref. #	Description	DB Part #
R1/S1	10K, 17-millimeter pot and switch	
R2, 3, 6	Four 1K, 1/4-watt resistors (br-blk-red)	
R4, 5	One 100-ohm, 1/4-watt resistor (br-blk-br)	
R7	470 1/4-watt resistor (yel-pur-br)	
R8, 9, 10	Three 10K, 1/4-watt resistors (br-blk-or)	
R11	100K, 1/4-watt resistor (br-blk-yel)	
R12	1M 1/4-watt resistor (br-blk-gr)	
C1, 5, 10	Three .1-microfarad, 50-volt caps	
C2, 3, 12	Three 10-microfarad, 25-volt vertical electro radial leads	
C4,8,14	Three 100-microfarad, 25-volt vertical electro radial leads	
C6	.22-microfarad, 250-volt metalized polypropylene	
C7	.1-microfarad, 50-volt cap	
C9	1-microfarad, 25-volt vertical electro cap	
C13	33-microfarad, 25-volt vertical electro radial leads	
C11	.01-microfarad, 50-volt disk (103)	
C15	1,000-microfarad, 25-volt vertical electrolytic capacitor	
C6	.22-microfarad, 250-volt metallized polypropylene	
C9	1-microfarad, 25-volt vertical electro cap	
C7	.1-microfarad, 50-volt cap (INFO#VG22)	
D1, 2	Two 1N4007 1-kilovolt diodes	
I1	LM386 operational amplifier *dual inline package* (DIP) *integrated circuit* (IC)	
I2	555 timer DIP IC	
Q1	PN2222 NPN small-signal transistor	
Q2	D40D5 high-voltage transistor	
TI	1.5K, 500-ohm audio interstage transformer	
M1	Electret microphone	
S2	4PST DIP switch	
PBOARD	4.2 × 1.5 × .1-inch grid perforated board, sized per chapter	
GR1	1/2-inch rubber grommet	
BUSCLP	1/4-inch Heyco clamp bushing	
CL1	9-volt battery clip with leads	
RJ11	RJ11 telephone plug and cable assembly	
WR20R	12 inches of #20 vinyl red wire for positive input	
WR20B	12 inches of #20 vinyl black wire for negative input	
ENC1	5 × 3 × 2-inch aluminum box and cover	

Chapter Twenty-Four

Long-Range FM Voice Transmitter

This project shows how to construct an electronic device that can transmit low-level sounds over a considerable distance (see Figure 24-1). It is intended for use as a security system when entry sounds are broadcasted over a standard FM radio, providing an early warning of an intrusion. It may be used to monitor children, swimming pools, sick and invalid people, and animals. The device can also be used for nature listening, which is made possible by the wide frequency response of the audio section, allowing good-quality reproducibility of rare bird calls, for example. An excellent application, when one is living in hostile areas, is to install the unit in one's home. This allows monitoring of the premises before possibly being surprised by a burglar.

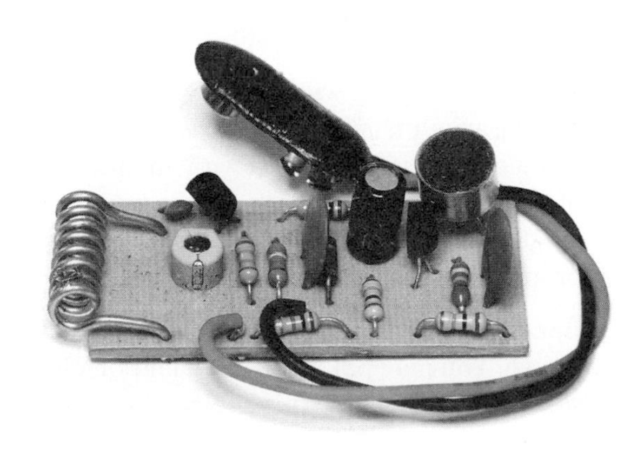

Figure 24-1 *Long-range FM voice transmitter*

The assembled transmitter is small enough to be placed at the foci of a sonic reflector, creating a highly directional wireless mike for targeting particular areas of interest. (The Audubon Society purchased 25 of these systems for monitoring bird and animal calls in mosquito-infested swamp areas.)

Circuit Description

This circuit shows how to build a super-sensitive, mini-powered FM transmitter consisting of a *radio frequency* (RF) oscillator section interfaced with a high-sensitivity, wide pass-band audio amplifier and capacitance mike with a built-in *field effect transistor* (FET) that modulates the base of the RF oscillator transistor (see Figure 24-2). Transistor Q1 forms a relatively stable RF oscillator whose frequency is determined by the value of coil L1 and tuning capacitor C4. The setting of C4 determines the desired operating frequency and is in the standard FM broadcast band with the tuned circuit design favoring the high end up to 110 MHz. Capacitor C2 supplies the necessary feedback voltage developed across resistor R3 in the emitter circuit of Q1, sustaining an oscillating condition. Resistors R1 and R2 provide the necessary bias of the base emitter junction for proper operation, while capacitor C1 bypasses any RF to ground fed to the base circuit. Capacitor C3 provides an RF return path for the tank circuit of L1 and C4 while

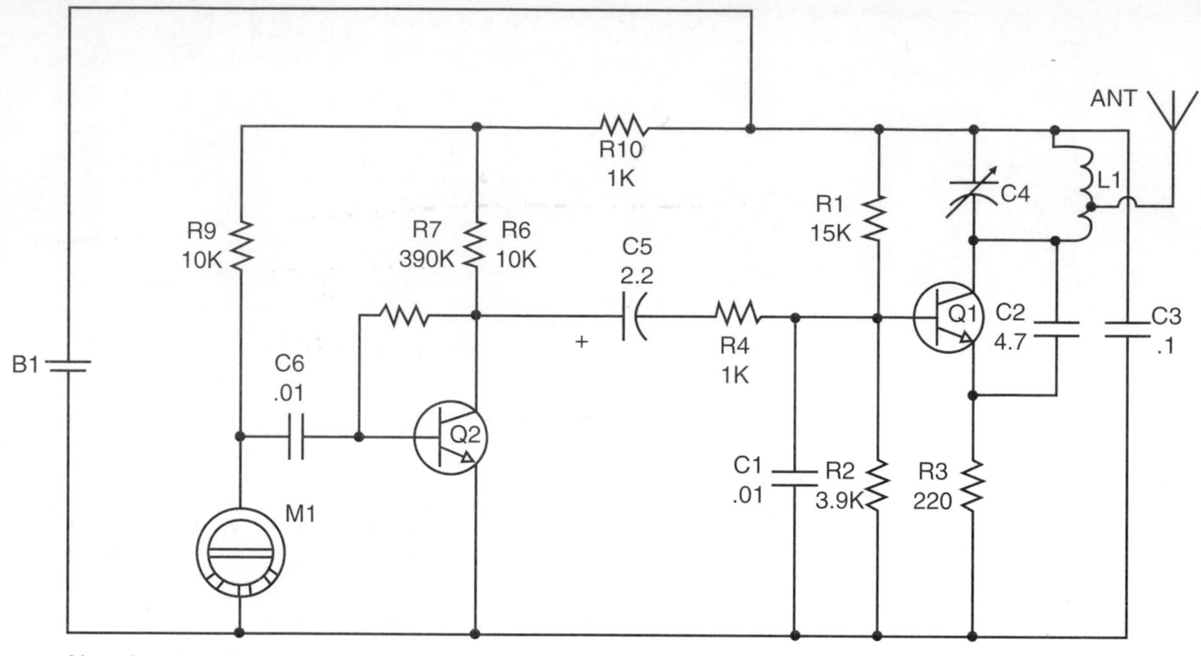

Figure 24-2 *Circuit schematic*

blocking the DC supply voltage fed to the collector of Q1.

The audio section utilizes a high-sensitivity capacitance mike (M1) and built-in FET transistor (field effect) and will clearly pick up all low-level sounds in the speech audio spectrum. The speech voltage developed across resistor R9 by M1 is capacitively coupled by C6 to the base of audio amplifier transistor Q2. You will note the base of oscillator transistor Q1 is biased by resistors R1 and R2. A signal voltage now developed across resistor R6 is capacity coupled through nonpolarized capacitor C5 to the base of Q1 through resistor R4. The gain of Q2 is controlled by the ratio of R7 and R6. The operating point is set to allow full excursion of the collector to the amplified signal. The amplified speech signal now causes FM and AM modulation of the oscillator circuit by slight shifting of the operating point of the base section. Resistor R10 decouples the oscillator and audio circuits and is necessary to prevent feedback and other undesirable effects.

When properly assembled, the circuit should produce crystal-clear quality when the receiver is properly tuned to the unit. Note that a shunt capacitor may be connected across the base lead of Q1 to reduce sensitivity. The circuit with the components listed operates best in the upper FM band. This is a clear spot without interference from FM radio stations. However, satisfactory performance is obtained above 110 MHz for limited range use. Use caution as this is the aircraft band and should not be used if near an airport (see Table 24-1).

Assembly Steps

Assemble the circuit as shown in Figure 24-3.

1. Cut a piece of .1-inch grid perforated board to a size of 2 × 1 inches.
2. Form L1 by tightly wrapping eight turns of #16 bus wire on a #8 wood screw. This produces an eight-turn coil with an inner diameter of approximately .135 inches and a length of about .625 inches. Insert it in the proper hole and solder as shown.
3. Insert trimmer capacitor C4 into the holes as shown. You may put this component on either side of the PC board. This choice is up to you and should be determined by the final packaging scheme you require. Three holes are available for the part with two of them electrically the same point. This is necessary as some

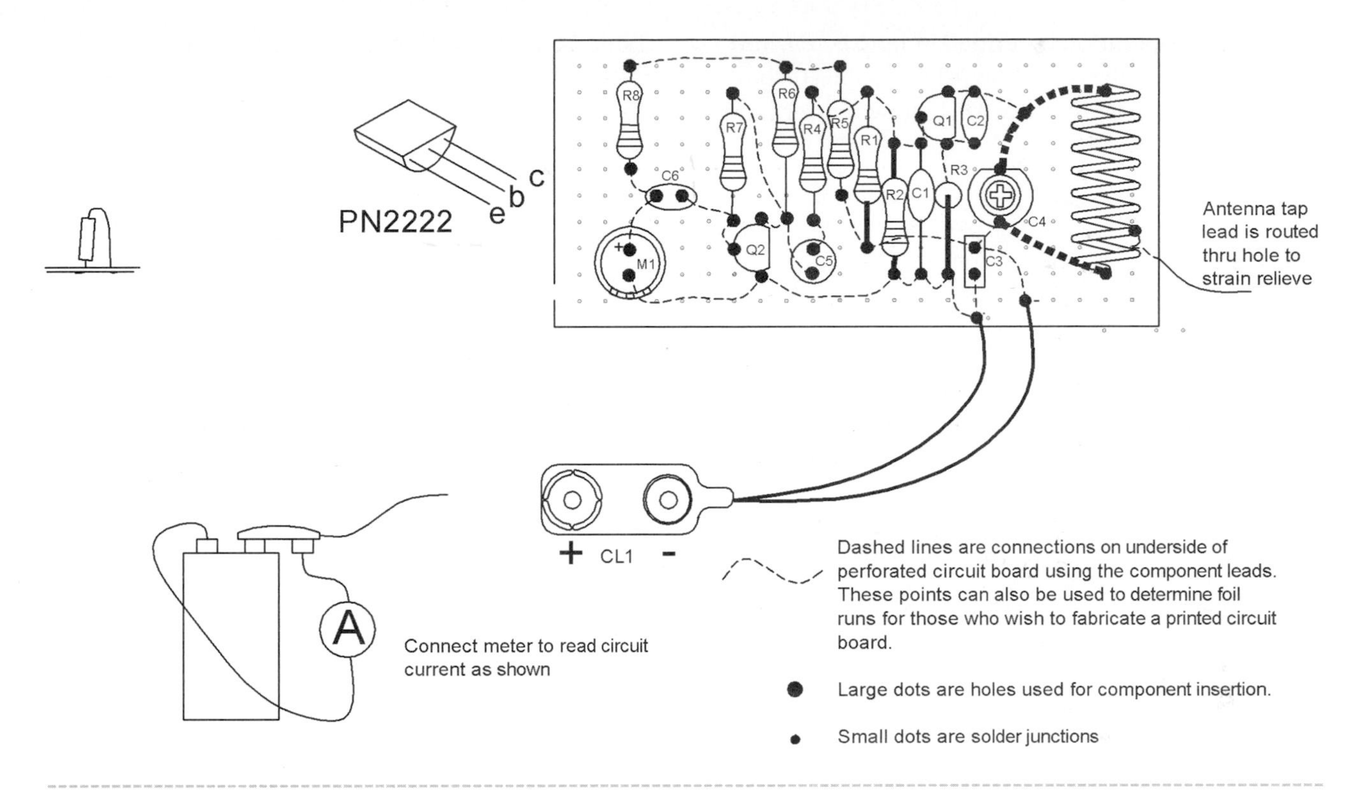

Figure 24-3 *Assembly board*

trimmers may have three pins. Make sure that the common pins are connected to the same electrical points with the odd pin to the other point.

4. If you are building from a perforated board, it is suggested that you insert components starting in the lower left-hand corner. Pay attention to the polarity of the capacitors with polarity signs and all the semiconductors. Route the leads of the components as shown and solder as you go, cutting away unused wires. Attempt to use certain leads as the wire runs or use pieces of the #24 bus wire. Follow the dashed lines on the assembly drawing as these indicate the connection runs on the underside of the assembly board.

5. Attach the external leads of the battery clip (CL1).

6. Double-check the accuracy of the wiring and the quality of the solder joints. Avoid wire bridges, shorts, and close proximity to other circuit components. If a wire bridge is necessary, sleeve some insulation onto the lead to avoid any potential shorts.

7. Tune an FM receiver to a fairly strong station at the high end of the band (108 MHz or higher). Turn up the volume and position the receiver 25 to 50 feet from the circuit.

8. If a multimeter or 50-milliamp meter is available, connect it in series with a battery lead. This can be done by removing one of the fasteners of the clip and connecting the meter to the free contacts, as shown in Figure 24-3. The meter should read 5 to 7 milliamps. Use a short piece of bare wire to touch the coil L1 a turn at a time, starting from the C3 end. Note that as you progress turn by turn away from C3 that the indicated meter current will drop or change.

9. If the device performs as follows with the battery connected, slowly rotate C4 with an insulated tuning wand until the station being received by the radio at approximately 108 MHz breaks out in audio feedback or is blocked out. It may be difficult at first to spot the signal as this adjustment is very touchy. Also note that several spots in the adjustment may be erroneous and will be weak and unstable. The correct signal will be strong, stable, and clear and can be verified by distance-of-transmission testing.

10. Once the desired setting of C4 is found, it should be marked with the frequency noted.

11. When operation is verified, it may be desired to enclose the circuit board into a plastic box along with the battery. This allows easier tuning and handling.

Notes

One of the things to watch for when using a device of this type is proper tuning. The adjustment capacitor, C4, is quite sensitive and requires only a slight movement to change the frequency, so always use a tuning wand. It is very easy when one is not familiar with this unit to tune to an erroneous signal. This phenomenon is likely to occur when the unit is close to the monitoring receiver. As previously mentioned, an erroneous signal will be weak, distorted, and unstable (often it is mistaken for the main signal and the unit blamed for poor performance), and the main signal will be strong, stable, and undistorted, if modulated. Several experiments in tuning the unit should be done before attempting to use it for the desired application.

Also, whenever possible, the unit should be used around 108 to 109 MHz, which is at the boundary of the aircraft band and upper FM broadcast band. When the approximate desired spot is found, final touchup tuning should be done at the receiver end for clarity. In most areas, these "upper" frequencies are clear and allow uninterrupted use in contrast to the lower ones in the FM band where a slight change in frequency from a clear spot results in the unit being drowned out by a strong broadcast station. Do not go above 108 MHz if near an airport or air traffic lane.

Most available FM radios can easily be detuned slightly to shift the dial readings down where 108 is 109. This is accomplished by carefully adjusting the "oscillator," the padding trimmer located on the main tuning capacitor, and "walking" a known station down the necessary megahertz or two. An antenna-peaking trimmer should now be adjusted for maximum signal at the high end.

Optimum performance will require a good-quality receiver with an analog slide-rule-type tuning dial. Digital tuned receivers will not work that well.

Table 24-1 Long-Range FM Voice Transmitter Part List

Ref. #	Description	DB Part #
R1	15K, 1/4-watt resistor (br-gr-or)	
R2	3.9K, 1/4-watt resistor (or-wh-red)	
R3	220-ohm, 1/4-watt resistor (red-red-br)	
R4, 5	Two 1K, 1/4-watt resistors (br-blk-red)	
R6, 9	Two 10K, 1/4-watt resistors (br-blk-or)	
R7	390K, 1/4-watt resistor (or-wh-yel)	
C1, 6	Two .01-microfarad, 50-volt disc capacitors	
C2	4- to 6-picofarad small silver mica disk (4.7)	DB# 4.7P
C3	.1-microfarad, 50-volt plastic capacitor	
C4	6- to 35-picofarad trimmer capacitor	DB# 635P
C5	2.2-microfarad, 25-volt vertical nonpolarized electrolytic capacitor	
CL1	Battery snap clip and leads	
Q1, 2	Two PN2222 NPN general-purpose transistors	
L1	Coil wound as described in assembly steps	DB# COIL8T
PBOARD	2 × 1 × .1-inch grid perforated board	
WR24BUSS	12 inches of #24 bus wire for wiring and antenna tap	
WR16BUSS	6 inches of #16 bus wire for making L1 coil	
M1	Special FET-bypassed microphone	DB# FETMIKE
PCXFM1		DB# PCXFM

Chapter Twenty-Five

FM Pocket Radio and TV Disrupter

Your TV disrupter, as shown in Figure 25-1, is designed and built on a piece of perforated circuit board that can fit inside a small plastic case. It operates using two standard, 9-volt transistor batteries in series. A unique pulsing circuit causes carrier deviation at 1,000 Hz, which creates a highly effective interfering device. The unit is tuned to a certain channel or frequency via a miniature tuning capacitor and then is controlled by a slight action of hand movement for further control.

This device is intended as a barroom joke and must not be abused. No attempts should be made to add an antenna because the range can be excessive and become a highly illegal device. When used as a harmless gag, such as in a bar, you can drive the bartender crazy because just as he is about to make an adjustment to a TV to get rid of the interference, you simply detune the device to make an adjustment to get rid of the interference and take the disrupter off the channel. When he thinks the problem is solved and continues his other duties, you hit it again. You can only begin to imagine the enjoyment (as a gag) of creating such havoc and being completely unnoticed.

Also, any FM radio station can be interrupted with a 1,000 Hz tone that is many times louder than the normal received station. Obviously, a fun-loving person can use his or her imagination and come up with many harmless uses for the device.

It must be stated that using a device of this nature could be considered illegal, and discretion should be used at all times because a good gag can become a hateful nuisance. Also note that when cable TV is used, it may be necessary to get as close to the set as possible for maximum effect.

The following plans show how a miniature, simple, inexpensive electronic circuit can disrupt TV and radio communications in a very limited area. The unit can also be used to blank out those tiresome commercials.

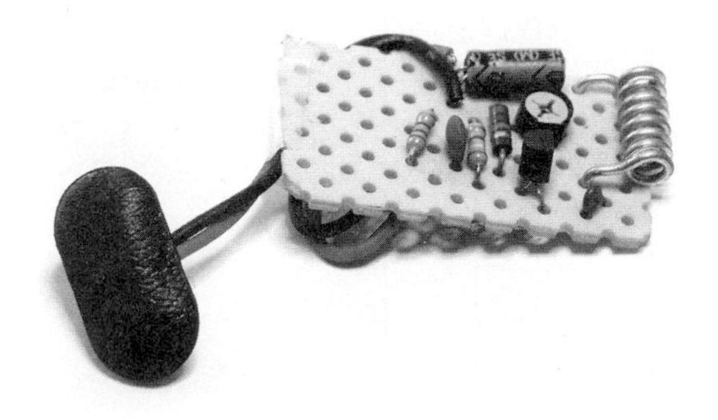

Figure 25-1 *Assembly board*

Circuit Theory

The I1 timer generates positive pulses that occur across R4. The frequency of those pulses is a function of R1 and C1 and consequently establishes the pulse repetition rate of the device. This frequency modulates the base of Q1, which is a variable oscillator whose frequency is determined by C6 (see Figure 25-2). See Table 25-1 for the parts list.

Assembly Steps

Lay out and identify the parts with the bill of materials. Insert the components as shown in Figure 25-3, starting from one end of the perforated circuit board, and follow the locations using the holes as guides. Use the leads of the actual components as the connection runs, which are indicated by the dashed lines. It is a good idea to trial-fit the parts before actual soldering. Always avoid bare wire bridges and globby solder joints, and check for cold or loose solder joints. Always pay attention to the capacitors with polarity signs and all the semiconductors.

To adjust for frequency, set the TV to channel 2 and turn C6 for maximum capacitance (for a preliminary frequency, when adjusting, it is advisable to use a lower voltage due to the possibility of interfering with other than TV or radio stations such as police or aircraft bands). Adjust L1 by compressing turns until the interference is noted on channel 2. Turn to other channels, turn on C4, and note the complete coverage. Channels above 6 are covered by the harmonic output of the device.

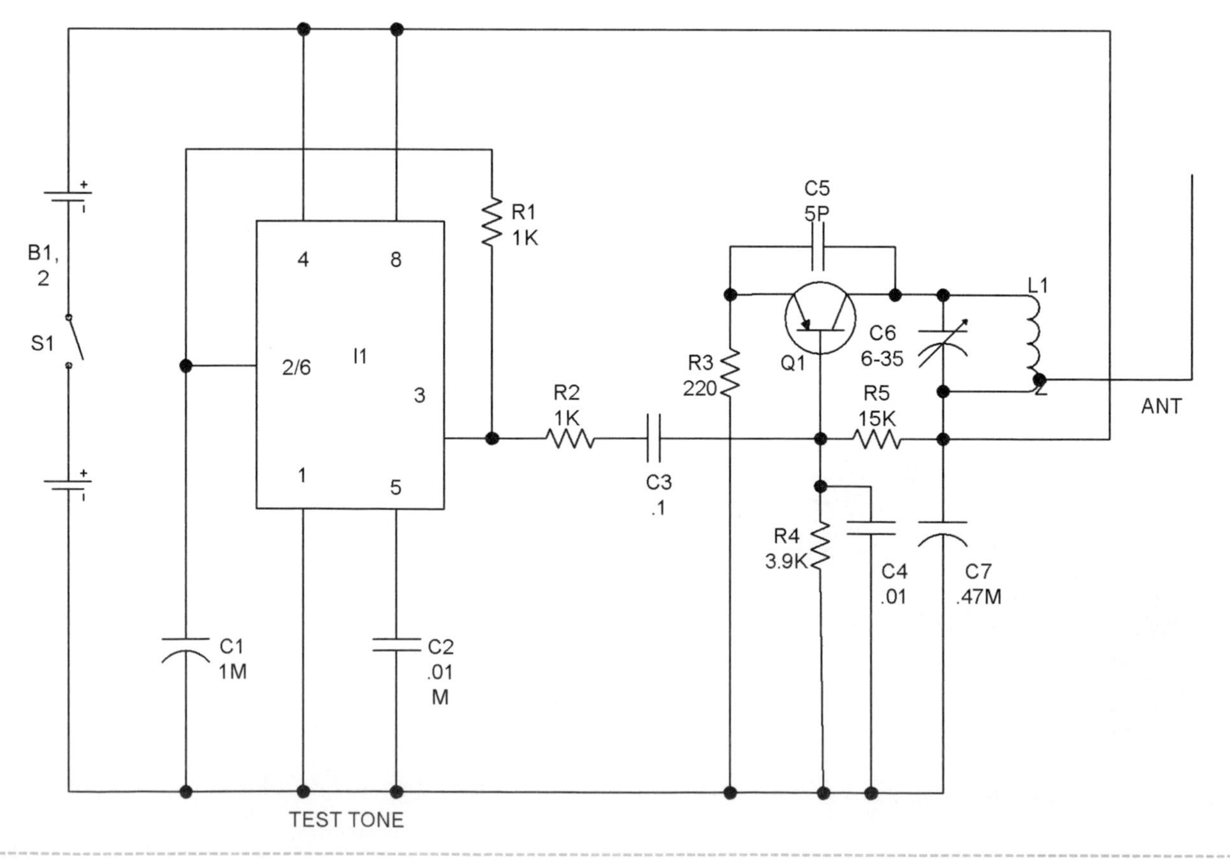

Figure 25-2 *Circuit schematic*

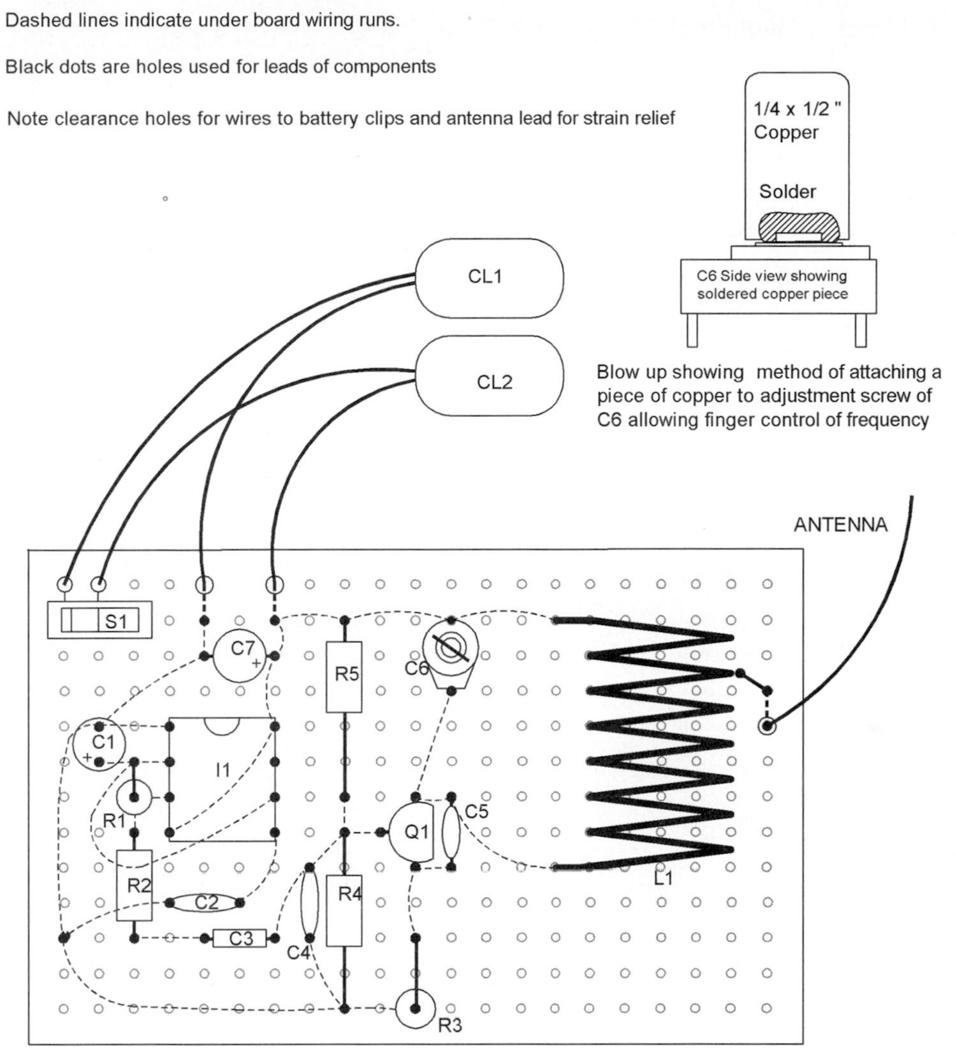

Figure 25-3 *Assembly board*

Several points of interest should be mentioned here, one being the ability to tune the unit on or off of a particular channel with just hand capacity (slight movement near L1); that is, the ability to detune the circuit with your hand creates a virtual capacitance. The other is the ability to also cover the FM radio band. This is accomplished by decompressing the coil (spreading the turns apart) and eventually adjusting L1 and C4 for the necessary coverage. It should be noted that all channels can be covered along with the FM band when properly adjusted. It may be necessary to remove a turn or two from L1 to a range on the FM radio stations.

Cable and satellite TV will often require the unit to be very close to cause an effect. There is no control over this, as the unit is designed to interfere with the actual frequency of the station being received, as was the case before cable and satellite services.

Table 25-1 FM Pocket Radio and TV Disrupter Parts List

Ref. #	*Description*	*DB Part #*
R1, 2	Two 1K, 1/4-watt resistors (br-blk-red)	
R3	220-ohm, 1/4-watt resistor (red-red-br)	
R3	100-ohm, 1/4-watt resistor (br-blk-br)	
R4	3.9K, 1/4-watt resistor (or-wh-red)	
R5	15K, 1/4-watt resistor (br-gr-or)	
C1	1-microfarad, 35-volt vertical electrolytic cap (green or blue can)	
C2, 4	Two .01-microfarad, 25-volt disk caps (103)	
C3	.1-meter, 50-volt ceramic cap (blue tablet .1)	
C5	6-picofarad, zero temp (4.7 disk)	DB# 4.7P
C7	.47-microfarad, 35-volt vertical electrolytic cap (green or blue can)	
C6	6- to 35-picofarad trimmer tuning Cap	DB# 635P
Q1	PN2222 NPN high-frequency, general-purpose transistor	
I1	555 *dual inline package* (DIP) timer *integrated circuit* (IC)	
S2	*Single pole, single throw* (SPST) slider switch	
CL1	Two-battery snap clip	
WR1	9 inches of #16 buss wire; wind seven turns on a pencil-sized object	
WR2	12 inches of #24 vinyl hookup wire	
PB1	2.5 × 1.5 × .1-inch grid perforated circuit board	
L1	Coil as shown	DB# COIL8T

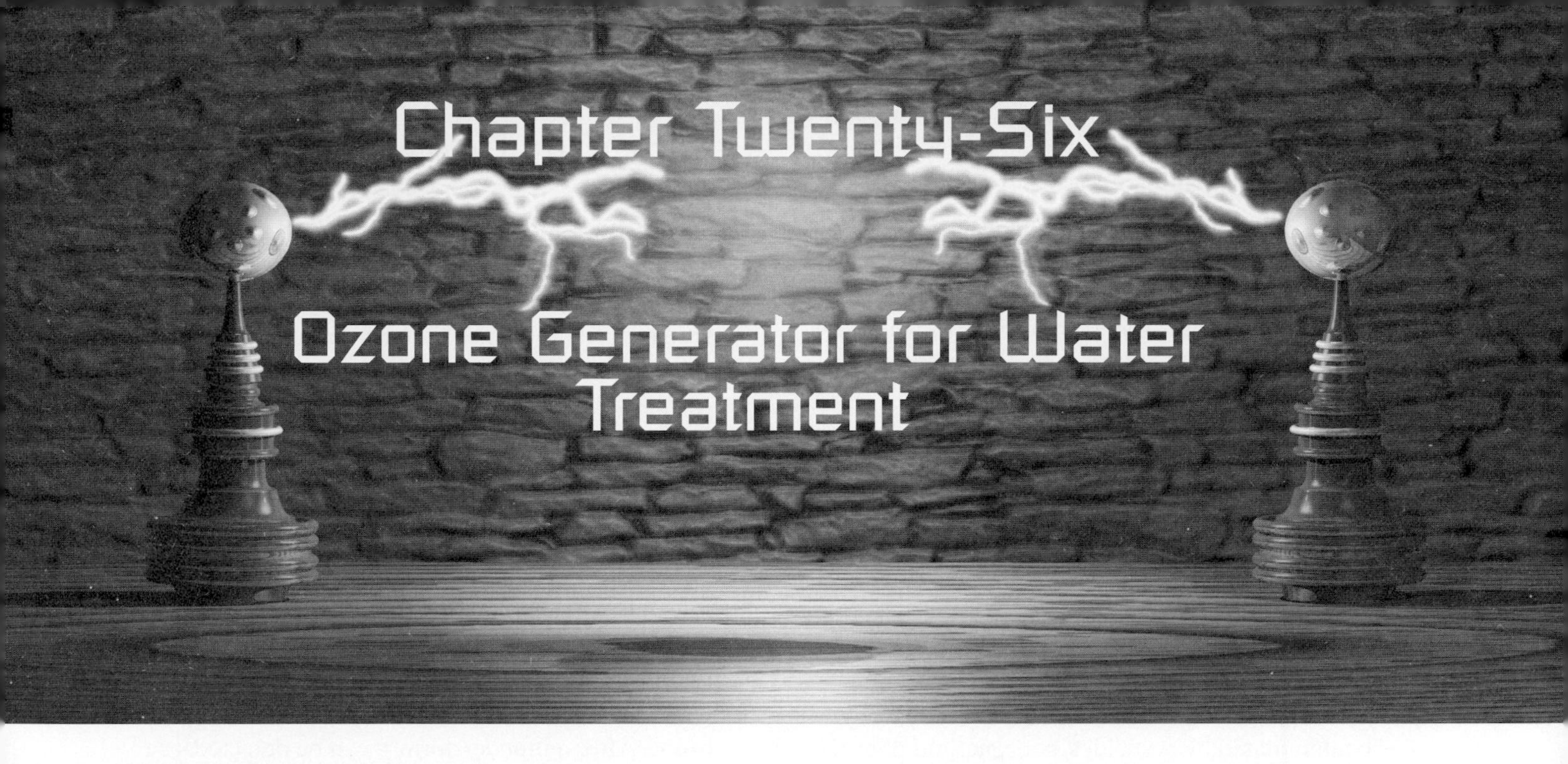

Chapter Twenty-Six

Ozone Generator for Water Treatment

This project, as shown in Figure 26-1, can very effectively purify water for use around the home or farm. The system will produce up to 5 grams per hour sufficient for swimming pools, laundries, and other medium-volume applications. The system covered here is shown in five different sizes and uses easy-to-construct modules requiring only a flow of air for operation.

This is an intermediate-level project requiring a basic electrical hookup. Expect to spend $50 to $250 depending on the size of the system chosen and the availability of the required air source. All parts are readily available, with specialized parts obtainable through Information Unlimited at www.amazing1.com, and they are listed in the parts list at the end of the chapter.

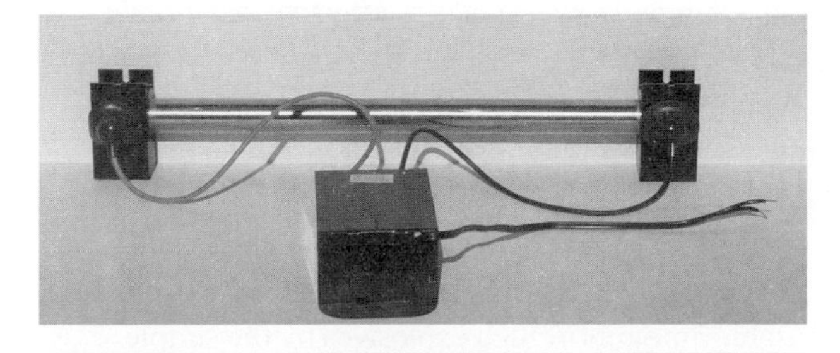

Figure 26-1 *The water treatment system*

Introduction to the Benefits of Ozone

Ozone is an unstable form of oxygen. Normal oxygen is diatomic (0_2), existing as two atoms of oxygen that form the molecule. Whereas ozone is tri-atomic (0_3), existing as three atoms of oxygen for the molecule. The tri-atomic form of oxygen is very unstable, wanting to lose the third oxygen atom and combine with whatever atom it can (oxidization). This property makes it the most active oxidizer known, with the exception of the very hazardous fluorine gas.

Ozone at normal pressures is colorless and produces a pleasing, fresh-air odor often like that after an old-fashioned thunderstorm. Under pressure, it becomes a bluish gas.

Ozone is also a very powerful bactericide. It is not affected by pH, as is chlorine, thus making it an excellent candidate for pools, spas, laundries, and general water treatment applications. It is many times more soluble in water, further enhancing its purifying effect. Ozone will combine with diatomic nitrogen N^2, forming nitrous oxide, 2NO. This oxide quickly combines with water, forming nitric acid, HNO_3. This is often a very undesirable effect when used with straight air. Pure oxygen greatly minimizes this effect and is often required in many applications. However, those requiring a supply of concentrated nitric acid for nitration may wish to consider ozone and air with a condensing apparatus to obtain this useful acid. A condensing apparatus takes vapor from the ozone gas along with water and nitrogen to form a liquid by contacting them with cooler surfaces. The liquid (now nitric acid) flows down into a collection container by gravity. In the same way the old timers made moonshine, this uses a simple and basic chemical reaction where $2O^3 + 2N^2 + 2H_2O$ yields $2H(NO^3) + H_2O + 2NO$.

Ozone for Water Applications

It is estimated that 20 percent of all ground water is contaminated by pesticides, benzene, and phenol derivatives along with other undesirable organic substances. Ozone can oxidize many of these compounds, along with deactivating many viruses and harmful bacteria. Ozone will also oxidize certain inorganic compounds, such as iron and manganese, making them more easily removable by filtration.

Chlorine and bromine are often the choice of disinfectant for swimming pools and spas. The effect of these halogens is often dictated by pH, temperature, and agitation. Extreme heat and agitation can produce chloroform, a very toxic carcinogenic. The *Environmental Protection Agency* (EPA) is already taking a look at these chemicals for this use.

Ozone-treated water will destroy fungus, mold, and many pathogens found in water when used for washing fruits and vegetables in packing lines. When discharged, ozone causes little change to the beneficial bacteria in sewage treatment facilities.

Additionally, freshly caught fish will last longer when washed with ozone-treated water, and ozone's oxidizing action can eliminate odors from stored cheese. Egg storage time is increased, and wine can be aged faster. The removal of odors produced by the bleaching of beeswax, starch, flour, straw, bones, and feathers are all aided by ozone treatment. Also, the grease and wax on cotton and wool fibers can be decomposed by ozone, and gray mold on the surface of fruits and vegetables can be controlled by ozone.

Ways to Generate Ozone

Ozone can be produced by an electrical discharge or by a high-frequency electromagnetic wave. A high frequency requires the wave to be in the ultraviolet spectrum where Planck's energy formula, $W = hc/v$, starts to become effective. This is where energy in a wave packet in ergs is equal to Planck's constant times the speed of light in centimeters divided by the wavelength in centimeters. It is this energy that causes the stable O_2 to break up and recombine with other O_2 to form unstable O_3. Germicidal lamps operating at 253.7 nanometers can produce ozone.

The method presented here utilizes the ozone-producing properties of an electrical discharge. We have all, at one time, smelled the by-products of ozone. After a thunderstorm, it can be detected, as well as on certain days where electrical activity is spawning a storm. A sparking electric discharge such as brushes on a motor will create ozone also.

Our method uses a metal tube with a conductor running down the center. The conductor and the tube are insulated by one another and must support the high frequency and high voltage necessary to create a corona without breaking down into an arc. The ozone produced in the tubular cavity must be made to flow using moving air for cooling and replenishing air to be converted.

This method, while producing usable ozone, has several disadvantages. First, normal air contains nitrogen that likes to combine with the ozone to produce nitrous oxides. Second, air contains water in the form of moisture. Without getting into basic chemistry, we know that water plus nitrous oxide forms nitric acid, which is corrosive and undesirable for air purification applications. However, a very effective way of making this acid is to allow the oxides to combine with water vapor or steam and condense in a cooling tube, producing concentrated nitric acid that can be used in the manufacture of high explosives. The 5 gram per hour system can produce enough nitric acid that, when mixed with battery acid (sulfuric acid), can produce usable amounts of high explosives by the simple nitration of many organic compounds.

Select the system you require from Figure 26-2. Note that each is complete with the mating power supply.

Wire up the system as directed in Figure 26-3. Obtain a suitable air supply and connect up the hoses as shown. Use ozone-resistant material for connections to the cell.

Turn on the air supply and note a bubbling from the output hose immersed under the water. Apply power to the high voltage and note a distinct smell coming from the hose when pulled out of the water. Reinsert and rig the hose end to stay under the water. The unit is now ozonating the water.

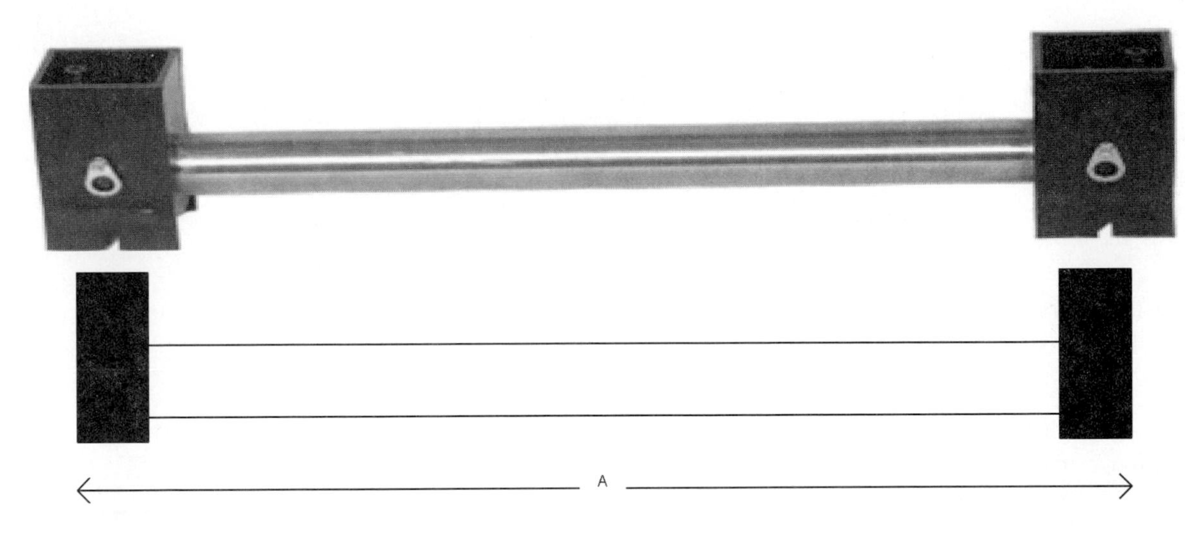

DATA BASE#	LENGTH A	OZONE OUTPUT/HR	AIR/O3 FLOW/MIN	INPUT V	INPUT I
OZONE300	182mm	100-500 mg/hr	.25-.55 CFM*	115 Vac	.2 Amps
OZONE800	282mm	500-1000 mg/hr	.35-.7 CFM*	115 Vac	.6 Amps
OZONE2000	382mm	1000-2500 mg/hr	1-1.5 CFM*	115 Vac	.9 Amps
OZONE5000	482mm	2500-5000 mg/hr	1.5-3 CFM*	115 Vac	l.2 Amps

* Multiply CFM by 28.3 to get Liters per minute

Each DATA BASE# system is complete with a matching power supply and ready to connect up as shown on figure 26-3

Figure 26-2 *System selection specifications*

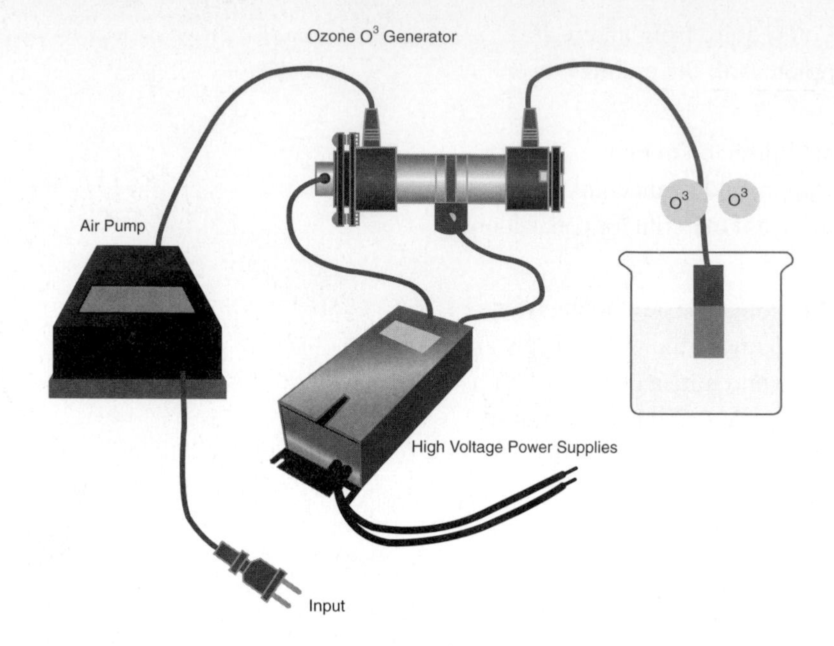

The high voltage power supplies used in these systems operate from normal 115 Vac 60 Hz. They are listed with the appropriate ozone cell on figure 26-2

The air pumps used can be that appropriate for the volume of the particular syste. A flow meter should be used when first setting up the system to verify proper air flow. Pure oxygen works the best and is available in cylinders with regulator valves and optional flow meters. Large aquarium air pumps may be used with the smaller systems while rotary vane air compressors are suggested for the larger systems. Note to multiply cubic feet per minute (CFM) by the factor 28.3 when converting to liters per minute.

Use ozone resistant tubing such as VITOR or HYPALON for connection to the cell

Use standard wiring codes for all 115 vac connections. The high voltage output wires from the high voltage power supply must be free from conductive objects and should be as short as possible. Do not twist together.

Figure 26-3 *Ozone system diagram*

Table 26-1 Ozone Generator for Water Treatment Parts List

Ref. #	*Description*	*DB Part #*
OZONE300	100 to 500 mg/hr system/182 mm/ 115 VAC @ .2 amps	DB# OZONE500
OZONE800	500 to 1,000 mg/hr system/282 mm/ 115 VAC @ .6 amps	DB# OZONE800
OZONE2000	1,000 to 2,000 mg/hr system/382 mm/ 115 VAC @ .9 amps	DB# OZONE2000
OZONE5000	2,500 to 5,000 mg/ hr system/482 mm/ 115 VAC @ .2 amps	DB# OZONE5000
HOSE	Connecting hoses, use ozone-resistant VITOR or HYPALON	
COMPRESSOR	Rotary vane compressor at required airflow	
CORD1	Three-wire power cord	
SWITCH	Two switches to control air compressor and high-voltage supply	

Chapter Twenty-Seven

Therapeutic Magnetic Pulser

This project shows how to build a battery-powered magnetic pulser that can provide relief from aches and pains, as well as improve upon many different health concerns and conditions. The unit is built on a small circuit board that can easily fit into a cigarette pack-size enclosure, along with its 9-volt battery or wall adapter (see Figure 27-1).

This magnetic pulser is designed so the magnetic north pole is applied to the body only. The coil strength is 10,000 gauss and it pulses at 1,000 times a second. A small switch can be used to turn the device off or you may simply unsnap the battery clip.

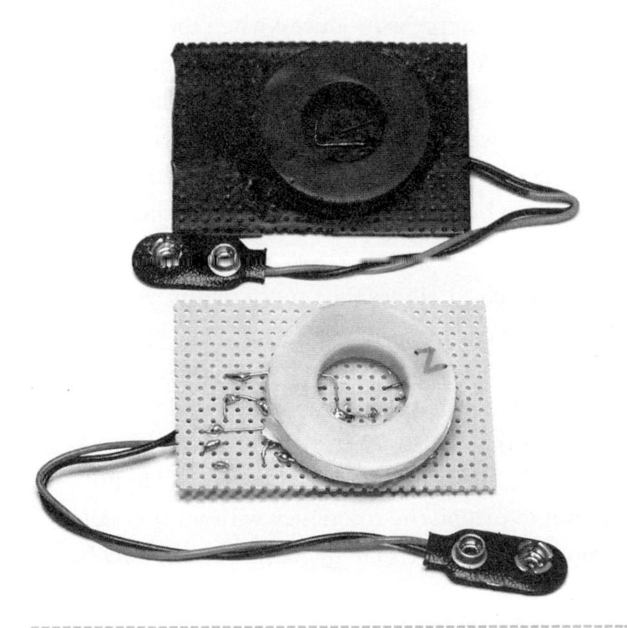

Figure 27-1 *The thermomagnetic pulser*

When an aluminum beer or pop can is held between your ear and this pulser, you hear a strong, distinct ringing as the can itself becomes the opposing polarity. If you hold the coil in front of a TV, you can see lines forming on the screen. On some TVs, the lines are visible when it's held as far away as 3 inches.

This magnetic pulsing action is more powerful and effective than traditional magnetic therapy. The pulsing coil takes magnetic therapy to a whole new dimensional level with many yet-to-explore exciting breakthroughs and results.

Permanent magnets, no matter how strong, will not produce the same results as this device's pulsed fields, which now produce induced back-emf currents. This requires a high-intensity, time-varying, magnetic impulse, not just a steady-state magnetic field.

Permanent magnets have their uses, but they work in different ways with different results. Equally important is the fact that we have "electric-powered" brains: All our thoughts and perceptions consist of complex networks of electrical signals and electromagnetic fields that pulse and sweep throughout the brain. So it then makes sense that harmonic electrical revitalization of the brain can harmonically influence your mental state and positively alter mental effectiveness.

Externally applied magnetic pulses to the lymphatic system, spleen, kidney, and liver help neutralize germinating, latent, and incubating parasites of all types, helping to block reinfection. This speeds up the elimination of disease, restores the immune system, and supports detoxification. The movement of the lymphatic system is essential in purifying, detoxifying, and regenerating the body, supporting the immune system, and maintaining health. Normally, the lymph is pumped by the movement of our body's electromagnetic field with vigorous exercise and physical activity. However, a clogged, sluggish, or weak lymphatic system prevents the body from circulating vital fluids and eliminating toxic wastes, thus weakening the immune system. It makes us vulnerable to infections and diseases. In order to be healthy, it is essential to keep the energy balanced and fluids moving so that the body's natural intelligence may operate at its full healing capacity. In addition, each cell must be enlivened with its own unique energetic frequencies and harmonic energy state and be harmonically connected to the life-force energy throughout the rest of the body.

Pulsed electromagnetic fields influence cell behavior by inducing electrical changes around and within the cell. Improved blood supply increases the oxygen pressure, activating and regenerating cells. Improved calcium transport increases the absorption of calcium in bones and improves the quality of cartilage in joints, decreasing pain dramatically. Acute and even chronic pain may disappear completely.

This device uses complex energy pulses of magnetic waves to stimulate certain body functions. It accelerates the production of vitally important hormones often providing miraculous effects. Magnetic pulsing aids in human growth hormone production and neotransmitter production. The results are remarkable, increasing vitality, sexuality, and accelerated healing. It also helps learning and the reduction of memory loss with reports of increased psychic ability for some people. The magnetic pulser also revitalizes and rejuvenates. As we get older, hormone production drops off considerably and is a reason why we age.

Users are experiencing faster healing of injuries, including bone fractures, carpal tunnel, and arthritis. The pulsed magnetic field stimulates blood flow and cellular respiration in the area applied. Reports have also documented that people's migraine headaches have ceased after magnetic therapy has been applied.

This pulser can be worn, held, or wrapped to any part of the body or held in place with a tension bandage. You can use your pulser while driving and on the go, rather than having to stop and apply it when sitting or lying down, as is necessary when using a 110-volt, standard magnetic pulser. The product is currently being sold only for experimental research and testing.

This is a simple, basic-level project requiring minimal electronic skills. Expect to spend $15 to $20. All parts are readily available, with specialized parts obtainable through Information Unlimited (www.amazing1.com), and they are listed in the parts list at the end of the chapter.

Circuit Description

Figure 27-2 shows a 555 timer (I1) connected as a free-running, astable pulse generator. The output pulses are nonsymmetrical with a ratio of low to high time of 10 to 1. Discharge resistor R3 and forward-conducting diode D2 control the low time, whereas charge resistor R2 and forward-conducting diode D1 control the high time. Timing capacitor C2 is common to both states of the pulse being alternately charged and discharged.

Magnetic energy is produced when the timer I1 turns on transistor Q1. The current now has the "on" time of the pulse to build up and then collapse in the flux coil L1 when Q1 is turned off. The stored energy is returned to the circuit by diode D3. The result is a steady train of magnetic pulses at 1,000 pulses per second.

The switch S1 allows the unit to be turned on and off and also allows switching in another range, dropping the pulse repetition to 100 per second. It is interesting to note that claims using specific pulse repetition rates of 20, 72, 95, 100, 125, 146, 440, 600, 660, and 727 have proven very beneficial. Do not attempt to extend the frequency without first determining the current rise in the coil. Current may be excessive and will require more turns as the frequency is lowered.

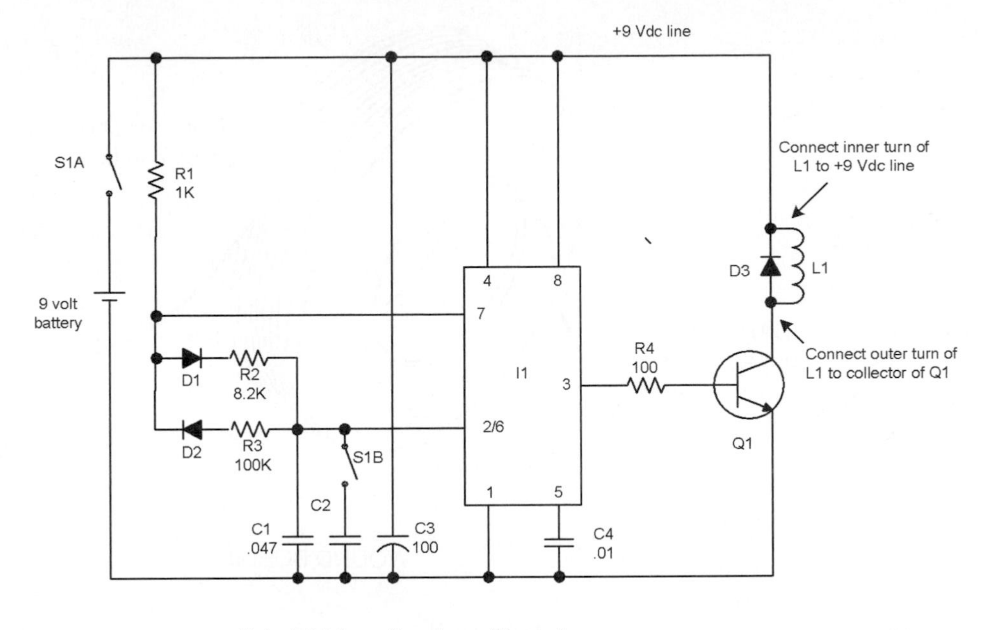

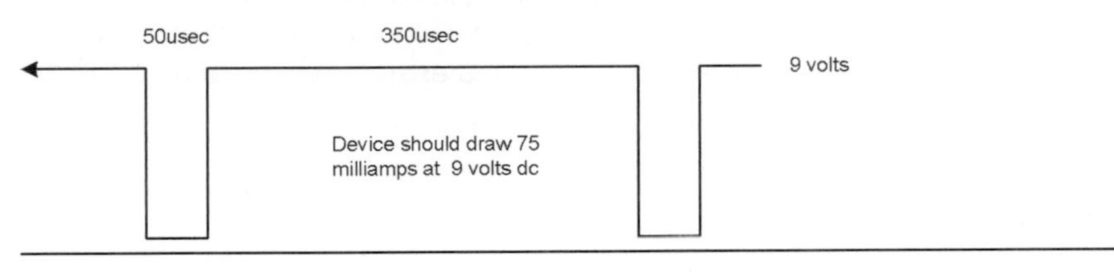

Figure 27-2 *Circuit schematic*

Assembly

1. Assemble the flux coil as shown in Figure 27-3 by first drilling a small hole adjacent to the inner diameter so that the "start" lead can exit the winding. Carefully wind 100 turns of #24 magnet wire in even windings. This is not necessary but makes for a neat-looking construction. Note to wind the wire so that its direction is the same as the figure with the leads routed as directed. This will guarantee the correct polarity as shown.
2. Cut a piece of .1-inch grid perforated board to a size of 2.6 × 1.8 inches, as shown in Figure 27-4. Locate and drill the holes as shown for the leads of battery clip CL1.
3. When building from a perforated board, it is suggested that you insert the components starting in the lower left-hand corner as shown. Pay attention to the polarity of the capacitors with polarity signs and all the semiconductors.

   Route the leads of the components as shown and solder as you go, cutting away unused wires. Attempt to use certain leads as the wire runs or use pieces of the #24 bus wire. Follow the dashed lines on the assembly drawing as these indicate the connection runs on the underside of the assembly board.
4. Attach the external leads from the battery clip and flux coil. Note the polarity of the leads on this part as indicated in Figures 27-2 and 27-4.
5. Double-check the accuracy of the wiring and the quality of the solder joints. Avoid wire bridges, shorts, and close proximity to other circuit components. If a wire bridge is necessary, sleeve some insulation onto the lead to avoid any potential shorts.
6. You are now ready to test the unit. Connect it to a 9-volt battery and note a current draw of 60 to 80 milliamps. Place the coil of the unit

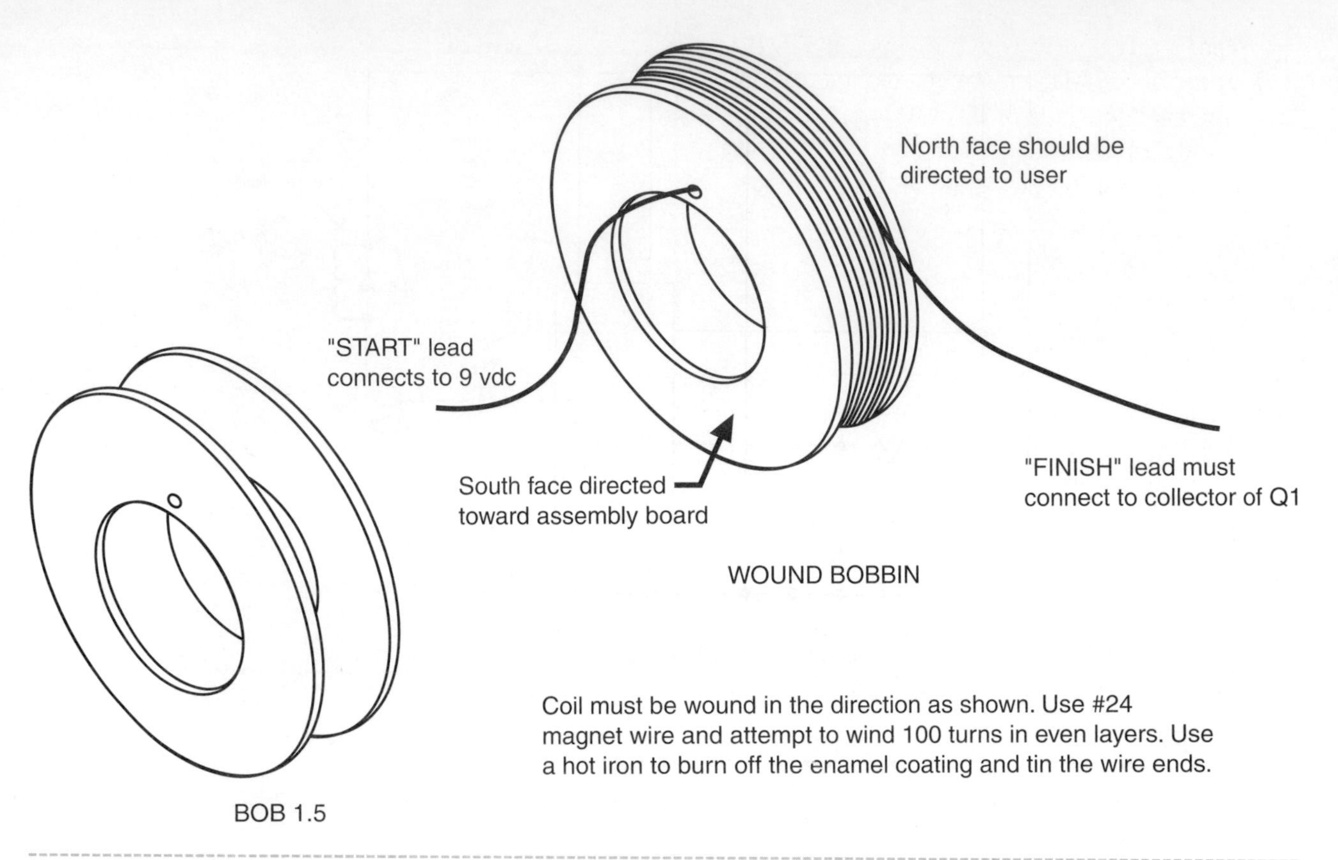

Figure 27-3 *Flux coil assembly*

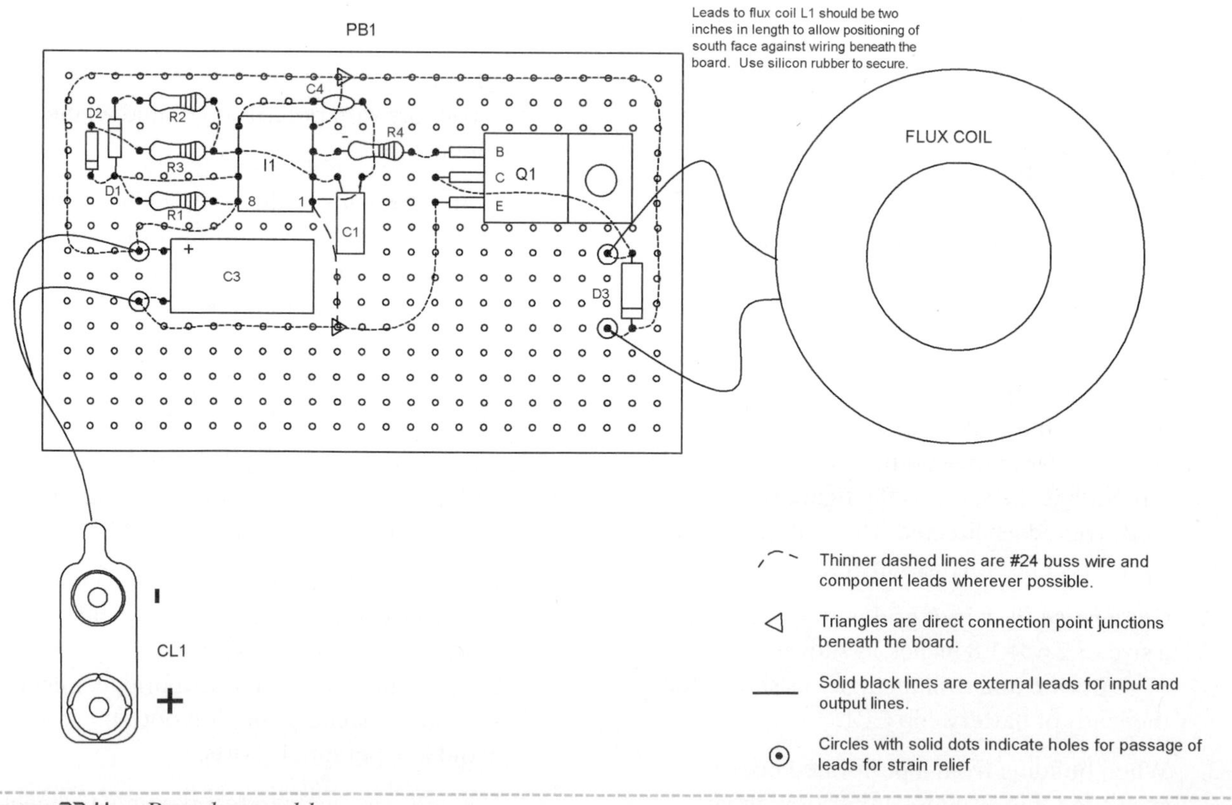

Figure 27-4 *Board assembly*

near a television screen and note the lines appearing. You may also put a metal can onto the coil head and place your ear to the can, noting a 1 to 2 KHz tone. If you have a scope, note the waveshape shown in Figure 27-2.

7. To use the unit, simply place the north face of the coil to the target area and allow exposure for several hours.

Table 27-1 Therapeutic Magnetic Pulser Parts List

Ref. #	*Description*	*DB Part #*
R1	1K, ¼-watt resistor (br-blk-red)	
R2	8.2K, ¼-watt resistor (gray-red-red)	
R3	100K, ¼-watt resistor (br-blk-yel)	
R4	100-ohm, ¼-watt resistor (br-blk-br)	
C1	.047-microfarad, 50-volt plastic capacitor	
C2	1-microfarad, 25-volt vertical electrolytic capacitor	
C3	100-microfarad, 25-volt vertical electrolytic capacitor	
C4	.01-microfarad, 50-volt disk capacitor (103)	
D1, 2	Two IN914 silicon diodes	
D3	1-kilovolt, 1-amp diode 1N4007	
Q1	TIP31 NPN TO-220 power transistor	
I1	555 dual inline package (DIP) timer	
BOB1.5	1.5 × .3 × .75-inch ID nylon bobbin (see Figure 27-3)	DB# BOB1.5
L1	Flux coil (see Figure 27-3)	DB# FLUXCOIL
PBOARD	2.6 × 1.8 × .1-inch grid perforated board as shown in Figure 27-4	
CL1	9-volt battery snap clip	

Chapter Twenty-Eight

Noise Curtain Generator

This unique circuit shown in Figure 28-1 is designed to produce a relaxing sound like that of a breaking surf. It is technically known as pink noise, being defined as a random distributor of equal sounds in the audible spectrum, favoring the higher-frequency end.

Pink noise has a spectral intensity that is inversely proportional to frequency for a given range. Therefore equal power is dissipated into a fixed resistance in any octave bandwidth in that range. White noise is random noise, which is typically thermal and shot noise, has a constant energy per unit bandwidth, and is independent of the central frequency at the band.

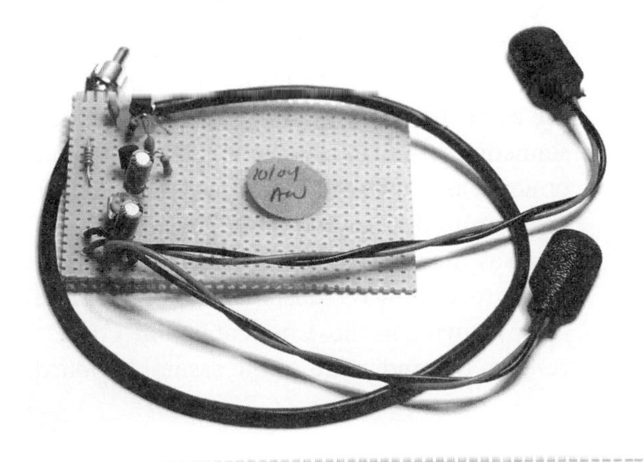

Figure 28-1 *The board-level generator*

This device is intended for use as a relaxing source of sound and also as a mask for certain types of surveillance equipment. Crashing surf is an excellent example of naturally produced pink noise that has a great relaxing effect. Due to the nearly equal amplitude of all the frequencies generated, the electronic equipment of an acoustical nature becomes easily overloaded and saturated, whereby normal voice sounds of varying amplitude and of limited bandwidth are much more finite and are unable to be processed by the equipment. In other words, the noise jams the microphone and audio preamplifier circuitry, thus rendering it unable to detect normal voices. (It should be noted that in certain instances a running shower will sometimes help simulate pink noise.)

The signal from the device is intended to be fed into the input of any amplifier, radio, or tape deck. It is easily connected to the center arm of the volume control if an input jack is not provided in a regular radio. (You will note the phono plug attached.) Also, the two batteries used are left on permanently, due to the fact that so little power is used.

The optional "speaker amplifier" reference schematic in Figure 28-2 is built on a piece of perfboard using standard audio-frequency wiring techniques. When used as a totally self-contained system, the electronics may be housed along with all batteries into a case or enclosure determined by the builder. This approach provides a convenient portable unit.

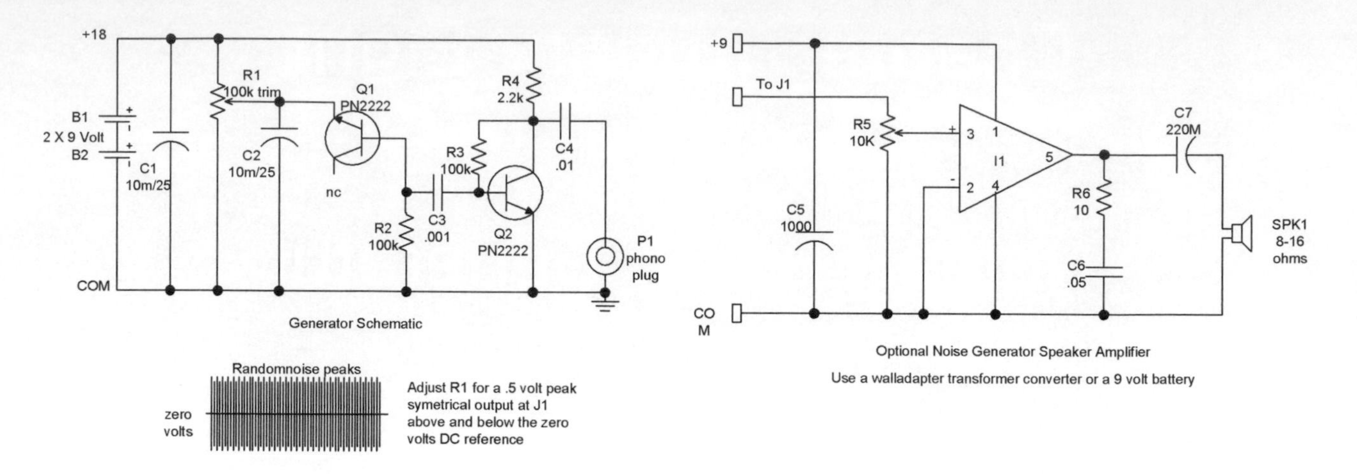

Figure 28-2 *Circuit schematic*

Circuit Description

Pink noise is a form of white noise that is the Gaussian distribution of all possible frequencies, the difference being that pink noise is more weighted to the audio spectrum. This form of noise has some very interesting properties, one being the ability to cause relaxation and a sense of well-being. Another property is that it provides a background that will completely mask an annoying device, rendering it unable to affect normal conversation. The device described is designed to work with any sound system with an audio input or is easily adapted to drive a loudspeaker by connecting it to a normal radio or building an optional speaker amp, as shown in the accompanying plans, making a complete system (see the parts list at the end of the chapter).

The circuit works in the following manner: A base emitter junction of transistor Q1 is reversed biased through a current-limiting resistor into breakdown (avalanche) condition. The random shot noise created is fed to the common emitter amplifier transistor, Q2, and to a filter, which in turn provides a low-level signal output. The unit module board and the batteries are placed in a plastic case with pieces of foam rubber or plastic. The outlet leads are fed out per the builder's requirements. The batteries are permanently installed and left on because the operating current is so low that there is no reason for a switch. Note that Jl is *not* required if using the optional speaker and amplifier circuit.

Assembly Steps

If you are a beginner it is suggested you obtain our *GCAT1 General Construction Practices and Techniques*. This informative manual explains basic practices that are necessary in the proper construction of electromechanical kits and is listed in Table 28-1.

1. Lay out and identify all the parts and pieces. Verify them with the parts list, and separate the resistors as they have a color code to determine their value. Colors are noted on the parts list.
2. Cut a piece of .1-inch grid perforated board to a size of 3 × 2.4 inches. Locate and drill the holes as shown in Figure 28-3 for the leads of the battery clips.
3. If you are building from a perforated board, it is suggested that you insert the components starting in the lower left-hand corner as shown in Figure 28-3. Pay attention to the polarity of the capacitors with polarity signs and all the semiconductors. Route the leads of the components as shown and solder as you go, cutting away unused wires. Attempt to use certain leads as the wire runs or use pieces of the #24 bus wire. Follow the dashed lines on the assembly drawing as these indicate connection runs on the underside of the assembly board.

4. Attach the external leads as shown in Figure 28-4. Notice the special note on the shielded cable.
5. Double-check the accuracy of the wiring and the quality of the solder joints. Avoid wire bridges, shorts, and close proximity to other circuit components. If a wire bridge is necessary, sleeve some insulation onto the lead to avoid any potential shorts.

Testing Steps

Insert the two battery clips and plug the output cable into the microphone or auxiliary jack of the existing system. Note the rushing sound and adjust the trimpot R1 to the desired effect. If you have a scope, make the adjustment as shown in Figure 28-2. If you build the speaker amplifier section, simply connect it as shown for a self-contained system suitable for nightstand use.

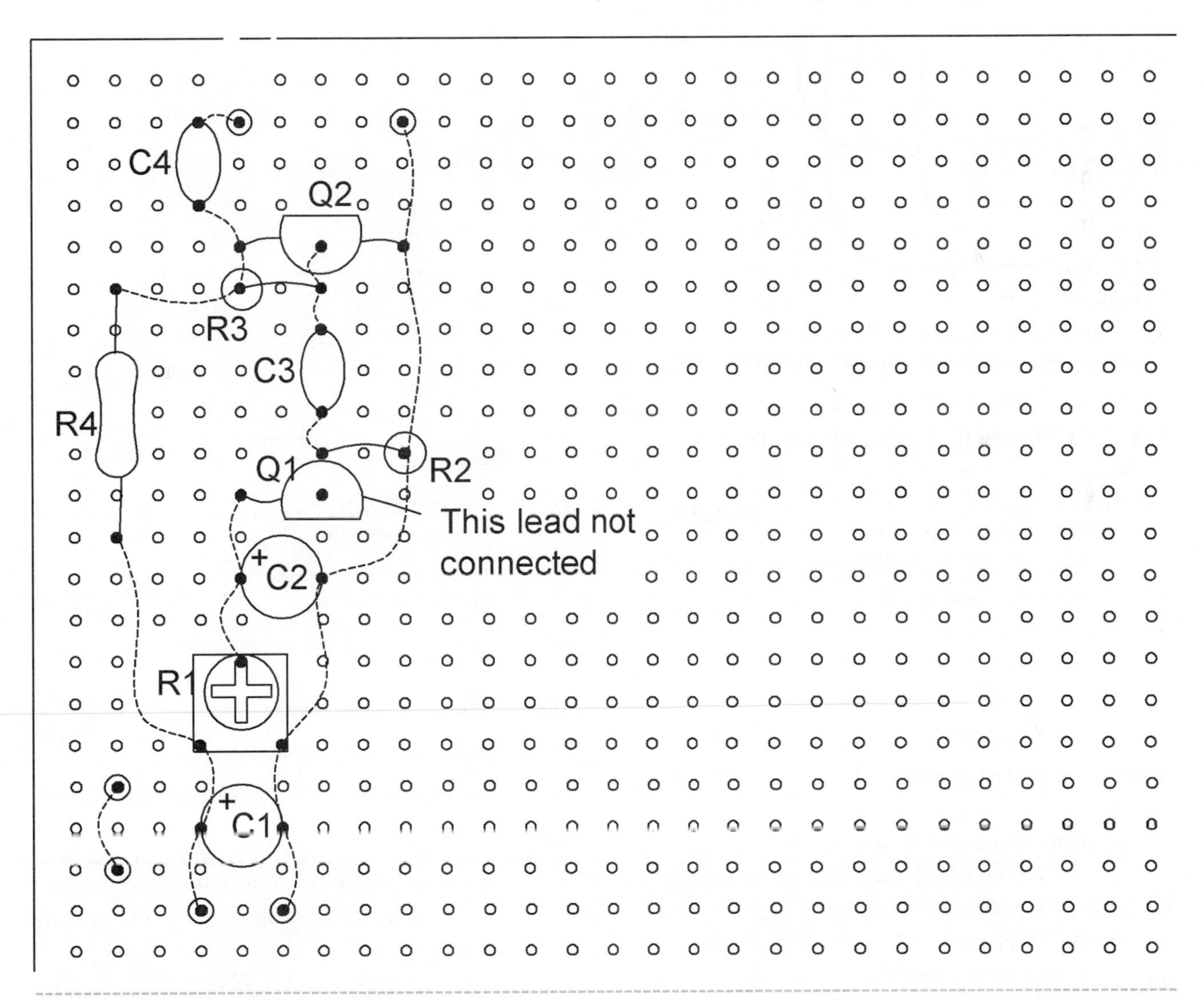

Figure 28-3 *Assembly board layout*

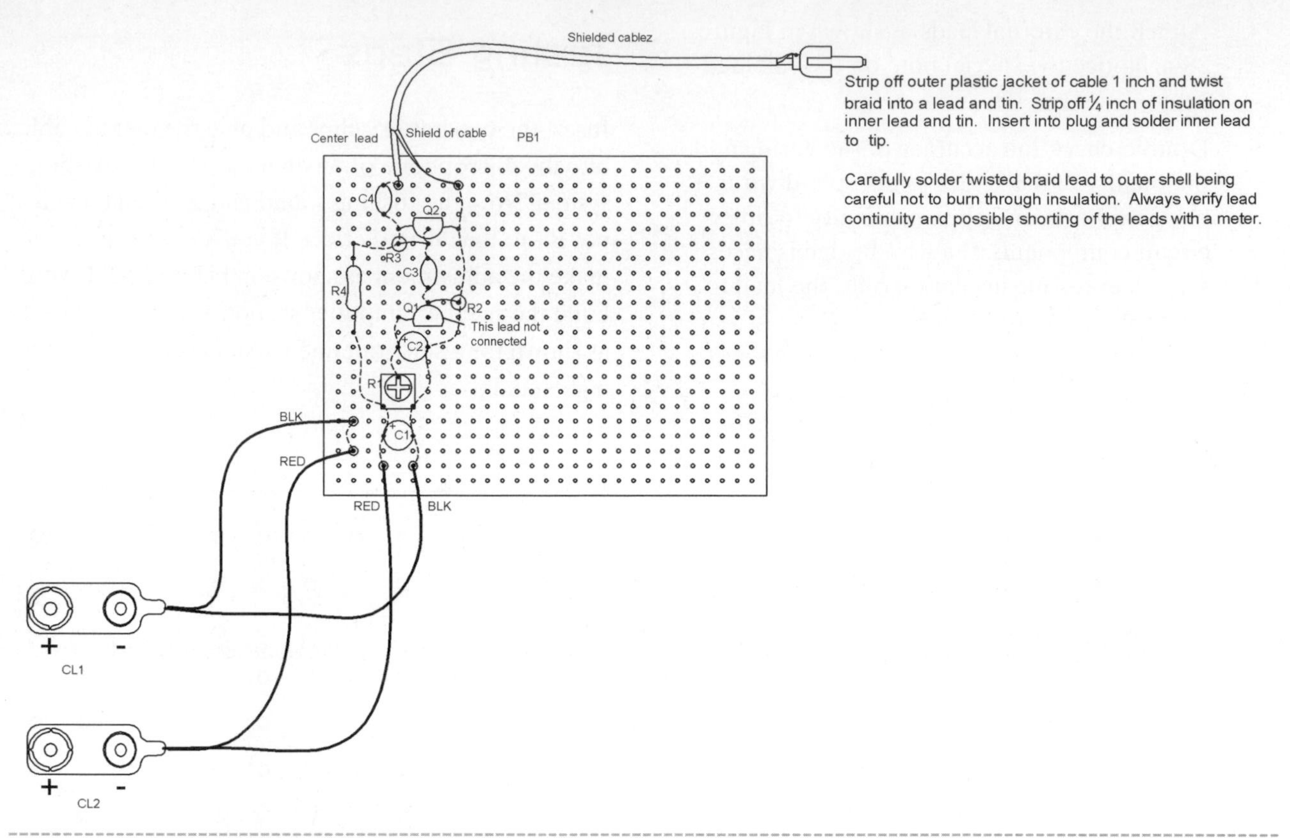

Figure 28-4 *Assembly showing external leads*

Table 28-1 Noise Curtain Generator Parts List

Ref. #	Description	DB Part #
R1	100K trimpot vertical mount	
R2, 3	Two 100K, ¼-watt resistors (br-blk-yel)	
R4	2.2K, ¼-watt resistor (red-red-red)	
R5	10K trimpot vertical mount	
R6	10-ohm, ¼-watt resistor (br-blk-blk)	
C1, 2	10-microfarad, 50-volt vertical electrolytic capacitor	
C3	.001-microfarad, 50-volt disc capacitor	
C4	.01-microfarad, 50-volt disk capacitor	
C5	1,000-microfarad, 25-volt vertical electrolytic capacitor	
C6	.047-microfarad, 50-volt plastic capacitor	
C7	220-microfarad, 25-volt vertical electrolytic capacitor	
Q1, 2	Two PN2222 NPN general-purpose transistors	
I1	LM386 dual inline package (DIP) operational amplifier	
PB1	2.25 × 3 × .1 × .1-inch grid perforated circuit board	
SPK1	Small 8- to 16-ohm speaker	
CL1, 2	Two 9-volt battery clips and leads	
P1	RCA phono plug	
SHC1	12-inch shielded microphone cable	

Chapter Twenty-Nine

Mind-Synchronizing Generator

Caution: This machine can trigger epileptic seizures and should be avoided at all costs by anyone who suspects even slightly that they are epileptic.

Your mind synchronizer is designed to be a completely self-contained unit with built-in batteries (see Figure 29-1). The system produces a variable-rate, flashing, monochromatic, directional, "shaped" light source (a light source with a variable ratio of off to on times), and a complementary synchronized audio pulser with tone-shaping control. The unit may be used so that a group can be exposed to light and sound pulses. For individual use, headphones can be used, along with special assembled light glasses. These are shown made from an existing pair of sunglasses. Optimum performance requires a quiet and reasonably low light condition, and distractions must be at a minimal. The mind-synchronizer can produce strange and bizarre hallucinations, as well as provide a sense of relaxation and well-being.

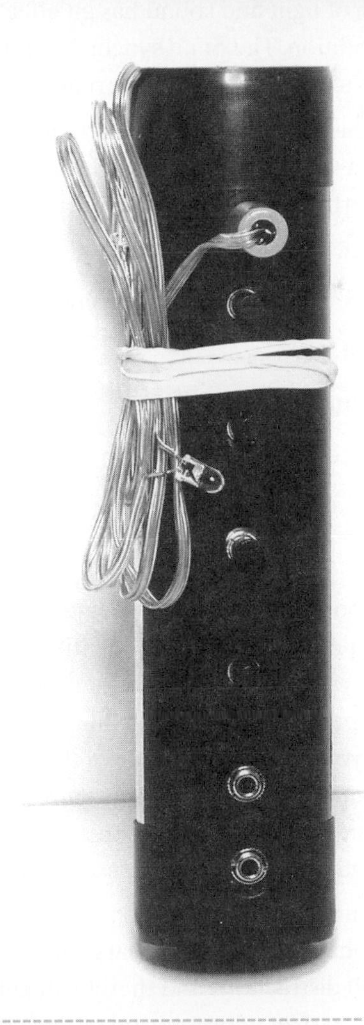

Figure 29-1 *Photograph of unit*

Information on Mind Control

Today's work environment demands that we have the ability to excel, to compete, and to resolve problems. Hence, the most important tool we possess is our mind. To optimize our potential, we now offer our low-cost mind-synchronizing unit.

After 25 years of research, Dr. Axel Bruck of Berlin, Germany, created a machine that can enhance, develop, and perfect the qualities within you in order to attain the goal of mental fitness. It has been found that external stimulation evokes a mental reaction in people that is either active or passive. The mind synchronizer can determine automatically the kind of person you are. An evaluation is performed via the means of a hand sensor using a technique called biofeedback, which is sensitive to the flow of blood. This knowledge is very important for providing optimal stimulation to the users.

Having determined the kind of person you are, the mind synchronizer now uses "programmed" light pulses and "shaped" sound to help you reach a required mental condition, be it relaxation or efficiency. It has been known for hundreds of years that an external rhythm of light and sound has an effect on your mental condition. The mind synchronizer is a modern version of a drum and a flickering fireplace.

Through extensive scientific measurements, experiments, and results, the mind synchronizer can improve your performance level by using the right stimuli:

- Deep relaxation to the point of sleeping
- Relaxation
- Equilibrium
- Activation of muscles doing physical movements
- Efficiency

Whether in sports, professions, or personal lives, the development of personality and identity is vital to personal success and satisfaction. The mind synchronizer trains and perfects your mental and psychic abilities so that you have a stable way of thinking with a healthy outlook. The mind synchronizer teaches you the technique of concentration, stress handling, focusing, visualization, relaxation, and self-awareness

When using the mind synchronizer, you learn to set a goal and concentrate on this goal completely. You eliminate all distractions and thereby maintain your concentration over a long period of time. By doing so, you improve your power of concentration. In the long run, you will think clearly, make better decisions, and deal more effectively with life itself.

The program of relaxation combats stress, tension, and direct pressure. Almost everything we do causes stress. Stress means that the body is in a chaotic condition. The opposite of stress is relaxation. The right kind of relaxation is an optimization of mind, concentration, learning ability, performance, and health.

In reality, a relaxed person saves more than half a billion heartbeats over a nervous or hectic person, according to Professor Prinzinger's theory of the Quiet Pulse. Professor Prinzinger performed much research into the various relaxed states of the body and the relationship of relaxation to heart rate and blood pressure. The body consumes less energy, organs are burdened less, and life expectancy increases.

The program of efficiency is a newly developed psycho-active technique: the indication of unconscious eye movement and the employment of rhythm displacement. Certain stimulating rhythms such as that of a strobing light can produce nausea when strobing at 20 repetitions per minute. Other stimulating rhythms of lower repetitions per minute can produce relaxation and the feeling of well being. In such a case, eye movement usually follows at the lower repetition rates. The combination of visual effects (flashing *light-emitting diodes* [LEDs]) with audio input (via a headset) produces high-efficiency performance. The mind synchronizer establishes physical and mental readiness to attain this goal of high performance.

The program of activation involves more optic, visual patterns than acoustic ones to provide "imagination journeys." To experience associations with your own history and your own personality, you will be able to discover your emotional condition and personality.

The program of equilibrium allows for meditation with your eyes open. You would observe attentively, simply, and exactly each individual light source and its changes. The mind synchronizer will bring you the healthy balance you desire through light and sound.

Mind machines are available from a wide variety of sources today. Often referred to as hemispheric synchronizers or cortical-frequency entertainment devices, they all basically perform one basic function: They lead the brain to synchronize both hemispheres and then lead those synchronized hemispheric frequencies to a specific target frequency. In order to understand why anyone would want to do such a thing, let's examine what frequencies the brain produces and what people experience when the brain is at these frequencies.

Generally, brain waves are broken down into four different levels. They are referred to as beta, alpha, theta, and delta. With respect to frequency, they are as follows:

Beta: 13 cycles per second or more

Alpha: 8 to 12 cycles per second

Theta: 5 to 7 cycles per second

Delta: 1 to 4 cycles per second

Profound changes take place in a person's conscious experience at each of these levels. The beta frequency is the "normal" waking frequency, the drive-in-the-city, go-to-work, drink-coffee frequency, and it can range as high as 40 cycles per second. It is *not* a relaxing frequency.

The alpha frequency is a realm of relaxed awareness. It has been shown in a concrete fashion that this frequency is associated with an increased capacity for learning, comprehension, and retention. It is in the alpha state that the best learning occurs.

Theta is a hypnotic state often associated with meditation, a zen-like state of waking dreaminess. It can be a profoundly relaxed state where internal associations and thoughts can form ideas with clarity that could never have emerged in the normal, high-noise environment of the beta state. This is also the range that many native shamans have been found to enter when engaging in healing trance work. It is a state of profound tranquility.

The delta state is much simpler to define; you're asleep. When your brain waves are in the delta range, you're in a sound and restful sleep.

A number of studies have been done that indicate that a flashing light has a profound impact on the entire brain, more so than a pulsing sound or feeling. When the brain is exposed for a time (as short as 2 to 5 minutes) to a flashing light at a speed close to the brain's existing operating frequency, the brain will begin to synchronize with the stimulation. Moreover, both hemispheres, which are often operating at different frequencies, will harmonize. This has also been shown clinically and statistically to be an aid to mental functioning. Many books are available on the market on this subject for anyone who would like more detailed information on these studies.

Operating a Mind Machine

Now with the previous information in mind, the operation of the mind machine in all of its forms is basically the same. Start the machine off at a frequency that is roughly equivalent to the operating frequency of your brain at the time. (Use your own feelings as a gauge. If you're frazzled, start at maximum frequency; if you're pretty mellow, start at about half-frequency.) Allow yourself about 3 to 4 minutes at this frequency, and then slowly begin bringing the frequency down, very slowly. You should take about 7 to 10 minutes to bring the control all the way down. When you reach a minimum, leave it there for about 5 to 7 minutes.

After you've gotten to the bottom range, you can either lead yourself back to normal, in which case you will feel what you may consider a "good normal." If you so choose, you can exit directly from the lowest setting. If you do, have taken your time getting to it, and stayed there for several minutes, you will notice that it may be very hard to begin to move. Your limbs may feel very heavy, as if your body were asleep. If you manage to get to your feet, you may notice that things won't feel as solid; they'll seem to have a rubbery quality to them. These sensations can be quite pleasant as long as you're prepared and expect them (and don't have to drive or operate heavy machinery afterward!).

The visual effects are the most stunning and can only be likened to the experience of ingesting certain mind-altering substances. As the frequencies change, you will notice colors. Yes, two little flashing red lights in front of your closed eyes will create the illusion of fantastic, swirling colors and geometric designs ranging from ultimate simplicity to incredible complexity. One effect will be observed as myriads of little shapes that form and begin to slowly spiral around your field of vision, first clockwise, then counterclockwise, then back again. You will see dreamy, smooth, cloudlike fields of frosty blue-white, deep blues, deep reds, electric greens, yellows, whites, and purples, all in balls, stripes, stars, and triangles, dancing and moving like the most intense kaleidoscope you can imagine. The imagery is absolutely fascinating and can transfix you with the dazzling display. You will be absolutely amazed and impressed with the fact that it is your brain that produces this light show.

This unit is designed to take you from a medium-level beta down to a low theta. You can use these states as you desire. You may wish to stop the frequency decrease early and remain in the alpha state before an exam, prior to entering a class, or when trying to learn something important. You can continue on down and enter the theta level and create healing imagery in your mind to assist your body's own natural healing processes.

Most of all, play with your unit and enjoy the show. Built right, it will provide many years' worth of reliable service. Through usage, you can become familiar with the settings and be able to use the machine in a way that's exactly right for you. Always leave yourself plenty of time to use the machine and at least a half-hour after to "normalize." Use the unit to help you drift off to sleep; you can actually lead your brain down into relaxation and turn the unit off after you've gone into a deep theta, and if you're ready to sleep away, it's only a short slide from theta into the peaceful slumber of delta.

Circuit Theory

Integrated circuit I1 is an astable oscillator with a pulse rate being controlled from approximately 3 to 40 reps per second by external pot R2. The output of Ul now triggers integrated circuit I2 being connected as a monostable circuit that controls these pulses' ratios of on to off time. This duty cycle control is via external pot R5. Light-emitting diodes (LEDl and LED2) are connected to the output pin 3 of I2 through brightness control pot R7. These are the diodes you install in the sunglasses or they can be a single ultrabright LED for group use. The result is a variable flash rate from 3 to 40 pulses per second with a variable duty cycle rate (ratio of on to off time).

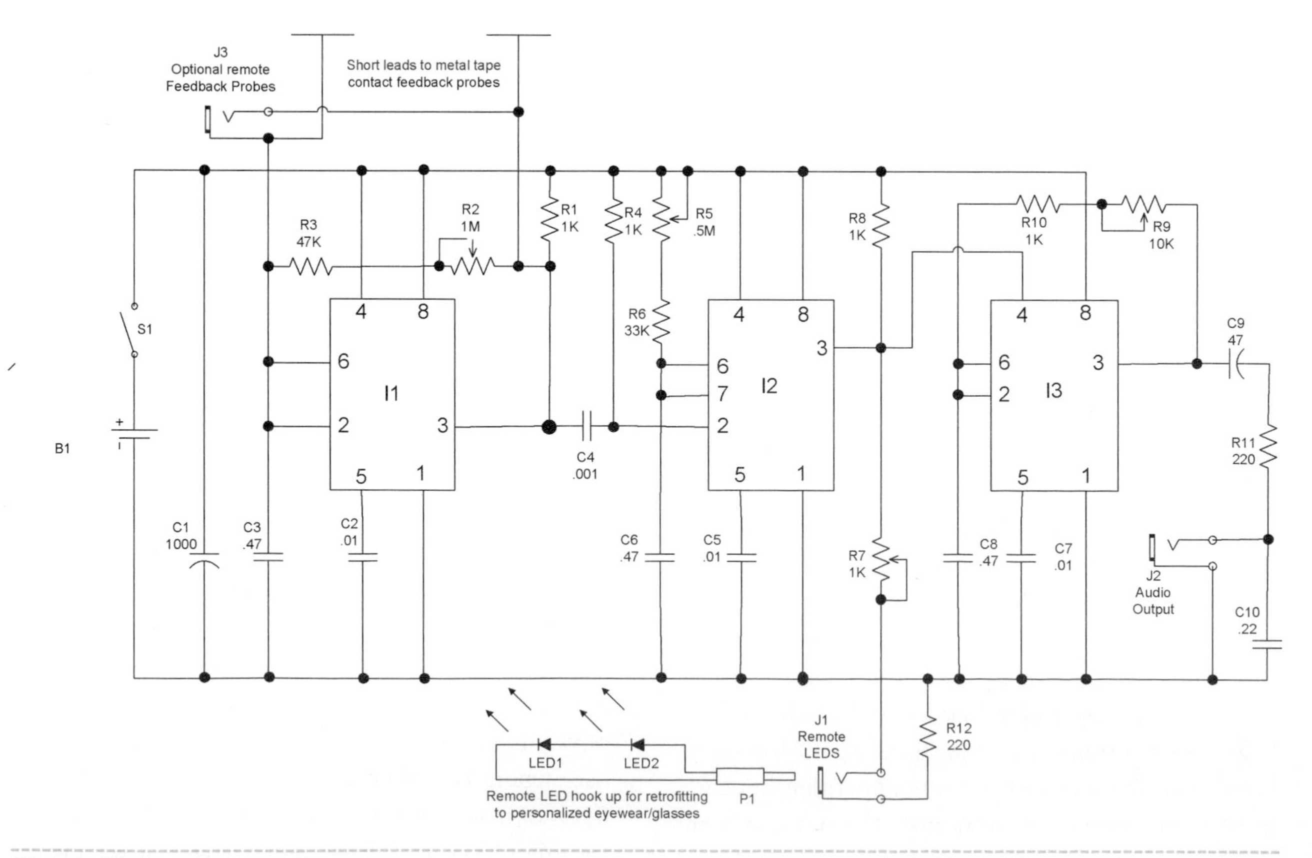

Figure 29-2 *Circuit schematic*

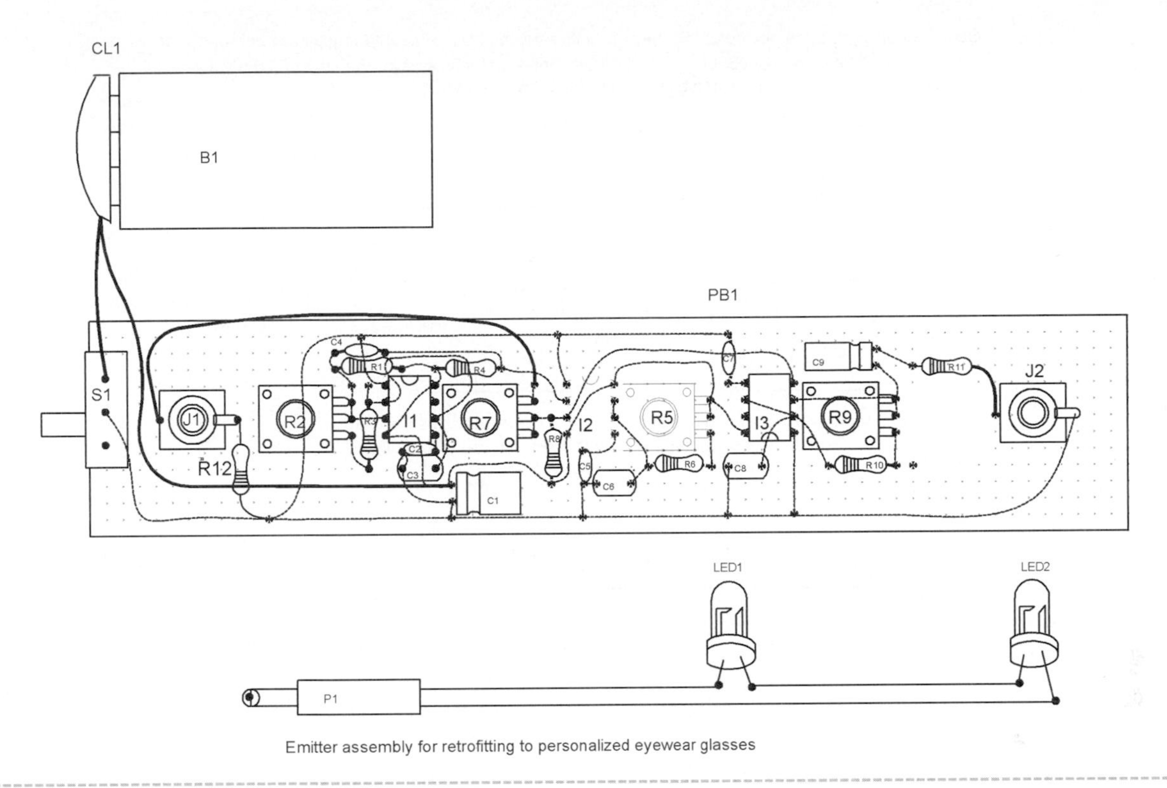

Figure 29-3 *Assembly board*

Note it is best to experiment with and record the duty cycle settings and their effects (see Figure 29-2).

The circuit also includes integrated circuit I3 connected as a variable-rate, sonically "shaped" audio oscillator. The output is via jack J2 driving a set of 32-ohm headphones.

The feedback circuit consists of two external pieces of metallic tape providing points of contact for the user's hand. Body resistance now partially controls the pulse rate of I1. An uptight situation provides lower contact resistance and consequently raises the pulse rate. As one relaxes, this resistance becomes higher and lowers the pulse rate. An external jack (J3) provides a connection for other bodily contact probes to the feedback system. These contact probes are left to the discretion of the user. This jack is omitted when using the external tape probes. Switch S1 controls power to the circuit and should be off when not in use to conserve the internal batteries.

Assemble the Board

If you are a beginner, it is suggested to obtain our *GCAT1 General Construction Practices and Techniques*. This informative manual explains basic practices that are necessary in proper construction of electromechanical kits and is listed in Table 29-1.

1. Lay out and identify all the parts and pieces. Verify them with the parts list, and separate the resistors as they have a color code to determine their value. Colors are noted on the parts list at the end of the chapter.

2. Cut a piece of .1-inch grid perforated board to a size of $6^1/_4 \times 1^1/_4$ inches, as shown in Figure 29-3. An optional PCB is individually available.

   The circuit is shown with the more challenging perforated circuit board often required for a science fair project. A PCB is also available and requires that you only identify the particular part and insert it into the respective holes as noted. The PCB is plainly marked with the part identification. Soldering is very simple as you solder the component leads to the conductive metal traces on the underside of the board.

   The perforated board approach is more challenging as now the component leads must be routed and used as the conductive metal traces. We suggest that the builder closely follow the drawings on this section and mark the actual holes with a pen before inserting the

Obtain a pair of sunglasses preferably the ones with side shields. Measure the distance between your eye centers. This is usually about 2½". Drill the two holes just a bit smaller than the LEDS to provide a secure fit. Press into the holes and glue if necessary to further secure.

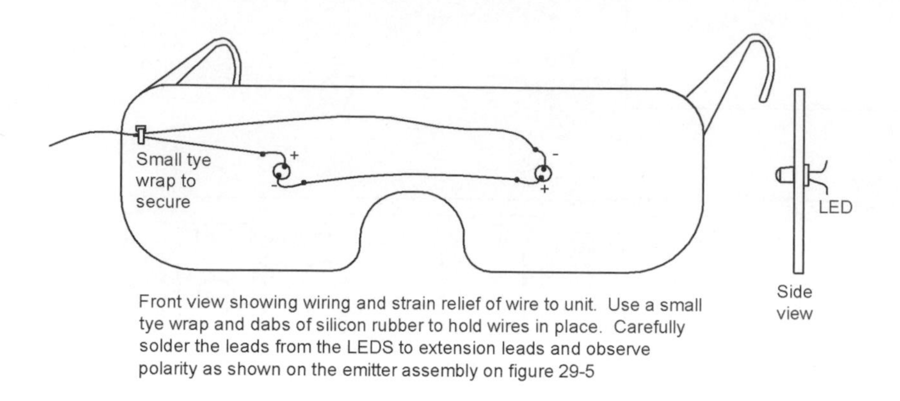

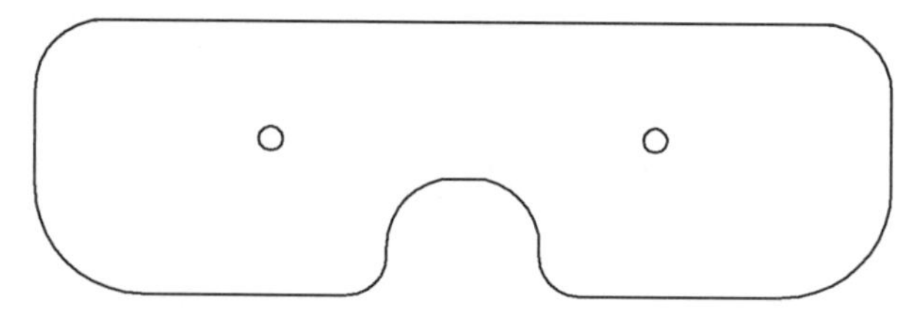

View towards the eyes. Secure the leads to the frames of the assembly to keep out of the way. You may also use pieces of duct or electrical tape to hold all in place.

Figure 29-4 *Assembly of glasses*

parts. Start from a corner, using it as a reference, and proceed from left to right. Note that the perforated board is the preferred approach for science projects as the system looks more homemade.

3. If you are building from a perforated board, it is suggested that you insert the components, starting in the lower left-hand corner, as shown in Figure 29-3. Pay attention to the polarity of the capacitors with polarity signs and all the semiconductors. Resistor RI2 is connected from the pin of 11 to the PC board. Note the control pots R2, R5, R7, and R9 and jacks J1, J2, and J3 should be positioned to closely match the holes, as shown in Figure 29-5. Route the leads of the components as shown and solder as you go, cutting away unused wires. Attempt to use certain leads as the wire runs or use pieces of the #24 bus wire. Follow the dashed lines on the assembly drawing as these indicate the connection runs on the underside of assembly board.

4. Attach the external leads to the battery holder as shown.

5. Double-check the accuracy of the wiring and the quality of the solder joints. Avoid wire bridges, shorts, and close proximity to other circuit components. If a wire bridge is necessary, sleeve some insulation onto the lead to avoid any potential shorts.

6. Assemble an LED to cable P1 as shown. Also, retrofit to the eyewear as per Figure 29-4.

7. Fabricate the enclosure tube (TUB1) as shown in Figure 29-5 and insert the assembly board and battery. Verify that the controls mounted on the assembly board will match the drilled holes with the dimensions as shown. Note the dimensions are shown for use with the optional PCB and may be changed. The unit is secured attaching the adjustment pots via nuts.

8. The final assembly is shown in Figure 29-6 and should be operated as instructed.

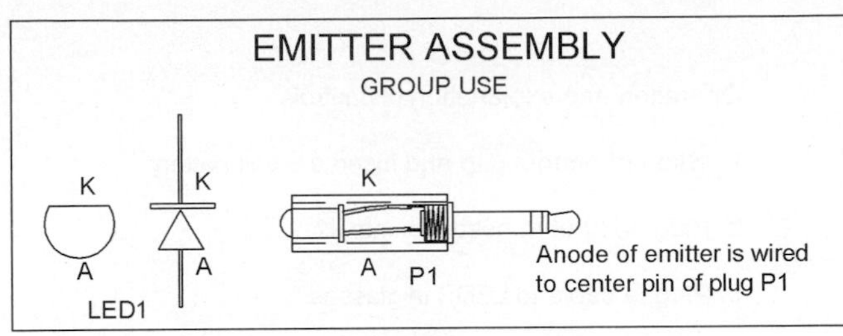

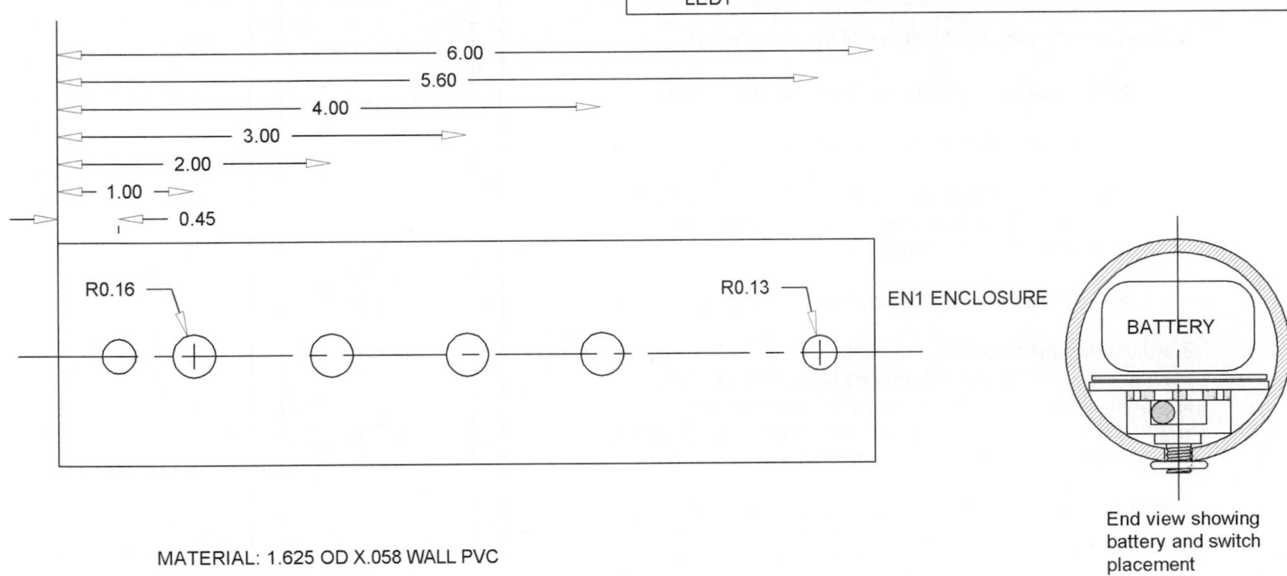

Figure 29-5 *Fabrication of enclosure*

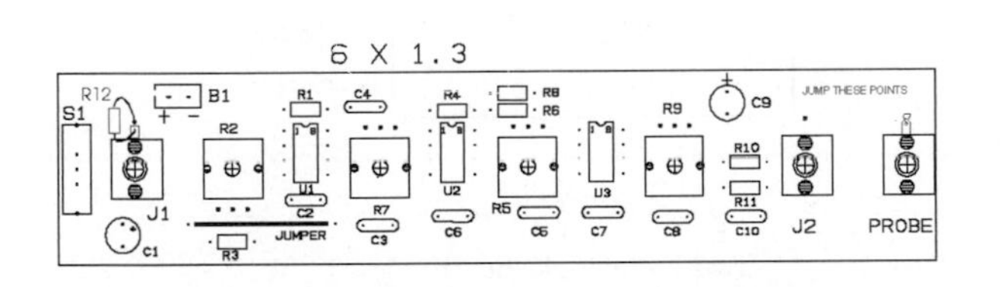

Figure 29-6 *Final assembly on PC Board*

Table 29-1 Mind-Synchronizing Generator Parts List

Ref. #	*Description*	*DB Part #*
R1, 4, 8, 10	Four 1K, ¼-watt resistors (br-blk-red)	
R2, 5	.5-1 meg potentiometer resistor	
R3	47K, ¼-watt resistors (yel-pur-or)	
R6	33K resistor (or-w-or)	
R7	1K pot resistor	
R9	10K pot resistor	
Rl1, 12	Two ¼-watt 220- to 330-ohm, resistors (red-red-br)	
Cl	1,000-microfarad, 25-volt vertical electrolytic capacitor	
C2, 5, 7	Four .01-microfarad, 50-volt disc capacitors	
C4	.001 mfd, 50-volt disk capacitor	
C3, 6, 8	Three .47-microfarad, 50-volt mylar capacitors	
C9	47-microfarad, vertical electrolytic capacitor	
C10	.22-microfarad, 50-volt mylar capacitor	
Il, 2, 3	Three 555 dual inline package (DIP) timers	
*LED1, 2	Two bright-red LEDs or use single, optional 15,000 mcd assembled as shown	
S1	Single pole, single throw (SPST) PC switch	
CLI	9-volt battery snap	
11, 3	Two 3.5 mm mono jacks	

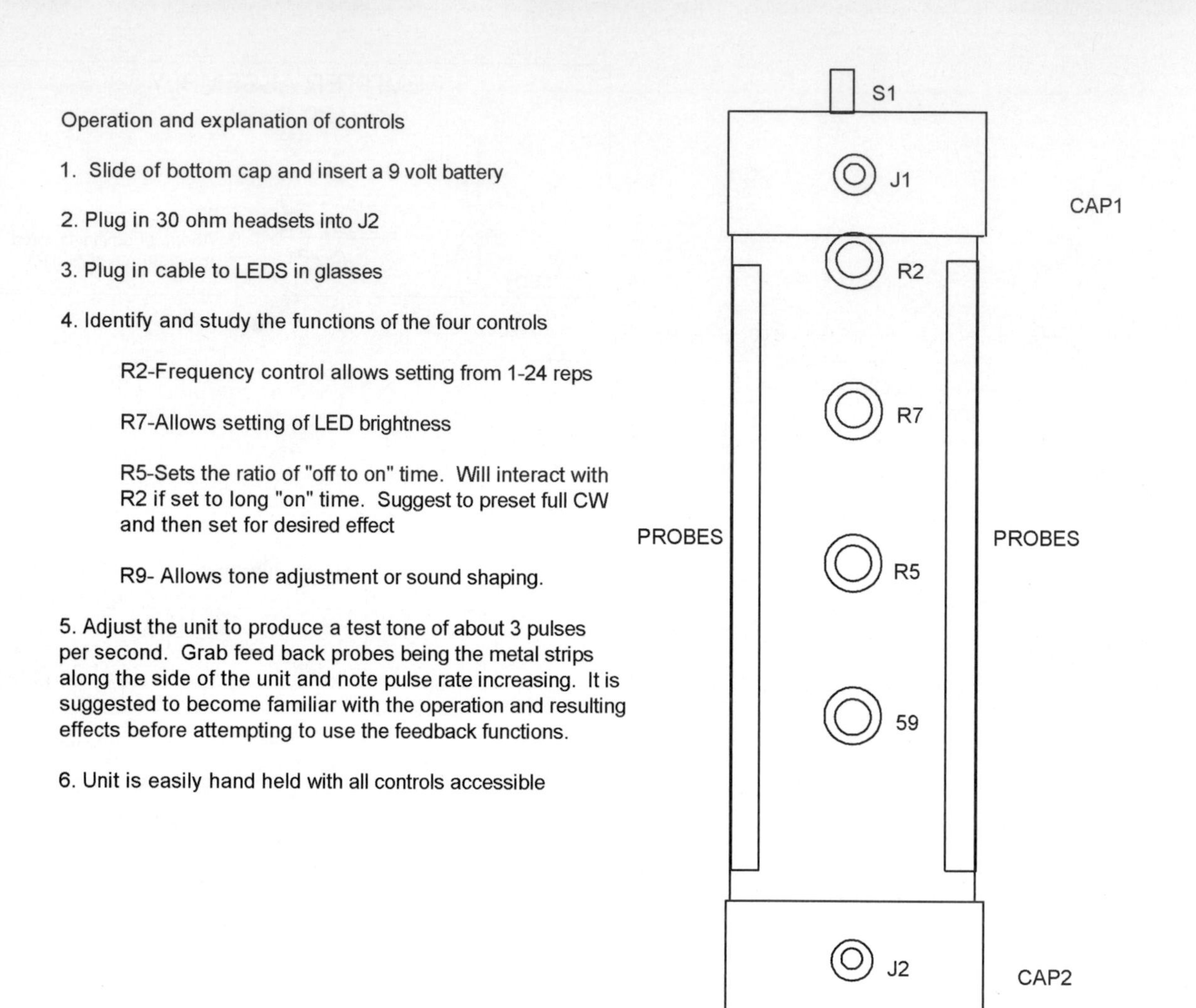

Figure 29-7 *Operation and Explanation of Controls*

J2	3.5 mm stereo jack
WR1	12 inches of #24 hookup wire
WR2	24 inches of #24 bus wire
WR3	48 inches of two-cord speaker wire
PBl	.1 × 6¼ × 1¼-inch grid perforated board
PCBOARD	Optional printed circuit board (PCB) with printing DB# PBMIND
EN1	Enclosure, sized at 1⅝ OD × 1 7/16 ID × 7 inches
CAl, 2	Two 1⅝ inch plastic caps #A 1⅝
P1	3.5 mm plug for LED cable to eyewear
HEADSETS	Stereo or mono, 32-ohm, cushioned headphones
TAPE	4 × ½-inch metallic silicon tape229

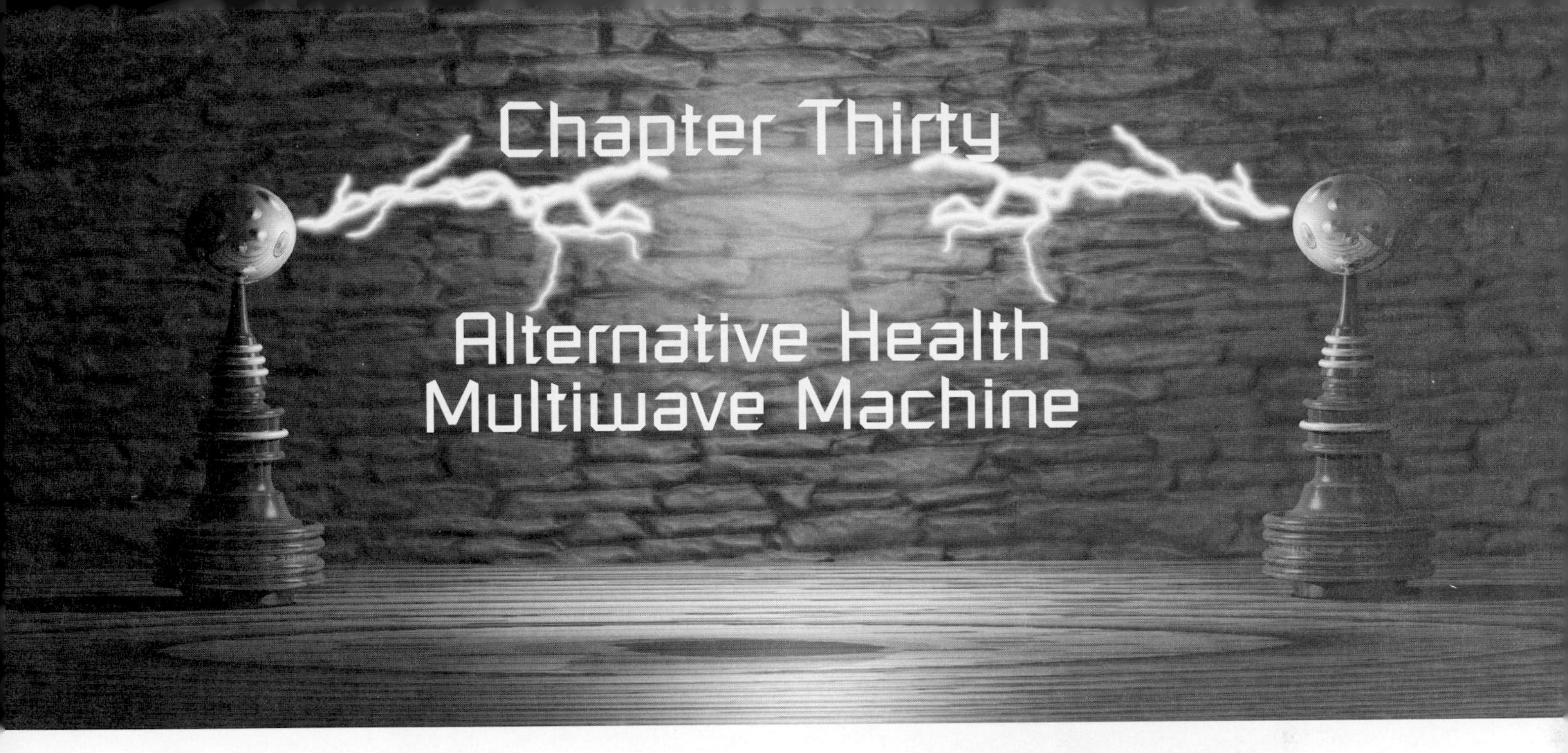

Chapter Thirty

Alternative Health Multiwave Machine

Figure 30-1 shows how to construct a Lakhovsky multiwave alternative health system. Dr. Lakhovsky was a Russian researcher who experimented with the effects of high-frequency electromagnetic radiation on various parts of the living organism. This controversial system is claimed to cure many health ills, including cancer. Its claim to fame is that all cells in a living body possess an intrinsic, vibrating resonant frequency that can be energized by an external means. Diseased cells appear to have a weaker and different frequency that can be brought back into step with the adjacent nondiseased cells. Antiaging appears to occur when cells are exposed to this full-spectrum electromagnetic vibratory energy. It is of special interest that this method of cellular regeneration has generated considerable interest in several research fields as a possible cure for cancer.

This project shows how to construct the multiwave radiating coil and antenna from readily available parts and pieces. System operation will require a working BTC30 12-inch-spark Tesla coil, as described in Chapter 14 of *Electronic Gadgets for the Evil Genius*. This Tesla coil is also available through plans, a kit, or as a complete and ready-to-use coil on the Information Unlimited Web site at amazing1.com.

It is suggested that the builder obtain the Lakhovsky handbook as listed in the parts list at the end of this chapter and study the methods of application and testimonials on the use of this controversial health machine before using it on one's self or others.

Figure 30-1 *The multiwave antenna*

Theory of Operation

Figure 30-2 shows a block diagram of the radiating coil and Tesla driver. The antenna coil contains a series of concentrically wound coil rings mounted in a flat plane. These coils all possess an inherent resonant

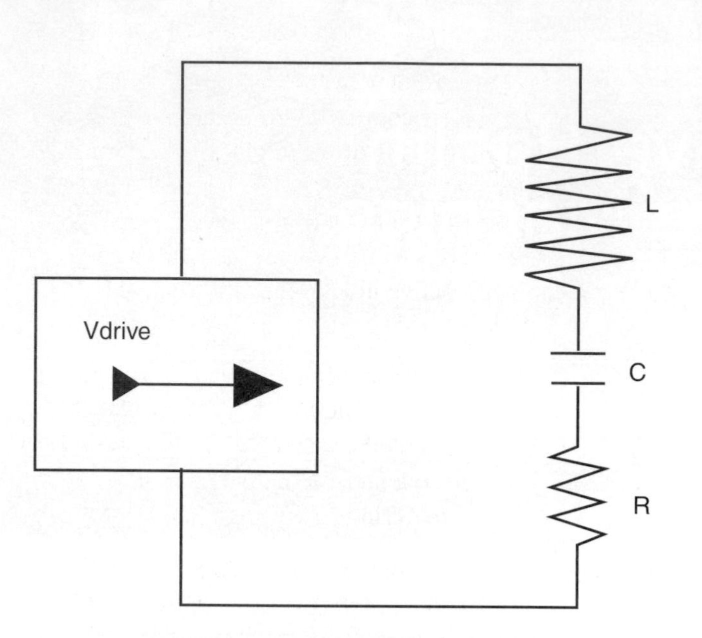

L1 to Ln represents the inductance of all of the coils.
C1 to Cn reprasents the capacity of all of the coils.
R1 to Rn is the radiation resistance of each section.
Frequency are the sum of the resonant sections ranging up to the gigahertz.

L is the sum of L1 to Ln
C is the sum of C1 to Cn
R is the sum of R1 to Rn and is mainly the radiation resistance.

The undamaged current wave from the output of the Tesla driver induces voltages at many frequencies into the multicoils of the antenna.

Figure 30-2 *Diagram of system*

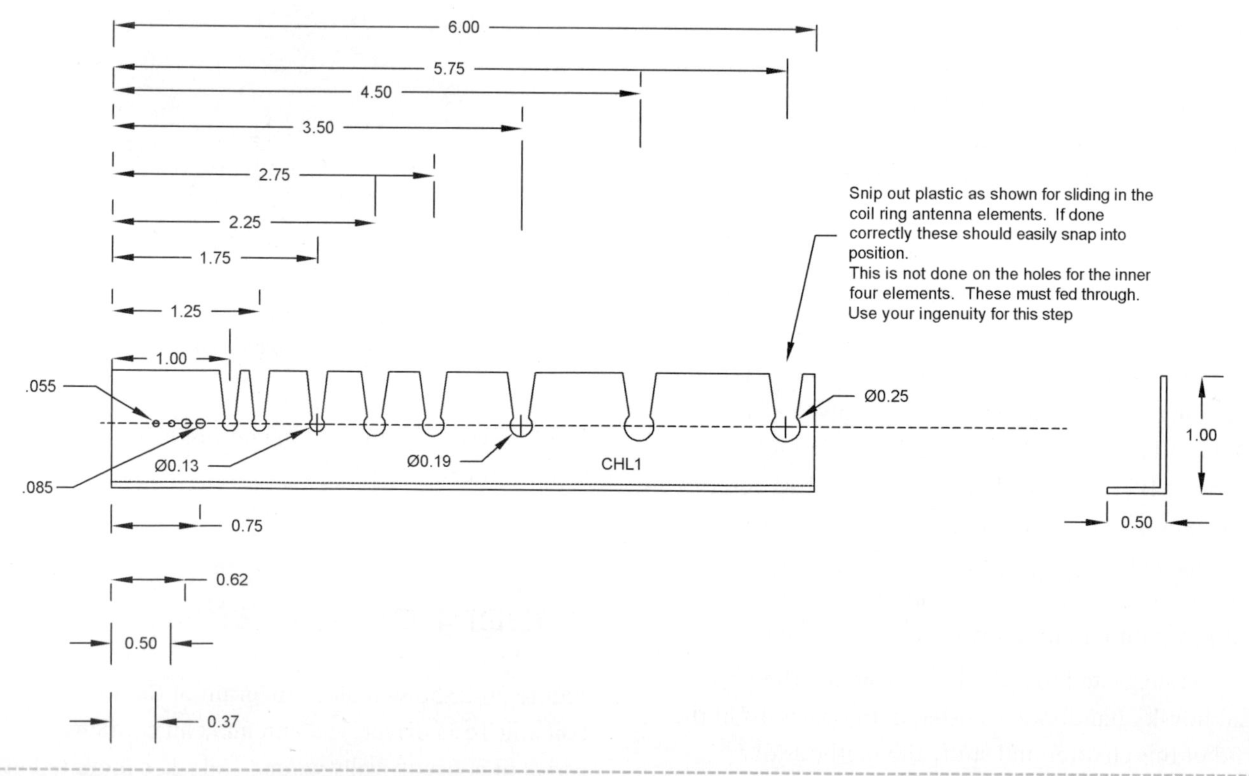

Figure 30-3 *Fabricating the four coil holders*

frequency as a function of their physical parameters. Harmonics are also produced and with all the coils oscillating they create a near-continuous spectrum of radio-frequency energy. A spark-driven Tesla coil is required for optimum results, as it generates many fast steps of current rich in harmonic content that drives the antenna coil.

Construction of the Multiwave Coil and Antenna

1. Create the four coil holders (CHOLD1) as shown in Figure 30-3. Use 1/16-inch Lexan polycarbonate.
2. Create the 12 × 12-inch main plate (MP1) from a piece of 1/16-inch Lexan sheet, as shown in Figure 30-4.
3. Mount the coil holders as shown and carefully drill the mounting holes. Use nylon or brass screws and nuts for mounting the holders to the plate. It is very important to have these holders as exact as possible or the wound coils will not fit properly in their predrilled slots.
4. Form the circular coil rings from the appropriate diameter copper tube and snap the tubing into place. Again attempt to form circles as perfect as possible for proper operation and esthetics. The inner coil rings are formed pieces of bus wire and are threaded through their respective holes. Note a 1/8-inch gap at the coil ends.
5. Attach a PVC cap (CAP1) using a small metal screw, solder lug, and nut. This screw is the high-voltage feed point to the antenna and is connected to a piece of wire soldered to the innermost coil ring (see Figure 30-5).
6. The antenna may be used with any suitable high-voltage, high-frequency generator. It is shown in Figure 30-6 with our BTC30 250-kilovolt Tesla coil, as described in Chapter 14 of *Electronic Gadgets*. It is available as a kit or completed system as noted in Table 30-1.

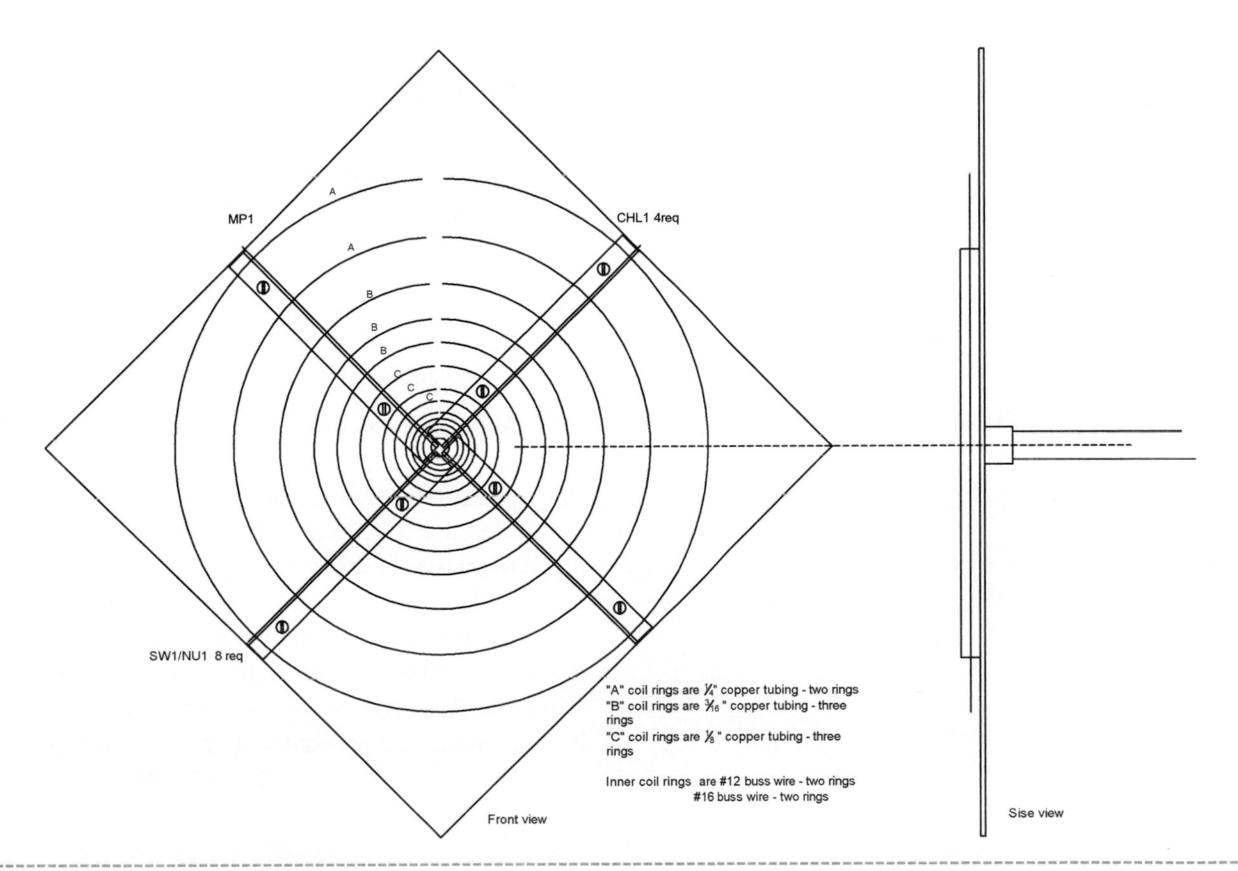

Figure 30-4 *Creating the base*

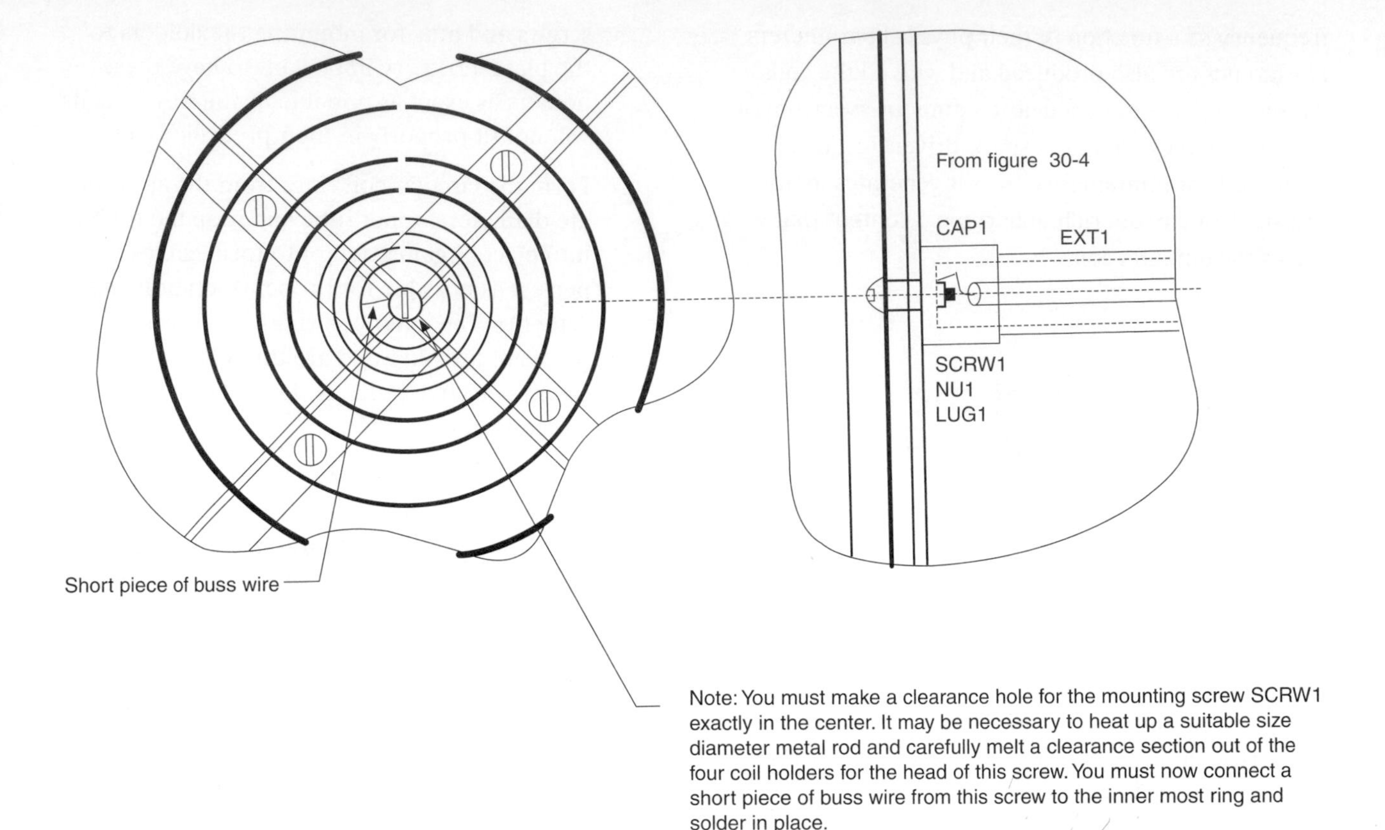

Figure 30-5 *Mounting and feeding details*

7. If you intend to use our coil, you will need to obtain a 3-inch slip cap (SLP1) and drill a hole for the extension arm tube (EXT1). You will also need the plastic stabilizing bracket (BRK1) required to keep the coil assembly upright.
8. The connection between the output of the Tesla coil and the antenna is made via the wire sleeved through the extension tube that connects to the mounting screw and exits through the rear to make contact with the Tesla coil output lead.

Operation

The following instructions reference the use of our 250-kilovolt Tesla coil as listed in Table 30-1 or as described in Chapter 14 of *Electronic Gadgets*.

1. Preset the spark gap on the Tesla coil to no more than 1/16 of an inch. Connect the tap lead for maximum inductance.
2. Turn on the switch on the Tesla coil and note all the antenna coil rings adjacently sparking and arcing between one another. Note that the outer ring may be grounded to the chassis to increase output activity.

At this point, it is recommended that you obtain the referenced book (LSK1), the *Lakhovsky Multiple Wave Oscillator Handbook,* as referenced in the parts list and familiarize yourself with the potentials, treatments, and experiments possible with this system.

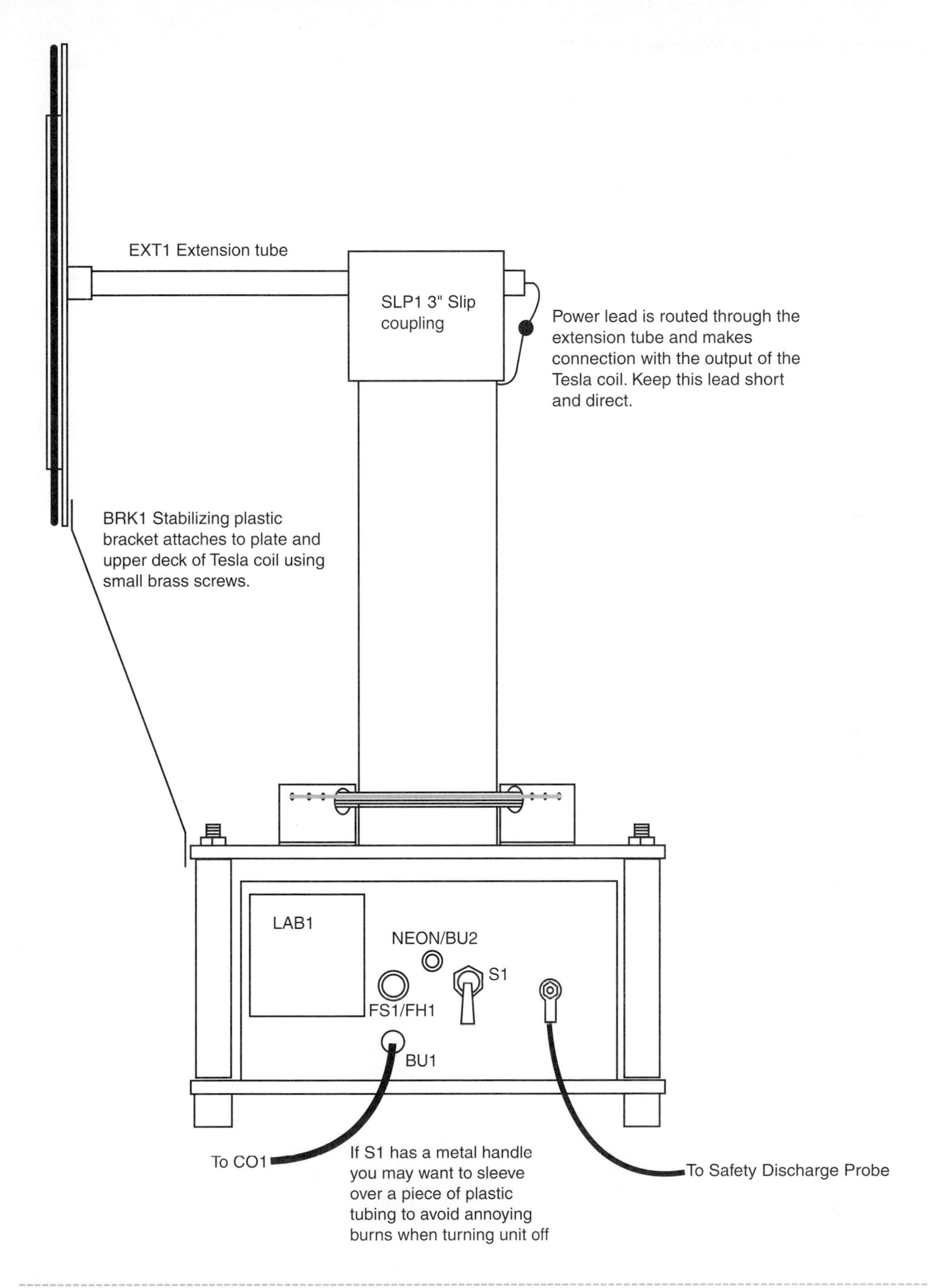

Figure 30-6 *Completed system*

Table 30-1 Multiwave Machine Parts List

Ref. #	Description	DB Part #
CHL1	Four coil-ring holders shown in Figure 30-3	DB# CHL1LAK
MP1	Mounting plate shown in Figure 30-4	
ARING	Two ¼-inch copper, soft drawn tubings	
BRING	Three 3⁄16-inch copper, soft drawn tubings	
CRING	Three ⅛-inch copper, soft drawn tubings	
INNERRING	Two #12 bus wire circular rings	
INNERRING SMALL	Two #16 bus wire circular rings	
SW1/NU1	106-32 × ½-inch brass or nylon screws and nuts	
SCRW1/NU1	6-32 × ¾-inch brass screw and nut	
LUG1	#6 solder lug	
WR20	24 inches of #20 bus wire	
BRK1	Stabilizing bracket shown in Figure 30-6	
CAP1	½-inch PVC slip cap	
SLP1	3 inches of PVC slip coupling	
EXT1	18 inches of ½-inch PVC tubing	
BTC3K	250-kilovolt Tesla coil generator kit and plans	DB# BTC3K
BTC30	250-kilovolt Tesla coil generator ready to use	DB# BTC30
LSK1	Book on Lakhovsky, 144 pages	DB# LSK1

Chapter Thirty-One

Mind Mangler

This board-level device, as shown in Figure 31-1, when properly assembled, can be a great prank. When placed in one's bedroom under normal light condition, it does nothing. As soon as the lights are turned off, it comes to life, producing pulses of high-frequency sound much like an insect. These control pulses can be timed to occur about every minute. When the unsuspecting victim attempts to locate it by turning on the lights, it ceases operation. Even using a flashlight will disable it. The long time between pulses also makes it very difficult to detect. Properly hidden, it can take a long time to locate and may turn a prank into a nasty situation.

Figure 31-1 *Board-level view of the mind mangler*

Circuit Operation

Figure 31-2 shows a timer, I1, as a free-running, astable pulse generator producing a symmetrical 10-second, off and on pulse controlling the second timer, I2. Timer I1 has a phototransistor Q1 connected in series with the reservoir capacitor C1. As long as there is light on Q1, C1 can never discharge due to the conductance of Q1. Therefore, pin 3 is at a low and I2 is disabled. When the lights go out, C1 now is allowed to discharge and pin 3 goes high, turning on I2 and producing the high-frequency sound to the output transducer TD1. The high-audio frequency is determined by resistor R3 and C3, and can be changed by altering these values. The output signal from I2 is stepped up by transformer T1 (see the parts list at the end of the chapter).

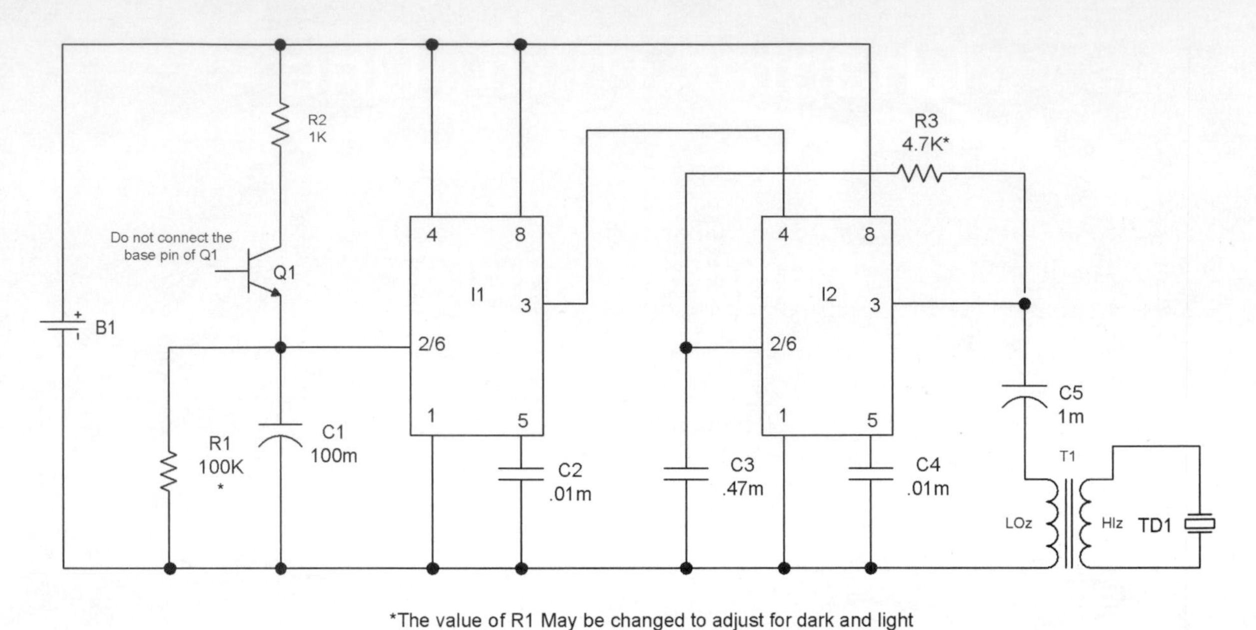

Figure 31-2 *Circuit schematic*

Construction Steps

1. Identify all the parts and pieces and verify them with the bill of materials.
2. Insert the components starting from one end of the perforated circuit board and follow the locations shown in Figure 31-3 using the individual holes as guides. Use the leads of the actual components as the connection runs, which are indicated by the dashed lines. It is a good idea to trial-fit the larger parts before actually starting to solder. Always avoid bare wire bridges, globby solder joints, and potential solder shorts. Also check for cold or loose solder joints.

Additionally, pay attention to the polarity of the capacitors with polarity signs and all the semiconductors. The transformer position is determined using an ohmmeter.

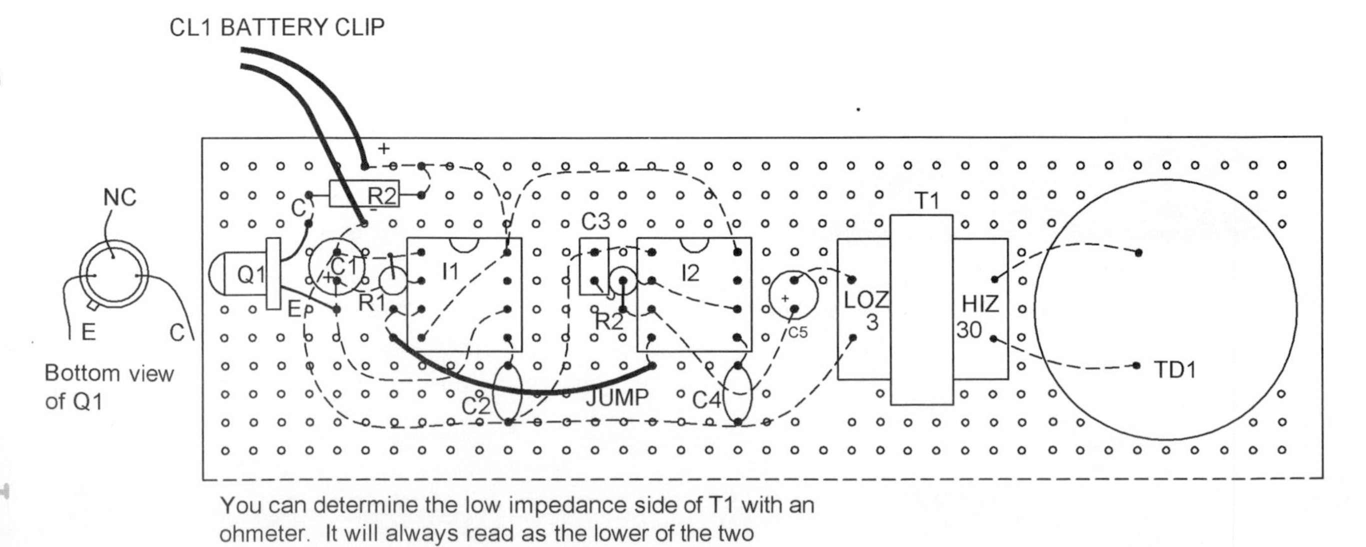

Figure 31-3 *Assembly board*

Testing

Verify the wiring accuracy and connection. Insert a 9-volt battery into the clip, cover the lens of Q1 or turn off the lights, and note the unit emitting a low-level, raspy, clicking tone like a sick cricket. Uncover the lens of Q1 and note the circuit shutting down. There will be a delay that is dependent on the level of light and dark. You may vary the response by changing the value of R1. Decreasing its value will require more light to shut down and vice versa.

Table 31-1 Mind Mangler Part List

Ref. #	Description	DB Part #
R1	100K, ¼-watt resistor (br-blk-yel)	
R2	1K, ¼-watt resistor (br-blk-yel)	
R3	4.7K, ¼-watt resistor (yel-pur-red)	
C1	100-microfarad, 25-volt vertical electrolytic capacitor	
C2, 4	Two .01-microfarad, 50-volt plastic capacitors (103)	
C3	.47-microfarad, 50-volt vertical electrolytic capacitor	
C5	1-microfarad, 50-volt vertical electrolytic capacitor	
I1, 2	Two 555 timer dual inline packages (DIP)	
Q1	L14P phototransistor	
CL1	9-volt battery clip	
TD1	Small piezo transducer	
T1	Small audio transformer, 8 to 1K	
PB1	4 × 1.5 × .1-inch grid perforated circuit board	

Chapter Thirty-Two

500-Milligram Ozone Air Purification System

Figure 32-1 shows a useful home device that generates ozone for eliminating odors, killing bacteria and mold, and disinfecting unsanitary areas such as pet boxes and bathrooms.

This project shows how you can build a system that could cost over $300 for less than $50. Its simple and basic design uses two 12-volt fans to blow air over a unique ozone-producing cell that is powered by a variable-output, high-voltage, high-frequency power supply. The power supply is featured in Chapter 7 of this book and is used in several other projects.

This is a beginner- to intermediate-level project requiring minimal basic electronic skills. Expect to spend $40 to $60. All the parts are readily available, with specialized parts obtainable through Information Unlimited (www.amazing1.com), and they are listed in Tables 32-1 and 7-1 at the end of this chapter and Chapter 7.

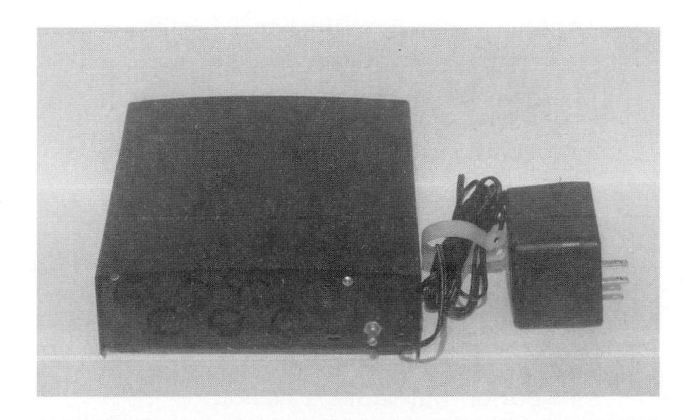

Figure 32-1 *Ozone purification system*

The Story of Ozone

Ozone is an unstable form of oxygen. Normal oxygen is diatomic (0_2), existing as two atoms of oxygen making up the molecule. Ozone is triatomic (0_3), existing as three atoms of oxygen for the molecule. The triatomic form of oxygen is very unstable, wanting to lose the third oxygen atom and combine with whatever atom it can (oxidization). This property makes it the most active oxidizer known, with the exception of the very hazardous fluorine gas. Ozone at normal pressures is colorless and odorless, yet it produces a pleasing, fresh-air odor as a result of the nitrous oxides it produces. The odor is usually noticeable after an old-fashioned thunderstorm. Under pressure, it becomes a bluish gas.

Ozone is also a very powerful bactericide. It is not affected by pH, as is chlorine, thus making it an excellent candidate for pools, spas, laundries, and general water treatment applications. It is many times more soluble in water, further enhancing its purifying effect. Ozone will combine with diatomic nitrogen, N_2, forming nitrous oxide, 2NO. This oxide quickly combines with water, forming nitric acid, HNO_3. This is often a very undesirable effect when used with straight air. Pure oxygen greatly minimizes this effect and is often required in many applications.

However, those requiring a supply of concentrated nitric acid for the nitration of many carbohydrates, such as cellulose, glycerine, hexamine, phenol, and toluene, may wish to consider combining ozone and air with a condensing apparatus to obtain this useful acid.

Ozone for Treating Air

Air pollution is one of the most serious environmental issues that we face. Visible and detectable smoke, dust, mildew, mold, and toxic odors, bacteria, pollen, and static electricity, along with the more elusive and invisible chemicals, have become a serious health threat. As buildings and homes are constructed with tighter air sealants for energy conservation, the problem will become even more severe. Often the effects from these problems include burning eyes, headaches, dizziness, depression, allergies, and general lethargy, all of which are usually attributed to colds and viruses. Air filters, such as the overpriced Sharper Image unit currently being advertised, are only partially effective, with some producing undesirable positive ions.

Ozone purifies the air from these undesirable pollutants by oxidization, as most pollutants readily combine with this highly reactive tri-atomic form of oxygen and break down into water and other non-toxic compounds. Ozone is produced artificially by electricity and is many times more antiseptically effective than oxygen.

Construction Steps

1. The system is shown using the high-voltage, high-frequency driver shown in Chapter 7. It is built on a printed circuit board or a more difficult perforated board. The PCB version is the easiest, as construction involves placing the correct components into the correct holes and soldering. The more difficult approach is also shown and uses a perforated circuit board. This is far more challenging and is intended for the experienced assembler. Figure 32-2 shows the board with the necessary changes for use with this circuit.
2. Create the base section, as shown in Figure 32-3. You can use sheet aluminum or bendable plastic sheets. Design a mating cover with a slight overhang over the front and rear panels. Use similar material and mount using #6 sheet metal screws along the lip of the base.
3. Fabricate the fan and cell mounting bracket as shown in Figure 32-4. Use clear 1/16-inch polycarbonate plastic and initially layout the piece, marking off the bend and cutting lines. You may fabricate separate pieces for the cell brackets and fans if you do not have access to the necessary equipment. Note dimensions and cuts must result in a single piece, as shown.
4. Cut out a piece of window screen large enough to cover the input air holes and secure it from the inside of the base enclosure using screws and silicon rubber cement.
5. Mechanically assemble the components, as shown in Figure 32-5. Verify all proper fittings and the clearance of components, especially the high-voltage points on the cell. Finally, wire everything as shown and verify the accuracy and any potential errors. Note: Figure 32-6 shows the wiring using the perforated board approach. Do not connect the 12 DC/1.5 adapter wall transformer at this time.
6. Obtain a 12-volt, 3-amp bench power supply with a voltage and current meter. Connect it to where the adapter's plus and minus wires go.
7. Connect a scope to the test points, as shown in Figure 32-5.
8. Preset the trimpot R1 to midrange and control pot R10 to full counterclockwise and click off.
9. Apply 12 volts from the bench supply and note that no current occurs.
10. Click on R10 and note a pulsing current of over 1 amp. Quickly adjust the trimpot to a dip in current, as noted on the meter. Our model was tuned correctly when the trimpot was set to one to two o'clock. The fans should both be in full operation and you should detect a faint, bluish glow coinciding with the pulsing. At no time should the heatsink tab on the power transistor Q1 be hot.

It will be necessary to construct the printed circuit assembly board as shown in chapter 7 with the following below list of revisions.

Wind L1 using 10 turns of #22 magnet wire. Air gap core with .007 (7mils) shims. Measured inductance should be between 15 -16 micro-henries.

Trimpot R6 is not used

Replace R5 with a wire jump

Replace C2 from a .0022 to a .0047 microfarad capacitor

Replace C6 from a .22 to a .47 microfarad 250 volt capacitor

Capacitor C4 is not used. insert a short piece of buss wire for a test point ground

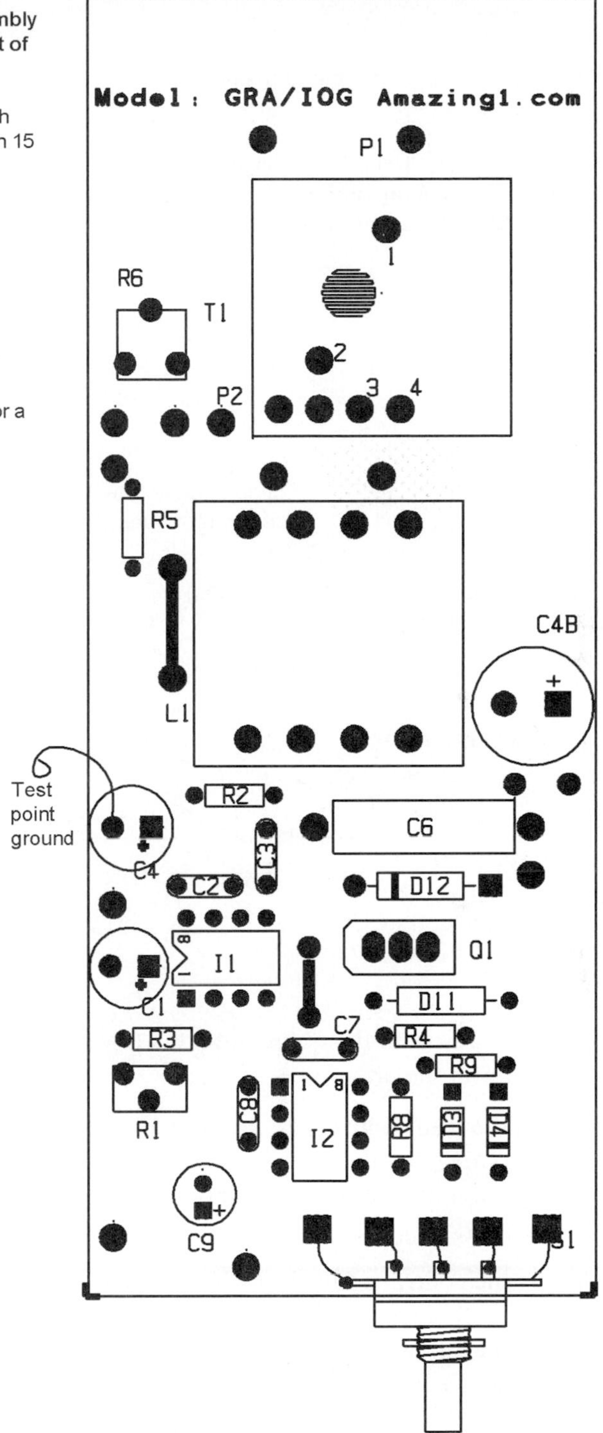

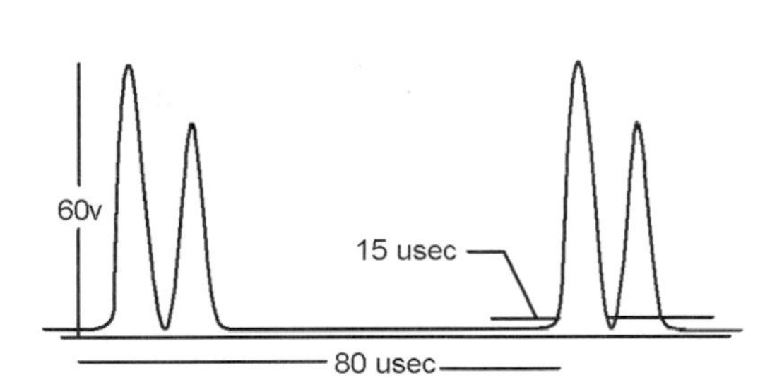

Correct wave shape measured at the drain of Q1 and test ground point

Figure 32-2 *PCB rework*

11. Turn R10 full clockwise and note the on time of the pulse increasing with the current but not exceeding 1.5 amps at these pulse peaks. Allow it to operate for 30 minutes and note ozone being produced, a soft, bluish, even glow on the cell, and the heatsink not getting too hot to touch.

12. Once this is verified, you can hook in the 12 VDC/1.5-amp adapter transformer. Allow it to operate and verify that the adapter does not get hot.

13. Attach the cover and place the unit in the target area.

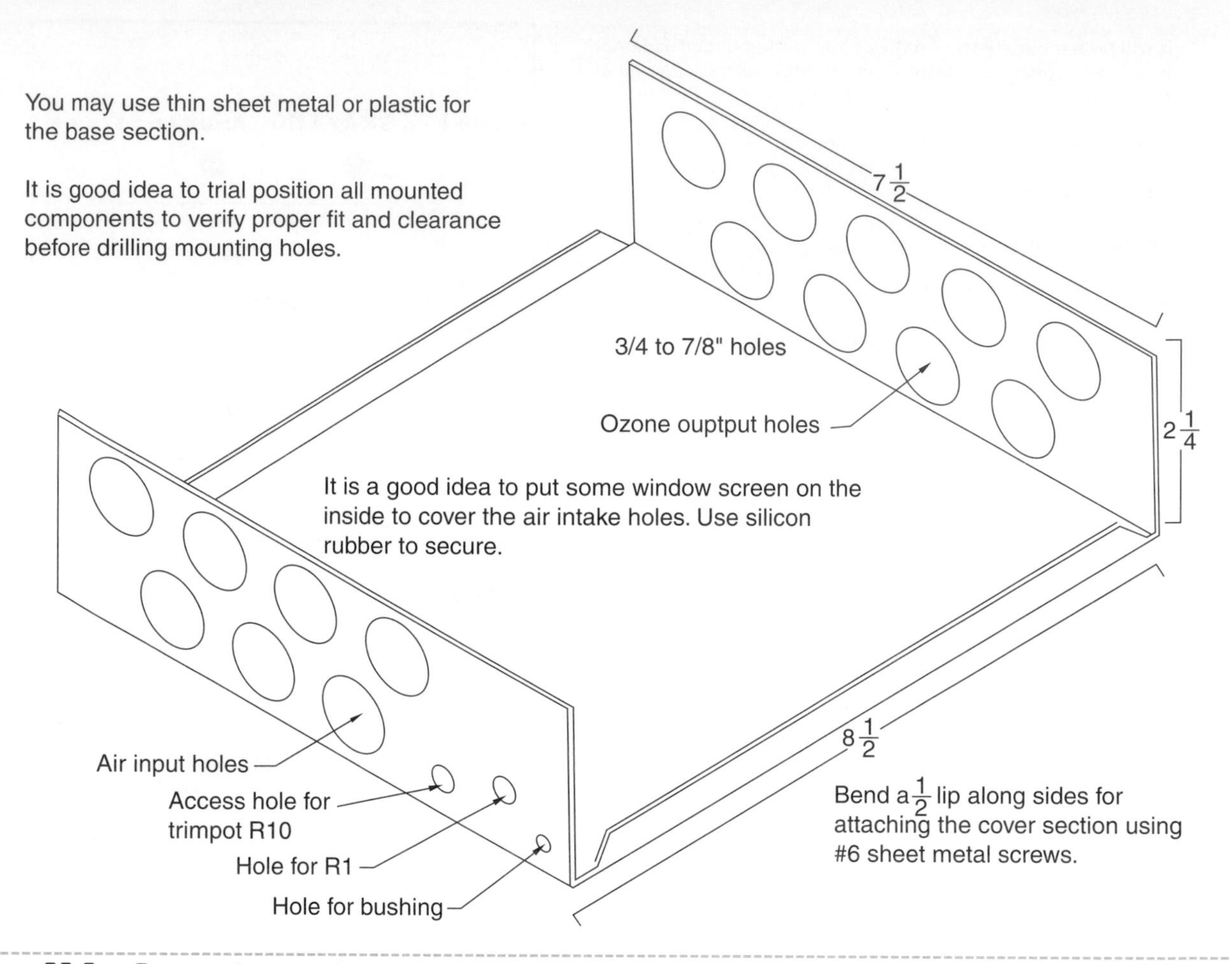

Figure 32-3 *Base section creation*

14. Place the unit with the output directed toward the center of the area.
15. Plug in the wall adapter and rotate the control switch to on. Note the fan rotating and freely turning without obstructions. It should produce a good air flow.
16. Adjust the control a full clockwise turn and note a smell of fresh air emitting from output ports. You will observe a purplish glow around the cell when viewed in darkness.
17. Allow it to run for several hours and adjust the control to where the fresh air smell is just detectable.
18. Check both the unit and the wall adapter for excessive heating. They should only be warm to the touch.
19. This unit may need cleaning from dust. Use compressed dry air or a soft cleaning brush.

Special Notes

You should never operate the unit when the odor becomes pungent or domineering. Ozone is a colorless, odorless gas composed of unstable diatomic oxygen. The smell of fresh air is not the ozone, but the result of secondary reactions with other chemicals.

The unit can be adapted to 12 VDC operation for vehicle or battery use. Simply remove the adapter and splice in some 12-volt connections. Use caution when observing the polarity. You may request a factory modification for this step.

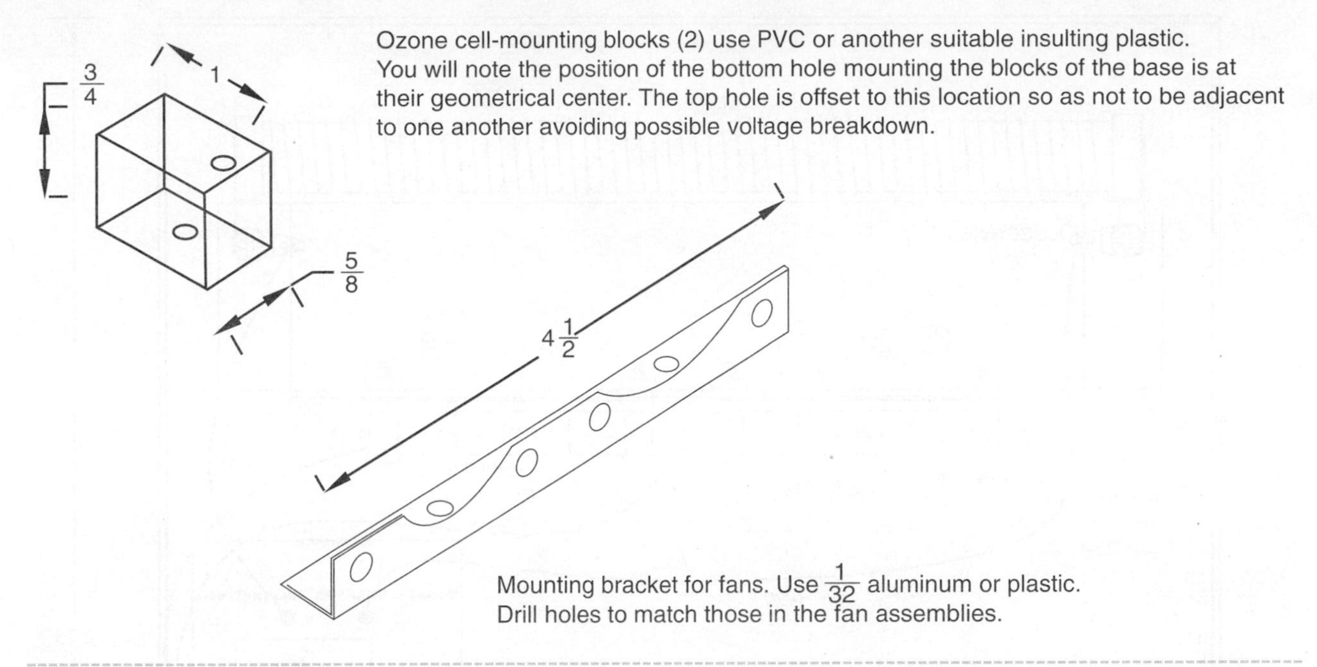

Figure 32-4 *Fabricated parts*

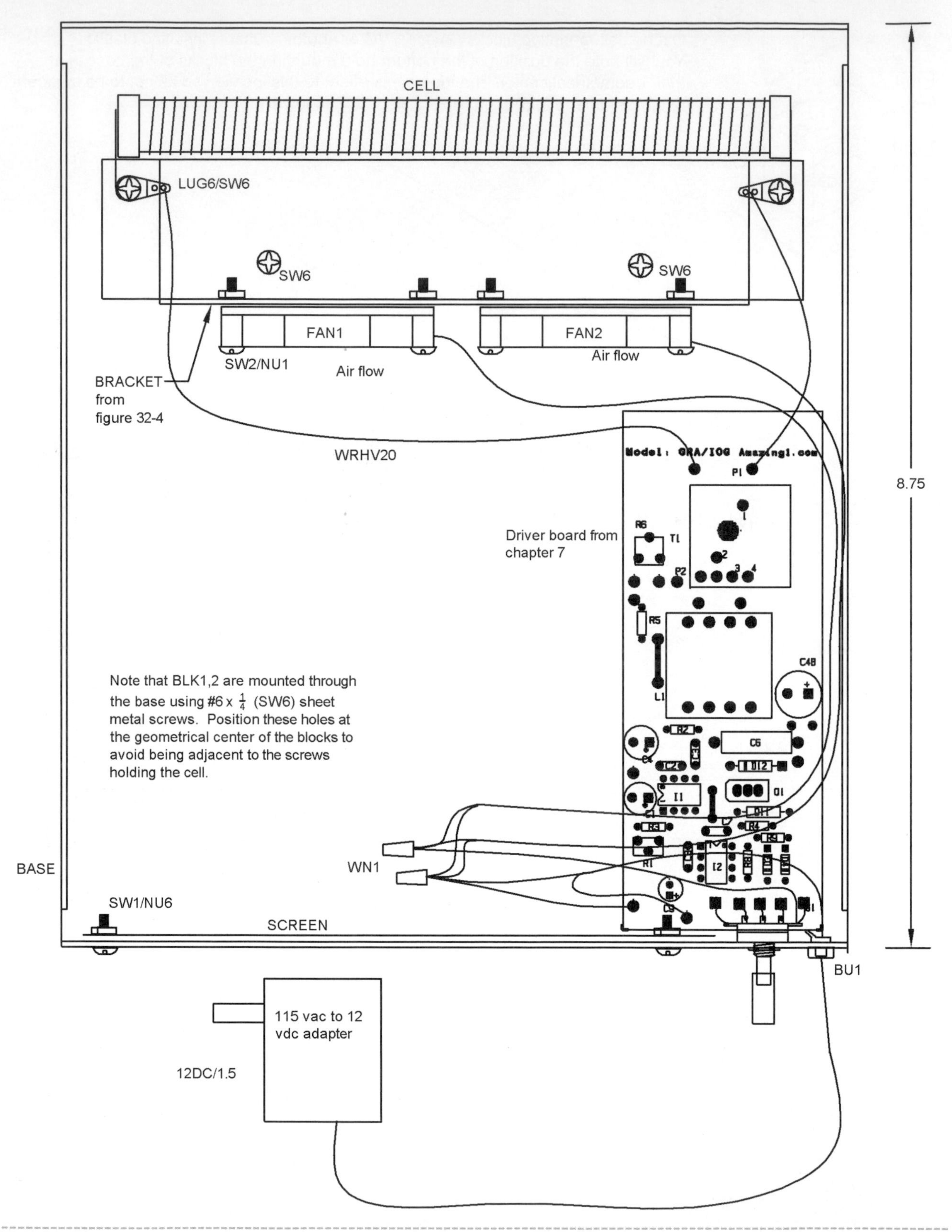

Figure 32-5 *Mechanical layout showing wiring when using the PCB*

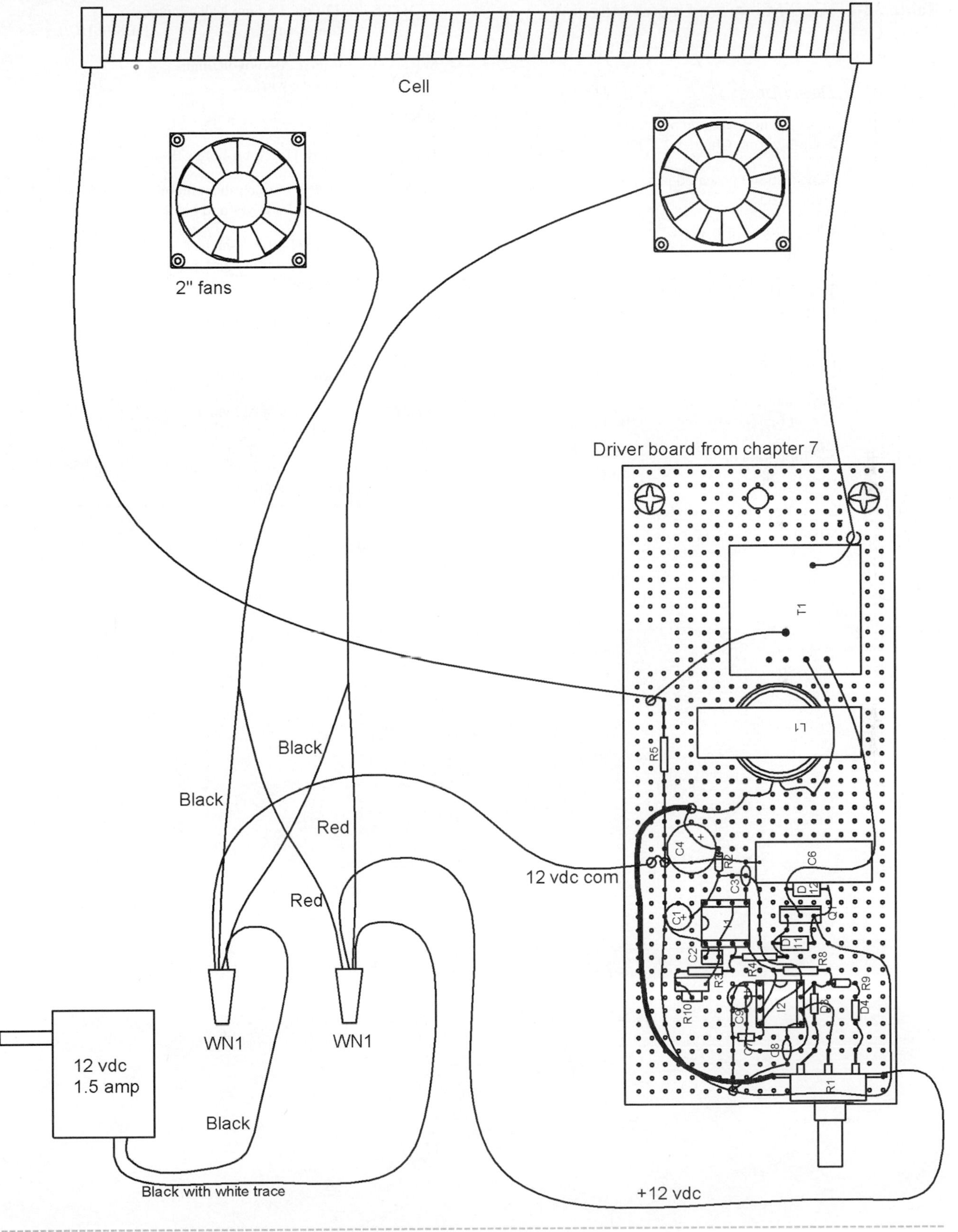

Figure 32-6 *Final wiring using a perforated board*

Table 32-1 500-Milligram Ozone Machine Parts List

Ref. #	*Description*	*DB Part #*
DRIVER	Modified high-voltage assembly from Chapter 7	DB# GRADRIV10
BASE	Base section fabricated per Figure 32-3	
COVER	Fabricated to fit BASE section	
BLOCKS	Two blocks fabricated per Figure 32-4	
BRACKET	Bracket fabricated per Figure 32-4	
FAN1, 2	Two small, 2 × 2-inch, 12-volt fans	DB# FAN2
CELL500	500-milligram ozone corona cell	DB# CELL500
SCREEN	2 × 5-inch piece of window screen	
12DC/1.5	12-volt, 1.5-amp wall adapter transformer	DB# 12 DC/1.5
WRHV20	6-inch, 20-kilovolt, high-voltage wire	
WN1,2	Two small wire nuts	
BU1	Small Heyco bushing #2P-4	
SW6	10 #6 × ¼-inch sheet metal screws for cover and blocks	
SW2/NU2	Four #4-40 × ¾-inch machine screws and nuts for fans	
SW1/NU1	Two #6-32 × ½-inch machine screws	
LUG6	Two #6 solder lugs	
WR22	12 inches of #22 vinyl hookup wire	
FEET	Four stick-on rubber feet for base section	

Chapter Thirty-Three

Invisible Pain-Field Generator

The invisible pain-field generator device shown Figure 33-1 is a handheld, battery-operated sonic, shock-wave generator that produces a *sound pressure level* (SPL) of up to 125 decibels (db) at 30 centimeters. The shock wave frequencies are user presettable at 25, 16, and 12 KHz. A sweep function is included in the circuitry where the selected center frequency varies between two set limits at an adjustable rate, providing a complex sonic signal that further enhances the effect.

Applications of the device can range from the routing out of agricultural pests in silage bins, chicken houses, grain bins, or wherever rats are a problem. The optimum frequency for this particular application is 16 KHz or just above that of human hearing. Many farmers use these devices for determining the effectiveness of pesticides, noting the reduction in infestation levels. Other applications include spooking birds and animals from unwanted areas and discouraging dogs and deer from decimating shrubs and ornamentals.

Figure 33-1 *Pain-field generator*

Adjustment of the frequency down to 12 KHz (within the range of human hearing) will produce extremely painful and annoying affects. The unit is excellent for disbursing crowds of potentially unruly people. A simple test is to point the device at an unsuspecting subject and momentarily push the button. You will notice a very positive response. Unfortunately maximum affect seems to favor younger women; older men seem to be less sensitive.

The unit is very directional when used outside, but it looses this property when used inside due to reflections from walls, ceilings, and furniture.

The unit is shown assembled into a small plastic enclosure that includes the electronics emitter transducer and the 9-volt battery. A removable plastic cover allows access in order to change the battery. Controls include a pushbutton emission control and sweep-activation pushbutton switches located on the top of the enclosure. The size of the main enclosure is 4 × 2 × 1 inches. The weight is only about 6 ounces with batteries installed

The unit can be used as a research tool for producing an effect on certain animals for their control and experimentation. The device has been successfully used for controlling certain dogs or other vicious animals by joggers or other outdoors enthusiasts. The unit has also been used in flushing out rats.

Caution: Caution must be used, as the effect on most people causes pain, headache, nausea, and extreme irritability. (Younger women are especially affected.) Do not, under any circumstances, point the unit at a person's ears or head at close range, as severe discomfort and possible ear damage may result. This also applies to dogs and other animals.

Include a caution label on the device to avoid exposure over 105 db for any continuous period of time. When using the device around people, you must be careful, as unjustified harassment is illegal and can result in prosecution.

This is an intermediate-level project requiring basic electronic skills. Expect to spend $25 to $50. All parts are readily available, and any specialized parts are available through Information Unlimited (www.amazing1.com) and are listed in Table 33-1 at the end of the chapter.

Circuit Description

A timer (IC2) is connected as an astable free-running multivibrator whose frequency is internally controlled by trimpot (R9). Resistor R8 selects the range limit of R9. The square-wave output of IC2 is via pin 3 and is directly coupled to power amplifier Q2. The drain of Q2 is DC biased through audio-frequency blocking chokes (L1A and L1B), providing a high impedance to the AC component of the signal.

DIP switches (S1,2,3) preselect the internal frequency for the required application and are shown in Figure 33-2. These switches select resonating inductors L2 and L3 connected in series with the output transducer TD1. The resonant action between the inherent capacity of TD1 and these selected inductances now produces a sinusoidal-shaped wave

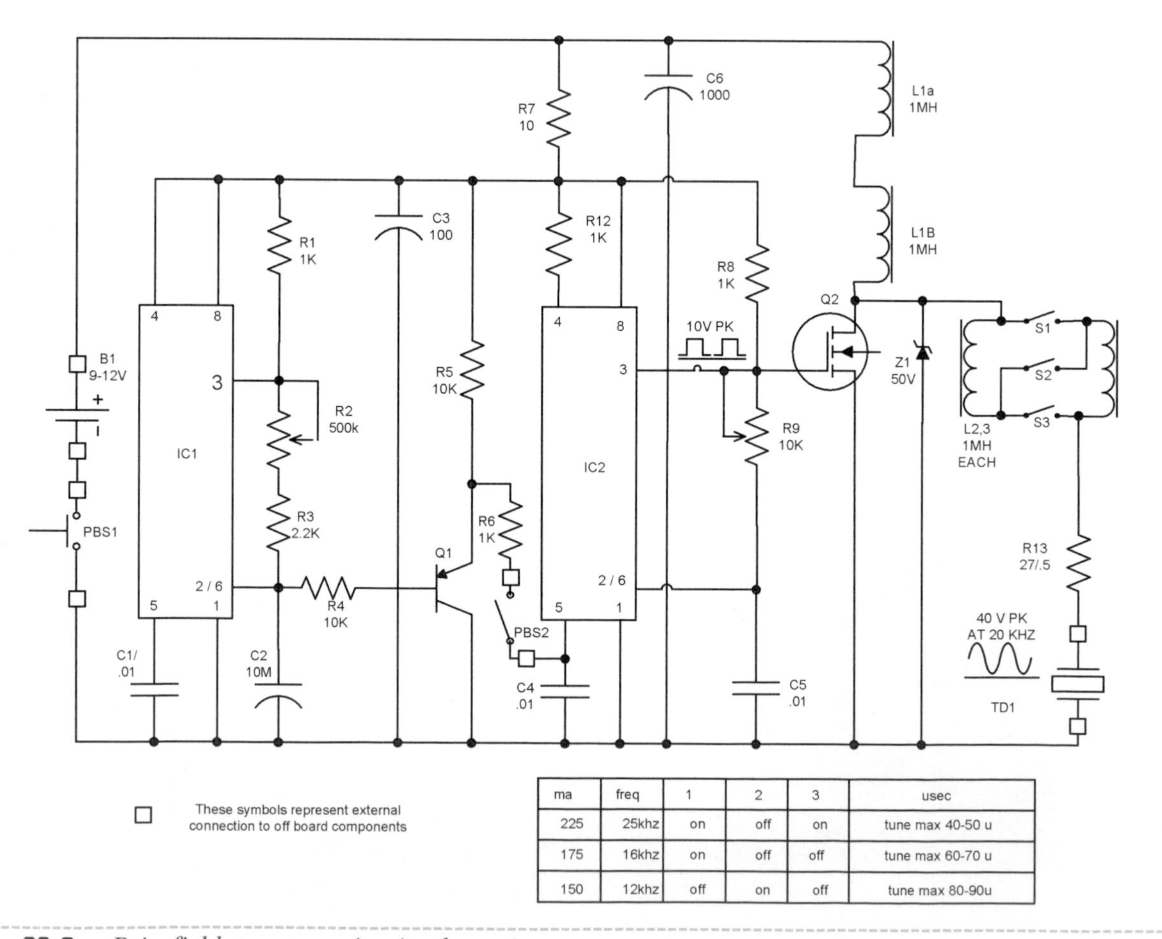

ma	freq	1	2	3	usec
225	25khz	on	off	on	tune max 40-50 u
175	16khz	on	off	off	tune max 60-70 u
150	12khz	off	on	off	tune max 80-90u

Figure 33-2 *Pain-field generator circuit schematic*

peaking around the upper limit of the tuning range. This *signal* waveform now has a peak-to-peak voltage several times that of the *original* square wave. Transducer TD1 now can take advantage of these peak voltages to produce the high sound-pressure levels necessary without exceeding the high RMS ratings of an equivalent voltage-level square wave. Zener diode Z1 clips any excessive overshoots across Q2.

Timer IC1 is similarly connected as an astable running multivibrator and is used to produce the sweeping voltage necessary for modulating the frequency of IC2. This sweep repetition rate is controlled by trimpot R2. Resistor R3 limits the range of this repetition time. C2 sets the sweep time range. Output for IC1 is via pins 6 and 2, where the signal ramp function voltage is resistively coupled to inverter transistor Q1 via resistor R6. The output of Q1 is fed to pin 5 of IC2 and provides the output modulation voltage necessary to vary the frequency as required. Note that the modulation signal is enabled by pushbutton PBS2.

Power to the system is via battery B1 and pushbutton PBSl. Capacitor C6 guarantees an AC return path for the output signal. Power to the driver circuits IC1 and IC2 is thru a decoupling network consisting of resistor R7 and capacitor C3.

Construction

1. Layout and identify all parts and pieces. Verify with parts list. Separate resistors, as they have a color code to determine their value. Colors are noted on the parts list.

2. If you are building from a perforated board, use Figure 33-3 as a parts placement guide and insert components starting in the lower left-hand corner as shown. Pay attention to polarity of capacitors and all semiconductors. It is suggested you use sockets for integrated circuits IC1 and IC2.

   Route leads of components, as shown, and solder as you go, cutting away unused wires. Attempt to use certain leads as the wire runs or use pieces of the #24 buss wire. Follow dashed lines on assembly drawing, as these indicate connection runs on underside of assembly board. The heavy dashed lines indicate use of thicker #20 buss wire, as this is a high-current discharge path.

   Please note that this circuit is very cramped, and it is suggested you obtain the optional printed circuit board shown in the parts list.

3. Attach external leads to components as shown in Figure 33-4, noting the individual lengths and twisted pairs.

4. Double-check accuracy of wiring and quality of solder joints. Avoid wire bridges, shorts, and close proximity to other circuit components. If a wire bridge is necessary, sleeve some insulation onto the lead to avoid any potential shorts.

5. Connect an ohm meter between CLI contacts and pushbutton PBSl. Note a reading of several thousand ohms. This may vary but should not indicate a short circuit. Preset trimpots R2 and R7 to midrange.

6. Connect a 9-volt battery and note a current of 200 to 250 milliamps when R9 is set to midrange. The meter can be connected across PBSl for this step.

7. Simultaneously press both PBS2 and PBSI and note a piercing sweeping tone coming from the transducer TD1. Turn R9 fully ccw and note the decrease in signal tone.

8. Frequency range should be approximately 10 to 25 KHz, with a sweep of approximately 1 to 20 times per second selectable by trimpot R2.

9. Check the waveshapes as shown at pin 3 of the output timer IC2. This is determined by the following formula: $F = 1/(1.57 \times R9 \times C5)$.

10. Check the waveshapes as shown across transducer TD1. This waveshape is approximately 40 to 50 volts, peak to peak, and approaches a sine wave. This is the approximate resonant frequency point of the selected values of series inductances within the inherent capacitance of the transducer. A voltage-peaking effect will be noted, as the frequency is varied by trimpot R9. As the frequency is varied off of resonance, this wave will severely distort but will not damage the circuitry. Note the chart in Figure 33-2 showing the selection

It may be a good idea to use SOCK 1, 2 sockets for IC1, 2 to avoid unsoldering if you make an error.

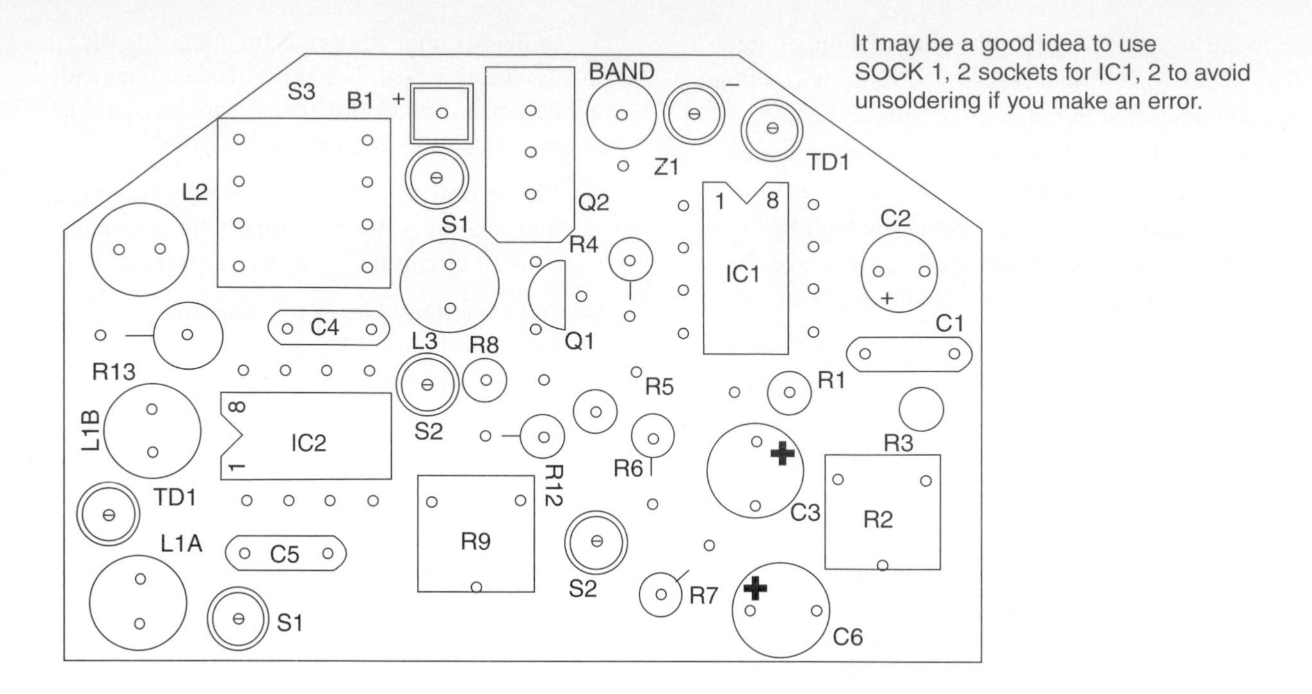

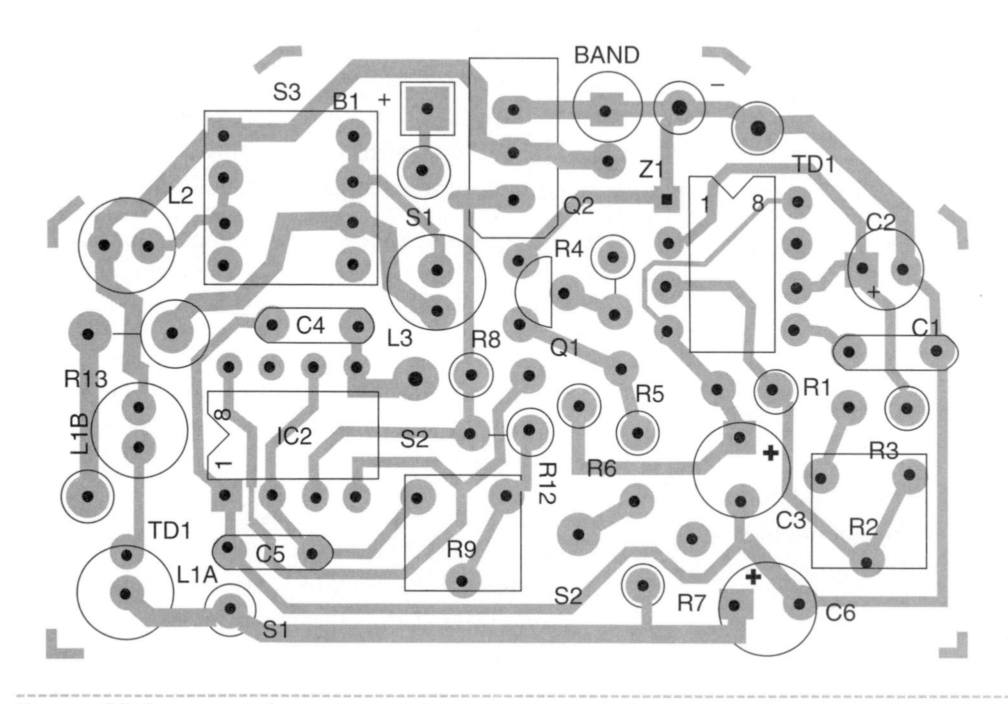

Figure 33-3 *Printed circuit board parts and wiring*

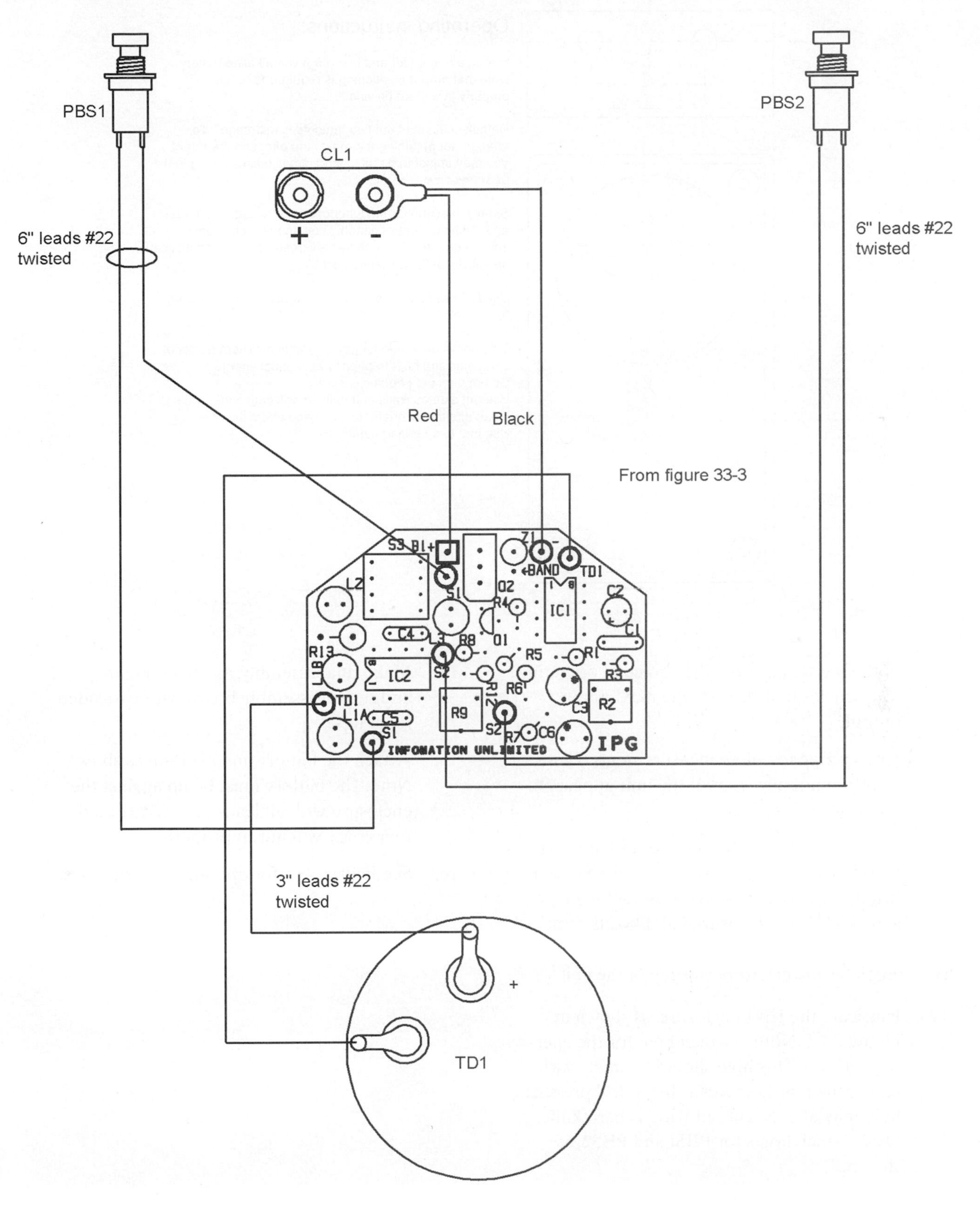

Figure 33-4 *External wiring*

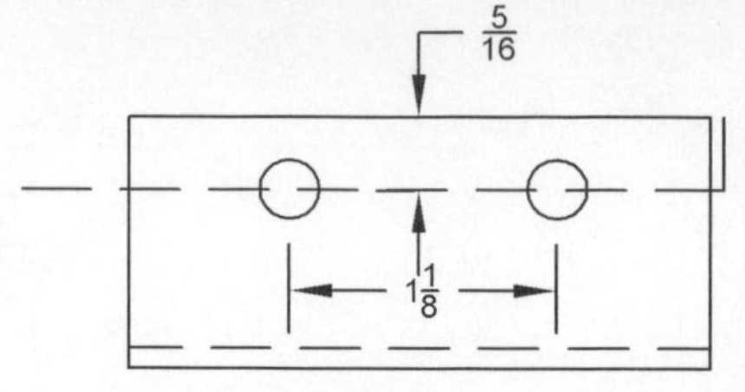

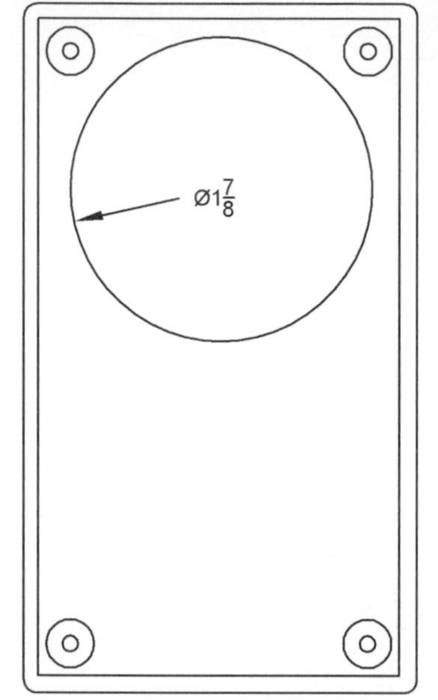

Operating instructions:

Remove rear cover and insert a 9 volt alkaline battery. Note that proper positioning is required for cover to properly fit without bowing.

Default settings of the two trimpots is midrange. To change for obtaining the maximum effect on the target you may adjust trimpots using a small plastic tuning tool or screw driver.

Simply direct unit transducer opening towards the target and push the control button. You should notice an immediate effect. The range will depend on the acoustical sensitivity of the target subject.

Use the sweep button to possible enhance the effect on the target.

This unit is designed for generating intermittent bursts of ultrasonic and high frequency acoustical energy.
Do not direct at people.
Use out side as walls and ceilings will cause the signal to loose directional characteristics and effect the user.
Use in 2 to 5 second bursts.

Figure 33-5 *Fabrication of plastic case*

of the inductances for the choice of operating frequencies that provide the maximum output.

Use a scope for these measurements. These steps are not necessary if the unit appears to function as described.

11. A sound pressure measurement of approximately 125 db was measured at 10 KHz at a distance of 16 inches on our model. Voltage across TD1 was measured at 40 volts peak.

This completes the electronic testing of the unit.

12. Fabricate the EN1 enclosure, as shown in Figure 33-5. Note 1 7/8-inch hole for the aperture of TDI. This hole should be made with a large punch or hole saw using a drill press. The hole may also be cut out with a sharp knife. Drill 1/4-inch holes for PBSl and PBS2, as shown.

13. Cut a piece of 2 1/8 × 2 1/8-inch window screen and secure it along with transducer TDI. Use RTV or suitable adhesive.

14. Make final assembly, as shown Figure 33-6, and secure assembly board with two-sided foam tape.

15. Attach the battery and position as shown. Note: The battery must be up against the enclosure with all leads properly routed or the rear cover will not fit properly.

16. See Figure 33-5 for operating instructions.

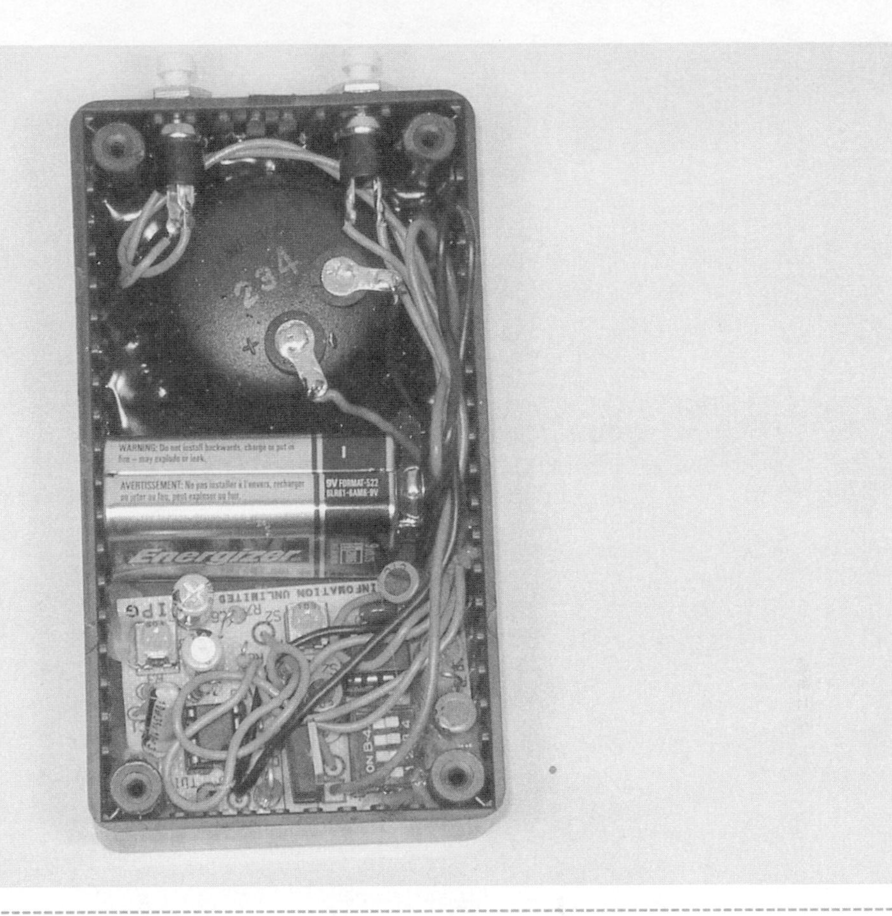

Figure 33-6 *Final assembly into EN1 enclosure*

Table 33-1 Invisible Pain-Field Generator Parts List

Ref#	*Description*	*DB Part #*
R1,6,8,12	Four 1 K, ¼-watt resistors (br-blk-red)	
R2	500 K vertical trimpot	
R3	2.2 K, ¼-watt resistor (red-red-red)	
R4,5	Two 10 K, ¼-watt resistors (br-blk-or)	
R7	10-ohm, ¼-watt (br-blk-blk)	
R9	10 K vertical trimpot	
R13	27-ohm, ½-watt resistor (red-pur-blk)	
C1,4	Two .01 mfd, 50-volt disc capacitors	
C2	10 mfd, 50-volt vertical electrolytic capacitor	
C5	.01 mfd, 50-volt plastic capacitor	
C3	100 mfd, 50-volt vertical electrolytic capacitor	
C6	1,000 mfd, 25-volt vertical electrolytic capacitor	
IC1,2	Two 555 timer DIP integrated circuits	
Q1	PN2907 PNP TO92 transistor	
SOCK1,2	Two 8-pin DIP sockets for above IC1,2 (Not shown)	
Q2	IRF540 Mosfet TO220	
L1AB,2,3	Four 1-millihenry inductors	DB# 1MH
Z1	50-volt, 1-watt Zener diode	
PBS1,2	Two pushbutton switches	
S1,2,3	3 or 4 section DIP switch	
CL1	Battery snap clip	
TD1	Special polarized transducer	DB# 1020A
PCB1	Printed circuit board	DB# PCIPG9
CASE	4 × 2 ⅜ × 1⅛-inch plastic enclosure	
SCREEN	2.5 × 2.5 inch piece of window screen	

Chapter Thirty-Four

Canine Controller

This very useful device, as shown in Figure 34-1, is intended for those who are bothered by nearby barking dogs. The control section is a modification of a circuit made by Bob Gaffigan in 1993. The project is intended to detect the dog's bark, which triggers a high-output, pain-field sonic generator, as described in Chapter 35, producing very uncomfortable sounds to the animal's sensitive hearing.

Figure 34-1 *The canine controller system with our sonic blaster*

The system may be covertly used without the dog owners knowing or it can even be tuned to bother the owners every time their mutt goes into a barking frenzy. The user can implement many options to discourage these constant barkers or give some of these inconsiderate owners a taste of their own medicine.

This is an intermediate-level project requiring basic electronic skills. Expect to spend $25 to $50. All the parts are readily available, with the specialized parts obtainable through Information Unlimited (www.amazing1.com), and they are listed in the parts list in Table 34-1 at the end of the chapter.

Device Description

A directional device picks up the dog's bark and processes it so that it turns on the sonic burst pain-field generator pointed in the offending animal's direction. The animal now experiences a very uncomfortable sound, much like a person would find the scratching of chalk on a blackboard. Eventually, the animal associates this sound with his barking, thus conditioning him to stop. The device contains adjustments that actually count the number of yelps so that it can be set not to trigger, thereby discouraging the animal from a normal barking trend that may be the warning of an intrusion or other important event.

The unit is easily mounted in any convenient location and operates from internal batteries, an external 12 volts, or 115 VAC via a wall adapter converter. The detection system contains adjustments for bark-level sensitivity, the number of barks detected before triggering, and the length of the triggered on time of the sonic burst. The pain-field generation section, as described in Chapter 35, has controls for the frequencies of the burst, sweep rate, sweep on/off, and main power control. The canine controller section is designed to easily connect to the existing jacks on the sonic generator section via a three-conductor cable. It may be mechanically attached as a single unit for a compact integrated system. The pickup microphone may even be placed at the focal point of a parabolic dish, providing very discrete selection of the target. As in any device such as this, many variables will enter into the equation that defines the overall performance. The suggested effective range will vary but can be up to 100 feet.

Special Note: In some circumstances, the owner is more to blame than the actual animal. As previously stated, the unit may be tuned to a frequency that the owner him- or herself finds very uncomfortable.

Circuit Description

Figure 34-2 shows the block diagram where the mic picks up the bark and amplifies it. A filter favors the assumed frequencies of the bark and feeds them to a bark counter. This circuit contains a selector switch to preset the necessary number of barks to trigger the bark period timer. The time of the bark period is also selectable by a switch. The bark period timer gates on the MOSFET switch transistor turns on the sonic generator, which drives the four transducers that now produce the directional, uncomfortable sound pain-field shockwaves.

Figure 34-3 shows the complete schematic of the controller section circuit, and the sonic pain-field generator section schematic is shown in Chapter 35. The bark is detected by microphone M1 and amplified by operational amplifier IC2A. The DC follower IC2D sets the midpoint bias, and resistor R2 sets the gain of this stage. The output level at this stage is controlled by potentiometer R5 and fed to amplifier/filter IC2C with a passband set by resistors R8 and R9 and capacitors C8 and C9. Schmidt discriminators

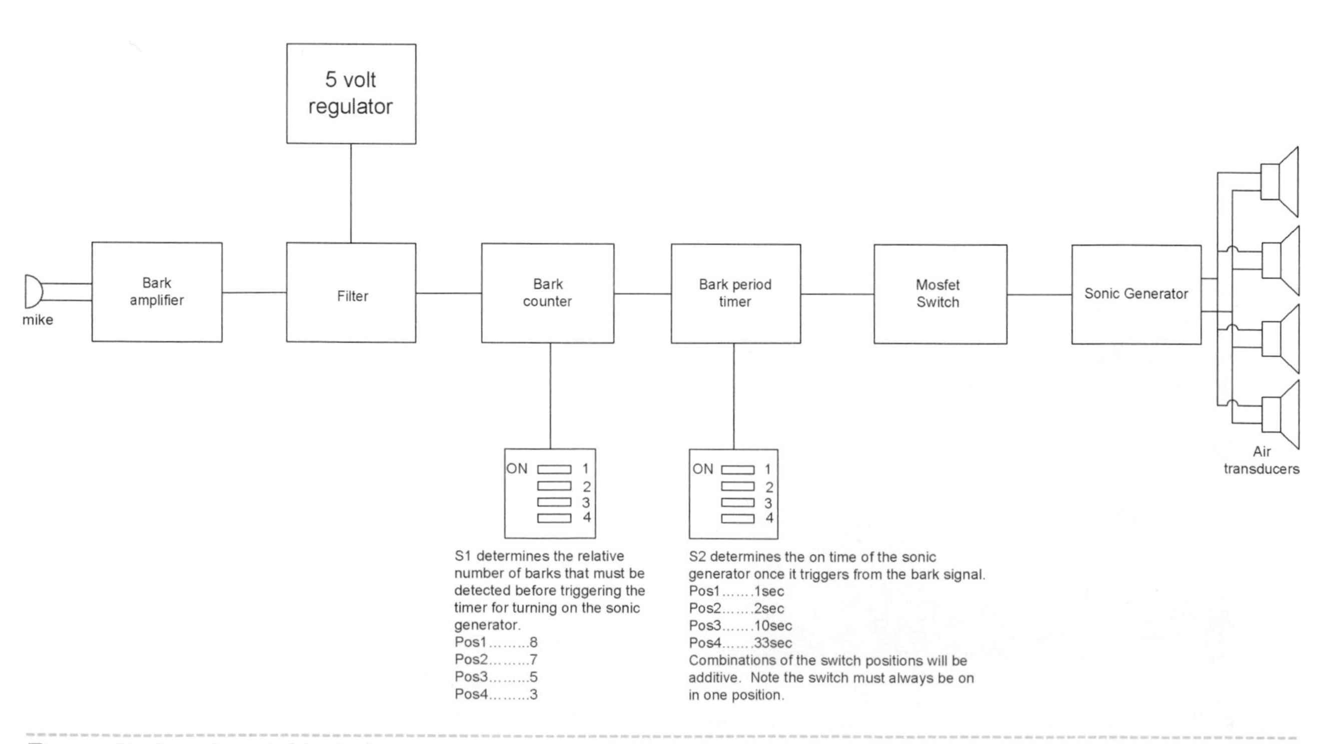

Figure 34-2 *Circuit block diagram*

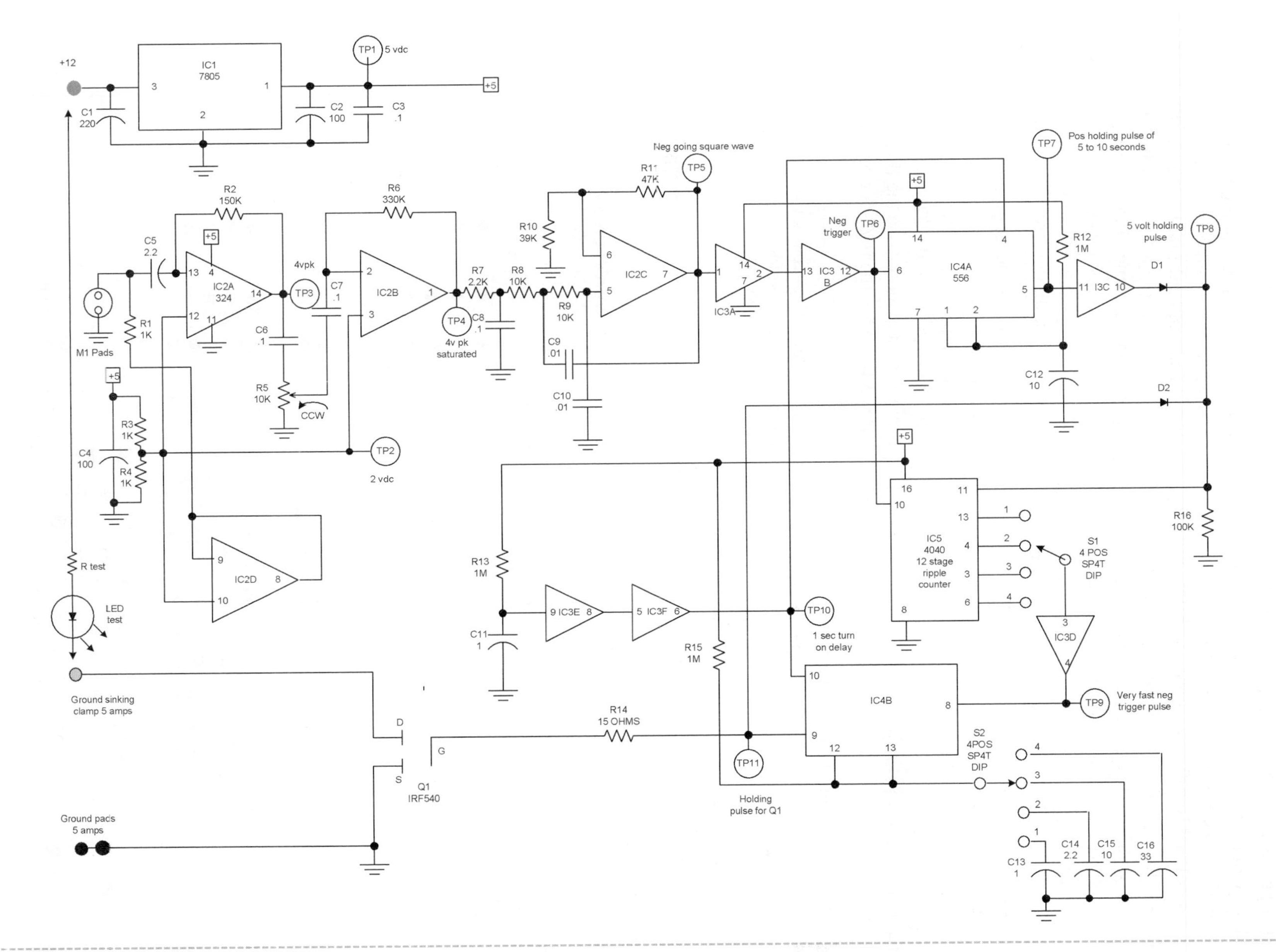

Figure 34-3 *Circuit schematic*

IC3A and IC3B trigger the signal, provide a defined level of the correct polarity to initiate timer IC4A, and enable ripple counter IC5 to count the barks within a time period determined by IC4A. If the number of barks exceeds a limit set by bark count switch S1, timer IC4B is initiated, turning on MOSFET switch Q1 for a predetermined amount of time selected by switch S2, enabling the sonic pain-field deterrent signal to be sent toward the target and then resetting. You will note that control of the sonic pain-field generator is made by sinking the negative return lead into the ground.

Circuit Assembly

1. Assemble the circuit board as shown in Figure 34-4. Note that Figure 34-5 shows the foil traces of the printed circuit. It is not recommended that you assemble this circuit on a perforated vector board unless you are an advanced assembler.

   Note that if you are building from a perforated or vector circuit board, it is suggested that you use the indicated traces for the wire runs and insert components starting at the lower left-hand corner. Pay attention to the polarity of the capacitors with polarity signs and all the semiconductors. It is a good idea to use sockets for all the integrated circuits.

   Route the leads of the components as shown and solder as you go, cutting away unused wires. Attempt to use certain leads as the wire runs or use pieces of the #26 bus wire. The heavy foil runs should use the thicker #24 bus wire, as these are the high-current discharge paths.

2. Double-check the accuracy of the wiring and the quality of the solder joints. Avoid wire

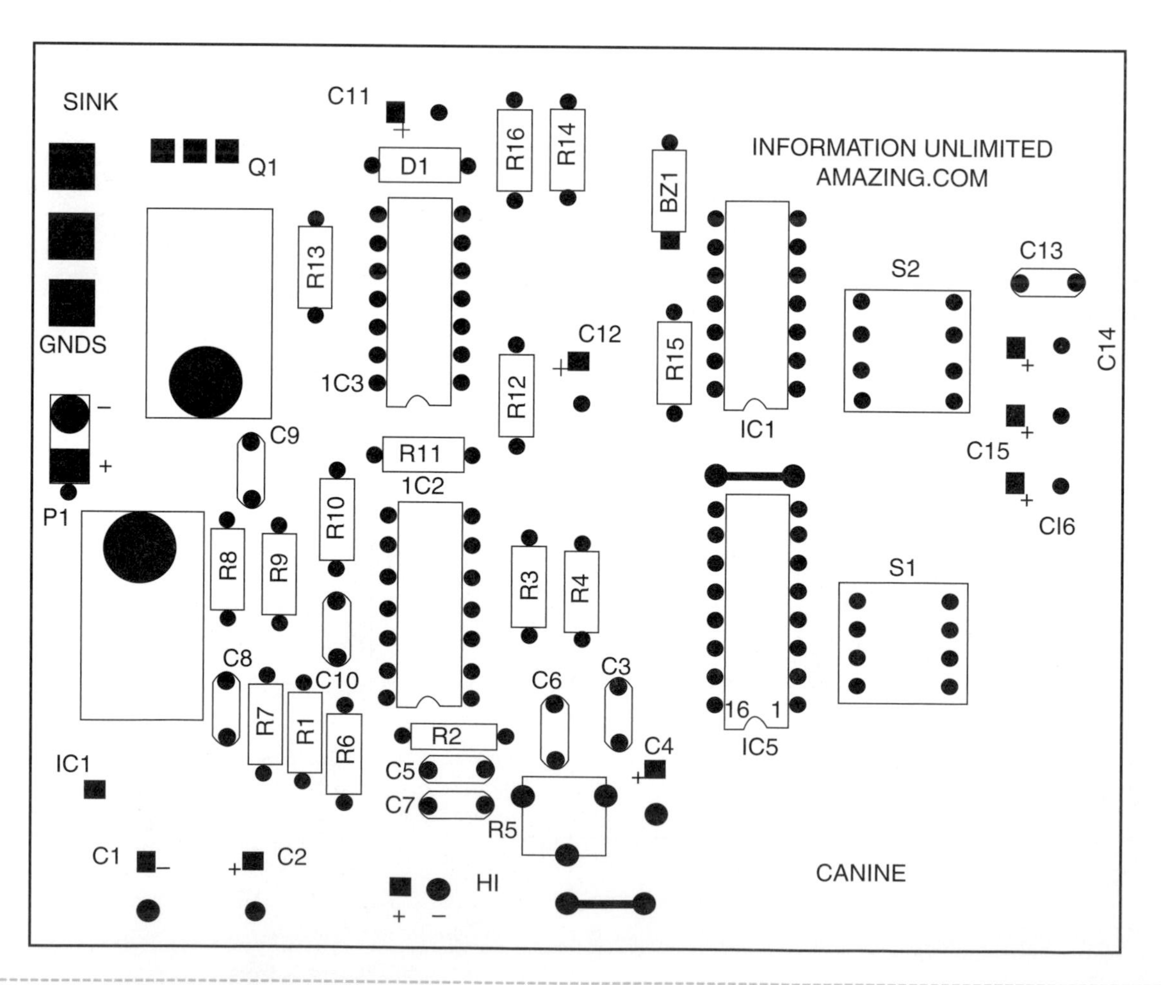

Figure 34-4 *Assembly of the circuit board*

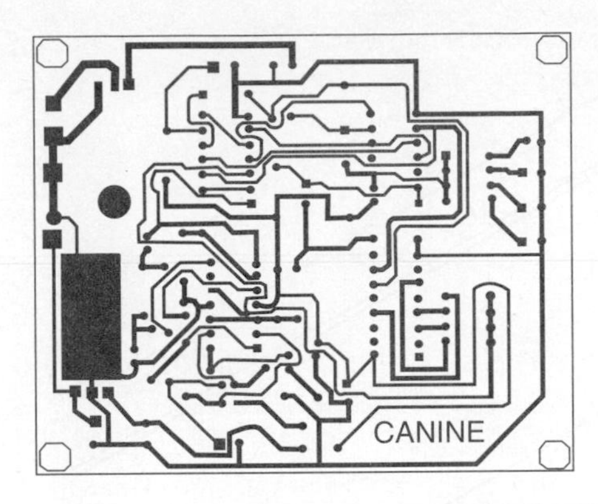

Figure 34-5 *Assembly circuit board traces*

bridges, shorts, and close proximity to other circuit components. If a wire bridge is necessary, sleeve some insulation onto the lead to avoid any potential shorts.

3. Connect the external components as shown in Figure 34-6. Use shielded microphone cable if you are not installing it on the actual board.

4. Fabricate the chassis as shown in Figure 34-7 from a piece of .035-inch 5052 bendable aluminum or plastic. If you use metal, you will need a piece of plastic material under the assembly board to prevent shorting the foil traces.

5. Create a mating cover and finally assemble everything as shown in Figure 34-8. Note the hole for the microphone with a screen. Wire the P1 plug and cable for interconnecting the controller and the sonic blaster using desired lengths per your requirements. Leave the other end with leads that will eventually connect to the sonic blaster.

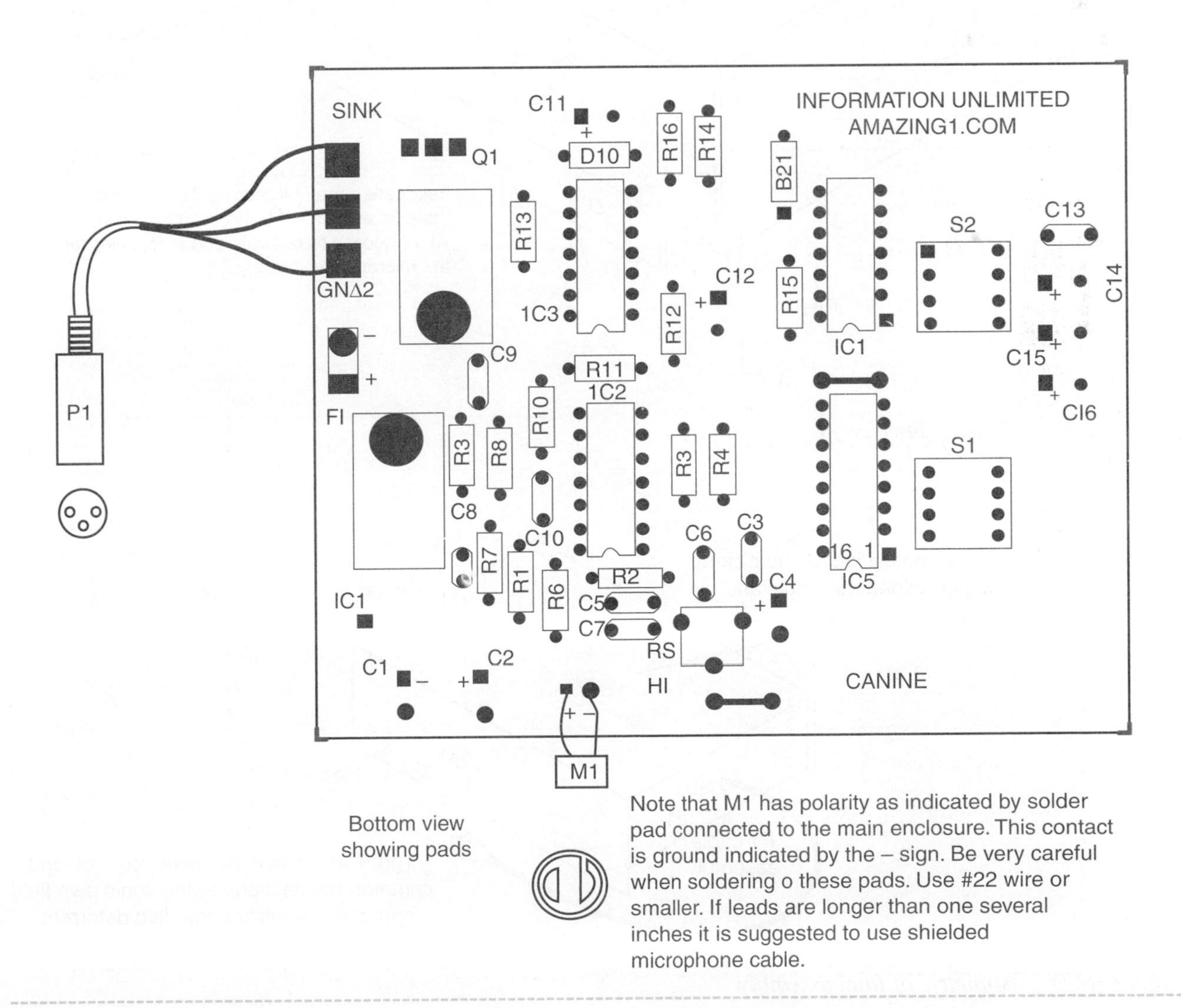

Figure 34-6 *External wiring*

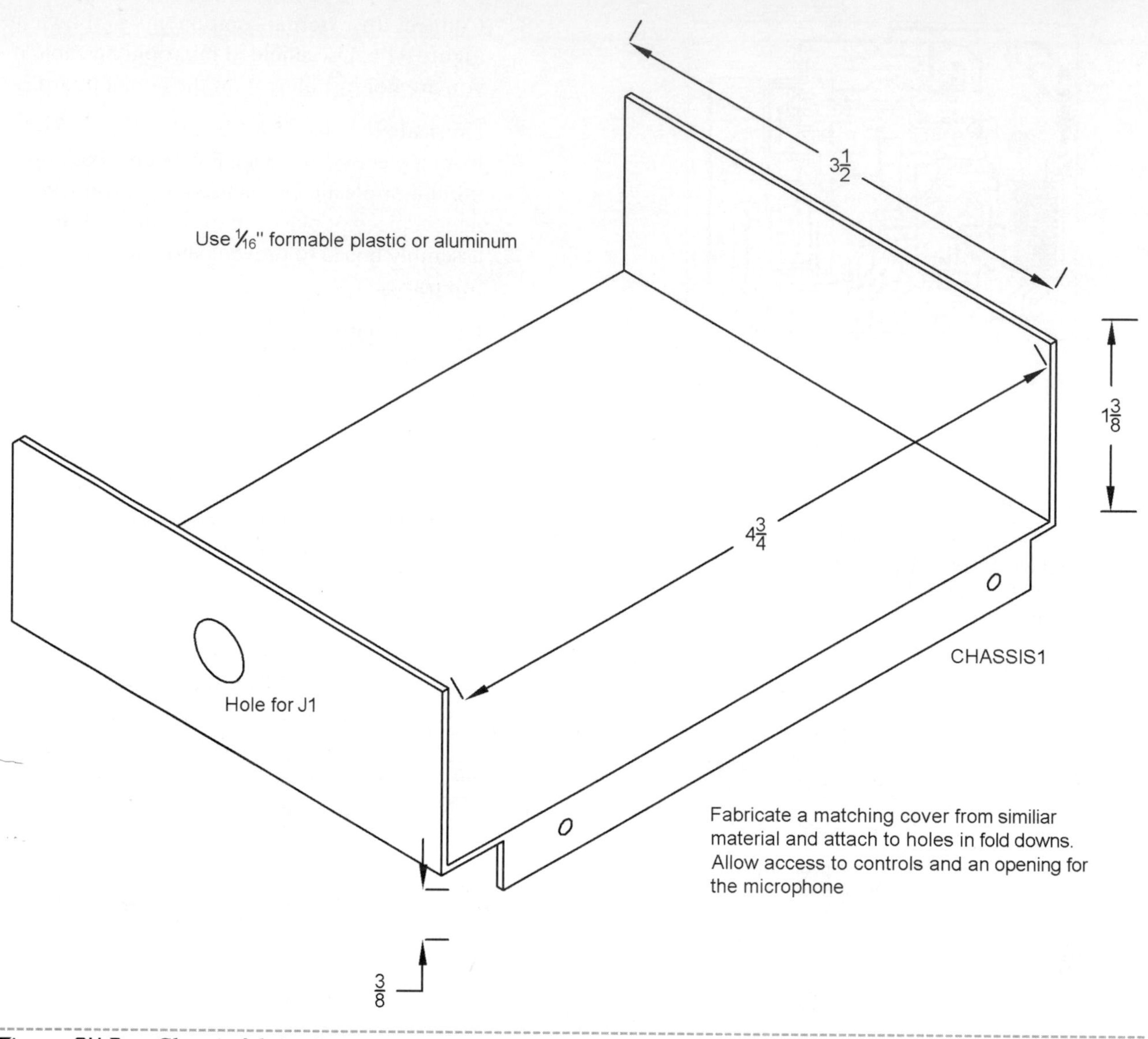

Figure 34-7 *Chassis fabrication*

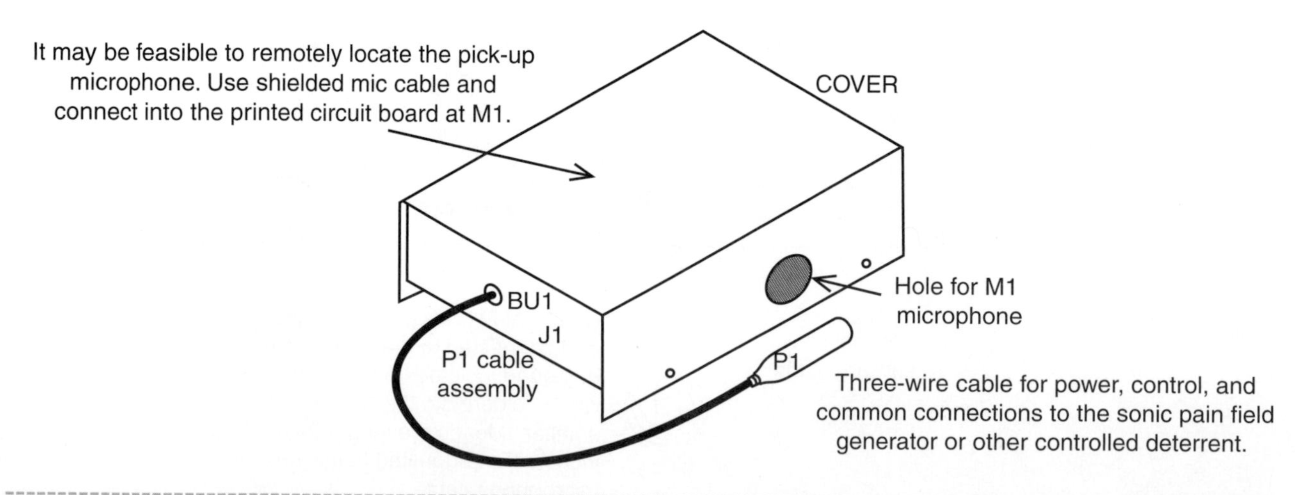

Figure 34-8 *Isometric of final assembly*

Testing the Circuit

1. Assemble the test jig, referring back to Figure 34-3. If the circuit is correct, you may go right to the connection of the test jig as shown and perform the following steps:

    a. Preset S1 to position 4 (the lowest number of barks for activation).

    b. Preset S2 to position 1 (the shortest time for the sonic generator).

    c. Preset R5 to midrange.

2. Apply a 12 VDC input and note a low-current draw of less than 10 milliamps, which is the same value as the quiescent operating current. C cell batteries will work approximately 30 days before requiring replacement.

3. Simulate a bark and note the test LED momentarily coming on. If this occurs, you may want to test the various positions of S1 and S2. Also verify the sensitivity control, R5.

4. If the circuit does not work, you may need a scope to test the various test points as shown in Figure 34-3. They should have the following values:

    TP1: +5 volts DC

    TP2: 2 to 3 volts DC

    TP3: 4-peak-volt audio level

    TP4: Saturated audio signal

    TP5: Negative-going pulse

    TP6: Negative-going square wave

    TP7: Positive 10-second holding pulse

    TP8: Positive holding pulse

    TP9: Very fast negative trigger pulse

    TP10: 1-second turn-on delay

    TP11: Holding pulse for Q1

10. Once the controller operation is verified, you may connect the system as shown in Figure 34-9. You will need a properly working sonic pain-field generator, as described in Chapter 35.

It is suggested that you experiment with the system before actually installing it. Once familiar with it, you may position the system for the best effect. You also have the option of adjusting the sonic generator when the dog starts barking so that it can also be very annoying to the animal owners when their mutt goes into an uncontrolled barking rage. Use your own judgment as you can get tagged for harassment!

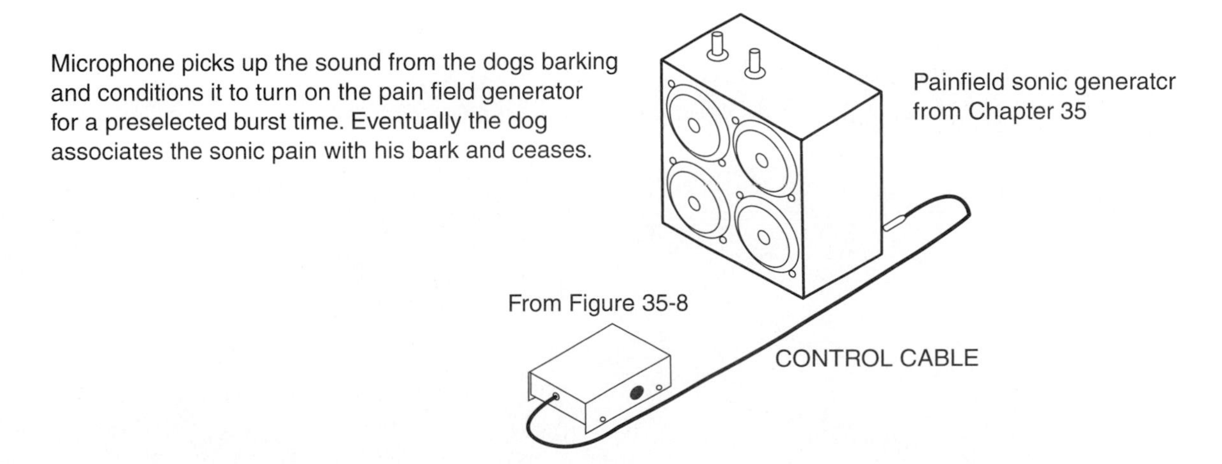

Figure 34-9 *Systems view*

Table 34-1 Canine Controller Parts List

Ref. #	*Description*	*DB Part #*
R1, 3, 4	Three 1K, ¼-watt resistors (br-blk-red)	
R2	150K, ¼-watt resistor (br-gr-yel)	
R5	10K pot and 12-volt switch, 17 mm	
R6	330K, ¼-watt resistor (or-or-yel)	
R7	2.2K, ¼-watt resistor (red-red-red)	
R8, 9, 10	Three 10 kilo-ohm, ¼-watt resistors (br-blk-or)	
R11	47 kilo-ohm, ¼-watt resistor (yel-pur-or)	
R12, 13, 15	Three 1M, ¼-watt resistors (br-blk-gr)	
R14	15-ohm, ¼-watt resistor (br-gr-blk)	
R16	100K, ¼-watt resistor (br-blk-yel)	
C1	220-microfarad, 25-volt vertical electrolytic capacitor	
C2, 4	Two 100-microfarad, 25-volt vertical electrolytic capacitors	
C3, 6, 7, 8	Four .1-microfarad, 50-volt plastic capacitors	
C5, 14	Two 2.2-microfarad, 50-volt vertical electrolytic capacitors	
C9, 10	Two .01-microfarad, 50-volt plastic capacitors	
C11, 13	Two 1-microfarad, 50-volt vertical electrolytic capacitors	
C12, 15	Two 10-microfarad, 50-volt vertical electrolytic capacitors	
C16	33-microfarad, 50-volt vertical electrolytic capacitor	
D1, 2	Two IN914 silicon diodes	
IC1	7805 5-volt regulator TO220	
IC2A, B, C, D	LM324 quad amp in dual inline package (DIP)	
IC3A, B, C, D, E, F	40106 hex Schmidt in DIP	
IC4	556 dual timer in DIP	
IC5	4040 complementary metal oxide semiconductor (CMOS) PLL phase lock loop in DIP	
Q1	100-volt IRF540 metal-oxide-semiconductor field effect transistor	MOSFET
M1	FET mic element	DB# FETMIK
S1, 2	Two four-position single-throw DIP switches	
PCCANINE	Printed circuit board (PCB)	DB# PCCANINE
CHASSIS1	Metal or plastic chassis as shown in Figure 34-7	
COV1	Cover fabricated as shown in Figure 34-8	
J1/P1	Three-pin chassis mount jack and mating plug	
WR3C	3-inch three-conductor cable for interconnecting	
PPF4K	Pain-field generator kit as described in Chapter 35	DB# PPF4K
PPF40	Pain-field generator assembled as described in Chapter 35	DB# PPF40

Chapter Thirty-Five

Ultrasonic Phaser Pain-Field Generator

This project, as shown in Figure 35-1, shows how to construct a moderately high powered sonic generator that can be used for tasks that range from animal control to discouraging personal encounters. It can be used as part of our laser property protection guard described in Chapter 5 or as the deterrent for use with the canine controller, as described in Chapter 34. The unit can generate a variable rate of complex waves from 5 to 25 kHz well into the ultrasonic range. These waves can be very painful or disorientating to animals and people, depending on where the controls of the unit are set.

This is an excellent device for use in animal control as well as a low-liability deterrent in anti-intrusion alarms and detection systems. This is an intermediate-level project requiring basic electronic skills. Expect to spend $50 to $75. All parts are readily available, with specialized parts obtainable through Information Unlimited at www.amazing1.com, and they are listed in the parts list at the end of the chapter.

Figure 35-1 *The phaser pain-field generator*

Basic Device Description

This chapter demonstrates how to create a variable-sweep-frequency, ultrasonic, pain-field generator capable of producing the equivalent of 400 watts of resultant power such as that obtainable from a conventional dynamic transducer system. This is possible due to recently developed piezoelectric ceramic devices. The high efficiency requires very little driving power, consequently resulting in a lightweight, portable, handheld device that is battery driven and capable of producing high sound pressure. It is a directional device and can be set up in a target area such as a field or garden to discourage any sound-sensitive animals. It is also capable of being powered by a car's 12-volt system and can be mounted on the hood or roof of the vehicle. Transducers are mounted in an array for a concentration of energy in one direction.

It should be understood that certain people subjected to different degrees of exposure are affected more than others, some to a point where they may vomit or experience severe headaches and cranial pains. Some people will experience severe pain in the ear, teeth, or lower head. Statistically, women and younger children are many times more sensitive to this device than the average male adult. With this in mind, the user must exercise consideration when testing and using the device for animal control, as many people will not be aware of the source of this pain and attribute it to a headache or other physical ailment. Also, certain people are affected mentally to a point of actually losing their tempers completely or becoming extremely quick-tempered. Some will experience a state of extreme anxiety when overly exposed. Therefore, consideration must be used at all times when testing or using this or similar devices.

It should also be noted that using the transducers in an array configuration may be damaging to hearing at close range. The array approach produces high sound-pressure density occurring on or near the output axis.

Circuit Theory

Figure 35-2 shows a timer, IC2, connected as an astable free-running multivibrator whose frequency is externally controlled by pot R9. Resistor R10 selects the range limit of R9, and capacitor C5 determines the frequency range of the device along with the previous resistors.

The square wave output of IC2 is via pin 3 and is directly coupled to power amplifier Q2. The drain of Q2 is DC biased through choke L1.

The square wave output signal is then fed into transducer TD1 in series with parallel combination of resonating coils L2 and L3. The resonant action between the inherent capacity of TD1, the C7 tuning capacitor, and the inductance now produces a sinusoidal-shaped wave peaking around 25 kHz or the

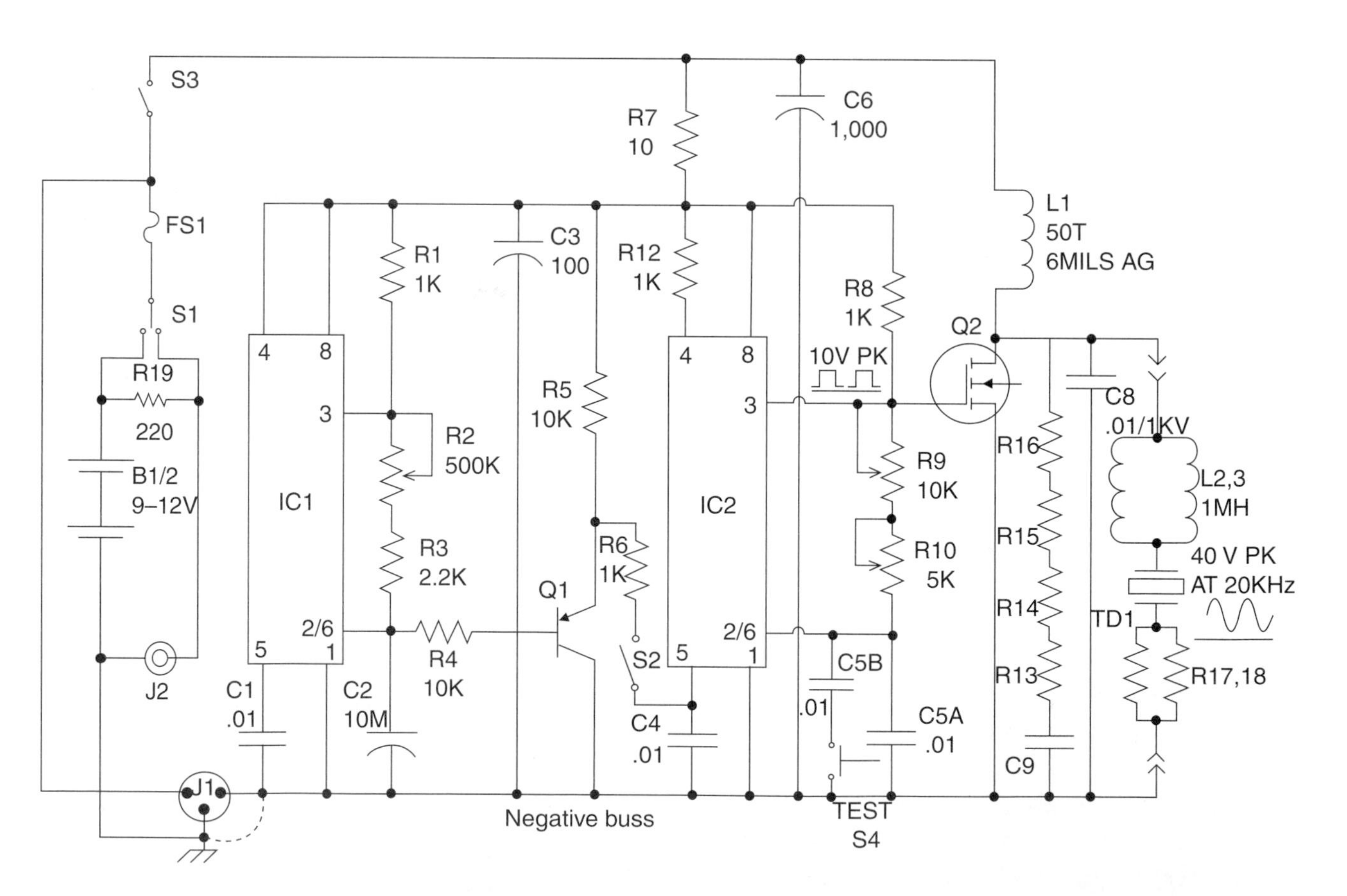

Dashed line indicates bypass connection when not using J1 for external control. You may also short these pins using a mating plug with pins connected together.

Note four transducers and associated networks.

Figure 35-2 *Schematic of the phaser pain-field generator*

upper limit of the tuning range. This signal waveform now has a peak-to-peak voltage several times that of the original square wave. Transducer TD1 now can take advantage of these peak voltages to produce the high sound-pressure levels necessary without exceeding the high *root means square* (rms) ratings of an equivalent voltage-level square wave.

Timer IC1 is similarly connected as an astable, running multivibrator and is used to produce the sweeping voltage necessary for modulating the frequency of IC2. This sweep repetition rate controlled by pot R2 and resistor R3 limits the range of this repetition time. Resistor R1 selects the duty cycle of the pulse, while capacitor C2 sets the sweep time range. The output for IC1 is via pins 6 and 2 where the signal ramp function voltage is resistively coupled to inverter transistor Q1 via resistor R4. The output of Q1 is fed to pin 5 of IC2 and provides the output modulation voltage necessary to vary the frequency as required. Note that the modulation signal is easily disabled via R2/S2.

Capacitor C6 guarantees an AC return path for the output signal. Power to driver circuits IC1 and IC2 is through a decoupling network consisting of resistor R7 and capacitor C3.

Power to the system is via internal battery B1 or an external 12 VDC such as from a vehicle. Three-way switch S1 selects between these two power sources or shuts the unit off in the center position. Resistor R19 provides a trickle charge to the internal battery when using an external source. Jack J1 allows a connection that is similar to our canine controller, turning on the unit when it detects a barking dog.

Construction Steps

If you are a beginner it is suggested to obtain our *GCAT1 General Construction Practices and Techniques*. This informative literature explains basic practices that are necessary in proper construction of electromechanical kits and is listed at the end of the chapter.

1. Lay out and identify all the parts and pieces, and check them with the parts list. Note that certain parts may sometimes vary in value. This is acceptable as all components are 10 to 20 percent tolerant unless otherwise noted.
2. Fabricate the heatsink bracket (HS1) as shown in Figure 35-3 from a .75 × 2 × .065-inch aluminum piece. Bend it 90 degrees at its midsection and drill a hole for the SW1/NU1 screw and nut. Attach it to Q2 as shown.
3. Assemble the L1 choke coil as shown in Figure 35-4 by wrapping 50 turns of #24 magnet wire on the nylon bobbin as evenly as possible. Leave 2 inches of leads for a connection to the circuitry. Assemble the E core as shown and shim each side with pieces of yellow cardboard strips of 3 mils each for a total of 6 mils. If you have an LCR bridge, measure 1 millihenry.
4. Assemble the PCB as shown in Figure 35-5. Note the two wire jumps and component polarity. Wire in inductor L1 and secure it to the board with *room temperature vulcanizing* (RTV) silicon rubber or another suitable adhesive.

   If you are building from a perforated or vector circuit board, it is suggested that you use the indicated traces for the wire runs and insert the components starting in the lower left-hand corner. Pay attention to the polarity of the capacitors with polarity signs and all the semiconductors.

   Route the leads of the components as shown and solder as you go, cutting away unused wires. Attempt to use certain leads as the wire

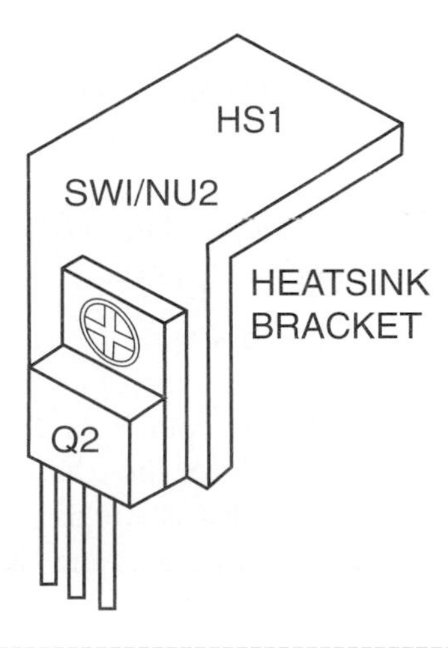

Figure 35-3 *Fabrication of heatsink*

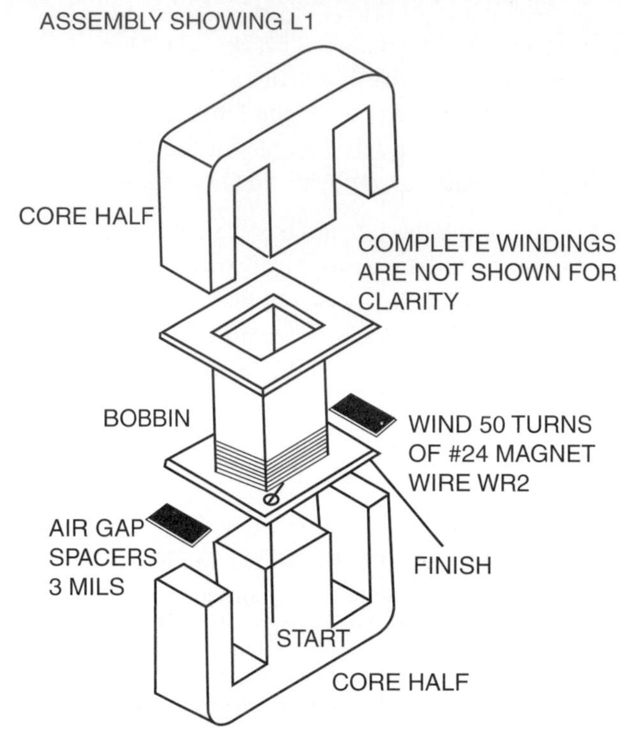

Figure 35-4 *Assembly of L1 inductor*

runs or use pieces of the #24 bus wire. The heavy foil runs should use the thicker #20 bus wire as these are the high-current discharge paths.

It is suggested that you use an 8-pin integrated circuit socket (SO8) for IC1 and IC2.

5. Double-check the accuracy of the wiring and the quality of the solder joints. Avoid wire bridges, shorts, and close proximity to other circuit components. If a wire bridge is necessary, sleeve some insulation onto the lead to avoid any potential shorts.

6. Fabricate the chassis and cover, as shown in Figure 35-6 from 1/16-inch Lexan plastic or aluminum. Use a large punch or circle saw to cut out the four holes for the transducers. Drill the remaining holes as you assemble it, verifying the proper clearances and sizes with components.

7. Complete the final assembly shown in Figure 35-7 and mount the assembly board, battery holders, switch S1, jacks J1 and J2, and fuse holder FS1. It is a good idea to trial-fit all the parts in this section before actually drilling or making any mounting holes. Note a piece of plastic (SPACER) that insulates the bottom solder connections of the PCB from shorting out to the metal chassis base. Also note resistor R19 across S1. Secure the battery holders using pieces of two-sided foam sticky tape (TAPE). Wire using #20 vinyl wire as shown.

8. Mount the transducers as shown in Figure 35-8 using SW2/NU2 6-32 × 1/2-inch screws and nuts.

   Twist the leads of the two inductors L2 and L3 together, effectively paralleling these components. Repeat this for resistors R17 and R18. You should end up with four sets of these parallel components. These components should be self-supporting by being connected to stiff pieces of bus wire. Take caution to observe any potential shorts. Solder the two 18-inch connecting leads and connect them to point P6 on the assembly board.

9. The final assembly is shown in Figure 35-9. Do not attach the cover until you have tested the unit.

Testing the Assembly

10. Turn the pots counterclockwise, set the switch to mid position (off), and install eight fresh batteries into the battery holders. If you have access to a bench supply of 12 VDC at 2 amps, it may be connected to J2 via a mating plug, eliminating the need for the batteries in the preliminary test. Short out the chassis ground and negative bus pins of J1. This jack is for connection to the canine controller module or other remote control devices.

Note: A variable bench supply capable of 12 VDC at 2 amps with a voltmeter and current meter can be a great convenience for the remaining steps and the testing of other similar circuits.

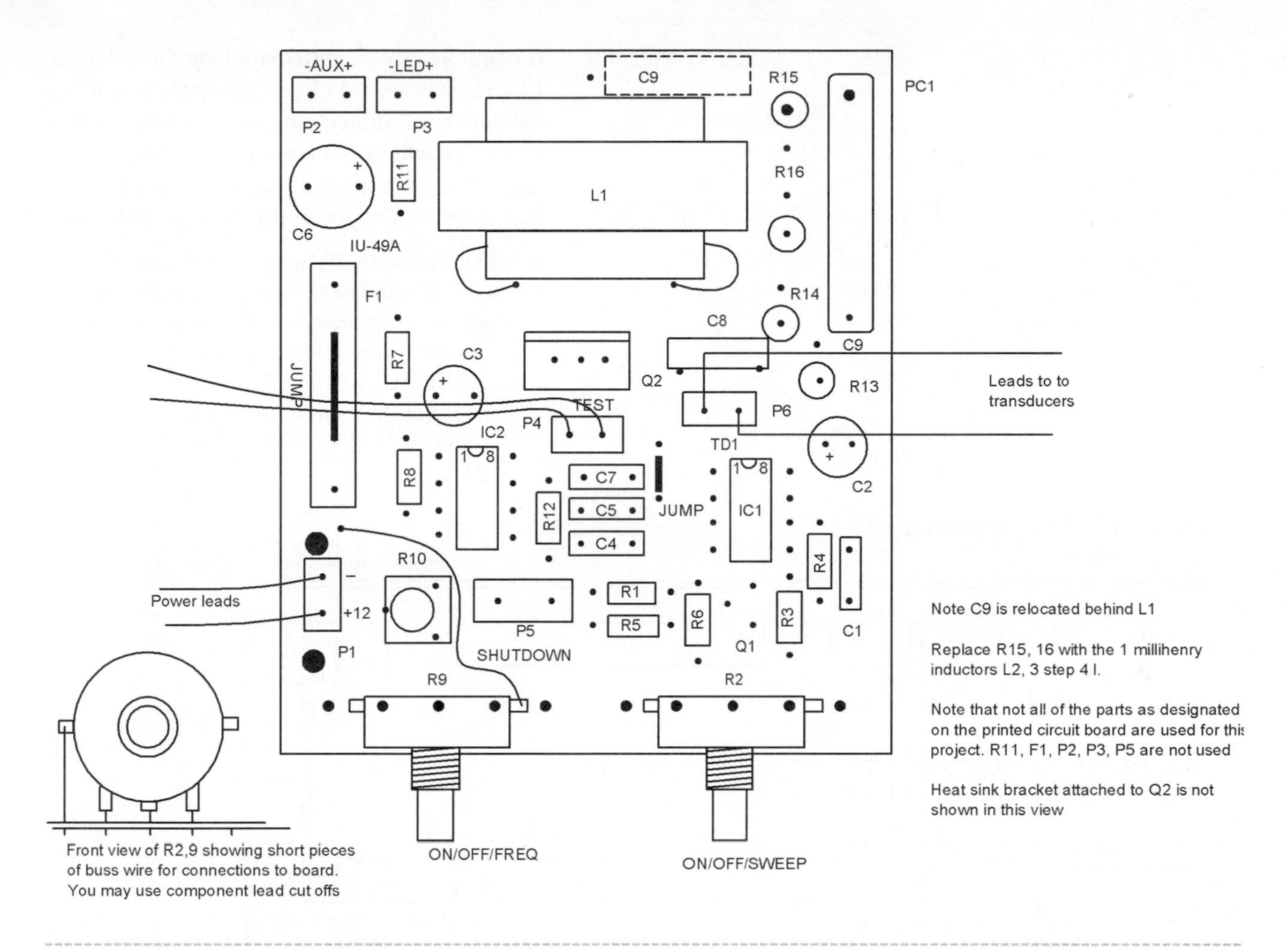

Figure 35-5 *Assembly of PCB*

11. Press push button S1 and note a loud, piercing wave coming from TD1. Measure a current of 500 to 600 milliamps in series with a battery or bench supply. Note that the switch position when up should be power supplied by internal batteries, and the down position being supplied by the external 12 VDC supply. This function may be reversed but should be noted, as the external position will automatically trickle charge the internal batteries.

12. Rotate R9 and note the frequency increasing to above the audible range; measure a current of 400 to 800 milliamps. You may preset the limit of the high range by the R9 setting. The normal factory setting is 25 kHz maximum with R10 full clockwise. Note the wave shapes shown in Figure 35-2 for those who have a scope.

13. Turn on sweep control R2/S2 and note the frequency being modulated by a changing rate as this control is adjusted. Use caution as certain sweep rates may cause epileptic seizures and other undesirable effects. Sweep rates between 7 to 20 per sec should be used with caution.

Basic Operating Instructions

14. Connect a 12-volt source capable of supplying 2 amps of current to designated leads or use internal *nickel cadmium* (NiCad) batteries that are type C cells. These will provide approximately 3½ hours continuous operation at slightly reduced output.

15. The battery-charging function occurs when the system is connected to an external power source and will charge the NiCad batteries at a 100-milliamp rate, taking about 14 hours to fully charge.

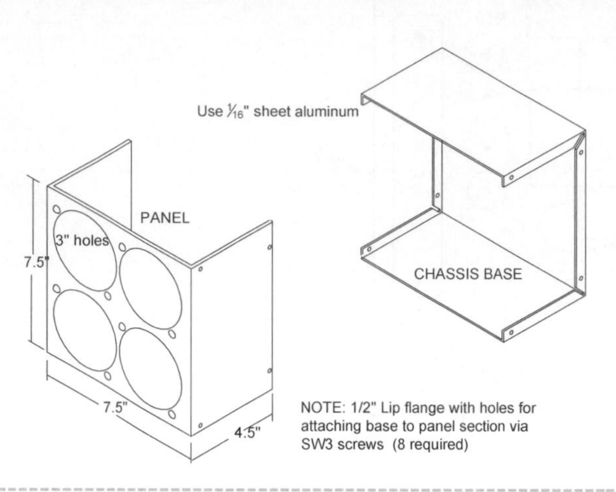

Figure 35-6 *Chassis fabrication*

16. If continued use is anticipated via the external 12 volts, it is a good idea to remove the internal batteries' connections as overcharge may occur that will ruin the batteries or actually cause them to explode. The recommended method of setting the controls is the following:

    a. Determine the frequency limits per the application. These types of devices have two basic applications. When used as an anti-intrusion device to discourage unauthorized entrance, adjustments are made for maximum human annoyance, usually with frequencies from 10 to 15 kHz.

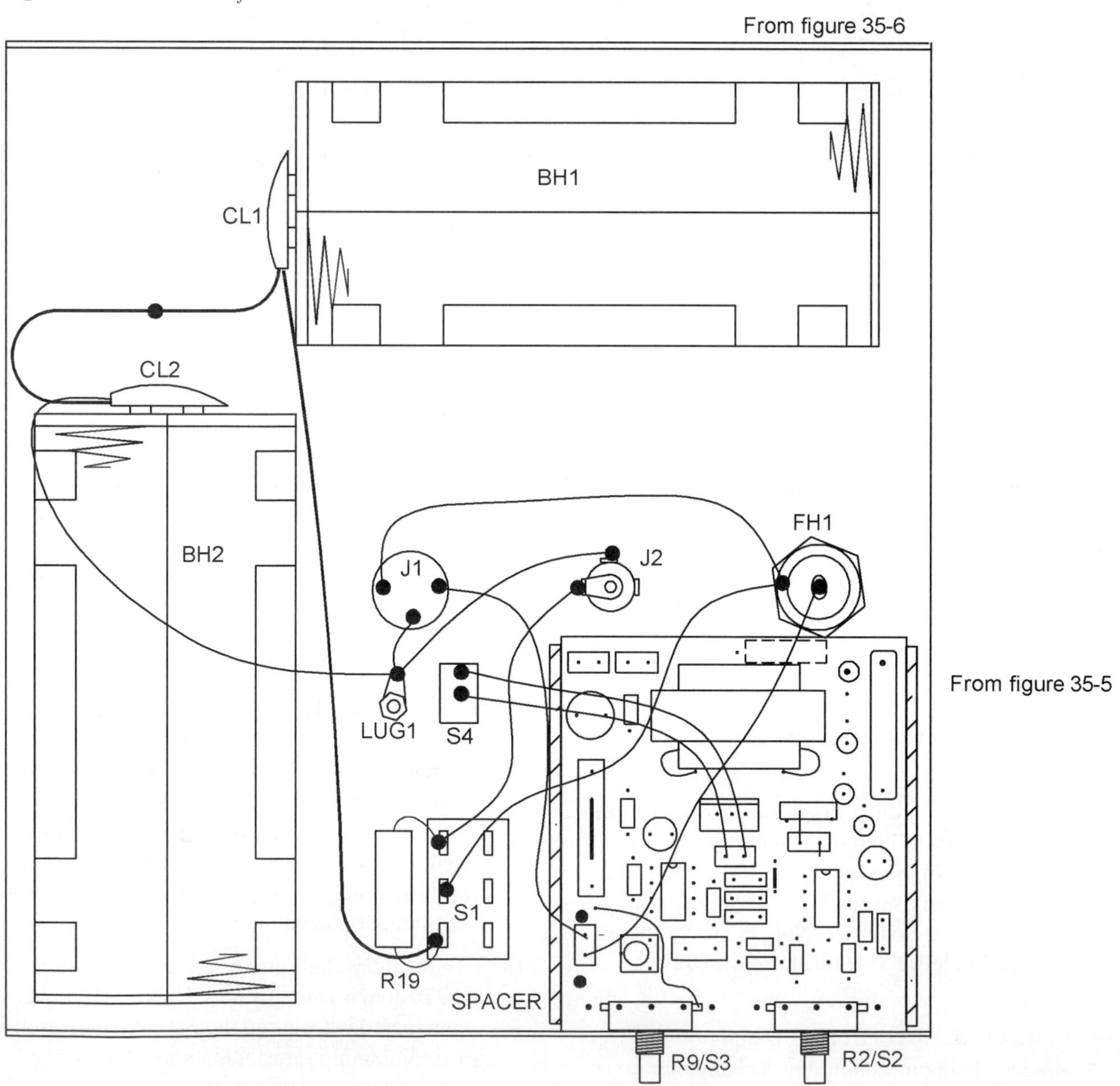

Figure 35-7 *Chassis wiring*

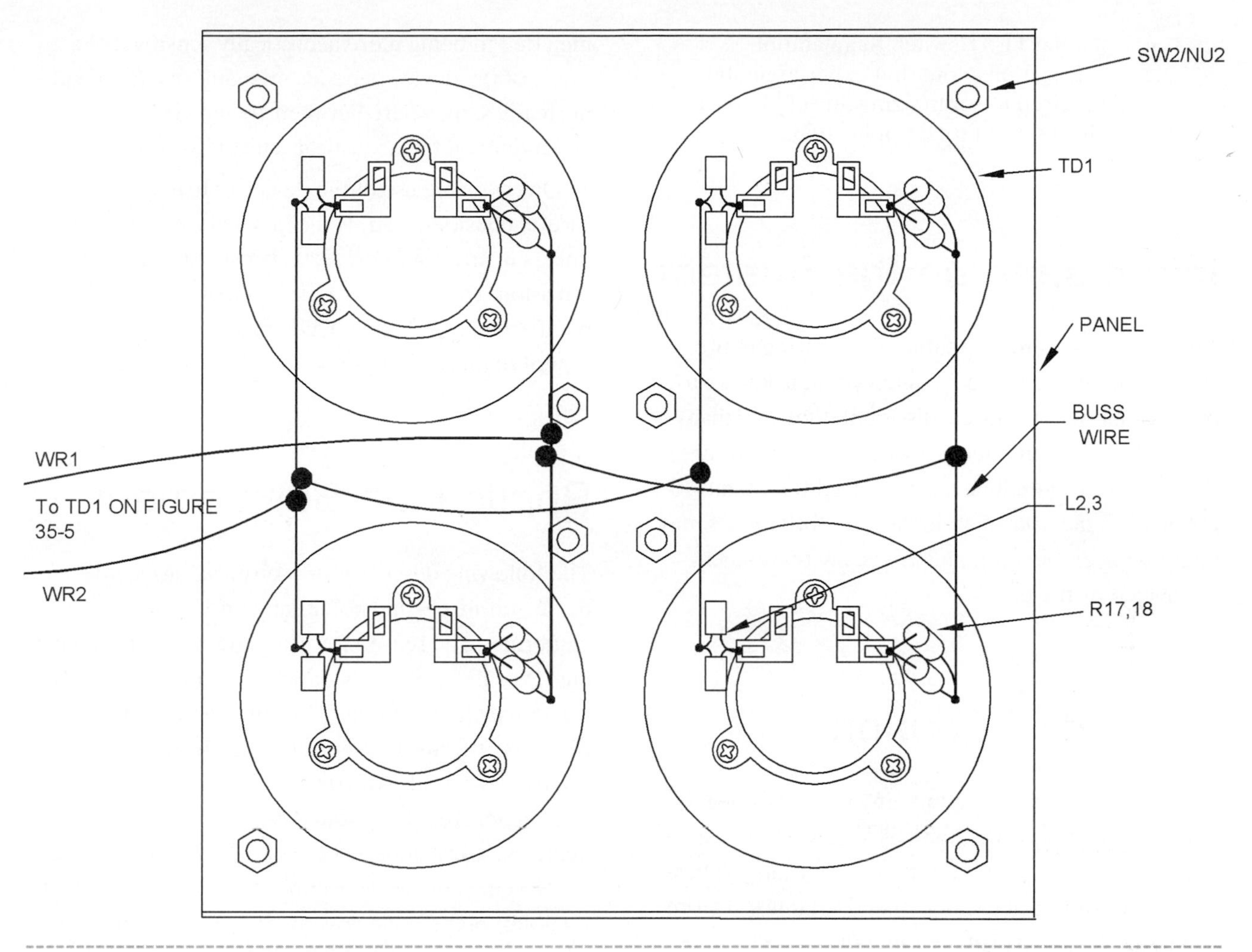

Figure 35-8 *Transducer panel wiring*

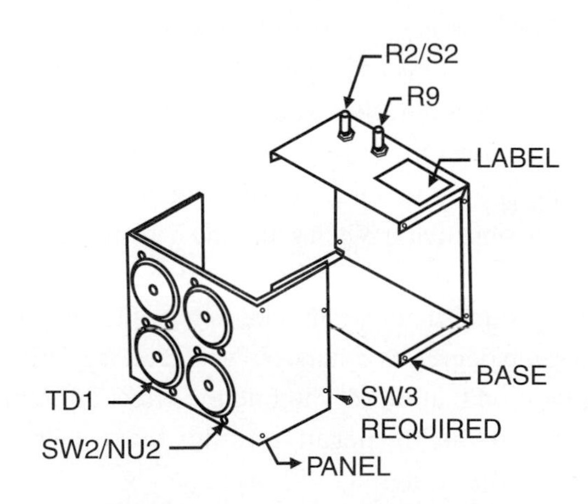

Figure 35-9 *Final assembly*

When used as a rodent device, we have found that the lowest tolerable frequency to humans usually has the greatest effect on the common species of rats. This is not always true but serves as a starting point in initially setting the adjustments. A frequency setting of 15 to 20 kHz for nonhuman living areas and 20 to 25 kHz for areas where people are present usually suffices for good rat control. Note that at no time should the unit be set higher than 25 kHz.

Dog control is usually around 18 kHz, but experimenting with the target animal may be necessary for optimum effect. Stubborn dog owners fall into the category of anti-intrusion if they will not cooperate.

b. Turn the SWP control off. Adjust the TONE control so that the output is detectable by the human ear or just above the point of annoyance.

c. Set the TEST switch to an audible verification. Note that experimentation is required for optimizing any effect on a given set of target applications.

Information on the System

Your phaser system is capable of operating in two modes. Mode one is at a frequency that is known to produce paranoia, nausea, disorientation, and many other physiological effects. Mode two allows using the system as an audible alarm to frighten off intruders or warn the user of an intrusion. Both modes may be used in combination and are easily rear-panel-controlled by the user.

A Word of Caution

Ultrasonics is a grey area in many respects when an application involves the control of animals or even a deterrent to unauthorized intrusion. It is always best to consult with local municipal and state laws before using this device to protect home or property. Remember many state laws lean more towards the right of the criminal rather than the victim.

General Information on Ultrasonics

Numerous requests have been made for information on the effect of these devices on people. First, let us make it clear that no device such as this should purposely be used on humans, and we discourage this use due to the possibility of acoustically sensitive people being highly irritated.

None of these devices has the ability to stop a person with the same effect as a gun, club, or more conventional weapon. They will, however, produce an extremely uncomfortable, irritating, sometimes painful effect in most people. Everyone will experience this effect to some degree. Unfortunately, younger women are much more affected than older men due to being more acoustically sensitive. The range of the device depends on many variables and is normally somewhere between 10 and 100 feet depending on the acoustical sensitivity of the target.

One possible use of the device (which deserves careful consideration) is the installation of all transducers in an area to protect against unauthorized intrusion. This would produce an irritating and painful feeling for the intruder, along with a condition of paranoia of not knowing what to expect next.

Application Supplement

The following describes an acoustical, ultrasonic device for dispersing potentially unruly crowds or gatherings. The reason for using the higher-frequency energy is that it is easily directed without being diffracted in all directions. Therefore, the use of directional-type transducers producing a front-to-back relative signal ratio of 20 to 30 dbs is easily achieved. The higher-frequency sound can produce nausea, headaches, and depression without the subject necessarily being able to determine the source. This effect no doubt presents an advantage to law enforcement when using these devices.

It is a known fact that the development of any type of demonstration in "psychological crowd control" follows the natural laws of growth and decay, in that a positive influx of people encourages a crowd growth, whereas a negative outflux creates a diminishing situation. One must realize that differences exist among most people regarding sound level and frequency sensitivity. When sampling a group or crowd as to the overall sensitivity of these devices, it is usually found that certain ones may not be affected to the same degree as others. However, taking into consideration that those who find the effect intolerable does generate the negative outflux that usually terminates the gathering.

This method, to this date, produces no permanent, lasting effects or symptoms and therefore becomes a more humane means of dispersing potentially unruly crowds. It should also be mentioned that younger people appear to be many times more sensitive to this effect. This usually goes hand in hand in most demonstrations, since younger people are more apt to become unruly.

This section encompasses many of the methods for applying this device via a permanent installation where transducers can be placed and adjusted to produce the necessary sound-pressure levels at distances to achieve the desired effects. This method could prove effective as an intrusion deterrent for a certain protected area.

Another method could be a handheld directional device similar to a megaphone where the amount of transducers used would be determined by the affects desired. This device would take into consideration the highly directional characteristics of this energy, being able to be directed at problem areas as a handheld portable system.

Another method for effective control would be clusters or arrays of transducers with these systems being mounted to vehicles via conventional roof mounts or on the hoods of these vehicles. Operation could take further advantage of the fair to good soundproofing found in most automobiles and vehicles could approach problem areas, taking advantage of close-proximity situations.

It has been found that certain frequencies also can be made to produce intense irritability in certain types of people. This method could justify the eventual use of more severe methods of restraint, assuming that the target subjects would eventually become more aggressive or violent, thus warranting this action. This is only to be considered as a potential extreme use of the device.

The objective here is to describe a similar method whereby the sound-pressure levels and frequencies can be obtained economically and efficiently, using state of the art methods and easily available parts, yet they can still be portable and easily handled for complete operational flexibility. Construction is based around the use of the piezoelectric tweeter transducers. These lightweight devices are easily driven and produce a conversion efficiency of over six times their electromagnetic counterpart. Being lightweight, economical, and easy to use when driving, they are an excellent candidate for this type of system. Higher driving impedance versus frequency is the inverse of the electromagnetic type.

The system, as described, contains several modes of operation. The first mode is a steady state of a certain frequency determined by the operator. This mode is termed *manual frequency control* and is adjusted by its appropriate control. The next mode is *low speed sweep*, which consists of the unit starting at a low frequency and automatically increasing to its higher limit where it again repeats itself. The remaining modes are controlled rates of this sweep, determined by the appropriate control referred to as *sweep rate control*. It is these controls that can be remotely operated from the inside of a vehicle when high-powered, outside arrays in clusters are used. As was mentioned before, certain frequencies combined with certain sweep rates can cause different degrees of effectiveness. It is not the objective of this information to analyze the potential types of effect and behavior versus control setting, but to describe the working system as a generator of these controlled sounds.

Table 35-1 Pain-Field Generator Parts List

Ref. #	Description	DB Part #
R1, 6, 8, 12	Four 1K, ¼-watt resistors (br-blk-red)	
R2/S2	500K, 1 meg pot/switch, 17 mm	
R3	2.2K, ¼-watt resistor (red-red-red)	
R4, 5	Two 10K, ¼-watt resistors (br-blk or)	
R7	10-ohm, ¼-watt resistor (br-blk-blk)	
R9/S3	10K pot/switch, 17 mm	
R10	5K horizontal trimpot	
R13–16	Four 4.7-ohm, 3-watt resistors (yel-pur-sil-gold)	
R17, 18, 19	Nine 120-ohm, 5-watt resistors (br-red-br); use two per transducer and one across S1	
C1, 4	Two .01-microfarad, 50-volt disk capacitors (103)	
C3	100-microfarad, 25-volt vertical electrolytic capacitor	
C2	10-microfarad, 25-volt vertical electrolytic capacitor	
C5A, B	Two .01-microfarad, 50-volt polyester capacitor	
C6	1,000-microfarad, 25-volt vertical electrolytic capacitor	
C8	.01-microfarad, 2-kilovolt disk capacitor	
C9	.22-microfarad, 250-volt polypropylene capacitor	
L2, 3	Eight 1 mh inductors (6080), 2 used per transducer, in place of R15 and R16 as marked on PC board	DB# 1MH
L1	Inductor: two Hitachi 30.48 E cores and mating bobbin assembled and wound L1 inductor	DB# PPPL1L1
Q1	PN2907 PNP GP transistor	
Q2	IRF530 or 540 N-channel metal-oxide-semiconductor field effect transistor (MOSFET)	
IC1, 2	Two 555 dual inline package (DIP) timers	
HS1	Heatsink bracket, created as shown	
SW1	6-32 × ¼-inch screw	
SW2	Eight 6-32 × ½-inch screw	
SW3	Eight #8 × ¼-inch sheet metal screw	
NU2	Nine 6-32 hex nuts	
FH1/FS1	Fuse holder and 2-amp fuse	
J1	Three-pin din chassis mount jack	
J2	DC 2.5 mm jack panel mounting	
CL1, 2	Two battery 9-volt clips	
BH1, 2	Two four-C-Cell holders	
S1	Single pole, single throw (SPDT) power switch	
S4	Small SPST toggle switch	
PC1	Printed circuit board (PCB) or use perforated vector board	DB# PCSONIC
TD1	Four polarized 130 db piezo transducer	DB# MOTRAN
WR1/2	36 inches of #20 vinyl hookup wire, red and black	
CHASSISBASE	Create as shown in Figure 35-6	
PANEL	Create as shown in Figure 35-6	
TAPE	8 inches of two-sided 1-inch foam tape for securing battery holders to base	
SPACER	3 × 3-inch piece of plastic sheet to insulate assembly board from metal BASE	
LABULTRA	Label, ultrasonic warning	

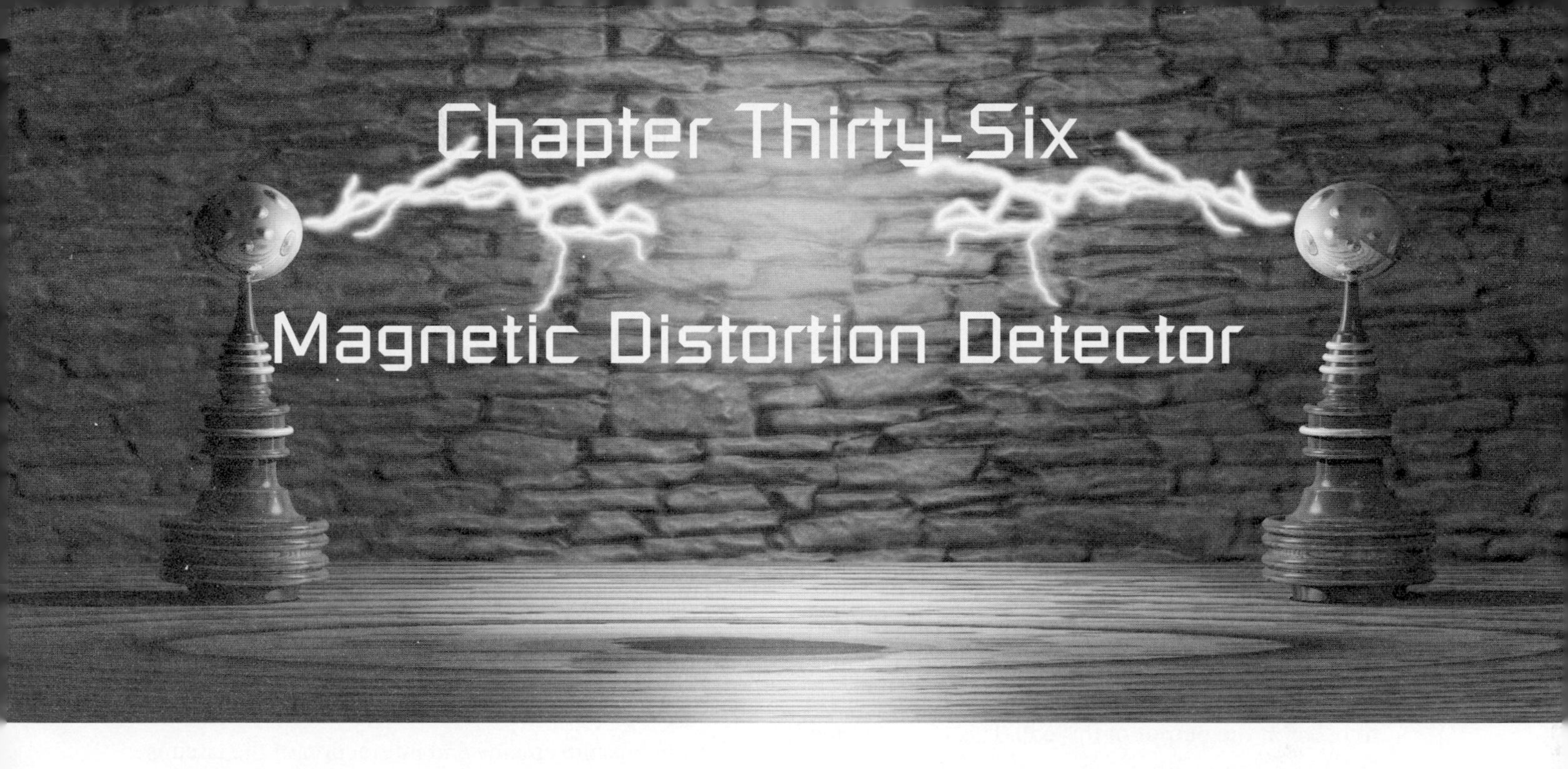

Chapter Thirty-Six

Magnetic Distortion Detector

The following project, as shown in Figure 36-1, describes an excellent, low-cost device that has amazing sensitivity in its ability to detect magnetic disturbances down into the microgauss. A microgauss is one millionth of a gauss. The earth's magnetic field is 500 milligauss where a milligauss is one thousandth of a gauss. The unit easily detects sunspot activity and auroras, and it can even be used as a vibration sensor that can detect a person's finger tapping on a rock. It can also detect moving aircraft and vehicles, as well as other changing electric and magnetic phenomena.

Actual detection is accomplished via an analog meter that responds to the relative strength of the changing magnetic fields. An audible alarm also allows the presetting of an activation level where it will alert the user of certain activity. An output port is included for connecting to a chart recorder.

The magnetic fields produced by many devices, both manmade and natural, are detectable by this system. Interesting results are produced when this unit is placed near anything electrical or magnetic such as a car, motor, or TV. The sensitivity is impressive when one sees the results of a 1-amp pulse of electrical current through a wire located a meter away from the unit. (This corresponds to a milligauss. Earth's magnetic field is .5 gauss). When the system is properly installed, it will detect passing automobiles, aircraft, or any moving, ferrous source that will cause the terrestrial magnetic field to distort. The unit is an excellent science fair project and can supply the amateur scientist with hours of knowledgeable entertainment in listening and detecting magnetic fields and their abnormalities.

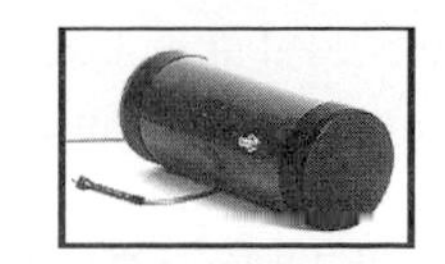

Figure 36-1 *The magnetic distortion detector*

Construction is shown to be as flexible as possible. It is built in two sections: the sensing head and the control box, which are interconnected via a length of shielded cable. The maximum usable sensitivity requires the sensing head to be placed away from power lines and other potential disturbances. The sensing head, which must be securely mounted to prevent interference from mechanical movement due to wind and vibration, can cause erroneous signals as it shifts position through the earth's magnetic field. The unit can also be made as a sensitive magnetic probe by installing the sensing head and the control box together, bearing in mind that the usable sensitivity will be decreased due to erroneous movement near magnetic materials when moving its position. The detection of field changes produces a moderately loud alarm along with a relative indication from an analog meter located on the panel. An optional jack is used for an X/Y chart recorder, and the system is fully battery powered.

This is an intermediate to advanced-level project requiring basic electronic skills. Expect to spend $45 to $65. All parts are readily available, with specialized parts obtainable through Information Unlimited at www.amazing1.com, and they are listed in the parts list at the end of the chapter.

Circuit Description

Figure 36-2 shows a coil (L1) consisting of many thousands of turns is wound on a 2-inch bobbin and placed over a soft iron core of about 1 to 2 feet in length. Any magnetic changes induce small voltages to occur at the output of this coil. This output signal is fed into a nano-amp, sensitive DC amplifier (I1), providing a current gain of 500 times the original. The iron core of L1 serves to concentrate the magnetic flux lines by offering a lower reluctance path than that of air. The input of I1 is protected by overvoltage and transients via clipping diodes Dl and D2. The output of I1 is forced to near zero via a multiturn trimpot (R4) for balancing and compensating any offset.

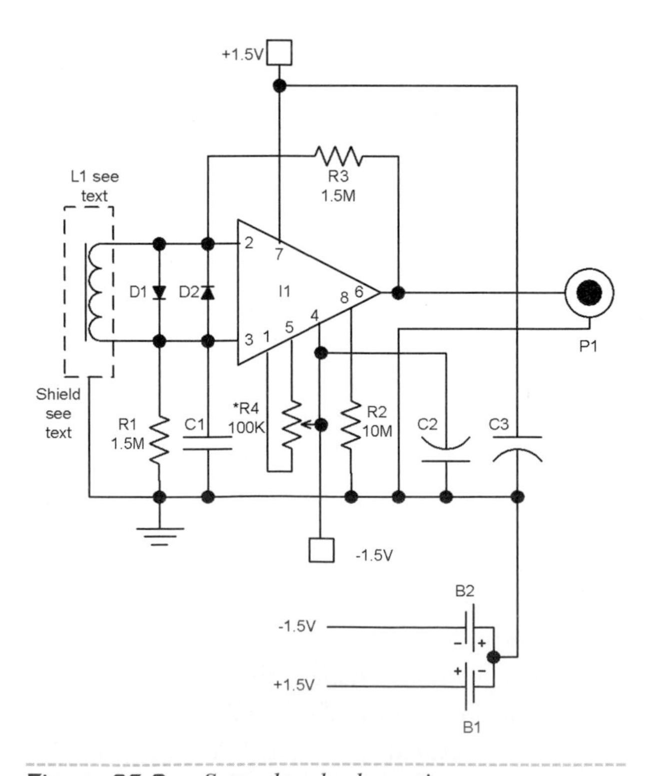

Figure 36-2 *Sense head schematic*

Power for I1 consists of batteries Bl and B2, which are left in the circuit permanently as current drain is very low and replacement can be on a yearly basis. It should be noted that if the sensing head is to be located in a hard-to-get area the builder might use D-size cells, as these will last that much longer.

The sense head output is fed to the control box via any reasonable length of shielded cable, WR10. As mentioned before, the sensing head should be located away from any potential magnetic interfering devices, such as anything electrical and large, ferrous objects.

Aluminum foil can be wrapped around the sense coil and is connected to the circuit common line. This reduces the capacitive pickup of unwanted signals.

Again, the sensing head and control box may be assembled together and used as a probe. This configuration may reduce the system's sensitivity due to erroneous movements and interaction of the circuits such as magnetic speakers and meters. This method is not encouraged due to these erroneous readings.

Figure 36-3 shows the control box providing another amplifier, I2, with a gain of 200 times the original amount, which gives the system an overall gain of 100,000. Full-scale meter deflection of the 50 ua meter (M1) occurs when a current of .5 nano-amp is obtained from L1. It is obvious that with the many thousands of turns on L1 that a detectable signal can occur with very minute flux changes, such as the result of changes in the microgausses. It should be noted that the earth's magnetic field is only .5 gauss.

I1 is connected as a conventional DC operational amplifier with its output DC coupled to I2 via a shielded cable. This cable is necessary to separate the sense head and the control box by at least 6 feet. This is necessary to prevent magnetic coupling from the speaker SPK1. Sensitivity control pot R5 controls the system gain and is usually set at maximum.

You will note that meter deflection pot R11 also is ganged with switch S1, which controls the +6-volt power to the control box section. R11 controls the deflection of meter M1 with diode D3 and capacitor C6, serving to smooth out and average the signal. The voltage level at jack J2 is also controlled by R11 and is an auxiliary jack intended for a chart or other recorder. The audible alarm activation point is controlled by pot R12 that is ganged with switch S2.

Switch S2 controls the −6 volt power to the control box.

The alarm is nothing more than a reflex-type oscillator consisting of transistors Q2 and Q3 with its feedback paths consisting of resistor R15 and capacitor C8. Trimpot resistor R16 is set for a reliable, clean-sounding signal.

Assembly Steps

1. Figure 36-4 shows coil L1 wound with thousands of turns of fine magnet wire. This may be difficult to do and it is suggested that the builder locate a high-voltage oil burner ignition, neon sign, bug killer, or other similar transformer and carefully remove the high-voltage secondary winding. The parameters of these transformers are usually around 5,000 volts at 10 to 20 milliamps. These are available from junkyards and other places. Many are rejected due to radio and TV interference and yet have intact secondary winding suitable for this project.

   The objective is to obtain one with as many turns as possible; however, most will work quite well. A quick test is to measure the DC resistance of between 5 and 20K across the coil. A relay coil of high impedance such as 5 to 10K may also work. The coil must have connecting leads and these should be secured with *room temperature vulcanizing* (RTV) adhesive, wax, or something similar to prevent breaking that may render the coil useless. A soft iron core of 1 to 2 feet is inserted through the coil and serves to offer a lower reluctance to the magnetic flux lines. It is these lines that, as they change in the core, induce a voltage due to magnetic induction. Note to shield the coil from electrical and capacitive pickup by wrapping it in aluminum foil and connecting the foil to the circuit ground. A special custom-made coil is available from Information Unlimited and is noted in the parts list.

2. Lay out and identify all the parts and pieces, and check them with the parts list. Note that some parts may sometimes vary in value. This is acceptable as all components are 10 to 20 percent tolerant unless otherwise noted.

3. Assemble the board by inserting components into the board holes as shown in Figure 36-5.

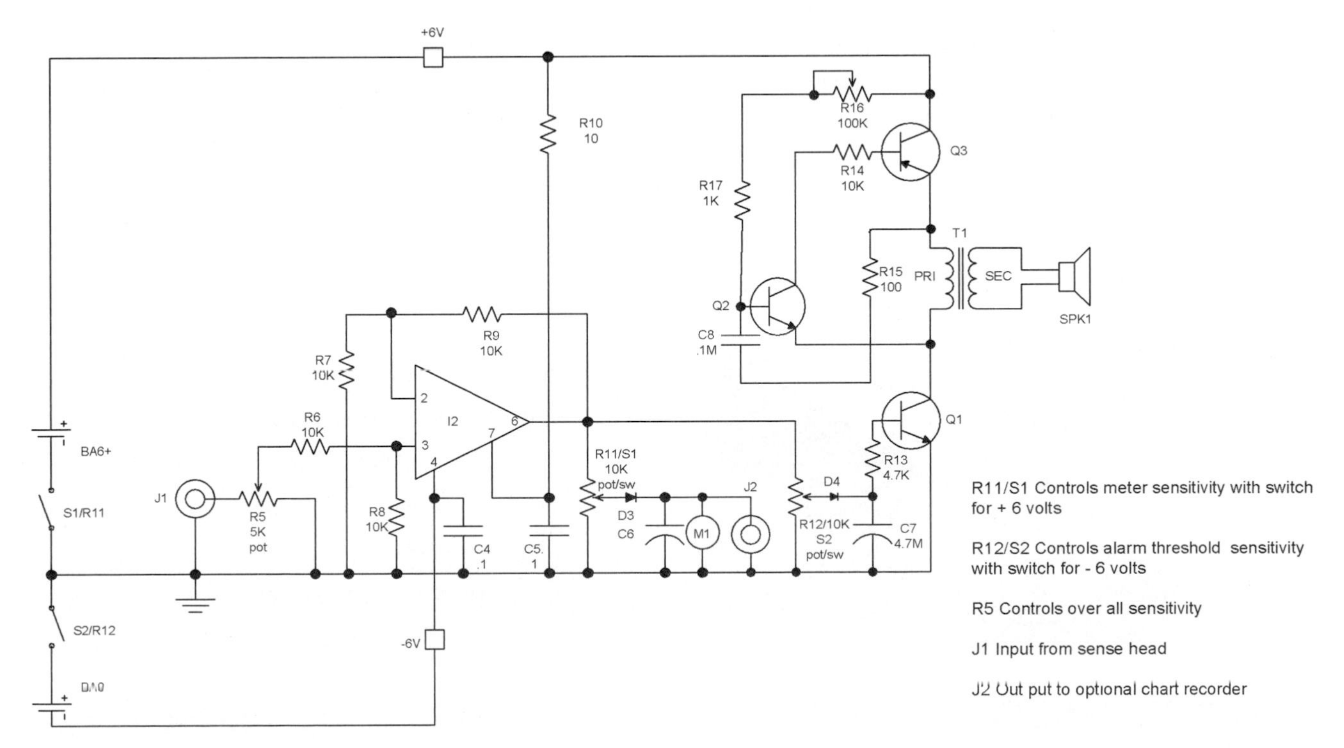

Figure 36-3 *Control box schematic*

Proceed from left to right, attempting to obtain the layout. Note the polarity of the components as indicated. Certain leads of the actual components are to be used for connecting points and circuit runs. Do not cut or trim them at this time. It is best to temporarily fold the leads over to secure the individual parts from falling out of the board holes for now. Solder as required and clip off excess leads. Some kits contain a length of insulated wire. This must be cut, stripped, and tinned according to where it is used.

4. Wire in the 6-foot length of shielded cable, being careful not to burn through the insulation and short out the cable.
5. Attach the battery clip HOLD1 using silicon rubber.
6. Attach the L1 coil to the PB1 perforated board as shown in Figure 36-5 and secure it with RTV adhesive. You will have to drill a hole for the iron core. The core should fit snuggly and can be glued in place in the final assembly.

Pretesting the Sense Head

1. Carefully check for wire and solder errors, and install fresh 1.5-volt AA batteries (B1 and B2). Connect a low-range voltmeter to the leads of the output cable and adjust the 25-turn pot (R4) for as close to a near-zero reading as possible. Finally, adjust the lowest range of the meter, preferably the 50 microamp range. This should be done outside away from power lines.
2. With the meter still connected, observe a reading when a piece of iron is moved near sensing coil L1 and moved about. If the piece of iron contains any magnetism, the effect will be enhanced.
3. Wrap the sense coil with several layers of aluminum foil and ground the foil to the negative circuit. Diodes D1 and D2 will require shading against light sources as the junction behaves like photodiodes at these low signal levels.
4. Create enclosure EN1, as shown in Figure 36-6, from an $8 \times 3^{1}/_{2}$-inch outer-diameter piece of plastic tubing.

   Note the small, $^{3}/_{8}$-inch access holes for adjusting R4 are drilled when all is in place. Create the brackets as required for mounting the sense head.
5. Install the assembly into ENl with caps CPl and CP2 as shown. Recheck it by connecting the meter across P1. Readjust R4 through the access hole if needed. The sense head should now be working correctly.

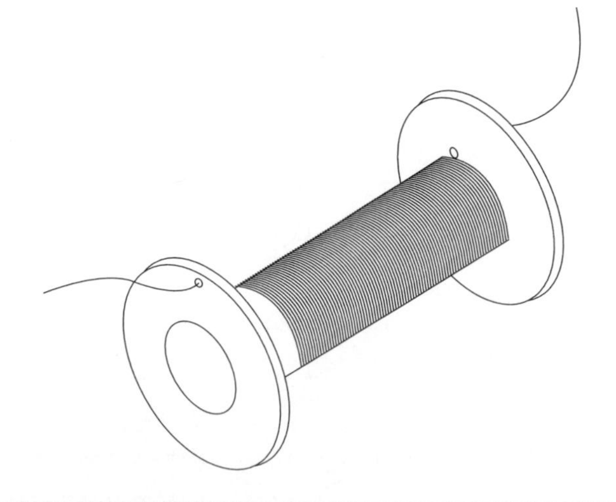

Figure 36-4 *Winding instructions for L1*

Strip off outer plastic jacket of cable 1 inch and twist braid into a lead and tin. Strip off 1/4 inch of insulation on inner lead and tin. Insert into plug and solder inner lead to lin tip.

Carefully solder twisted braid lead to outer shell being careful not to burn through insulation. Always verify lead continuity and possible shorting of the leads with a meter.

Dashed lines are connections on underside of perforated circuit board using the component leads. These points can also be used to determine foil runs for those who wish to fabricate a printed circuit board.

• Small dots are holes used for component insertion.

⊙ Circles with dots are holes that may require drilling for component mounting and points for external connecting wires, indicating strain relief points. It is suggested to drill clearance holes for the leads and solder to points beneath board.

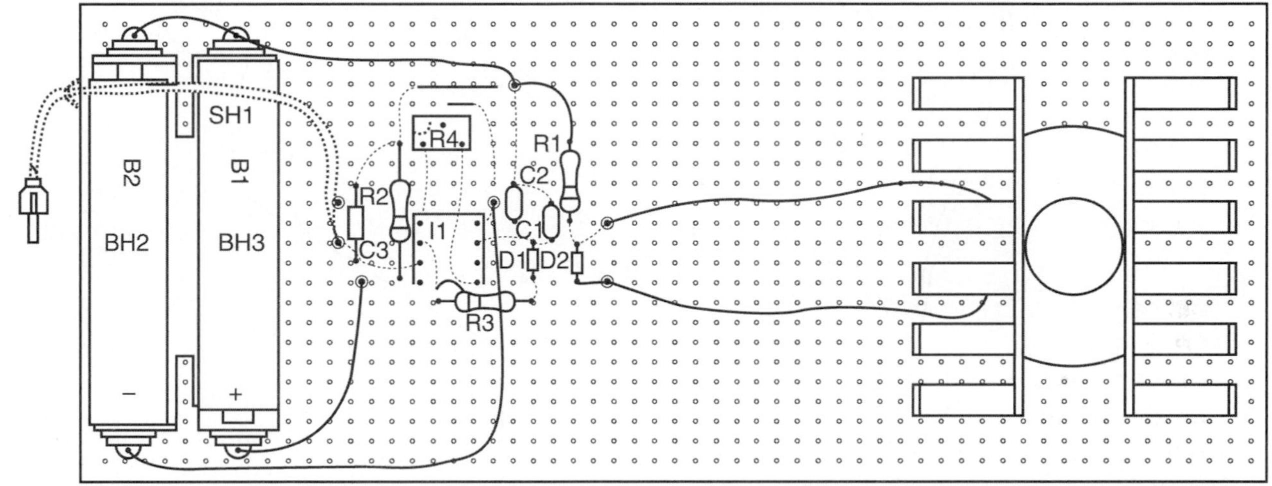

Note: The sense head as shown in this sketch is wound from a transformer bobbin. Any plastic bobbin 2"x1" will usually suffice. See Figure 36-4.

Figure 36-5 *Assembly of the sense board section*

Construction of the Control Box

1. Assemble the board as shown in Figure 36-7. Observe the proper position of I1 and the polarity of the diodes, transistors, and C6 capacitor. Note that Q3 is a PNP PN2907 transistor. Note the position of Tl regarding the primary and secondary winding. The secondary will read very low resistance when compared to the primary. It is a good idea to use a socket for the integrated circuit.
2. Create the enclosure base section (BASE) from sheet aluminum, as shown in Figure 36-8, with holes for meter Ml, speaker SPK1, R11/S1, R12/S2, R5, and holes for J1 and J2. Assemble the components as shown. Note the battery holders secured via two-sided tape, TAl.
3. Attach the interconnecting leads, as shown in Figure 36-9. Twist the leads whenever possible and keep them short and direct. Note the ground lugs on J1 and J2.
4. Connect the leads from the board assembly to the components in the box.

Control Box Pretesting

1. Position the board, but do not secure it at this point. Turn S1 and S2 off and insert four AA batteries into BH2 and BH3.
2. Turn on both switches, unsnap one side of the battery clip, and insert a milliamp meter. Measure .5 milliamps (−6 volts) at BH2 and .8 milliamps (+6 volts) at BH3. Reconnect the battery clips.
3. Turn the R16 trimpot to midrange, R11 full counterclockwise, R12 clockwise, and note the meter remaining at zero and the alarm

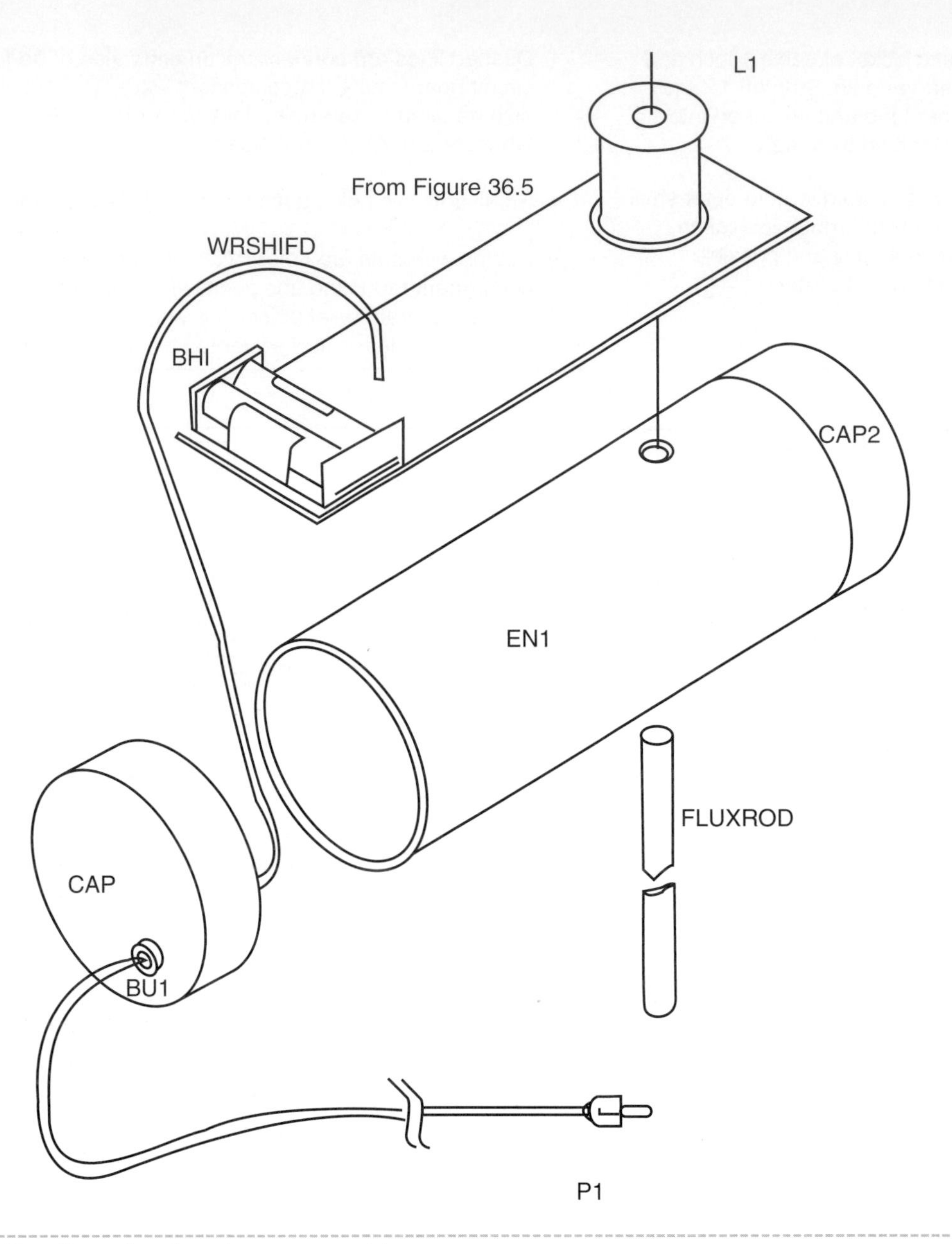

Figure 36-6 *Sense head enclosure assembly*

remaining silent. Now turn R11 clockwise and note the meter deflecting to full scale.

The following steps will require the sense head located away from electrical wiring and other potential sources of magnetic interference. The distance should be at least 5 feet from the control box to avoid feedback from the meter and speaker magnets.

4. Turn both R11 and R12 full clockwise. Plug in P1 from the sense head J2. Note the meter may pin, or jump off scale, and the alarm may sound. Have a friend slowly rotate R4 on the sense head to attempt to bring the meter into range. You may have to desensitize the circuit by turning R11 counterclockwise until the meter reads midrange and turn R12 to deactivate the alarm as it can be annoying during testing. Note that R5 is an overall sensitivity control and should be turned fully on in most applications for maximum sensitivity. The meter and alarm can be ranged by their respective controls, R11 and R12.

When properly set, the reading will constantly fluctuate and is the result of sunspots or other phenomena distorting the earth's magnetic field.

Dashed lines are connections on underside of perforated circuit board using the component leads. These points can also be used to determine foil runs for those who wish to fabricate a printed circuit board.

- Small dots are holes used for component insertion.

⊙ Circles with dots are holes that may require drilling for component mounting and points for external connecting wires indicating strain relief points. It is suggested to drill clearance holes for the leads and solder to points beneath board.

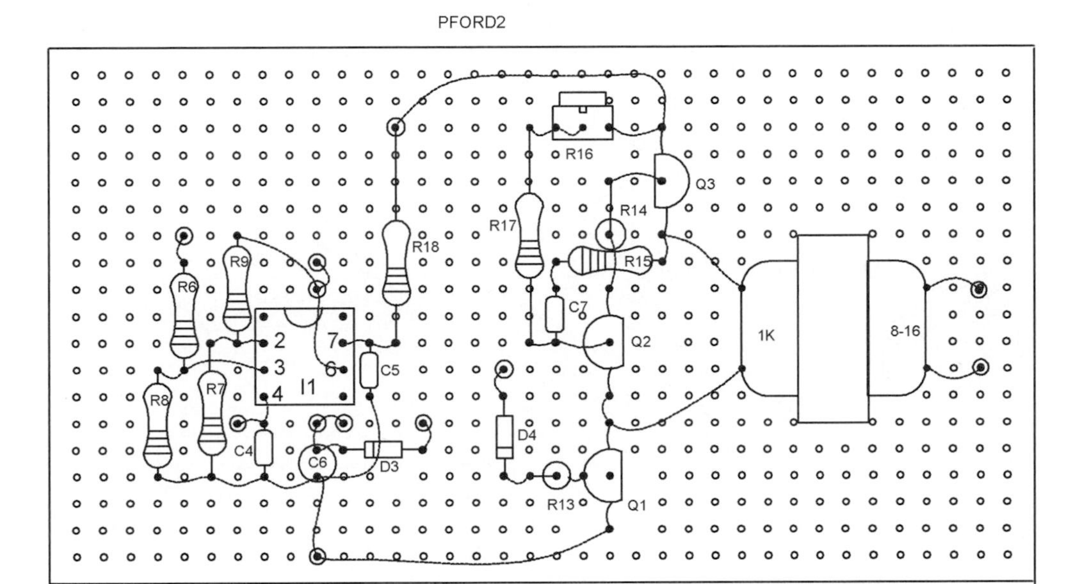

Figure 36-7 *Assembly control board*

If you choose a magnetically noisy location, you will have an added signal as the result of power lines, motors, and other magnetic disturbance. Note that if R11 cannot bring the meter into range without overly desensitizing it, it may be necessary to internally readjust R4 for proper ranging.

5. Preset R14 at midrange and adjust R12 until the alarm sounds. Readjust R16 for a clean, desirable-sounding tone if necessary.

Appreciating the sensitivity of this unit requires operation in a magnetically quiet location where all the controls can be advanced fully. Under the conditions of a low magnetic background, the unit can detect automobiles, manmade devices, and natural magnetic disturbances. Also, simply moving a piece of iron near the sensing head should produce an indication. (Note low-level-detection sensitivity may be difficult during the peak of the sunspot cycle due to the high background signal fluctuations.)

Jack J2 may be used to drive an XY recorder when observing sunspots or other time-related disturbances.

Note: No steps were taken to filter out 60 HZ disturbances, as it appears that many uses of this device involve the detection of power lines and other related equipment.

Applications

The uses of this device are obviously only appreciated by those who can understand its sensitivity to a minute change in magnetic flux. The unit can be made to detect the motion of large steel bodies as they distort the existing earth's magnetic field. Many applications and experiments are therefore possible with the device. Any high-impedance headphones

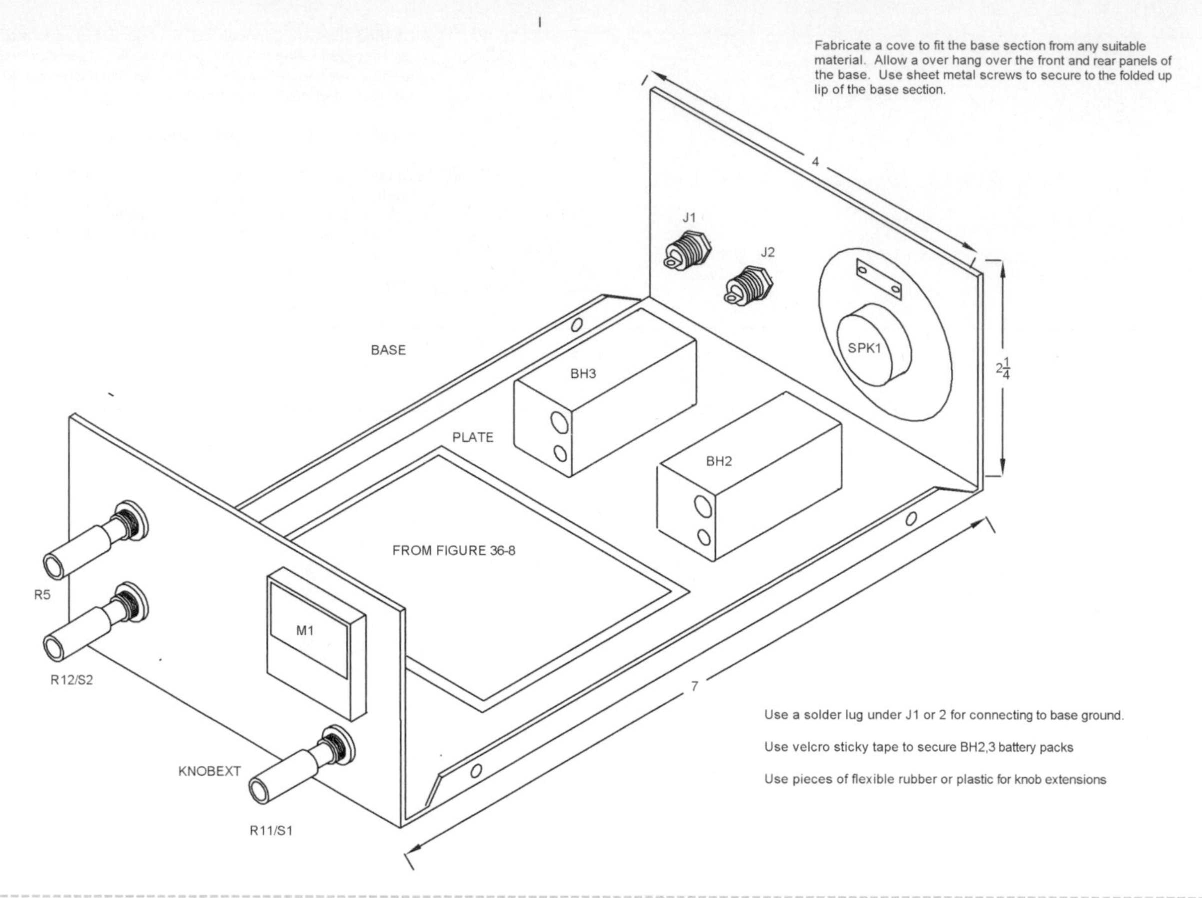

Figure 36-8 *Control box base section fabrication and layout*

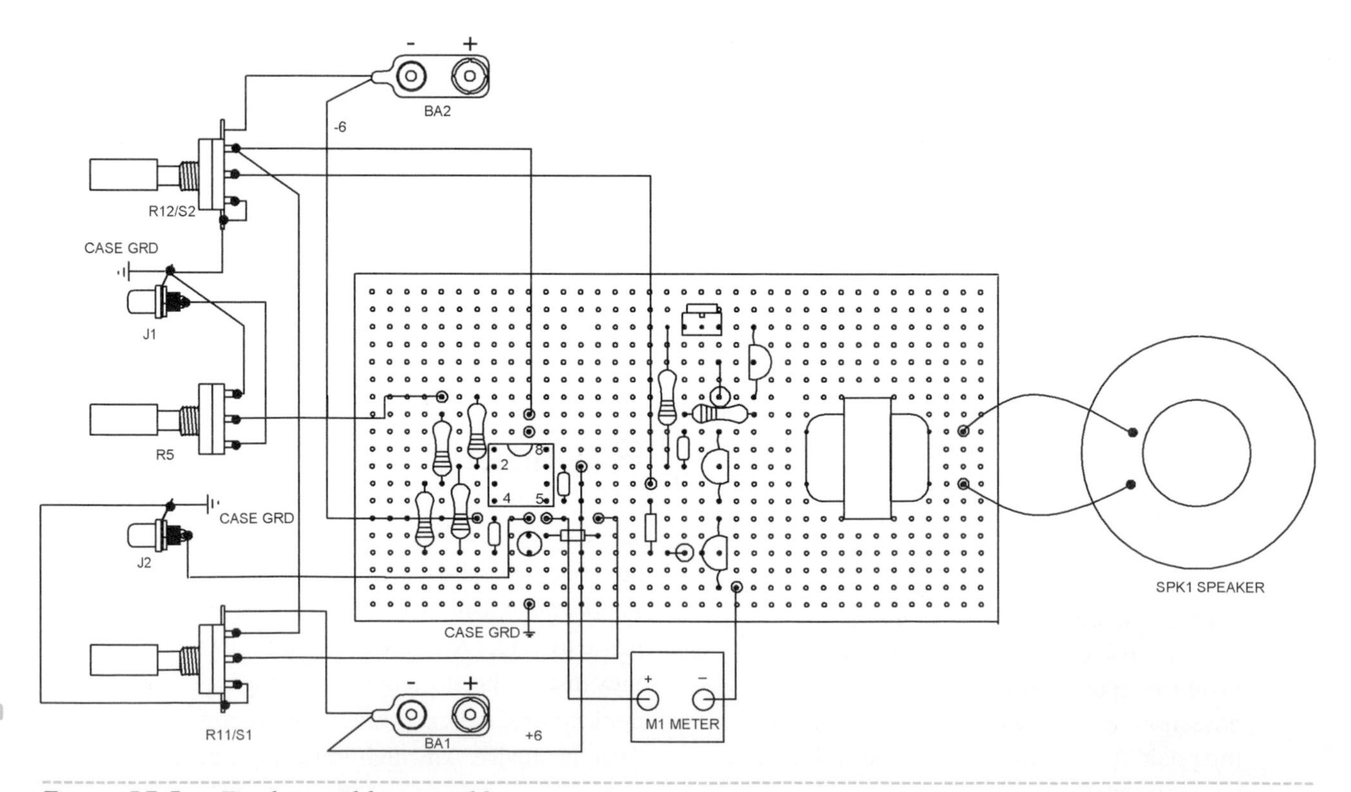

Figure 36-9 *Final assembly control box*

can be plugged into J2 where the experimenter can actually "listen in" and hear the field changing.

One verifying test that is easily done to determine the system's ability to detect the earth's magnetic field is the following. Rotate the sensing head with short, quick jerks in a clockwise direction. Note the meter pinning, and as rotation is continued a point will be found where the meter will respond less. This nulling point should occur with the fluxrod pointed in a north/south direction. If rotation is continued, it will be found that the short, quick jerks must occur in a counterclockwise direction to obtain a positive meter movement. This is similar to an electrical generator using the earth as its field magnet and the sensing head core as the moving armature.

Optional Ultra-Sensitive Vibration and Tremor Detector

You may easily convert this device to a vibration detector by simply modifying the sense head so it can be mounted to a rigid object such as a rock. The trick now is to place a magnet adjacent to the end of the sense coil, L1. The magnet should be held in place by elastic or other means whereby it is mechanically isolated relative to the sense coil. It should be very close to the sense coil for optimum sensitivity.

The theory is simple. Positioning a powerful magnet near a coil with many turns produces a signal. It is now the relative motion between this magnet and the coil that produces a voltage by the familiar equation $E = Nd0/dt$, where E is volts, N is the coil turns, 0 theta is the flux of the magnet, and t is the rate of relative motion in seconds. The combination of the high-resultant signal volts and the amplification sensitivity of the circuit provides amazing results.

Many variations are possible with the common objective being to obtain the highest reading for a given tremor. This unit will easily detect earthquakes occurring anywhere on the planet; however, the builder must experiment with motion dampening and other schemes to differentiate between manmade vibrations in order to isolate these signals. This is an excellent, low-cost project for any interested individual to learn by experimentation.

Table 36-1 Magnetic Distortion Detector Parts List

Ref. #	Description	DB Part #
R1, 3	Two 1.5M, ¼-watt resistors (br-gr-gr)	
R2	Two 10M, ¼-watt resistors (br-blk-bl)	
R4	100-kilo-ohm, 10-turn vertical trimmer	
R5	5K linear pot	
R6, 7, 8, 9, 14	Five 10K, ¼-watt resistor (br-blk-or)	
R10	10-ohm, ¼-watt resistor (br-blk-blk)	
R11/S1, 12/S2	Two 10K linear pots and switches	
R13	4.7K, ¼-watt resistor (yel-pur-red)	
R15	100-ohm ¼-watt resistor (br-blk-br)	
R16	100-kilo-ohm vertical trimmer	
R17	1-kilo-ohm, ¼-watt resistor (br-blk-red)	
C1, 2, 3, 4, 5, 8	Six .1-microfarad, 50-volt plastic capacitors	
C6	1-microfarad, 50-volt vertical electrolytic capacitor	
C7	4.7-microfarad, 50-volt vertical electrolytic capacitor	
I1	LM4250 nano-amp operational amplifier in a dual inline package (DIP)	
I2	LM741 operational differential amplifier in DIP	
D1, 2, 3, 4	Four IN914 small-signal diodes	
Q1, 2	Two PN2222 NPN GP silicon transistors	
Q3	PN2907 PNP GP silicon transistor	
TI	1K to 8-ohm, small audio transformer	
M1	50 uamp panel meter	DB# METER50S
SPK1	2-inch, 8- to 16-ohm speaker	
J1, 2	Two RCA panel-mount phono jacks	
P1, 2	Two RCA mating phono plugs	
CL1, 2	Two 9-volt battery clips	
SH1	3 feet of single-conductor, shielded mike cable	
WIRE24	4 feet of #24 vinyl stranded hookup wire	
BH1	Two-AA-battery holder	
BH2, 3	Two 4-AA-battery holders	
BU1	Small, plastic feed-through bushing	
L1SENSE	Sense coil per Figure 36-4	DB# SENSE
PBORD1	6¼ × 2½-inch, .1 × .1-inch grid perforated vector board	
PBORD2	4½ × 2-inch, .1 × .1-inch grid perforated vector board	
EN1	7½ × 3-inch schedule 40 PVC tubing	
CAP1	Two 3½-inch black plastic caps	
TAP1	4 inches of 4 × 1-inch strips of Velcro sticky tape	
KNOBEXT	4 inches of 3⁄16-inch inner-diameter plastic or rubber tubing for shafts	
FLUXROD	6 inches of ¼ × 20-inch steel-threaded rod and two mating nuts	
BASE	Created as shown in Figure 36-8 from .063 aluminum	
COVER	Cut to fit BASE section	
PLATE	5 × 2½-inch piece of insulating plastic to mount under board	

Chapter Thirty-Seven

Body Heat Detector

This project, as shown in Figure 37-1, shows how to construct a useful device that can detect human or animal motion up to several hundred meters. It provides a narrow field of view and can be used for paths and walkways. Farmers use these devices to detect when a predator comes into range, taking the necessary action when the alarm sounds. It may also be used as a medium-sensitive thermal detector for lab and research use. The device is an excellent science project demonstrating medium-sensitivity detection of relatively low temperatures.

This is an intermediate-level project requiring basic electronic skills. Expect to spend $30 to $45. All the parts are readily available, with specialized parts obtainable through Information Unlimited at www.amazing1.com. These are listed in the parts list at the end of the chapter.

Figure 37-1 *Body heat detector*

Description

Your body heat detector is a device capable of sensing an object at a temperature increase slightly over ambient. It is this property that allows it to detect a living body at a considerable distance. The sensitivity of the device corresponds to a black body wavelength of 10 microns, corresponding to a temperature of 100° Fahrenheit. Black-body radiation defines color temperature in wavelengths below what is visible to the naked eye. All objects above absolute zero radiate energy as a function of their temperature to the fourth power. This radiation decreases in wavelength as the radiating body becomes hotter. This radiation will consist of the visible color spectrum for a given range of temperatures. Hotter bodies will show visible color such as blue being on the hotter side of red. A body hot enough to glow red is around 900 degrees F. Black bodies radiate in the infrared (invisible) and lower as the generating temperatures are near ambient being above 70 degrees F. All objects above absolute zero radiate energy as a function of the fourth power of temperature, and body heat energy is in the infrared region at around 100° degrees F.

The applications for this device are numerous and as an example you could a create motion detection system for home security. These are normally short-range, wide-angle detection systems and are limited in performance. Our system would be mounted similar to an optical scope on a rifle or a tripod. A living target now can easily be ferreted out in darkness or under moderate cover due to heat reflections. A predator, as it approaches its target victim, is easily detected, alerting the farmer or rancher to take the next step of action. Garden pests, as they emerge from their burrow, and downed game can easily be detected by the device.

The ability to detect hot spots is an extremely useful function. An example of sensitivity can easily be demonstrated by the following. Obtain a piece of cardboard or plastic approximately 6 × 12 inches. Place your hand on one end of this piece for several seconds to produce a hot spot of your hand print. Then place the piece in front of the unit, noting a detected signal increase on the surface previously touched by your hand due to the heat stored. (Don't touch the other end of the piece). Note that as time continues, the hot spot dissipates into the object thereby reducing the temperature differential and the signal pickup on the cardboard piece becomes evenly distributed. The differences in temperature detection capability is in the millidegrees. You can now realize the many useful applications of this unit.

Body Heat Detector Modes

Your body heat detector operates in two modes:

- **MODE A:** This mode is used to detect moving bodies such as animals, people, and so on. It is intended for sensing intrusions along pathways and into areas where a narrow field of view is necessary at long ranges (100 meters or so). This mode is the most stable of the two and rejects steady background sources of thermal energy. Detected bodies, however, must be moving.
- **MODE B:** Operation in this mode allows the detection of stationary hot spots and is most useful with a "cold background" such as at night. It is useful for ferreting out thermal sources such as hidden persons and animals. Hot object spots and areas are easily detected, but unfortunately this mode is so sensitive that often signals appear to be erroneous and mask the target object.

Both modes of operation require a camera tripod for reliable and simple operation. Very often a panning action can be used to reject or detect the desired targets.

Circuit Description

The heart of the detector is a pyroelectric detector (PYR1) capable of detecting very minute temperature changes. This device is responsive to temperatures of 100° Fahrenheit equivalent to approximately 10 microns. Unfortunately, its response speed is relatively slow, peaking at about .1 Hz. This is shown in Figure 37-2.

The pyroelectronic effect utilizes dual crystals of lithium tantalum oxide (LiTa 02). These crystals are polarized in an electric field below the Curie point. Electrodes attached to the crystal become electrically charged and eventually neutralize due to the surroundings. When the neutralized crystal is exposed to an infrared signal, its temperature changes, producing a charge on the electrodes and hence a detected signal. This signal is amplified by a built-in, low-noise *field effect transistor* (FET) and becomes the output. It is important to note that only a changing infrared signal can produce output; therefore, a chopper must be used for stationary sources. A chopper is nothing more than a mechanical wheel that has vanes to periodically interrupt the heat radiation signal to the pyrodetector, simulating a moving object. A moving source obviously produces a changing signal. The dual crystals will also cancel out signals produced from sunlight as they are oppositely polarized in the detector.

The output of the detector is connected to the input of a low-frequency, dual-stage amplifier and filter (U1) with a gain of approximately 2,500. This amplifier is designed to respond to a frequency of .1 to 10 Hz, peaking at about 1 Hz. This matches the response of the pyrodetector and the signal commonly generated by a living, moving body in the field of detection.

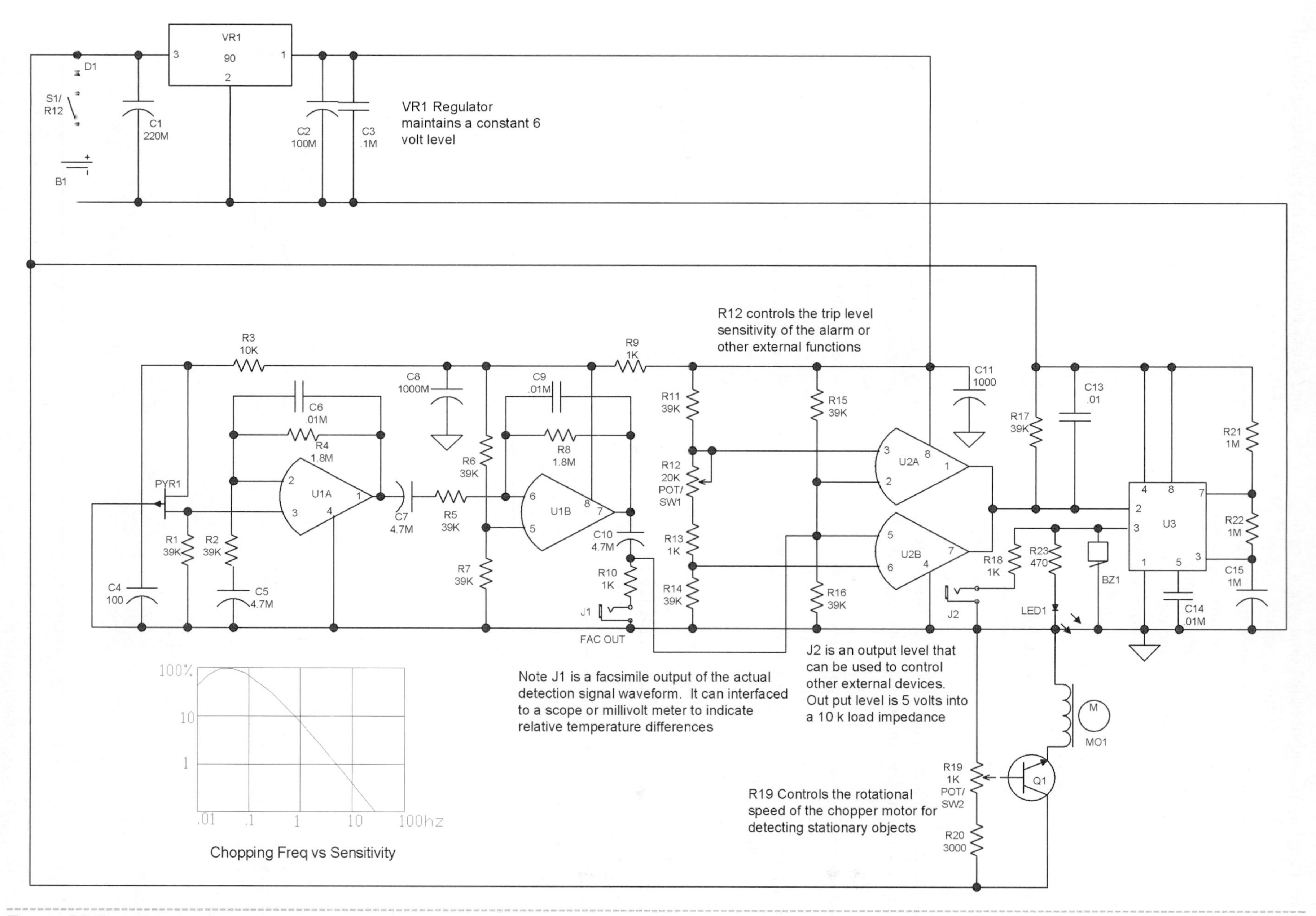

Figure 37-2 *Circuit schematic*

The changing AC output of the dual-signal amplifier (U1A and U1B) is capacitive coupled to output jack J1. It is this signal that is measured by an external AC millivolt meter and provides an analog indication of signal strength. This signal may be further processed for ultra-low-sensitive detection or maybe demodulated where the resulting DC level can control a VCO for audible detection via a changing tone. This AC signal is also fed to the dual plus and minus ("window") comparator (U2A and U2B). A threshold control (R12) presets the trip voltage or sensitivity of activation. The output of these two comparators is "anded" together and fed to the trigger input of the timer (U3). The output of U3 now activates audible buzzer BZ1 and visual indicator LED1. The buzzer on time is controlled by timing resistors R21 and R22 and capacitor C15.

An optional motor speed control for the chopper consists of Q1 and speed control pot R19. Jack J2 provides a 5-volt level through 1K resistor R18 and is used to trigger other systems.

Assembly Instructions

1. Lay out and identify all the parts and pieces, and verify them within the parts list at the end of the chapter.
2. Examine the PC board, as shown in Figure 37-3 and 37-4.
3. Proceed to assemble the device by inserting and soldering all the fixed resistors.
4. Insert and solder the capacitors, carefully noting the polarity on all the electrolytics: C1, C2, C4, C5, C7, C8, C10, C11, and C15. The other capacitors are not polarized.
5. Insert and solder in the semiconductors (D1 and VR1) and Q1. Note the polarity.
6. Identify and note the position marks on the integrated circuits U1, U2, and U3. Insert and carefully solder them, avoiding solder shorts between the pins. It is suggested to always use sockets for these components.
7. Identify the pots R12 and R9 and solder them in place. Note the switch section, S1, of R12. These connections are made using short pieces of cut leads to the pads directly beneath. Note these pots must be properly installed to allow final alignment with the PANEL plate.
8. Insert the wire jumps using short pieces of the cut leads of components that were used priorly.

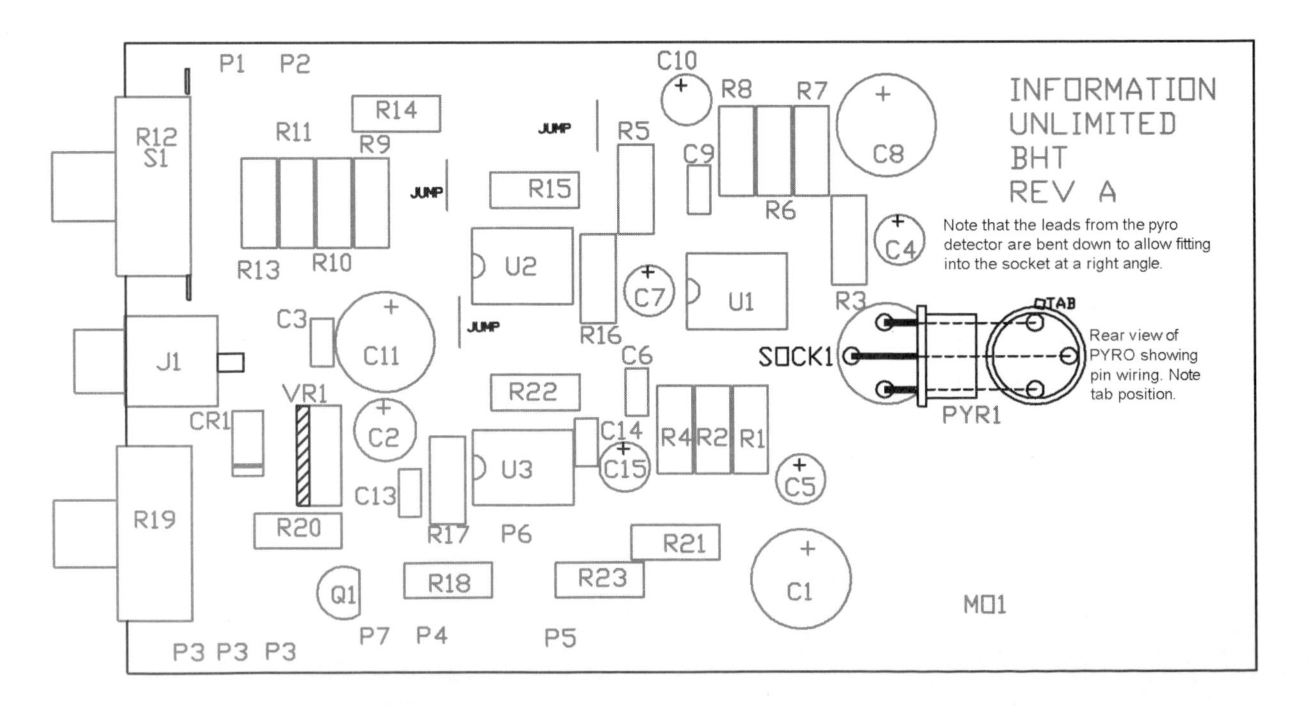

Figure 37-3 *PCB wiring and assembly*

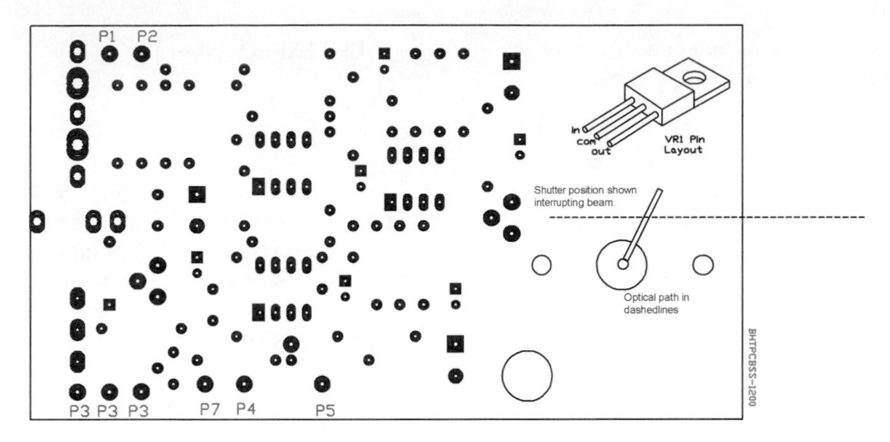

Figure 37-4 *Pad and connections*

9. Insert J1 and note the rear connection lead requiring a short piece of bare wire to connect to the board pad directly beneath.
10. Insert the SOCK1 pyrodetector socket as shown. Carefully form the three leads of the pyrodetector at 90°. Do not sharply bend it. Note the proper leads, and position the tab as shown. This part must be at the optical center of the system for proper operation.
11. Connect the BZ1 buzzer using its leads, as shown in Figure 37-5.
12. Connect output control jack J2 and indicator light-emitting diode (LED1) using a 3-inch length of #22 vinyl wire. Note the cathode lead of LED1 directly connecting to the ground lug of J2. Note these parts eventually will be mounted to the panel.
13. Connect the leads from the chopper motor MO1 and battery clip CL1 as shown.
14. Create the panel, as shown in Figure 37-6. Attach it to the PC board via the two pots and the jack. The panel and PC board should be flush to one another. Attach the remaining J2 and LED1 to the panel. Note to use a bushing for the LED.
15. Create the shutter flap (FLAP), as directed in Figure 37-6.
16. Recheck all wiring for shorts, accuracy, and correctness of parts. Also check for solder joints. You are now ready to bench test the assembly.
17. Connect a 9-volt supply to the battery clip.
18. Rotate R12 to on, and note the shutter flap rotating with R19 full counterclockwise. It should slow down as R12 is turned in a clockwise direction. This rotation should be variable in the range of 10 *revolutions per second* (rps), down to stalling at less than 1 rps. The motor will stall if there is any obstruction.

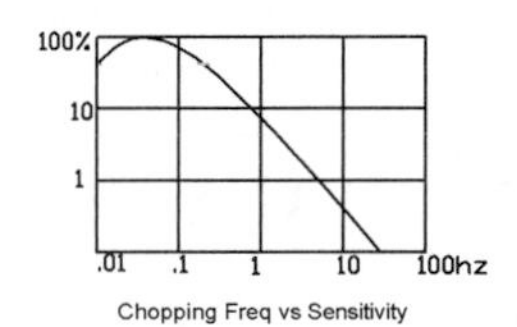

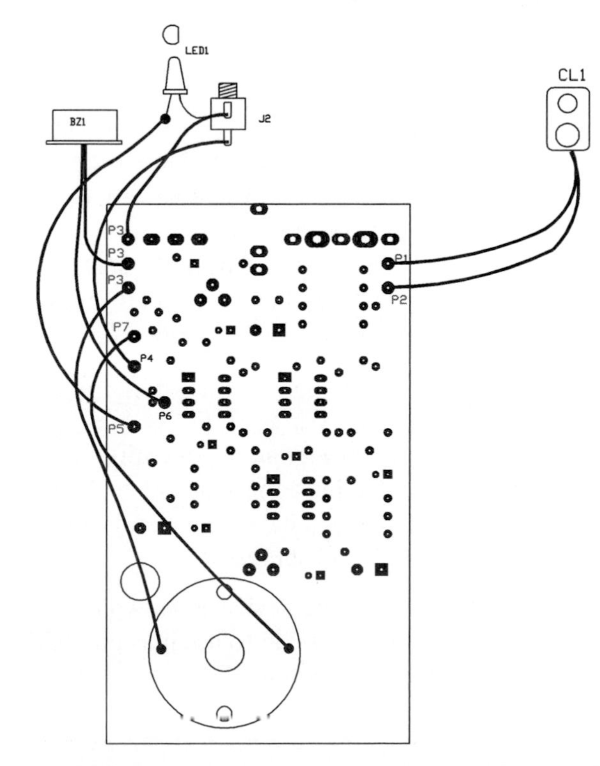

Figure 37-5 *Final wiring to external components*

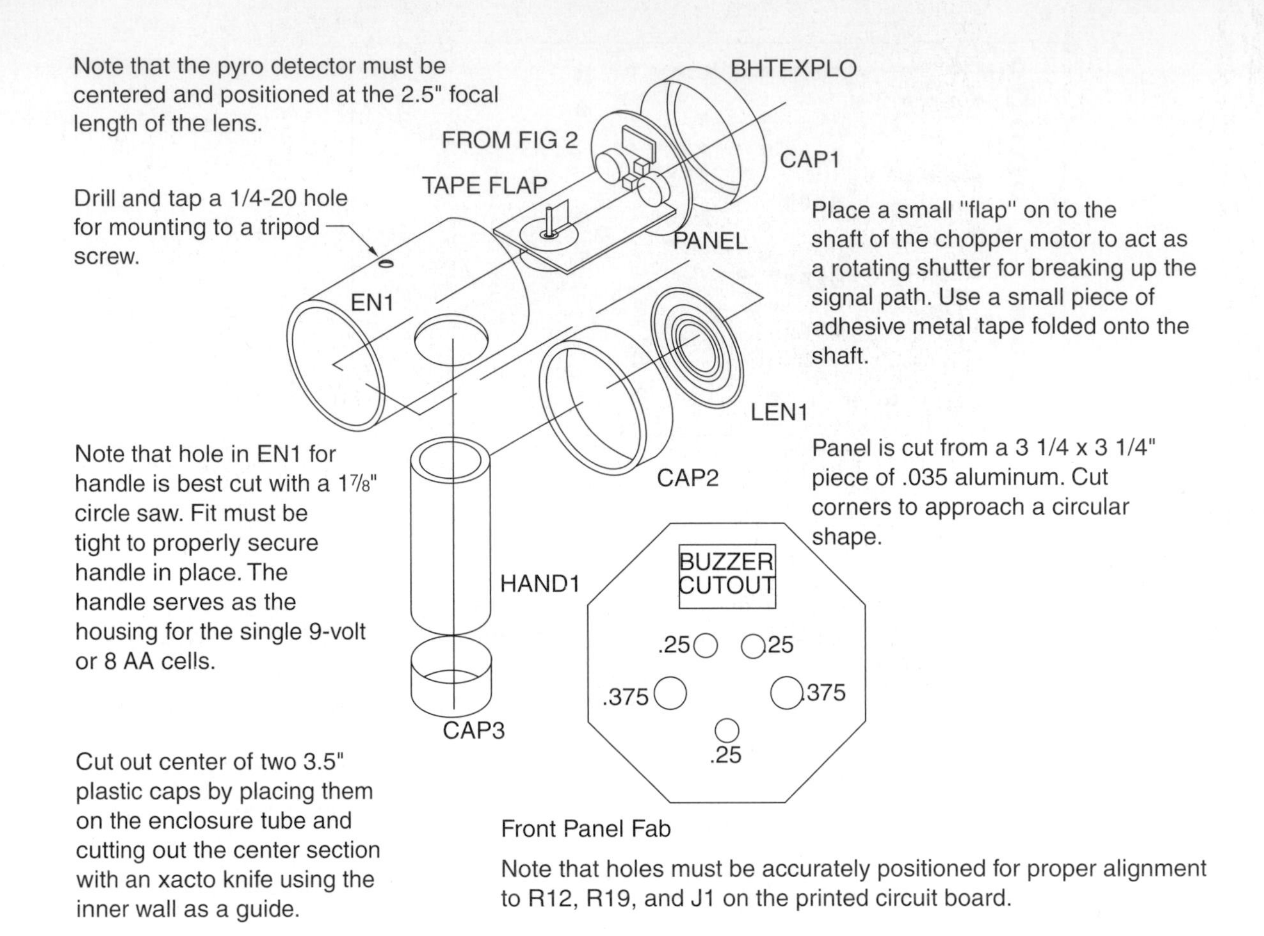

Figure 37-6 *Final housing setup*

Make sure the flap is clear and not touching the bearing bottom.

19. Connect a scope to J1 and note a changing level as your hand or another warm object is waved in view of the detector. Note the signal responding to a stationary source with the shutter flap or to some form of motion with the shutter flap stopped. Check the sensitivity control pot R12 and then verify a DC output level at J2, occurring when a body is detected. This function is for control of external alarms and alerts.
20. Check that the buzzer and light emitter activate when detecting a thermal source. Use your solder iron as a heat source and experiment with different sources.
21. Finally, assemble the device as shown in Figure 37-7 and proceed to use it as outlined in the instruction section.

Operating Instructions

1. Turn all controls fully counterclockwise to off.
2. Remove the battery compartment cover and slide out the eight-cell holder. Insert fresh batteries and reinstall.
3. Rotate the chopper control fully clockwise. View the shutter through the access hole and note how it is positioned in the figure.
4. Slowly turn the sensitivity control clockwise until the alarm sounds. If the unit does not respond, try moving your hand in front of it, noting its response. Turn the control back, just to the point where the unit does not respond.
5. Turn the chopper control until the unit starts to respond. You are now measuring the thermal background as indicated by the periodic sounding of the alarm as the shutter interrupts the heat path.

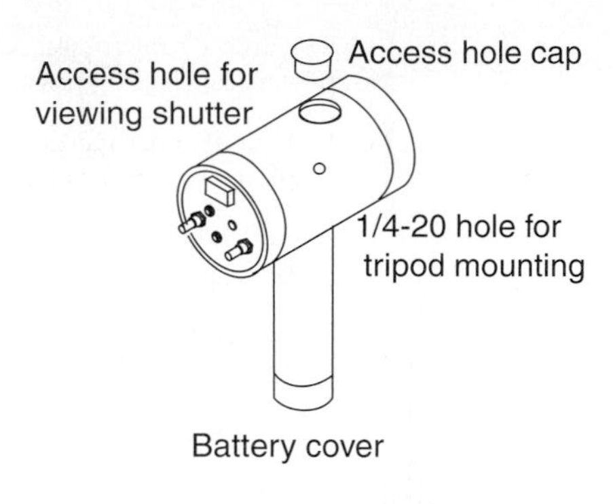

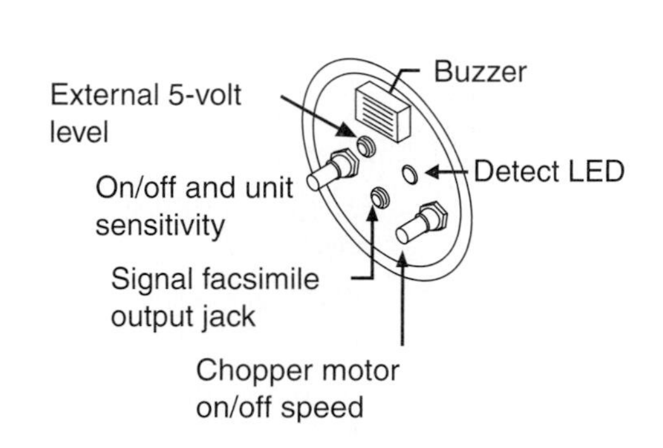

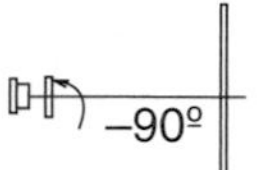

View through access hole showing shutter position relative to pyro detector optical axis and front lens.

Figure 37-7 *Final assembly view*

6. Connect a standard 3.5-millimeter plug into the 5-volt level jack and measure a positive 5-volt level every time the alarm responds. This output is through a 1K resistor and is intended for triggering other alarms or external functions.
7. Connect a standard 3.5-millimeter plug into the signal facsimile jack and connect it to a scope input set on the AC millivolt scale. Note the changing voltage level indicating the slightest amount of differential thermal energy. Check this level with and without the chopper and try it with various thermal sources, noting the voltage levels. This output port is intended for the lab technician desiring to explore low-level thermal sources. It requires a scope or a sensitive AC millivolt meter capable of measuring 10 to 100 millivolts.

The previous steps are performed without considering the thermal stabilizing of the unit or its immediate surroundings. Maximum sensitivity will not be obtainable under these conditions.

8. Select a target area and mount the unit on a tripod, using the 1/4-20 mounting hole located on the side of the unit. This scheme will place the handle in the horizontal plane (off to the side) but should not pose a problem. The access hole will be opposite the handle.
9. Point the unit at the target area and allow it several hours to thermally stabilize. Do not touch the enclosure and try not to get too close to the front of the system, as this can destabilize the system for low-level detection.
10. Point and pan the unit at some known source to verify the anticipated range of operation. Experiment with the chopper speed or no shutter to obtain the optimum results for your application.
11. Handheld operation will vary considerably due to the many variables in the thermal background. We can only suggest that you get accustomed by trial and error to determine your application.

Final Notes

The access hole for the chopper shutter is necessary in order to prevent the blockage of the thermal path when the chopper motor is off. There is no guarantee that the shutter will not randomly stop at the wrong place, blocking the detector and preventing normal operation. In such cases, the shutter must be manually moved with a pencil or some object.

The unit must be temperature stable to optimize maximum sensitivity. This usually takes about an hour if the unit has been recently handled or placed near radiant sources.

For actual body detection, the best performance will occur on a cold night. This does not mean that the unit will not work on a warm night, but that it might be less sensitive.

The detection of stationary objects may vary with chopper motor speed and background temperature. The maximum response of the pyrodetector is around .1 chops per second. This is very slow for many applications and we suggest increasing it to 1 to 3 per second.

The field of view may be narrow due to the Fresnel lens used, but you may change to other suitable lenses of polyethylene. Also, a panning action allows zeroing in on certain target subjects.

Table 37-1 Body Heat Detector Parts List

Ref. #	*Description*	*DB Part #*
R1, 2, 5, 6, 7, 11, 14, 16, 17	Ten 39K, ¼-watt resistors (or-wh-or)	
R3	10K, ¼-watt resistor (br-blk-or)	
R4, 8	Two 1.8-megohm ¼-watt resistors (br-gray-gr)	
R9, 10, 13, 18	Four 1K, ¼-watt resistors (br-blk-red)	
R20	4.7K, ¼-watt resistor (yel-pur-red)	
R21, 22	Two 1-megohm, ¼-watt resistor (br-blk-gr)	
R12/S1	20K pot and 12 VDC switch	
R19	500 to 1-kilohm pot	
R23	470-ohm, ¼-watt resistor (yel-pur-br)	
C1	220-microfarad, 25-volt vertical electrolytic capacitor	
C2, 4	Two 100-microfarad, 25-volt vertical electrolytic capacitors	
C3	.1-microfarad, 50-volt plastic capacitor	
C5, 7, 10	Three 4.7-microfarad, 25-volt vertical electrolytic capacitors	
C6, 9, 13, 14	Four .01-microfarad, 50-volt plastic capacitors (103)	
C8, 11	Two 1,000-microfarad, 25-volt vertical electrolytic capacitors	
C15	1-microfarad, 25-volt vertical electrolytic capacitor	
PC1	Printed circuit board (PCB)	DB# PCBHT
SOCK1	Three-pin TO5 socket	
Q1	PN2222 NPN transistor	
U1	LM358 dual-operational-amplifier integrated circuit	
U2	LM393 dual-comparator integrated circuit	
U3	LM555 timer integrated circuit	
VR1	7805 5-volt regulator TO220 package	
MO1	Low-voltage chopper motor	DB# CHOPMOT
PYR1	Pyroelectric detector	DB# PYR1
BZ1	Piezoelectric buzzer	
LED1	High-brightness light emitter	
J1, 2	Two 3.5-millimeter panel-mount jacks	
CL1	Battery clip and 9-inch leads	
PANEL	3.25 × 3.25-inch aluminum as shown	
EN1	6 × 3.5-inch OD schedule 40 PVC tubing as shown	
HANDLE	6 × 1.875-inch schedule 40 PVC tubing as shown	
CAP1, 2	Two 3 ½-inch plastic caps, as shown in Figure 37-6	
CAP3	2-inch plastic cap for retaining batteries in handle section	
LEN1	3-inch Fresnel polyethylene lens	DB# FRELEN3
BU1	Lens holder for LED1 or use a ⅜-inch bushing	
BHOLD1	8-AA-cell, 4 × 2 battery holder with snaps	
WR20B	24 inches #22 black vinyl hookup wire	

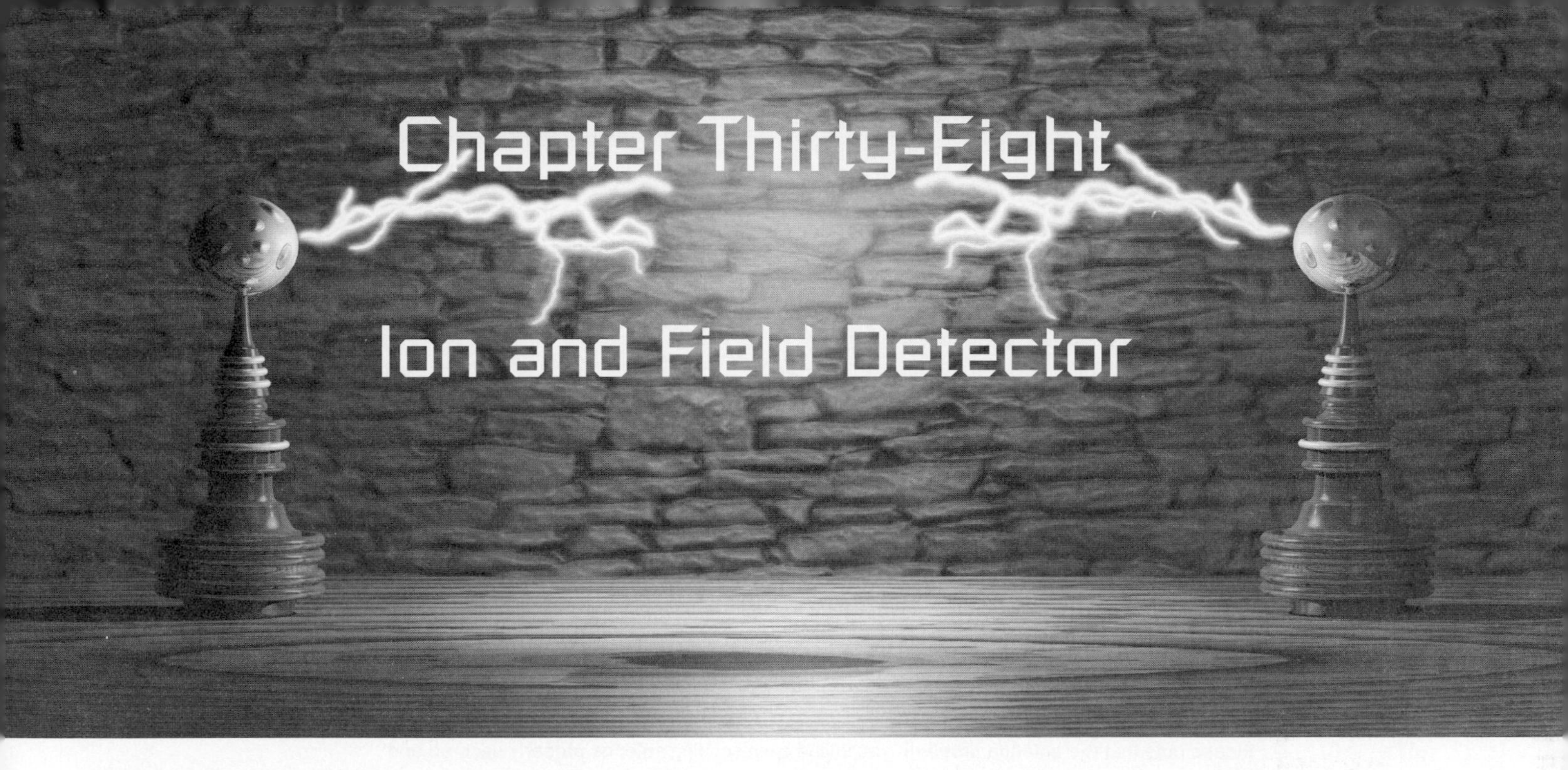

Chapter Thirty-Eight

Ion and Field Detector

This chapter focuses on a device capable of detecting and indicating the relative amount of free ions in the air (see Figure 38-1). It can determine the output of ion generators, high-voltage leakage points, static electrical conditions electric field gradients, and any other application where the presence of ions or a measurement of their relative flux density is required. It can also be used as a very sensitive lightning indicator, showing dangerous electric field levels.

Figure 38-1 *Ion and field detector*

The unit is handheld and enclosed in a plastic enclosure about the size of a king-size pack of cigarettes. A sensitivity control with an on/off switch, a high flux indicator lamp, and a meter are located on the front panel. A telescoping antenna exits a hole in the top of the unit and is used as a moving probe or a stationary pickup point for these charge fields. Grounding the unit is a strip of metallic foil on the outside of the enclosure that contacts the user's hand or may be directly connected to earth ground via a wire. Please note that the performance of this unit can be seriously affected by high humidity conditions.

Circuit Description

As shown in Figure 38-2, ions are accumulated on the collector probe ANT1 and cause a minute, negative base current to flow in transistor Q1 connected as a PNP Darlington pair with transistor Q2. Capacitor C1 and resistor R1 form a time constant to eliminate any rapid fluctuations. The response of the time constant can be changed for various applications and is discussed in Figure 38-2. Diodes D1 and D2 prevent excessive voltage from destroying Q1 by clamping any transients to ground. The collector of Q2 is DC

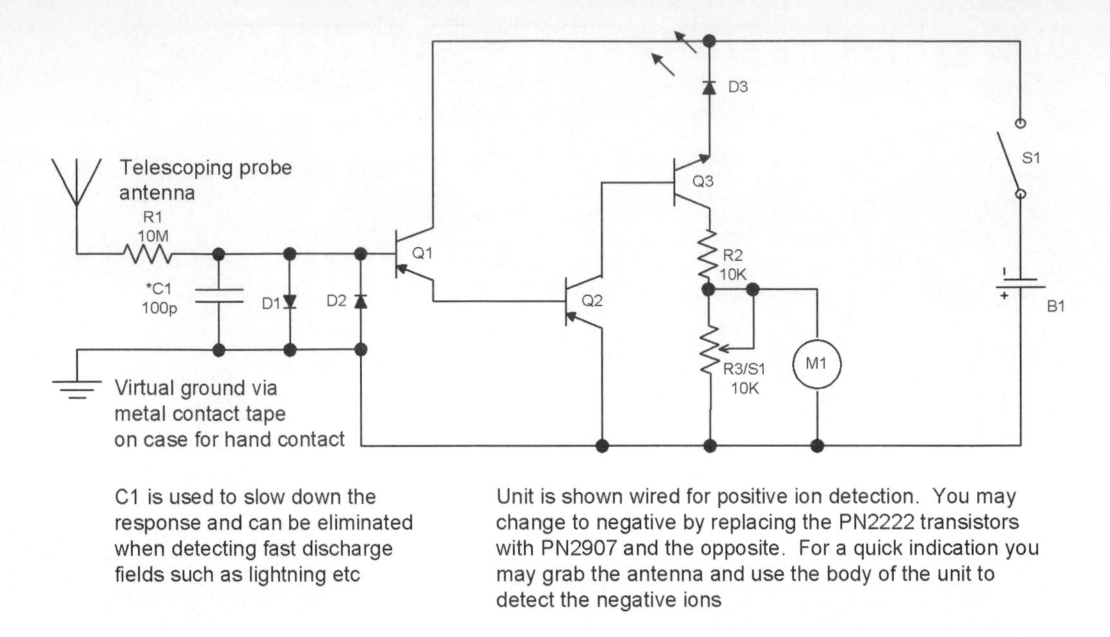

Figure 38-2 *Circuit schematic*

coupled to the base of NPN transistor Q3 in series with current-limiting resistor R2 and meter sensitivity pot R3. Current flowing through Q3 now causes a voltage to develop across R3, driving the indicator meter M1. *Light-emitting diode* (LED) D3 is connected in series with the emitter of Q3 and serves as a visual indicator of strong charge fields. It should be noted that the meter serves only as a relative indicator of ion flux and is not totally that linear.

A threshold voltage, as a result of the base emitter, drops off Q1 and Q2 is only noticeable at almost non-detectable levels. It should be noted that in order for the unit to operate properly, some sort of ground is usually required. Metallic tape is used as shown and provides contact to the operator's hand, providing a partial ground. If the unit is placed remotely, it should be earth grounded to a water pipe, for example.

Measuring positive ions simply involves holding the antenna, using the body of the unit as the probe, or reversing the polarity of the transistors. A *double pole, double throw* (DPDT) switch may be optionally installed for polarity reversal, allowing reversal at the flick of a switch. Be aware that a ceramic or quality switch must be used to avoid leakage in high humidity conditions.

Construction Steps

1. Lay out and identify all the parts and pieces. Verify them with the parts list, and separate the resistors as they have a color code to determine their value. Colors are noted in the parts list at the end of the chapter.
2. Cut a piece of .1-inch grid perforated board to 2 × 1 inches. Locate and drill the holes as shown in Figure 38-3. If you are building from a perforated board, it is suggested that you insert the components starting in the lower left-hand corner as shown. Pay attention to the polarity of all the semiconductors. Route

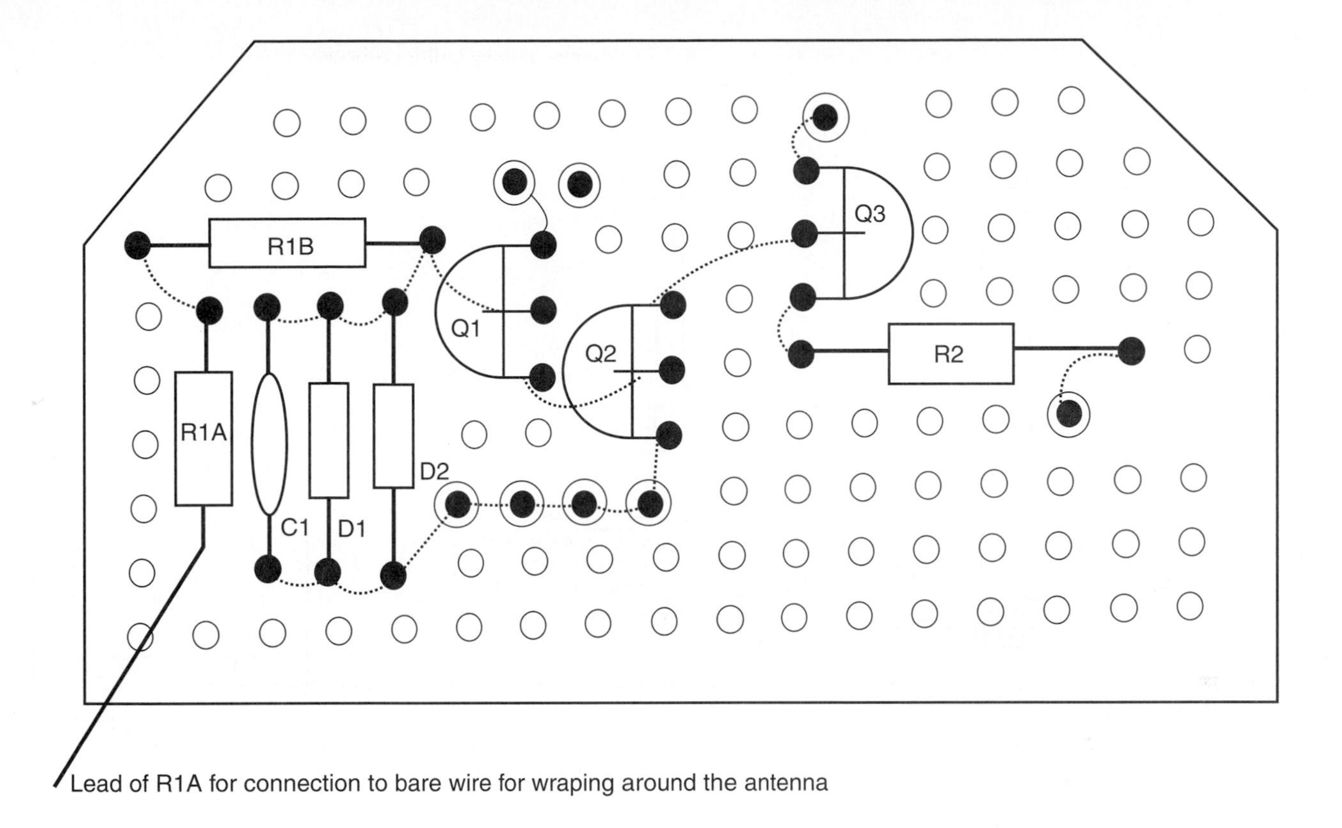

Figure 38-3 *Assembly board wiring*

the leads of the components as shown and solder as you go, cutting away unused wires. Attempt to use certain leads as the wire runs or use pieces of the #24 bus wire. Follow the dashed lines on the assembly drawing as these indicate the connection runs on the underside of the assembly board.

3. Attach the external leads and components as shown in Figure 38-4.
4. Double-check the accuracy of the wiring and the quality of the solder joints. Avoid wire bridges, shorts, and close proximity to other circuit components. If a wire bridge is necessary, sleeve some insulation onto the lead to avoid any potential shorts.
5. Create the plastic case, as shown in Figure 38-5. Note the 1-inch meter hole centered and located approximately 1 1/2 inch from the top as shown. Holes for the switches are centered and located 1 inch below the meter line. Location is not critical, but make sure everything is clear and fits together.
6. Place a strip of foil tape as shown, noting a hole for the connection to the common circuit

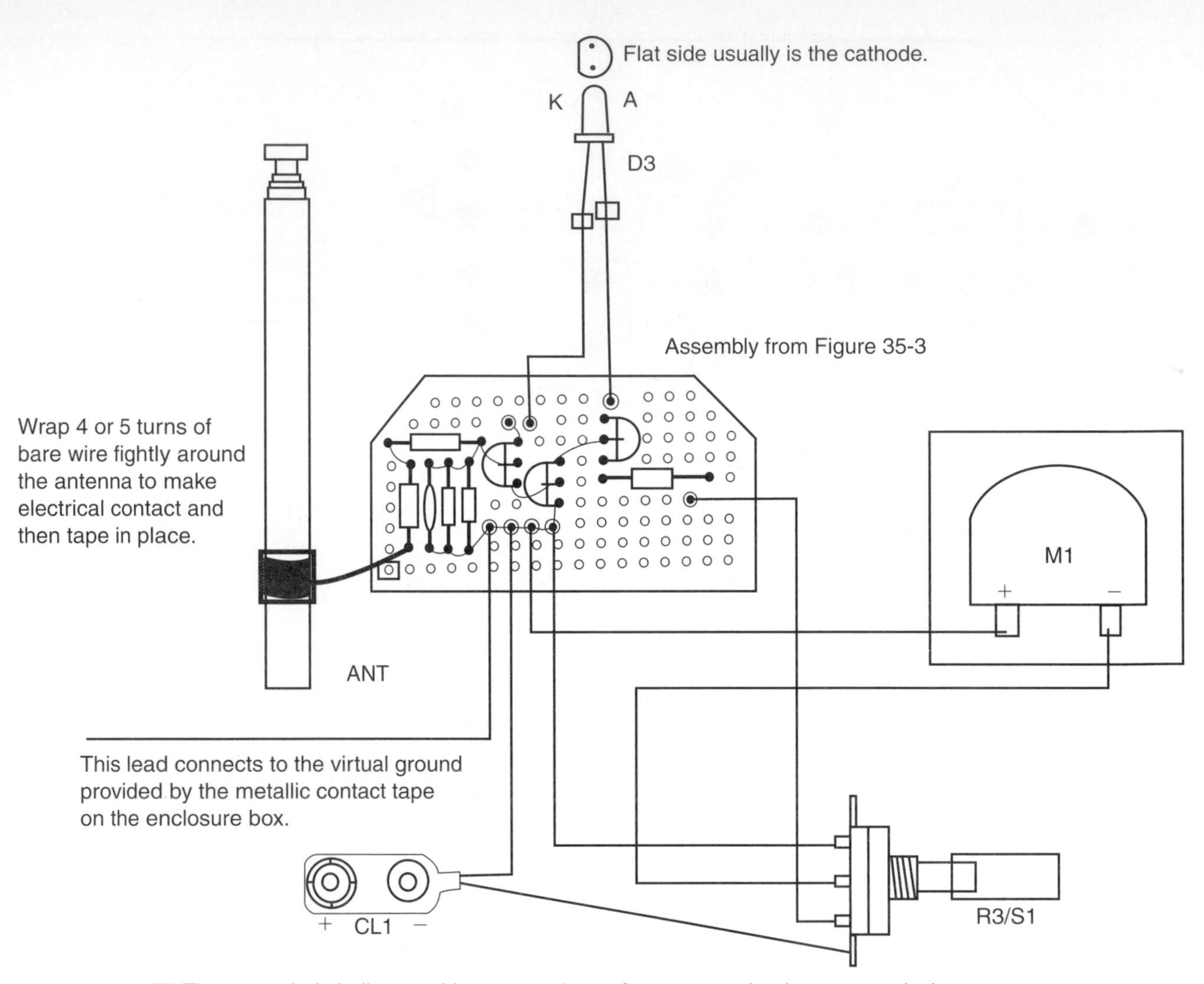

Figure 38-4 *Final wiring and assembly*

point. Secure the assembly board with double-sided tape after circuit operation has been verified.

7. Install battery B1 as shown and turn on the control. Rotate R3 fully clockwise and note the meter showing a slight indication. This is due to transistor and circuit leakage and should not be considered an indication of the ions. If meter sensitivity responds in the opposite way, you may have to reverse the wires across the outer terminals of R3 or the meter leads.

8. Extend the antenna about 6 inches and adjust R3 fully clockwise. Obtain a plastic comb and run it through your hair or rub a plastic object with a dry cloth. Place it near the probe and note the meter indicating a strong charge and the diode lighting. This effect may be reduced if the air is damp or moist.

9. Make the final assembly, as shown in Figure 38-6.

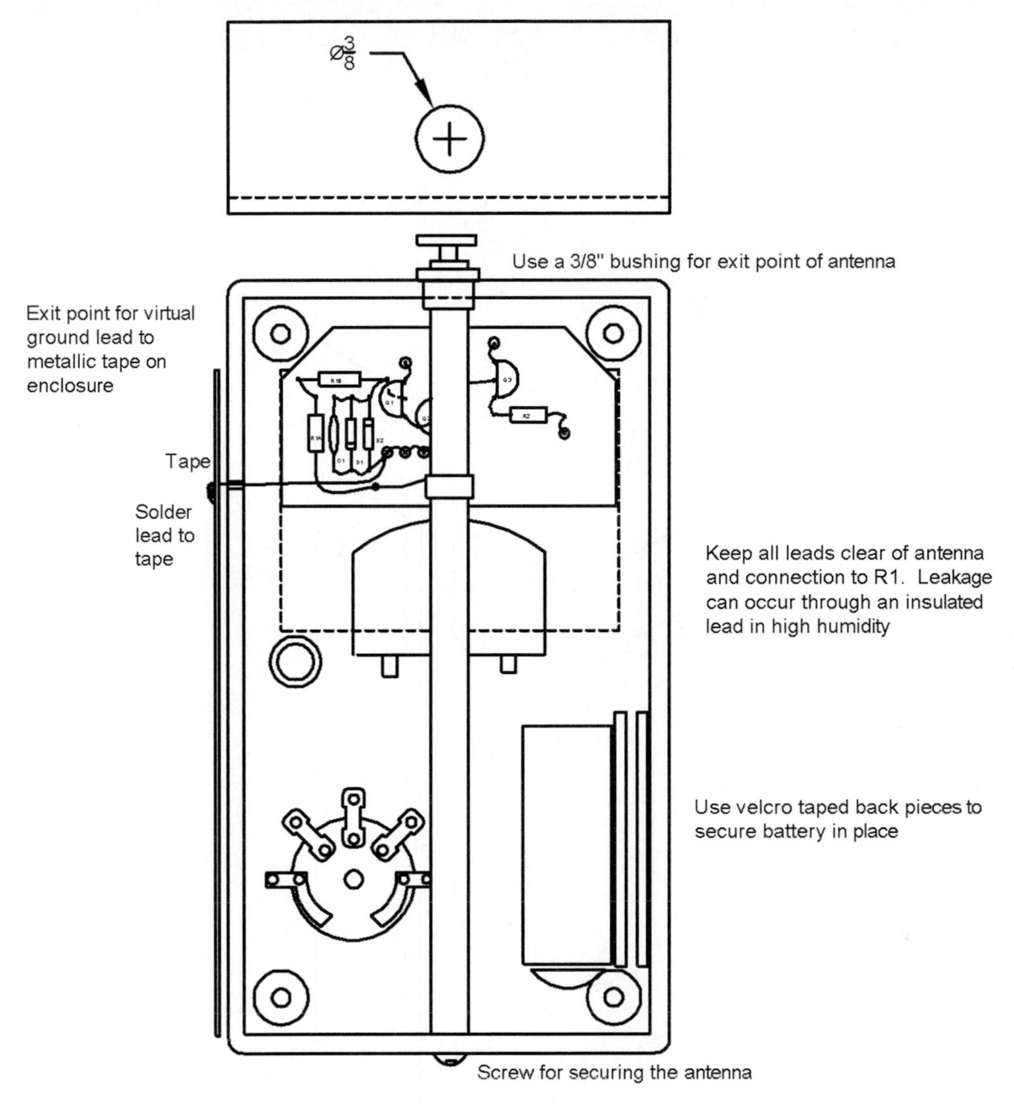

Figure 38-5 *Final assembly*

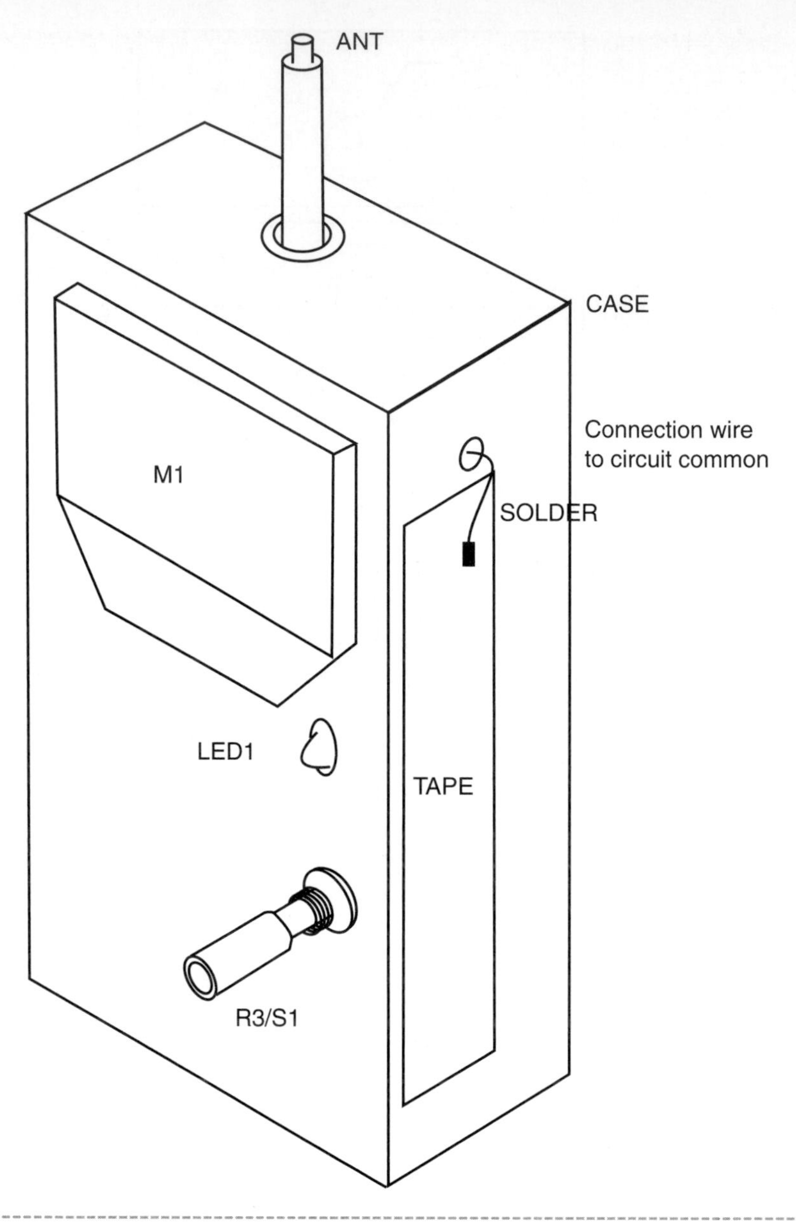

Figure 38-6 *Completed unit*

Applications

Your ion detector is a very, very sensitive ion-measuring device. It can be used for relative measurement but is not designed for absolute measurement.

For quick indications of the presence of a negative ion field, the unit may be handheld and actually used to determine where the source is. The sensitivity of this device can be realized by the simple experiment of running a plastic comb through one's hair and laying it near the probe, as described above.

Those who are familiar with the metallic leaf electroscope will soon realize the advantage of portability and sensitivity. When used for indicating or testing the relative strength of ion sources, the unit should be hard wire grounded for the best results. Adjustments to an ion source to determine the ion output may be made, noting the meter reading and then readjusting R3 to bring the meter reading to scale.

A very interesting phenomenon will be noted when using this device for detecting residual ion fields, the shielding of ions, field direction, static charges, resultant polarity, and the intensity of static charges, as well as a host of others. The unit is an invaluable tool for determining the output of ion generators, any purifiers, and the presence of dangerous static electricity situations. Many sources of charged particles soon become apparent when using the device. People's clothes, fluorescent lighting, plastic containers, and certain winds all will indicate a charge.

Please note that the antenna probe on your unit is a telescoping antenna properly secured and electrically isolated. It is important to remember that any type of leakage around the input of Ql can reduce the sensitivity. You may want to coat the circuitry with a good-quality varnish. Make sure the unit is dry and clean before sealing. The unit is shown wired for positive ion detection. You may change it to negative by replacing the PN2222 transistors with PN2907 and vice versa. For a quick indication, you may grab the antenna and use the body of the unit to detect the negative ions.

Table 38-1 Ion Detector Parts List

Ref. #	*Description*	*DB Part #*
R1	10 meg, ¼-watt resistor (br-blk-blue)	
R2	10K, 1-watt resistor (br-blk-or)	
R3/S1	10K pot and 12 VDC switch	
C1	100-picofard disk capacitor (see Figure 38-2)	
D1, 2	Two 1N914 signal diodes	
D3	Colored LED	
Q3	Two PN2222 NPN transistor (see note on reversing)	
Q1, 2	Two PN2907 PNP transistors	
M1	100-micro-amp, small panel meter	DB# METER50S
PB1	1 × 2-inch piece of .1-inch grid perforated circuit board	
ANT	Telescoping antenna	DB# ANT1
CL1	9-volt battery clip	
WR24	24 inches of #24 vinyl hookup wire	
CASE	4⅜ × 2⁷⁄₁₆ × 1¼-inch plastic box with lid	
TAPE	4 × ½-inch metal tape	
WR24B	12 inches of #24 bus wire	

Chapter Thirty-Nine

Light Saber Recycling Stand

This project is intended for those who have built or purchased a light saber from Information Unlimited or who have built one from our first book as directed in Chapter 21 in this series of three.

The saber stand recycles your saber where the plasma display travels up the tube, appearing to evaporate into open space (see Figure 39-1). It then retracts into the hilt and continues repeating until power is removed. The effect is very relaxing and provides an excellent display piece or prop for a bar, a point-of-purchase display, an attention getter, a conversation piece, or just a nightlight for a bedroom.

Operation is powered from a 12-volt wall adapter and is completely safe for children. This is an intermediate-level project requiring basic electronic skills. Expect to spend $20 to $30. All the parts are readily available, with specialized parts obtainable through Information Unlimited (www.amazing1.com), and are listed in the parts list at the end of the chapter.

Circuit Description

Figure 39-2 shows a 12-volt wall adapter (T1) being fed into a voltage regulator chip (I2) where a regulated 9 volts is applied to the remaining circuitry. Capacitors C1 and C2 smooth out the voltage waveform. An astable pulse generator using a timer (I1) provides an uneven "on to off" time determined by the ratio of resistors R2 and R3. The output pulse repetition rate is a function of timing capacitor C3 with

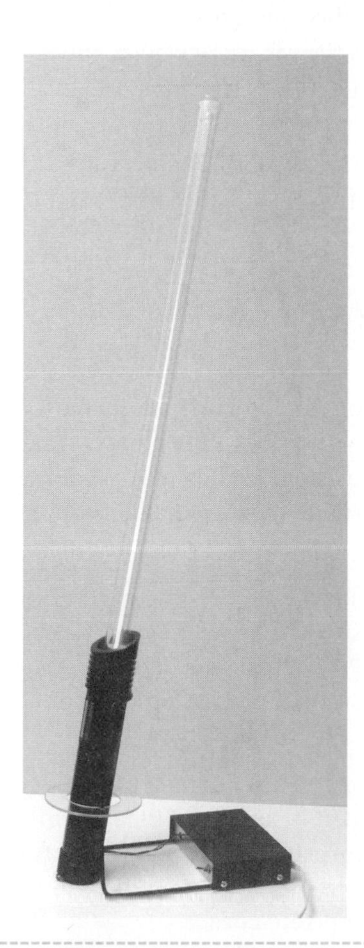

Figure 39-1 *The saber recycling stand*

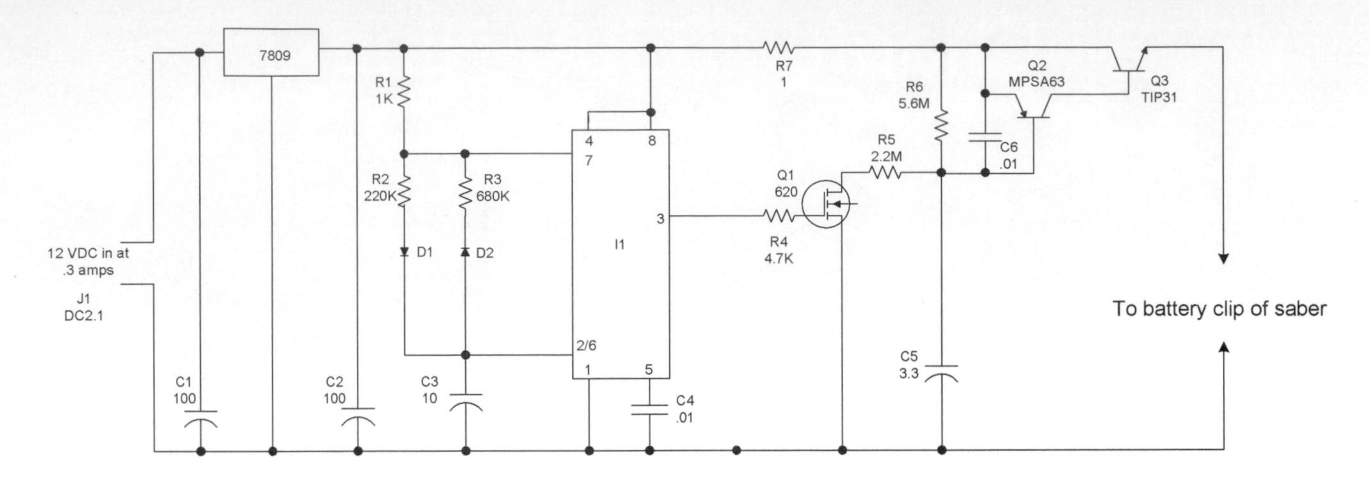

Figure 39-2 *Circuit schematic*

output at pin 3 of I1. The pulse output serves as a gate for MOSFET switch Q1, creating a charging and discharging of low-leakage capacitor C5. A quasi-linear voltage ramp now appears across C5, controlling the current through a high beta Darlington transistor (Q2) in turn controlling the main-current pass transistor Q3. It is this ramping up and down of current that now powers the saber electronics, allowing the plasma to flow up and down, all automatically. The cycling action is set by the circuit to travel up a half-second and down 3 seconds. This timing can be changed to some extent by the selection of R2 and R3.

Assembly

1. Lay out and identify all the parts and pieces. Verify them with the parts list, and separate the resistors as they have a color code to determine their value. Colors are noted on the parts list.
2. Cut a piece of .1-inch grid perforated board to a size of 3.1 × 1.8 inches, as shown in Figure 39-3. Locate and drill the holes as shown to relieve any strain on the input and output leads.
3. When you are building from a perforated board, it is suggested that you insert the components starting in the lower left-hand corner, which is used as a reference, and proceed to the opposite corner.

   Pay attention to the polarity of the capacitors with polarity signs and all the semiconductors.

   Route the leads of the components as shown and solder as you go, cutting away unused wires. Attempt to use certain leads as the wire runs or use pieces of the #24 bus wire. Follow the dashed lines on the assembly drawing as these indicate the connection runs on the underside of the assembly board.
4. Attach the external leads for battery clip CL1 and the leads to input power jack J1. *Important!* The battery clip CL1 must be wired in reverse by connecting the red lead to the circuit ground and the black lead to the emitter of Q3.
5. Double-check the accuracy of the wiring and the quality of the solder joints. Avoid wire bridges, shorts, and close proximity to other circuit components. If a wire bridge is necessary, sleeve some insulation onto the lead to avoid any potential shorts.

Parts Fabrication

1. Create the chassis shown in Figure 39-4. This is attached to the BASE section using four #6 sheet metal screws.
2. Create the base section from a piece of plastic or aluminum, as shown in Figure 39-5. Note

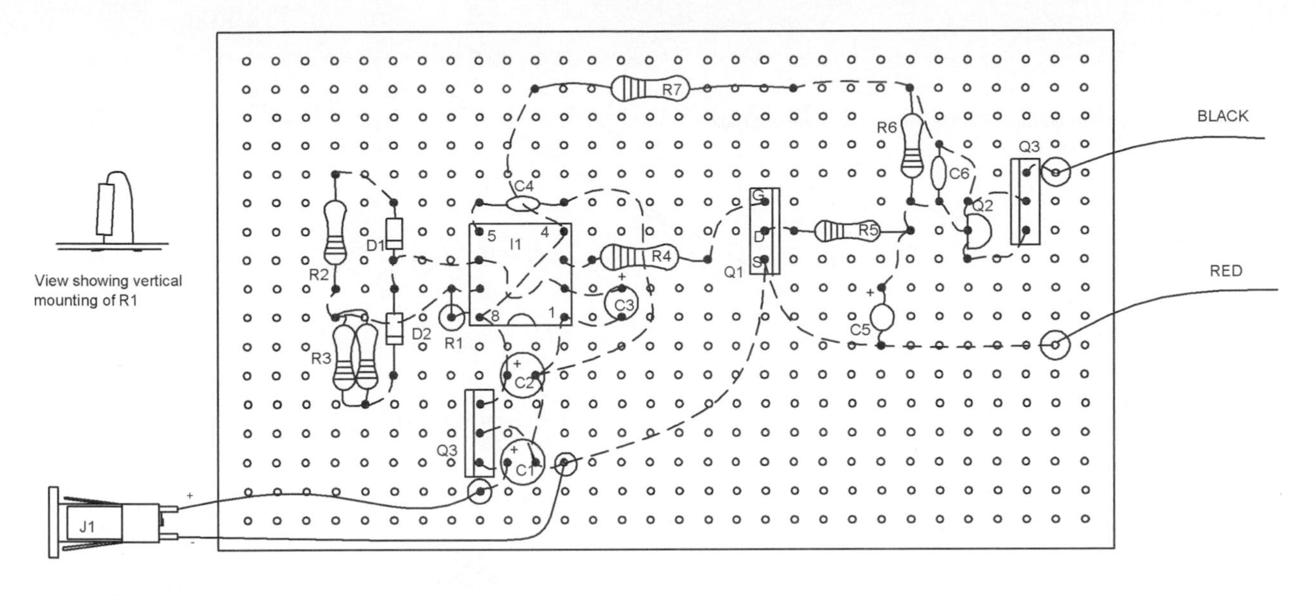

Notes

The looped dashed lind connecting pins 2 and 6 of I1 must have a sleeve of insulated tubing to prevent shorting the lead between pins 4 and 8.

Dashed lines are connections on underside of perforated circuit board using the component leads. These points can also be used to determine foil runs for those who wish to fabricate a printed circuit board.

Small dots are holes used for component insertion.

Large dots are solder junctions

Circles with dots are holes that may require drilling for component mounting and points for external connecting wires indicating strain relief points. It is suggested to drill clearance holes for the leads and solder to points beneath board.

Figure 39-3 *Board assembly*

the four holes for mounting the chassis section, two holes for the holder section, and two holes for the tie wraps.

3. Create the holder from a piece of 2³/₈-inch, schedule 40 gray PVC tubing. Note the slight angle cut for aesthetics being congruent with the light saber when in the stand. This piece is mounted to the base using two #6 × ³/₈-inch sheet metal screws.

4. Form the frame as shown in Figure 39-6 from ³/₁₆-inch cold roll steel rod or an equivalent. The frame must be formed to provide a stable assembly with the saber in place in the stand. The longer 36-inch blade unit is the most critical, requiring the most attention. Attempt to follow the dimensions as shown and determine the base angle as a final adjustment with the saber in place.

   The frame is secured in place by inserting the horizontal members through the corner holes in the chassis section. The front section of the frame is secured in place by two small tie wraps, as shown in Figure 39-5.

10. Finish the assembly, as shown in Figures 39-7 and 39-8.

Operation

Remove the cap at the rear of the saber hilt and remove both batteries. Take out one of the battery clips and snap it into the mating clip of the saber stand. Position the saber in the stand and verify its mechanical stability. Plug in the wall adapter and turn on the switch. The plasma should immediately go to the end and then drop down into the hilt. Recycling should commence and continue until power is removed.

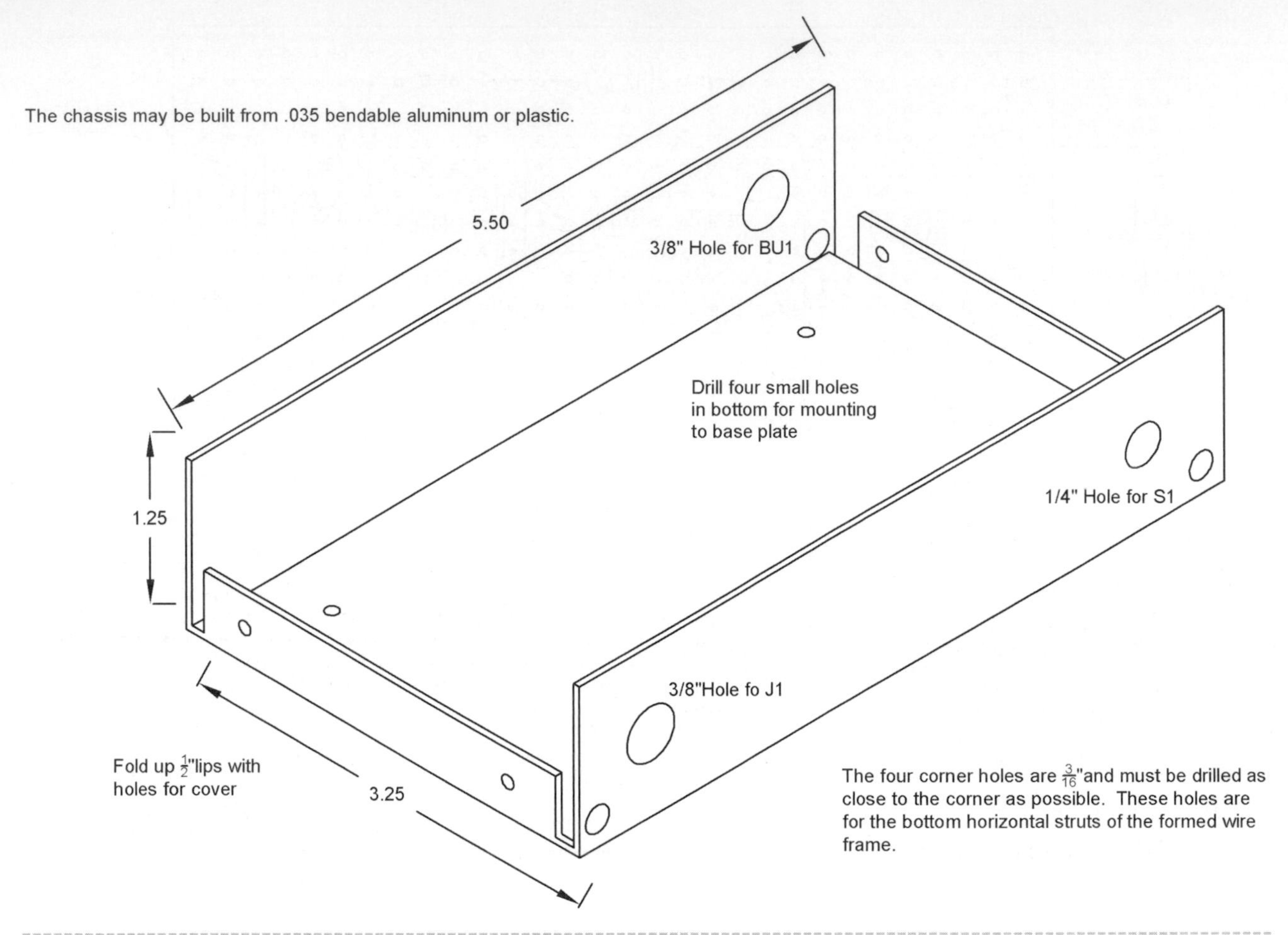

Figure 39-4 *Chassis fabrication*

Light sabers are available from Information Unlimited at www.amazing1.com in 24- or 36-inch lengths in colors of phasor green, photon blue, starfire purple, neon red, or solar yellow.

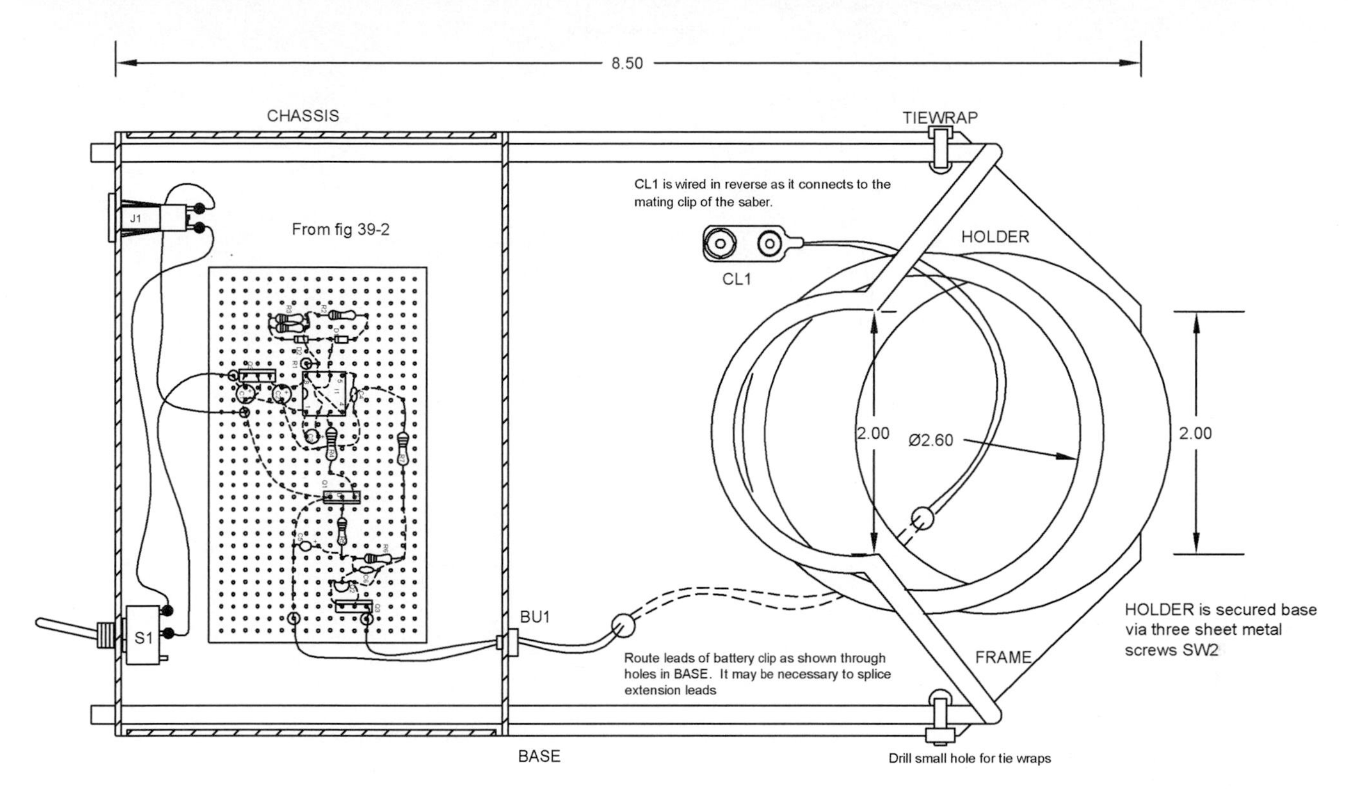

Figure 39-5 *Base fabrication*

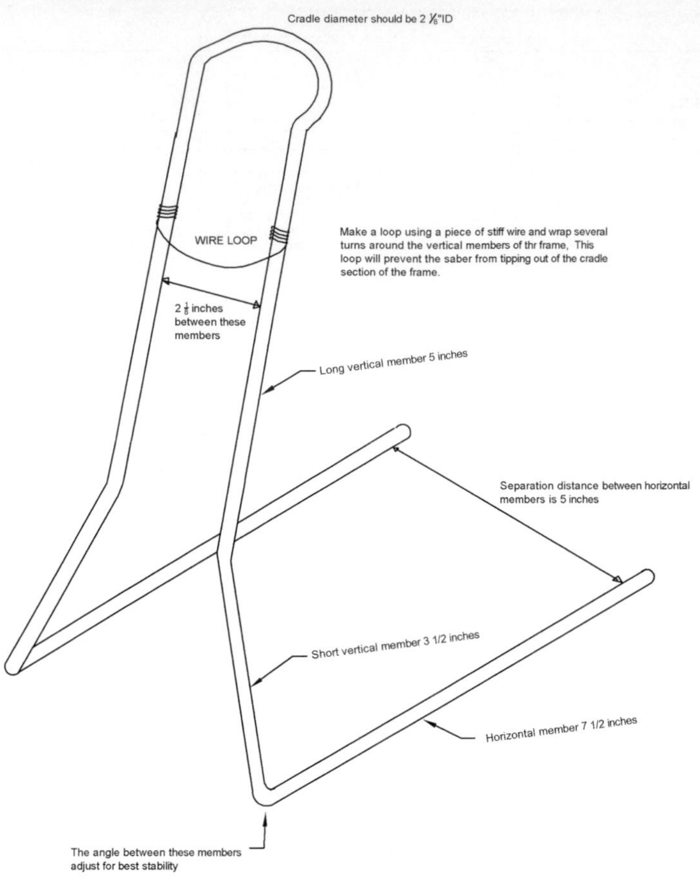

Figure 39-6 *Frame section*

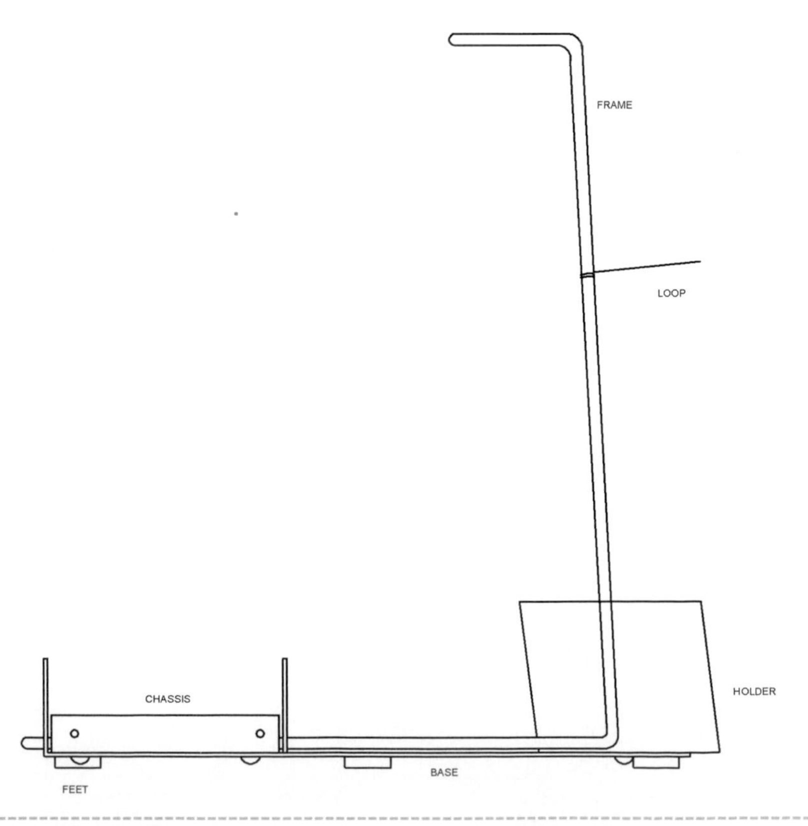

Figure 39-7 *Side view of assembly*

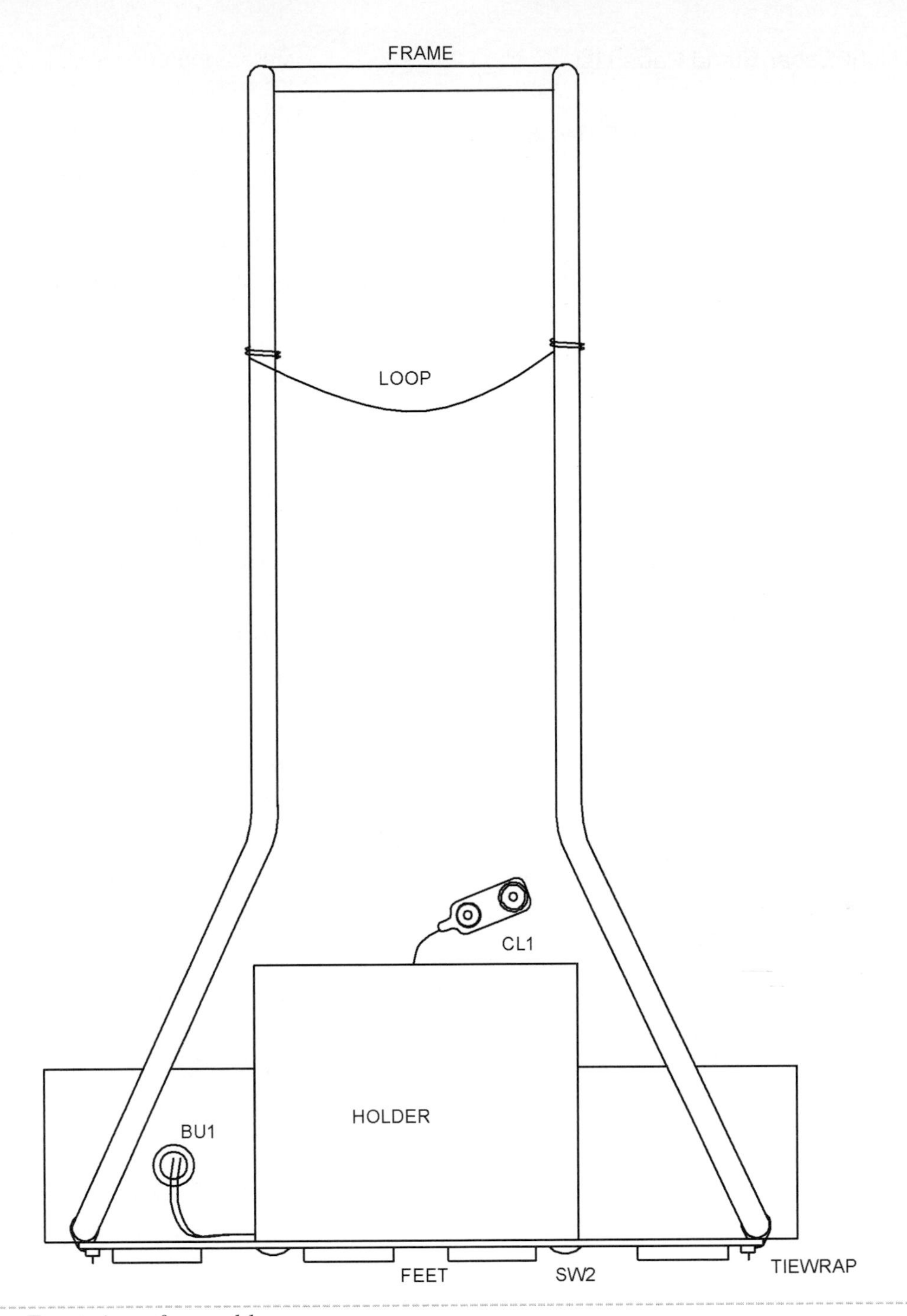

Figure 39-8 *Front view of assembly*

Table 39-1 Light Saber Stand Parts List

Ref. #	Description	DB Part #
R1	1K, ¼-watt resistor (br-blk-red)	
R2	220K, ¼-watt resistor (red-red-yel)	
R3	680K, ¼-watt resistor (blue-gray-yel)	
R4	4.7K, ¼-watt resistor (yel-pur-red)	
R5	2.2 m, ¼-watt resistor (red-red-gr)	
R6	5.6 m, ¼-watt resistor (gr-blue-gr)	
R7	1-ohm, ¼-watt resistor (br-blk-gold)	
C1, 2	Two 100-microfarad, 25-volt vertical electrolytic capacitors	
C3	10-microfarad, 25-volt vertical electrolytic capacitor	
C4, 6	Two .01-microfarad, 50-volt disc capacitors	
C5	3.3-microfarad, 35-volt tantalum capacitor	
D1, 2	Two IN914 silicon diodes	
I1	555 dual inline package (DIP) timer integrated circuit	
I2	7809 T0220 9-volt regulator	
Q1	IRF620 metal-oxide-semi-conductor field effect transistor (MOSFET)	
Q2	MPSA63 PNP Darlington transistor	
Q3	TIP31 T0220 NPN transistor	
J1	Snap-in DC panel, 2.5 mm jack	
S1	Single pole, single throw (SPST) small toggle switch	
PB1	3.5 × 2 .1 × .1-inch grid perforated board; trim to size	
CL1	Battery clip with 12-inch leads or extend shorter leads	
TIEWRAP	Two small tie wraps	
CHASSIS	Create as shown in Figure 39-4 from .035-inch aluminum	
BASE	Create as shown in Figure 39-5 from .035-inch plastic	
HOLDER	Create from a length of 2⅜-inch schedule 40 gray PVC	
FRAME	Create as shown in Figure 39-6	
FEET	Six large, rubber stick-on feet	
SW2	Six #6 × ⅜-inch sheet metal screws	
BU1	Two ⅜-inch plastic bushings	
LIGHT SABER	36- or 24-inch saber in blue, green, red, or purple	

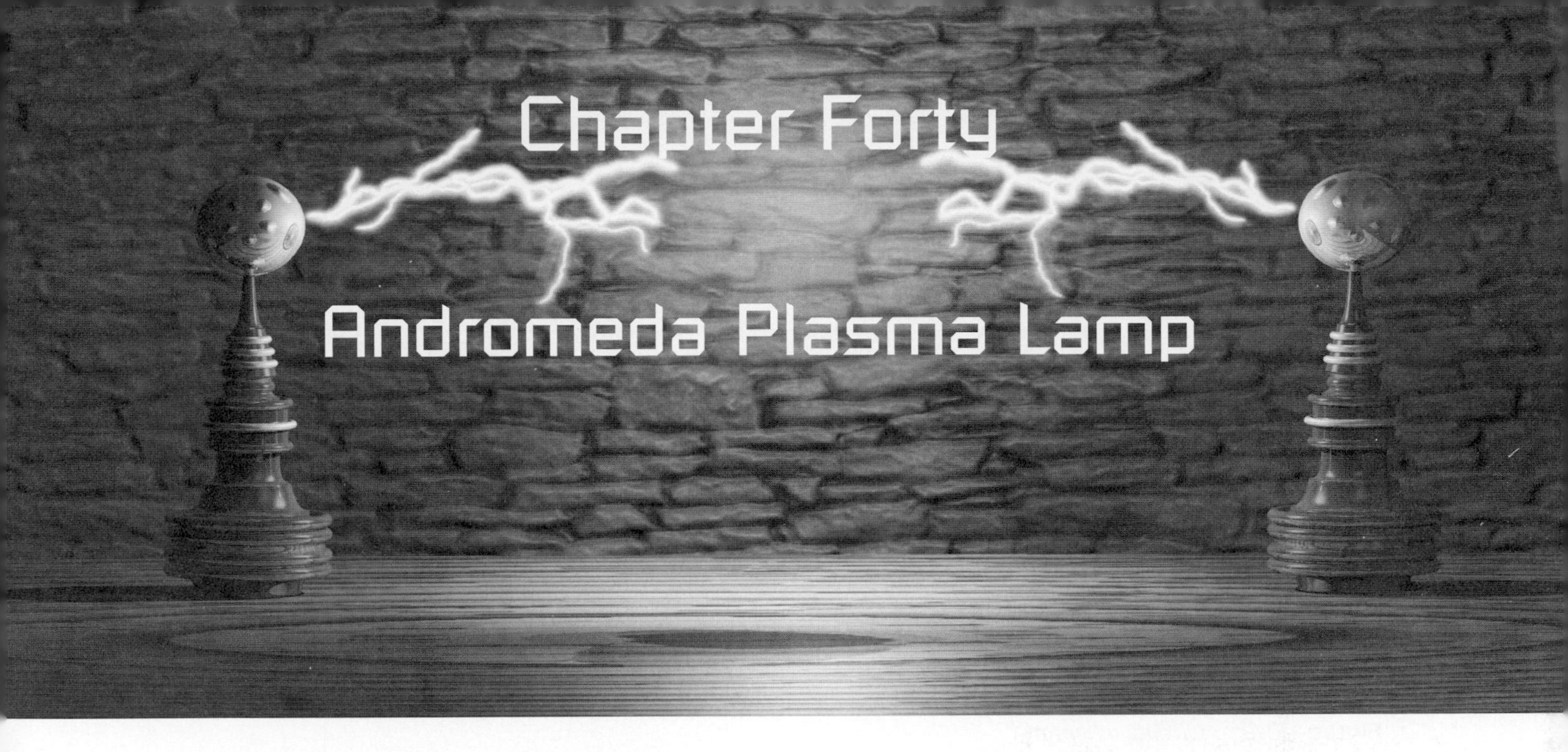

Chapter Forty: Andromeda Plasma Lamp

This project shows how to build a unique type of outside lamp suitable for illuminating driveways, pools, gardens, walkways, docks, or any place where colorful or functioning illumination is desired. The lamps are patented by the author and are currently in preproduction.

The lamp's features are high efficiency as five individual lamps can be powered by a single, conventional, 12-volt wall adapter as the lamps require only 3 watts each. The lamps provide a high amount of useful illumination and must not be compared to the low-light, solar-powered accent lamps.

Another feature of the lamp is that it never needs replacing and should last a lifetime. Other lamps such as incandescent ones are always burning out filaments, whereas conventional fluorescents darken and fail to start due to metallic sputtering.

The third feature is that the lamps are available in pure spectral colors of green, blue, purple, yellow, red, and cool and warm white. These are vivid colors and not the result of filters or paint. Their color is the result of the true spectral emission.

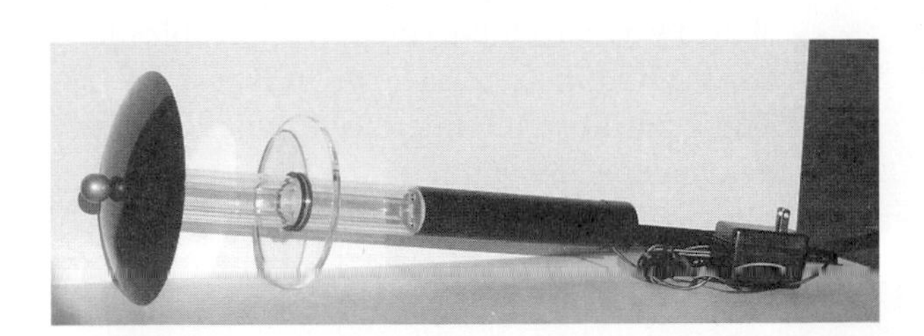

Figure 40-1 *The Andromeda plasma lamp*

This is an intermediate-level project requiring basic electronic skills. Expect to spend $25 to $35. All the parts are readily available, with specialized parts obtainable through Information Unlimited at www.amazing1.com, which are listed in the parts list at the end of the chapter.

Circuit Description

Transistor Q1 forms a self-oscillating circuit where current rises in the collector, COL, winding transformer T1 and inducing a voltage in the feedback (FB). This also causes a winding that keeps Q1 on until T1 saturates and the feedback voltage can no longer be induced. The collector current rapidly falls and forces the base to go negative further, turning Q1 off. When the collector current falls to zero, Q1 is turned on by bias-set resistors R1 and R2, which must charge up timing capacitor C2. The setting of trimpot R1 determines the power output by controlling the number of cycles per second, and capacitor C1 bypasses any residual signal voltage to ground (see Figure 40-2).

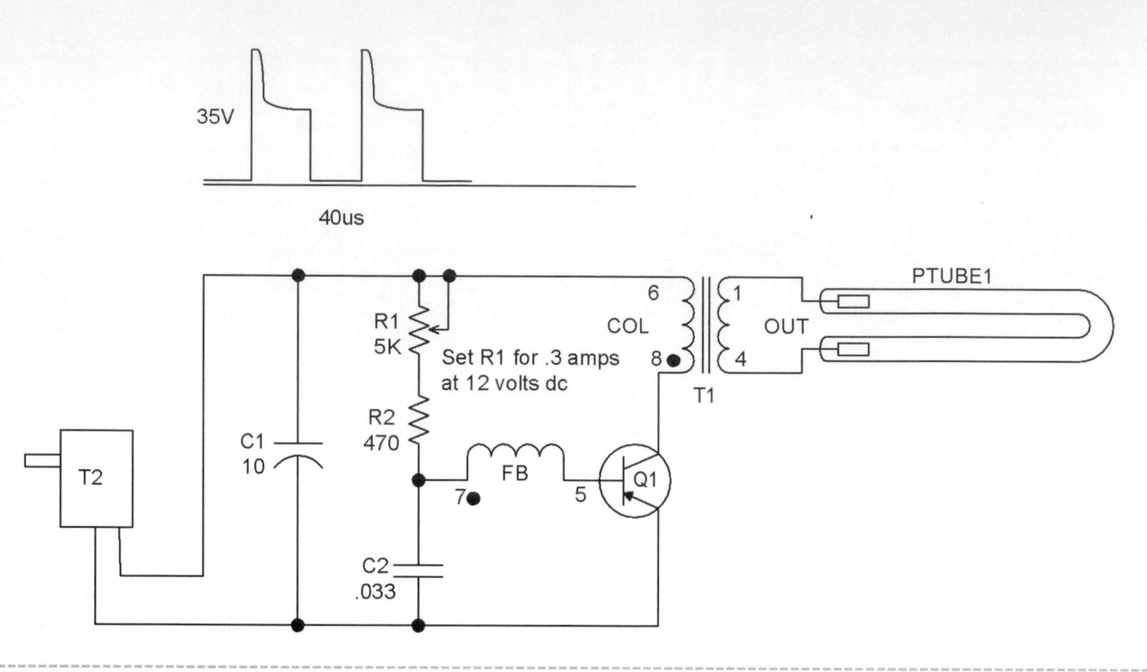

Figure 40-2 *Circuit schematic*

Assembly Board Wiring

1. Cut a piece of .1 × .1 grid vector circuit board to 1 × 4.5 inches, as shown in Figure 40-3.
2. Whenever you are using a perforated board, it is suggested that you insert components starting in the lower left-hand corner. Pay attention to the polarity of the capacitors with polarity signs and all the semiconductors. Route the leads of the components as shown and solder as you go, cutting away unused wires. Attempt to use certain leads as the wire runs or use pieces of the #24 bus wire. Follow the dashed lines on the assembly drawing as these indicate the connection runs on the underside of the assembly board.
3. Double-check the accuracy of the wiring and the quality of the solder joints. Avoid wire bridges, shorts, and close proximity to other circuit components. If a wire bridge is necessary, sleeve some insulation onto the lead to avoid any potential shorts.
4. Attach leads from the plasma tube PT1, being careful not to stress the glass enclosure. Note the small tie wraps (TY1) securing the tube to the perforated board. Again, do not cause any stress to the glass. Make sure that when tightening the wraps they do not stress the tube. When soldering, you must not allow the glass or metal seal to get too hot. Leave at least a half-inch lead between the tube and the connection. Use flux and make the solder joints quickly.
5. Connect the input power leads to a 12-volt DC source and note the lamp lighting. Quickly adjust trim pot R1 to 300 milliamps of current. The lamp should be very bright and the heatsink tab on Q1 should only be warm. Allow it to run for several hours and check adapter transformer T2 for excessive heating.
6. Create the EN1 enclosure tube and the STK1 mounting stake, as shown in Figure 40-4. Note the two mating holes for attaching the stake and the 5/16-inch hole for the rubber tubular bushing for routing the input power leads.
7. Obtain a top canopy (CANOPY7). The one shown was selected to mate with the plastic baffle (BAFFLE5) as shown in Figure 40-5. These parts are intended for aesthetics only and may be any combination that appears pleasing to the builder. A good source is a

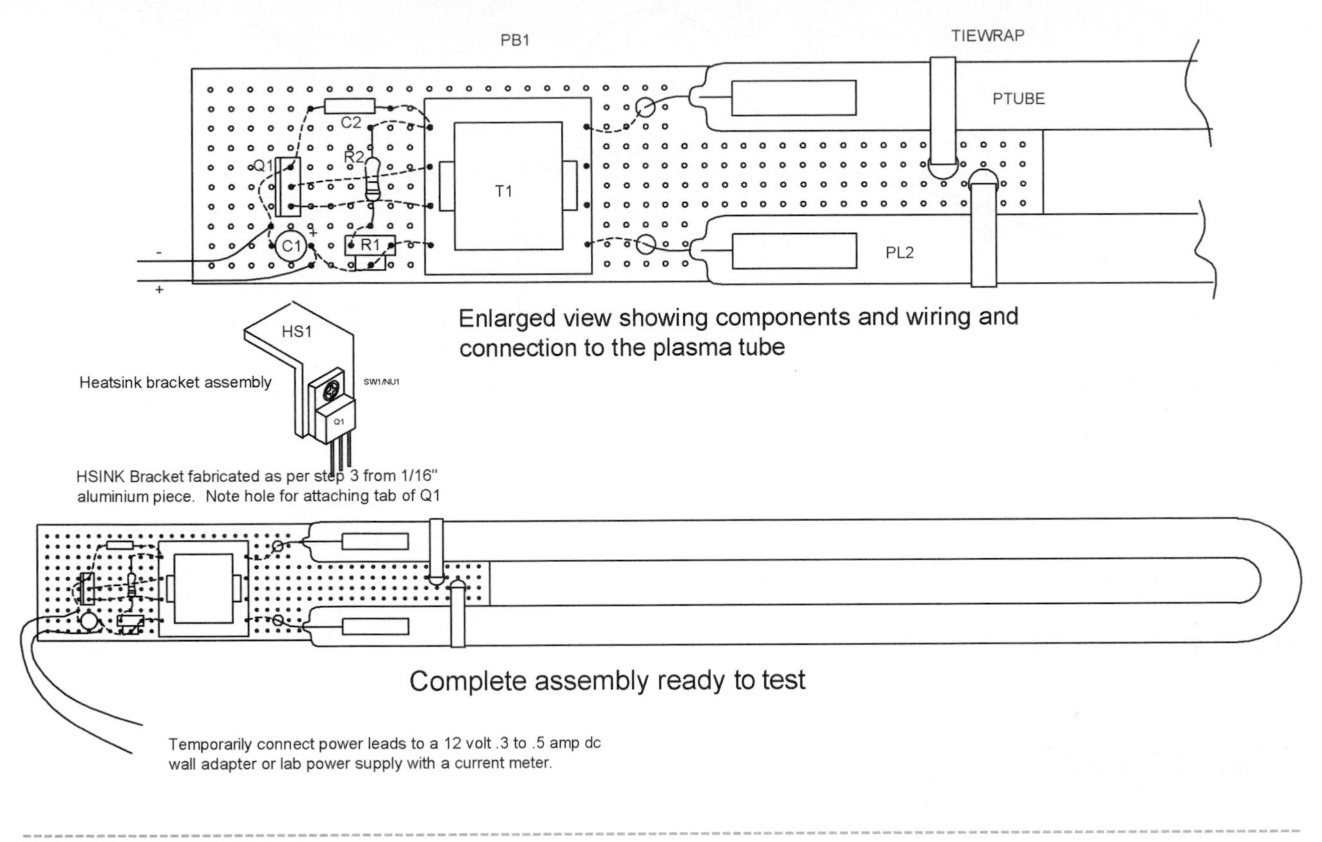

Figure 40-3 *Assembly board*

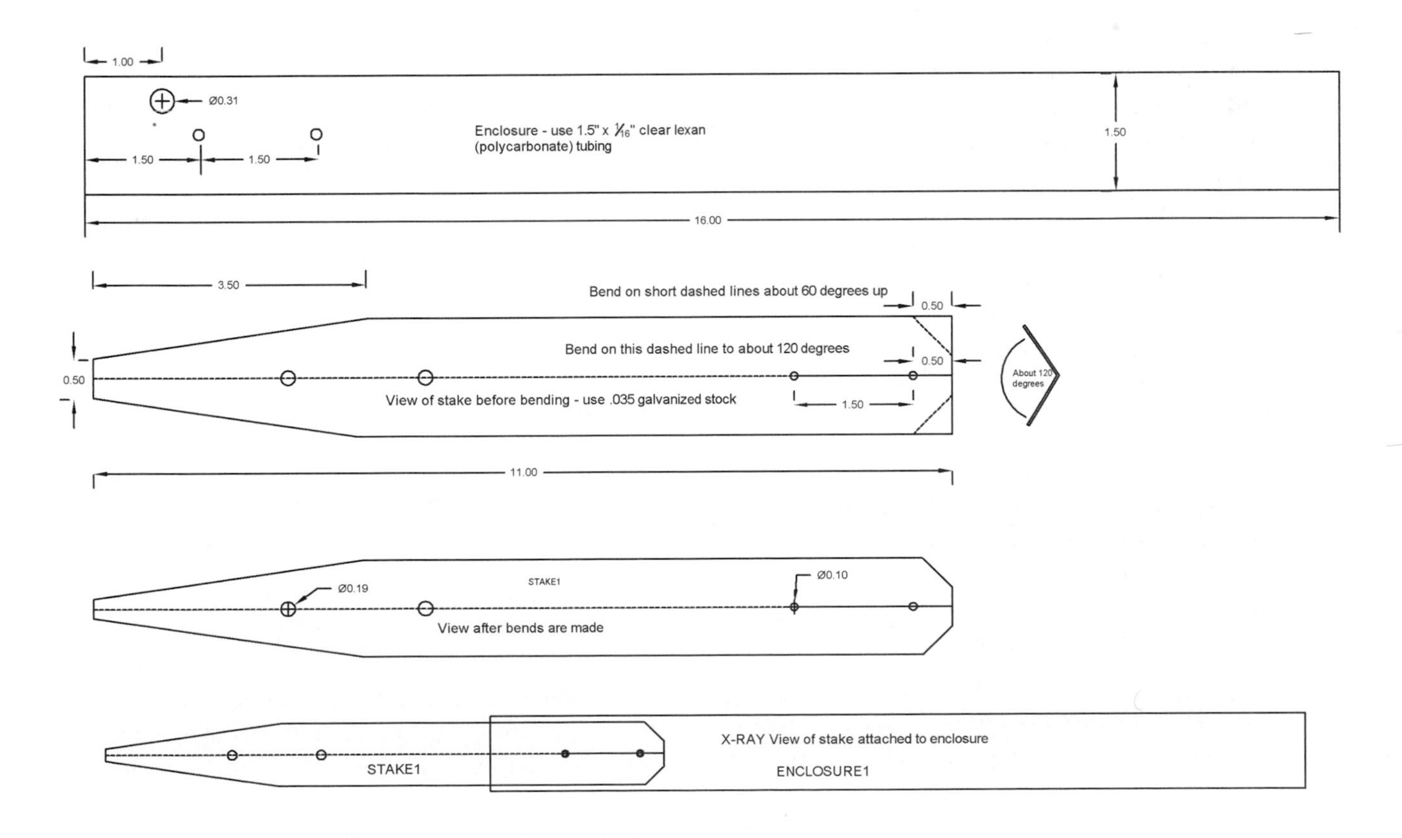

Figure 40-4 *Creating the mounting stake and enclosure*

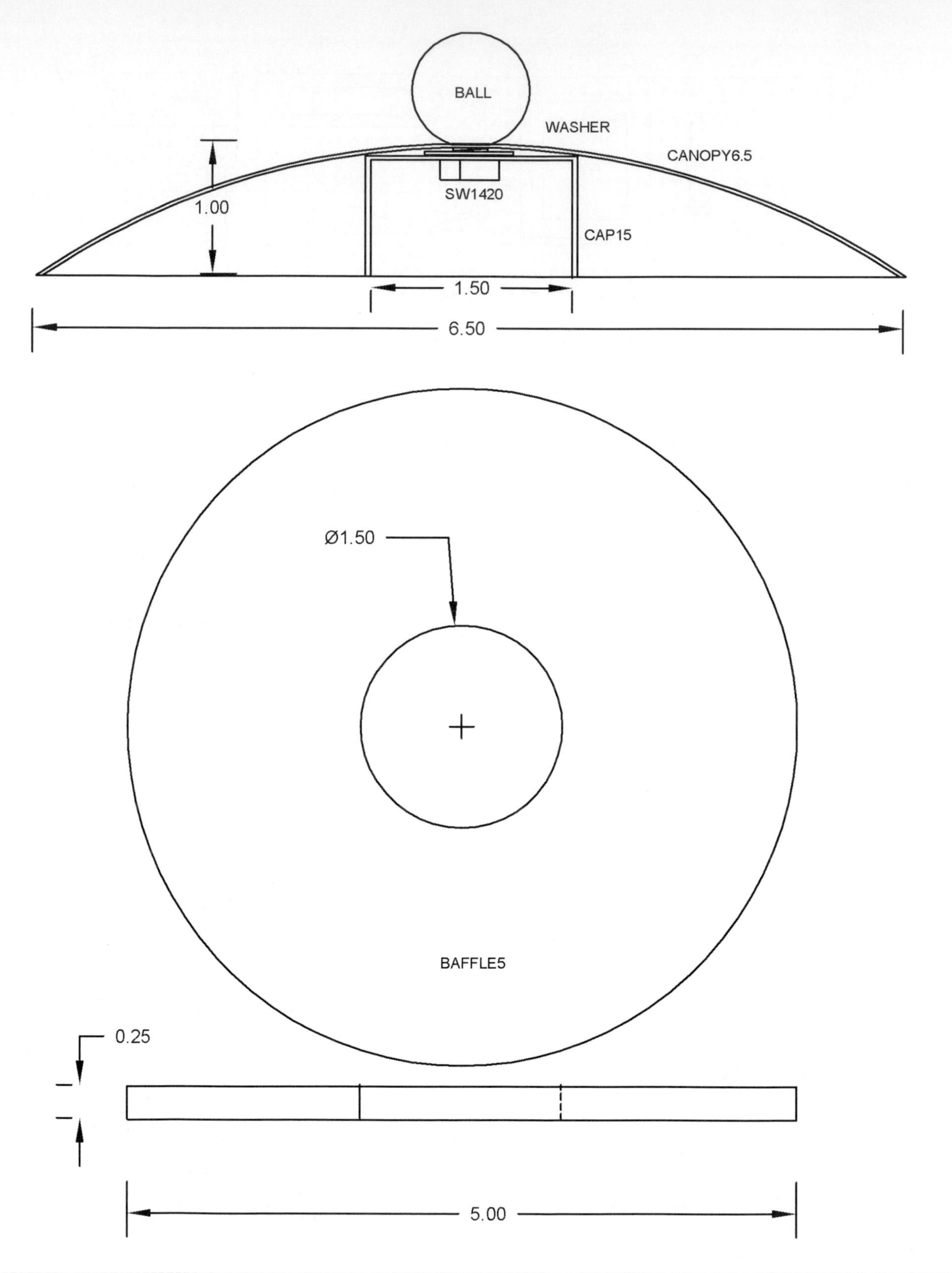

Figure 40-5 *Canopy and baffle design*

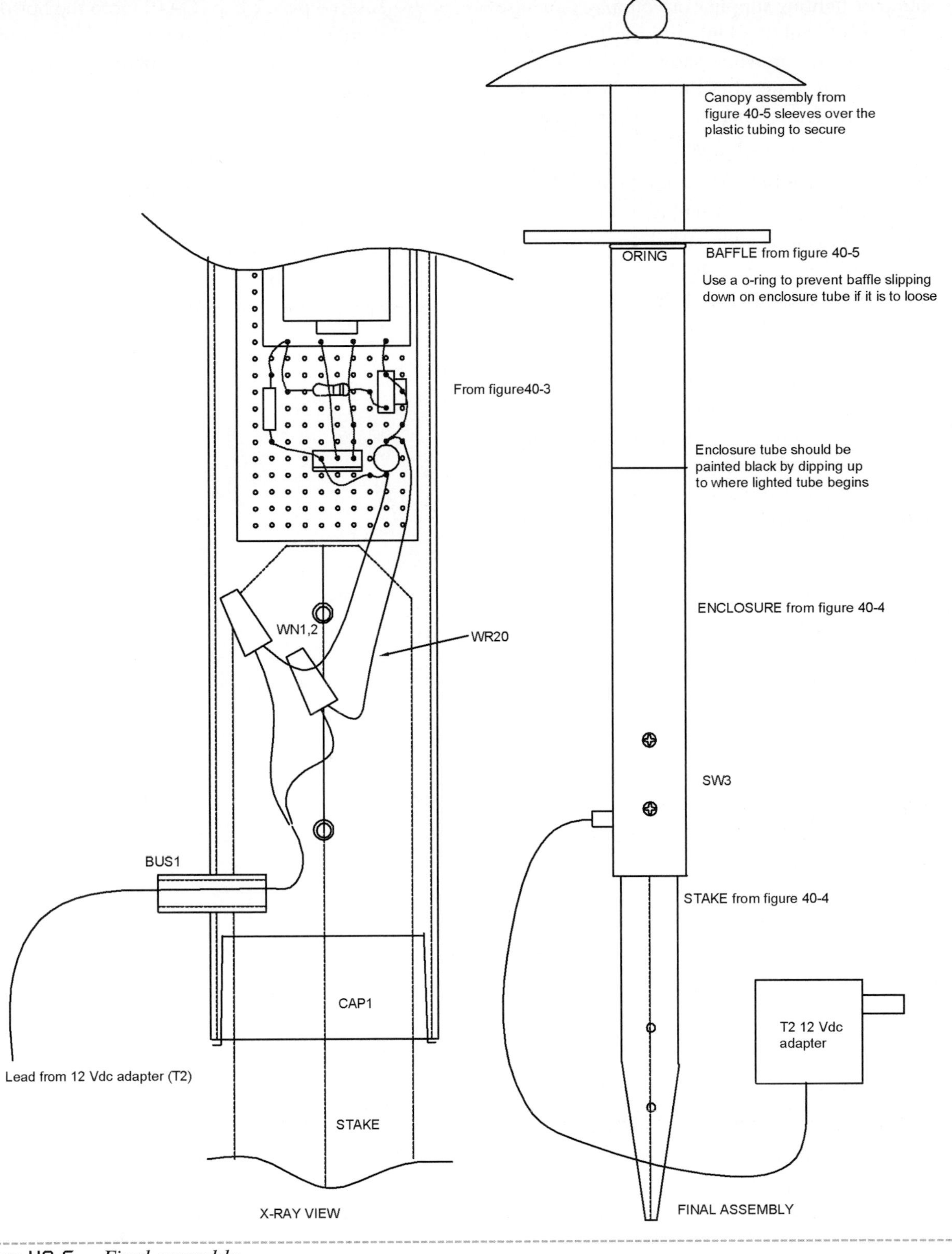

Figure 40-6 *Final assembly*

lamp or lighting supplier or you may purchase it from Information Unlimited, as noted in the parts list. Our drawings show what we are using in our design, but this may be altered to fit the builder's choice or his or her available sources.

8. The final assembly is shown in Figure 40-6. Note that the entire assembly from Figure 40-3 should easily slide into the enclosure tube and abut onto the folded-over corners of the stake. Thread the leads from wall adapter transformer T2 through bushing BUS1 and pull them out through the top end of the enclosure tube. Wire nut the power leads from the main assembly with the leads from T2. Observe the polarity or circuit damage will result. Gently slide them into the enclosure, taking up the wire slack by gently pulling the leads through BUS1.

9. Insert a plastic cap (CAP1) into the bottom section of the enclosure tube. Seal it around the stake using silicon rubber. This step prevents bugs and moisture from getting into the assembly.

Lamps, when built as shown, are fully weatherproof and can be left on indefinitely under snow, in the rain, or in any weather, short of a twister. Five lamps can all be operated from a single 12-volt, 1.5-amp wall adapter listed at the end of the chapter. Interconnecting wiring can be #18 zip cord due to the low-current consumption.

Table 40-1 Andromedia Plasma Lamp Parts List

Ref. #	*Description*	*DB Part #*
R1	5K vertical trimmer	
R2	470-ohm, ¼-watt resistor (yel-pur-br)	
C1	10-microfarad, 50-volt vertical electrolytic capacitor	
C2	.033-microfarad, 50-volt plastic capacitor (332)	
T1	Inverting transformer	DB# ANQP2824
T2	12 VDC, 300-milliamp wall adapter transformer	DB# 12 DC/.3
PTUBE-C	Cool white/blue/green/purple/red/yellow plasma tube	DB# PTUBE-COLOR
Q1	PN3055 NPN power T0-220 transistor	
PBOARD	1 × 4.5 × .1-inch grid perforated board	
HS1	1.5 × 1 × ¹⁄₁₆-inch aluminum heatsink per Figure 40-3	
WR20	12 inches of #20 vinyl hookup wire	
WR24BUSS	12 inches of #24 buss wire	
CANOPY6.5	6.5-inch, round metal canopy, as shown in Figure 40-5	DB# CANOPY65
BAFFLE	5 × .25-inch, round, clear plastic baffle	DB# BAFFLE5
BALL	⅞-inch brass ball with a hole tapped for a ¼-20 bolt	DB# BALL78
STAKE1	Metal stake created as shown in Figure 40-4	
ENCLOSURE	Plastic tube created as shown in Figure 40-4	
SW1	6-32 ¼ Phillips screw	
NUT	6-32 keep nut	
SW3	Two #6 × ¼-inch sheet metal screws	
SW1420X1	¼-20 × 1-inch hex bolt for attaching the ball (see Figure 40-5)	
TIEWRAP	Two tie wraps used per Figure 40-3	
CAP1	Plastic cap used per Figure 40-6	
CAP1.5	1.5-inch plastic cap shown in Figure 40-6	
WN1, 2	Two small wire nuts	
BUS1	⁵⁄₁₆ OD × ³⁄₁₆ ID × 1-inch length of rubber tubing (see Figure 40-6)	
ORING1.25	1¼-inch O-ring for holding the baffle in place per Figure 40-6	
5 LAMP ADAPTER	12 VDC, 1.5-amp adapter for five lamps	DB# 12/1 .5

Index

D

E

F

G

H

I

K

L

M

N

O

P

Q

R

S

T

Z